Springer Proceedings in Physics

Volume 192

The series Springer Proceedings in Physics, founded in 1984, is devoted to timely reports of state-of-the-art developments in physics and related sciences. Typically based on material presented at conferences, workshops and similar scientific meetings, volumes published in this series will constitute a comprehensive up-to-date source of reference on a field or subfield of relevance in contemporary physics. Proposals must include the following:

- name, place and date of the scientific meeting
- a link to the committees (local organization, international advisors etc.)
- scientific description of the meeting
- list of invited/plenary speakers
- an estimate of the planned proceedings book parameters (number of pages/ articles, requested number of bulk copies, submission deadline).

More information about this series at http://www.springer.com/series/361

H. Paul Urbach · Guangjun Zhang
Editors

3rd International Symposium of Space Optical Instruments and Applications

Beijing, China June 26–29th 2016

Springer

Editors
H. Paul Urbach
Optics Research Group
Delft University of Technology
CJ Delft
The Netherlands

Guangjun Zhang
Southeast University
Nanjing, Jiangsu
China

ISSN 0930-8989 ISSN 1867-4941 (electronic)
Springer Proceedings in Physics
ISBN 978-3-319-49183-7 ISBN 978-3-319-49184-4 (eBook)
DOI 10.1007/978-3-319-49184-4

Library of Congress Control Number: 2016956845

Printed on acid-free paper

This Springer imprint is published by Springer Nature
The registered company is Springer International Publishing AG
The registered company address is: Gewerbestrasse 11, 6330 Cham, Switzerland

The original version of the book was revised: Spell errors in the authors' names have been corrected. The erratum to the book is available at 10.1007/978-3-319-49184-4_50

Contents

Study of System Effectiveness Evaluation for Optical Imaging Reconnaissance Satellite Based on Fuzzy Theory

Liu Tao

Abstract Evaluation of system effectiveness for optical imaging reconnaissance satellite (OIRS) is a tough issue which is multi-level and multi-factor. According to key parameters of OIRS, a system effectiveness evaluation model for OIRS is proposed in this paper which includes 4 levels and 13 main factors. The OIRS model was verified by using fuzzy comprehensive evaluation (FCE) method and some typical satellites (Geoeye-1, WorldView-1/2/3, Pleiades and Quickbird-2) were used as evaluating cases. The results showed that the OIRS model was effective. This model and FCE method are very useful for development research and planning of optical imaging satellites.

Keywords Fuzzy theory · Reconnaissance satellite · System effectiveness evaluation

1 Introduction

System effectiveness evaluation of optical imaging reconnaissance satellite (OIRS) is very complex, which is a multi-level and multi-factor issue, and covers the objective and subjective factors; moreover, quantitative factors have different units.

Weapon System Effectiveness Industry Advisory Committee (WSEIAC) defined systems effectiveness as a measure of the extent to which a system may be expected to achieve a set of specific mission requirements [1]. WSEIAC also defined the system effectiveness as a function of the system's availability, dependability and capability which is often called the ADC method. To some extent, ADC method has authority, and many researchers have proposed several types of weapon system effectiveness evaluation model based on ADC method, however, if the number of sub-systems becomes too large, the analytical application of the WSEIAC framework becomes prohibitive [2].

L. Tao (✉)
Beijing Institute of Space Science and Technology Information, Beijing, China
e-mail: lttaotao@yeah.net

H.P. Urbach and G. Zhang (eds.), *3rd International Symposium of Space Optical Instruments and Applications*, Springer Proceedings in Physics 192,
DOI 10.1007/978-3-319-49184-4_1

Analytic Hierarchy Process (AHP) is also a popular technique for evaluating system effectiveness. However, there are many questionable theoretical issues in the AHP technique, such as scale misinterpretation, comparison matrix eigenvalue evaluation, and rank reversal problems [3, 4].

A system effectiveness evaluation index architecture of OIRS based on fuzzy theory is proposed in this paper, and it shows the fuzzy comprehensive analysis can be widely used in the evaluation of system effectiveness of OIRS.

2 The Architecture of Evaluation Index

The architecture of system effectiveness evaluation index is shown in Table 1, which is composed of system capacity, cost, reliability, and survivability. These four top-level indexes are chosen because the first three indexes are widely used in weapon system or satellite effectiveness evaluation [5, 6], and survivability is a key component of resilient space architecture which has been mentioned since 2012 [7, 8].

Procedures for building the architecture of system effectiveness evaluation index are shown is Fig. 1. A very important procedure is the last step. If a result can not

Table 1 The architecture of evaluation indexes and weights of the indexes

Top level index	Second level index	Third level index	Fourth level index
System capacity (0.65)	Image characteristics (0.7)	PAN resolution (0.35)	–
		IR resolution (0.3)	–
		Swath (0.1)	–
		Spectrum character (0.15)	PAN imaging capacity MS imaging capacity Hyperspectral imaging capacity
		Location accuracy without GCP (0.1)	–
	Observation timeliness (0.1)	Average revisit time (0.8)	–
		Data rate (0.2)	–
	Agile ability (0.2)	Maximum agile angle (0.5)	–
		Angular velocity (0.5)	–
Cost (0.1)	–	–	–
Reliability (0.1)	Lifetime (1)	–	–
Survivability (0.15)	On orbit maneuverability (0.65)	–	–
	Anti-interference capacity (0.35)	–	–

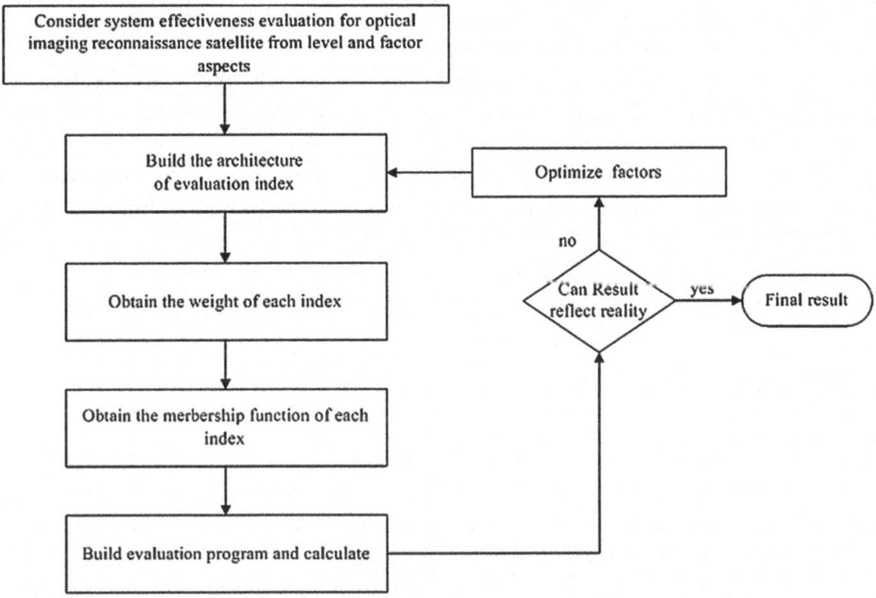

Fig. 1 Flow chart for building the architecture of system effectiveness evaluation index

reflect obvious diversities of different generations of OIRS in reality, we will optimize evaluation factors, and repeat procedures.

In the final architecture, the value of system capacity is determined by five subcapabilities: image characteristics, observation timeliness and agile ability. For example, image characteristics consist of resolution, swath, spectrum character, and location accuracy. Increasing spatial resolution and swath is general trend for development of OIRS. Multi or hyper-spectral capabilities allow OIRS to easily distinguish target characteristics. Good location accuracy ensures images from OIRS are as closely aligned as possible to a known coordinate on the surface of the earth.

3 Methodology

In this section, we present an algorithm for evaluating OIRS using fuzzy comprehensive evaluation method which is composed of expert scoring method and fuzzy evaluation method.

V is the evaluation set (valuation scale), which can be partitioned into a five-component subset in proposed model. $V = (v1, v2, v3, v4, v5)$ = (Excellent, Good, Medium, General, Poor). U is the evaluation criteria. There are four top-level criteria described above. Furthermore, assume that the weights of the criteria supplied by our expert panel are represented by a weighting vector A. R is fuzzy matrix, and is obtained by the membership functions (MF). B is result set, and is calculated by $A * R$, where * is an operator and represents the matrix multiplication. b_j, component of B, is a subordinate degree of the evaluation object associated with corresponding V_j. In other word, b_j specifies the extent to which j can be regarded as a result and one can simply choose the V_j of corresponding maximum b_j as final evaluation result. However, to make full use of information brought by the B, a degree parameter vector C should be used. The value of $B * C$ is our final result.

4 Empirical Illustration

An illustrative example of evaluating system effectiveness for OIRS is provided to demonstrate that the fuzzy method provides a good evaluation result. Several OIRS satellites are selected as the study cases (Geoeye-1, WorldView-1, WorldView-2, WorldView-3, and Quickbird-2 were largely funded by U.S. NGA's (National Geospatial-Intelligence Agency) contracts, so they belong to OIRS in this research.) (Table 2).

The weight of each evaluation index is given in Table 1, obtained by expert scoring method, as shown in (1). Where n is the number of experts we invite, and n_j is the weight scored by one expert.

$$weight = \sum_{j=1}^{n} n_j/n \quad (1)$$

The most important step of fuzzy evaluation method is to define the MF of each evaluation index. There are quantitative and qualitative indexes for types of evaluation indexes. Some quantitative indexes belong to the smaller the better type, such as spatial resolution, location accuracy without GCP. Some quantitative indexes belong to the bigger the better type, such as the swath.

Taking Geoeye-1 satellite as an example, the trapezoid MF of panchromatic resolution is written as below. We utilize boundary values of the National Imagery Interpretability Rating Scale (NIIRS) shown in Table 3 to define fuzzy values (a, b) of the MF of PAN resolution.

$$Y_1(x) = \begin{bmatrix} 1 & 0 \le x \le a_1 \\ \frac{b_2 - x}{b_2 - a_1} & \mathrm{a}_1 < x < b_2 \quad (\mathrm{i} = 1) \\ 0 & \mathrm{x} \ge \mathrm{b}_2 \end{bmatrix} \quad (2)$$

Table 2 Parameters of satellites as study cases

Name	PAN/m	IR/m	Swath/km	Spectrum character	GCP/m	Revisit time/d	Data rate (Mbit/s)	Agile angle (°)	Angular velocity (°/s)	Cost/$10^8$$	Lifetime/y	M.	A.
Geoeye-1	0.41	Nil	15.2	2	5	2.6	740	60	2.4	4.835	7	Nil	Nil
WorldView-1	0.5	Nil	17.6	1	4	1.7	800	45	4.45	4.73	7.25	Nil	Nil
WorldView-2	0.46	Nil	16.4	2	3.5	1.1	800	40	3.86	4.63	7.25	Nil	Nil
WorldView-3	0.31	Nil	13.1	2	3.5	<1	1200	40	4.45	5.2	7.25	Nil	Nil
Pleiades	0.7	Nil	20	2	8.5	2.1	465	60	3.4	4.955	5	Nil	Nil
Quickbird-2	0.65	Nil	16.5	2	23	2	320	45	1.17	2.75	5	Nil	Nil
Kanopus-V1	2.1	Nil	23	2	40	17	300	40	0.3	0.17	5	Nil	Nil
SPOT-6	2	Nil	60	2	35	2	300	45	2.5	1.875	10	Nil	Nil
Landsat-8	15	100	185	2	65	16	384	15	0.05	8.5	10	Nil	Nil

Infrared (IR) represents thermal infrared in this paper. For Spectrum character, *1* represents PAN, *2* represents PAN+MS, *3* represents PAN+MS+ Hyperspectral imaging capacity. GCP means Ground Control Point. Revisit time is measured under the following conditions, 40°N latitude and 1 m GSD

M represents maneuverability. *A* represents anti-interference capacity

Table 3 The national imagery interpretability rating scale

Level	Spatial resolution (m)
1	>9
2	4.5–9
3	2.5–4.5
4	1.2–2.5
5	0.75–1.2
6	0.4–0.75
7	0.2–0.4
8	0.1–0.2
9	<0.1

$$Y_i(x) = \begin{bmatrix} 0 & x \le a_{i-1} \\ \frac{x-a_{i-1}}{b_i-a_{i-1}} & a_{i-1} < x < b_i \quad (i = 2, 3, 4) \\ 1 & b_i \le x \le a_i \\ \frac{b_{i+1}-x}{b_{i+1}-a_i} & a_i < x < b_{i+1} \\ 0 & x \ge b_{i+1} \end{bmatrix}$$

$$Y_5(x) = \begin{bmatrix} 0 & 0 \le x \le a_4 \\ \frac{x-a_4}{b_5-a_4} & a_4 < x < b_5 \quad (i = 5) \\ 1 & x \ge b_5 \end{bmatrix}$$

Then, we set the following relationships, $a_1 = 0.1$ m, $b_2 = 0.2$ m, $a_2 = 0.4$ m, $b_3 = 0.75$ m, $a_3 = 1.2$ m, $b_4 = 2.5$ m, $a_4 = 4.5$ m, $b_5 = 9$ m. 0.41 m resolution of Geoeye-1 is substituted into the (2), so the fuzzy vector of resolution, r_{111}, is (0, 0.9714, 0.0286, 0, 0).

Similar steps are repeated to calculate fuzzy vectors of IR resolution, swath, spectrum character and location accuracy. These parameters are belonging to quantitative factor, so trapezoid MF is also used for them. Then, fuzzy matrix of image characteristics r_{11} is obtained, and is shown below.

$$r_{11} = \begin{pmatrix} 0 & 0.9714 & 0.0286 & 0 & 0 \\ 0 & 0 & 0 & 0 & 1 \\ 0 & 0 & 0.52 & 0.48 & 0 \\ 0.7 & 0 & 0 & 0 & 0.3 \\ 0.75 & 0.25 & 0 & 0 & 0 \end{pmatrix}$$

The result vector b_{11} is obtained using weight vector of image characteristics (0.35, 0.3, 0.1, 0.15, 0.1) to multiply by above fuzzy matrix, and the result is (0.1450, 0.3650, 0.0100, 0.0780, 0.4020). b_{12} for observation timeliness and b_{13} for agile ability are also obtained by same method, and then, b_1 is composed of b_{11}, b_{12} and b_{13}.

On-orbit maneuverability and anti-interference capacity belong to qualitative indexes. Military OIRS such as U.S. Keyhole satellites may have stronger on-orbit maneuverability than commercial OIRS because they take more fuels for both

Table 4 Final results

Sat name	The result set, B	Final score
Geoeye-1	(0.1376, 0.3834, 0.1106, 0.0355, 0.3329)	5.9146
WorldView-1	(0.1551, 0.3974, 0.0655, 0.0519, 0.3302)	5.9906
WorldView-2	(0.1858, 0.3552, 0.0906, 0.0437, 0.3247)	6.0672
WorldView-3	(0.2073, 0.3609, 0.0633, 0.0212, 0.3472)	6.1199
Quickbird-2	(0.1219, 0.1725, 0.2251, 0.1565, 0.3240)	5.2233
Pleiades	(0.0987, 0.2835, 0.2459, 0.0719, 0.3001)	5.6171
Kanopus-V1	(0.1318, 0.0217, 0.1923, 0.1915, 0.4627)	4.3371
SPOT-6	(0.2319, 0.1853, 0.1263, 0.1110, 0.3456)	5.6933
Landsat-8	(0.2001, 0.0600, 0.0409, 0.0771, 0.6219)	4.2787

Fig. 2 A part of computer program for calculating the system effectiveness of OIRS

```
% part of program
[data,text]=xlsread('data.xls','A2:O10');
[m,n]=size(data);
%weight
A=[0.65,0.1,0.1,0.15];
A1=[0.7,0.1,0.2];
A11=[0.35,0.3,0.15,0.1,0.1];
for i=1:m
r111=Pan_res(data(i,1));
r112=IR_res(data(i,2));
r113=Swath(data(i,3));
r114=Spec(data(i,4));
r115=Gcp(data(i,5));
r11=[r111;r112;r113;r114;r115];
B11(i,:)=A11*r11;
B1(i,:)=A1*[B11(i,:);B12(i,:);B13(i,:)];
end
for i=1:m
B(i,:)=A*[B1(i,:);B2(i,:);B3(i,:);B4(i,:)];
end
%to print result
```

orbital maintenance and maneuver for required operational tasks. To simply the problem, we may arbitrarily define that commercial OIRS do not have On-orbit maneuverability, and its fuzzy vector is set by (0, 0, 0, 0, 1).

The final results are obtained by the result set of system effectiveness multiply by C, $(10, 8, 6, 4, 2)^T$, shown in Table 4. Results show that the WorldView-3 satellite launched in 2014 get the highest score in this research; Landsat-8 which is not OIRS but can be used for military purposes gets the lowest score. These results are consistent with fact, which show that this evaluation method is effective. Furthermore, our algorithm can be automatically operated, and a part of program is shown in Fig. 2. This program can easily obtain data from an Excel file, and has potential in development planning of optical imaging satellites.

5 Conclusion

This paper has addressed an evaluation problem of system effectiveness for optical imaging reconnaissance satellite. Future research should include the development of the most suitable membership function; developing different system effectiveness models for different tasks (communication satellites, positioning and timing satellite, missile warning satellites, and environment monitoring satellites, etc.); developing an index model for evaluating whole space capabilities in country level.

References

1. US Army Material Command, Engineer Design Handbook - Systems Analysis and Cost-effectiveness, Document AMCP 706-191, Washington:, pp. 2–18, 1971.
2. Weapon System Effectiveness. http://uir.unisa.ac.za/bitstream/handle/10500/1527/03Chapter3.pdf.
3. R. Whitaker. "Criticisms of the Analytic Hierarchy Process: Why they often make no sense," Mathematical and Computer Modelling, vol. 46, pp. 948–961, 2007.
4. L. Warren, "Uncertainties in the Analytic Hierarchy Process," http://dspace.dsto.defence.gov.au/dspace/bitstream/1947/3553/1/DSTO-TN-0597%20PR.pdf.
5. S. Chen. "Evaluating weapon systems using fuzzy arithmetic operations," Fuzzy Sets and Systems, vol. 77, pp. 265–276, 1996.
6. M. Elhady, "Remote sensing satellite system overall effectiveness analysis and modeling," IEEE Aerospace Conference, pp. 1–10, 2014.
7. DoD Directive 3100.10, "Space Policy," pp. 12, 2012.
8. M.T. Trussell1, M.N. Martin, R.M. Ewart, "Resilience architecting and methodology development for the space enterprise," AIAA SPACE 2015 Conference and Exposition, pp. 1–11, 2015.

Analysis of Platform and Payload Integrated Design Technology for Optical Remote Sensing Satellites

Yanfeng Yao

Abstract Driven by the goal of high resolution, high agility and low cost, the integrated design technology of satellite platform and payload considering payload as center is an important trend in the development of advanced optical remote sensing satellites. The paper investigates the typical applications of integrated design technology in remote sensing satellites. The technical features of the technology, such as integrated satellite configuration, integrated thermal control, integrated electronic system and integrated micro-vibration suppression are also discussed. On this basis, it puts forward the development trends of the platform and payload integrated design technology, which include multidisciplinary design optimization taking into account structural-thermal-optical performance, integrated electronic system based on reconfigurable modules, focus on improving the ability of micro-vibration suppression and innovation of satellite development process.

Keywords Remote sensing satellite · Platform · Payload · Integrated design

1 Introduction

With the development of remote sensing satellite technology, space optical remote sensor payload is developing towards the direction of long focal length and large aperture. The traditional platform and payload subdivision design mode will lead to an increase in the volume, weight and inertia, which is difficult to meet the demands of high resolution, high agility and low cost of advanced remote sensing satellites. The integrated design of platform and payload considering payload as center can break through the limitations of traditional separate service and payload modules, realize the whole satellite design optimization from system level, and significantly reduce satellite weight and development costs. Platform and payload integrated

Y. Yao (✉)
Institute of Telecommunication Satellite, China Academy of Space Technology, Beijing 10094, China
e-mail: yaoyanfeng21@163.com

H.P. Urbach and G. Zhang (eds.), *3rd International Symposium of Space Optical Instruments and Applications*, Springer Proceedings in Physics 192,
DOI 10.1007/978-3-319-49184-4_2

design involves multiple disciplines, requires overall consideration of various factors such as optical, mechanical, electrical, thermal, etc. Configuration, layout and thermal control of satellite should be designed around remote sensing sensor tightly. Meanwhile, the electronic system also needs to be integrated designed, achieving the goal of unified scheduling and management of satellite resources.

The current advanced optical remote sensing satellites generally adopt the platform and payload integrated design concepts, such as Ikonos-2, WorldView, GeoEye-1 and Pleiades. The paper investigates the typical applications of integrated design technology in remote sensing satellites. Furthermore, the technical features of integrated design are also discussed from four aspects. Finally, it puts forward four trends of the platform and payload integrated design, which expects to provide a reference for the development of the technology.

2 Typical Satellites Applied Integrated Design

2.1 Ikonos-2

The Ikonos-2 satellite launched successfully in 1999 is the world's first 1 m-resolution commercial earth observation satellite, developed by Space Imaging Corporation [1]. The design idea of integration and high rigidity is adopted in the satellite. As shown in Fig. 1, the remote sensor is embedded in the platform, and the structure of the satellite is strengthened by using the plate and rod structure. Star tracker and gyro are directly mounted on the remote sensor to achieve high attitude accuracy. Other platform equipments are also installed on the remote sensor, saving the redundant structure of platform. The rigid fixed support solar arrays are designed to reduce the influence of the flexible appendages on the attitude maneuver. The surface of the remote sensor is coated with multi-layer thermal insulation materials to reduce the influence of the space environment on the temperature control, and a thermal door is installed on the top of the remote sensor to reduce the heat leakage.

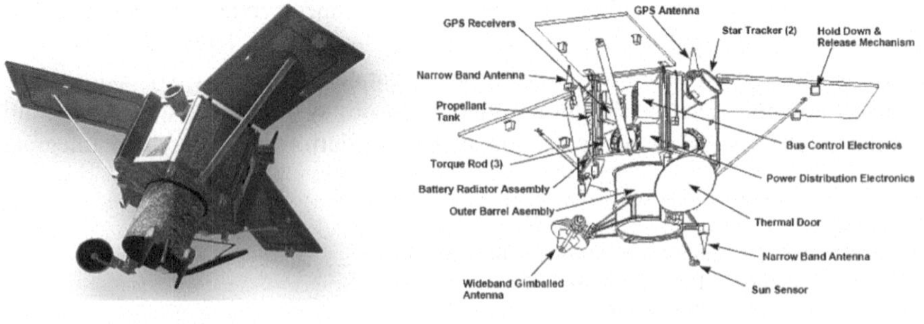

Fig. 1 Ikonos-2 satellite

2.2 WorldView Series

WorldView series satellites are developed by DigitalGlobe Inc. including WordView-1/2/3 on orbit. The design of WorldView satellites fully reflects the integrated idea of high stiffness and small inertia. As shown in Fig. 2, the composite structure of beam and plate is adopted to ensure the rigidity of the satellite. The electronic equipments are mounted on the outer surface of the platform. CMG module with vibration isolation devices is installed at the bottom of the hollow platform cabin, while the spherical tank and optical telescope are mounted successively on the top of the CMGs. The top of the telescope is also equipped with a thermal protection door, while special radiators are arranged for the high heat consumption equipments like focal plane unit, battery and solid state memory [2]. Another highlight of the satellite is the advanced integrated electronic systems [3].

2.3 Pleiades

Pleiades is the high resolution earth observation satellite developed by the CNES, including Pleiades-1A/1B, representing the most advanced technical level of France in the earth observation field. The focus of Pleiades design is the agile imaging and

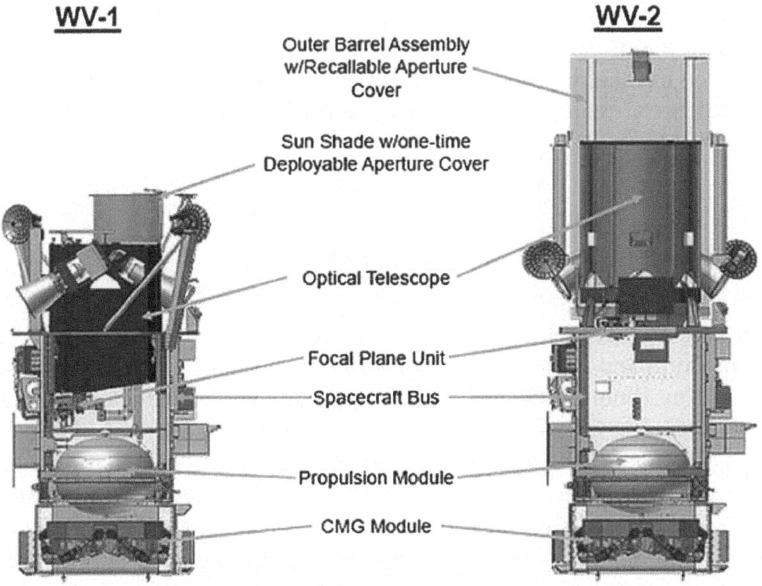

Fig. 2 WorldView-1/2 satellite

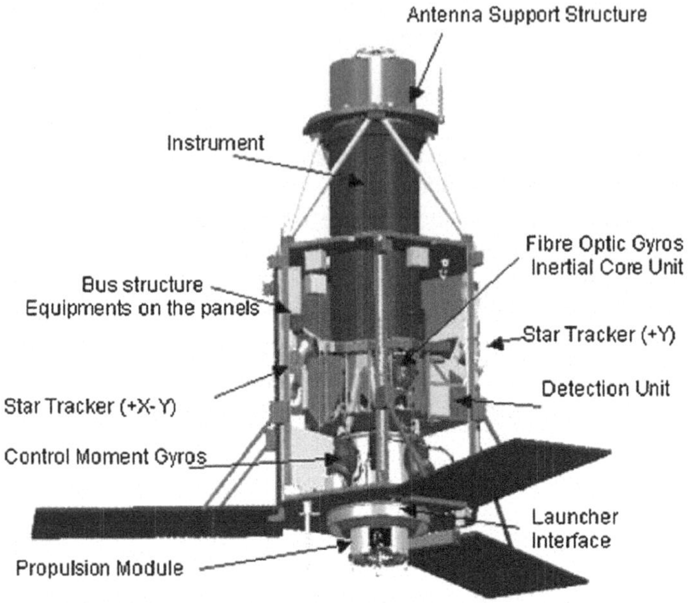

Fig. 3 Pleiades satellite

high positioning accuracy. As shown in Fig. 3, the satellite has adopted an integrated rigid structure similar to Ikonos-2. Meanwhile, latest technologies are introduced in the design of satellite including fiber optic gyro, star tracker, CMG and advanced navigation receiver. Pleiades uses the integrated design of electronic system, which can realize the scheduling and management of the whole satellite resources [4, 5].

3 Technical Features of Integrated Design

Through investigation and analysis, the ultimate goal of the integrated design of satellite platform and payload is to coordinate the design from the system level to achieve the maximum efficiency of the payload, which include configuration, layout, thermal control and electronic system design. In addition, the integrated design makes the satellite more compact, the influence of micro-vibration of reaction wheel, CMG, etc. on the imaging of remote sensor can't be ignored and micro-vibration suppression measures must be taken. In summary, the integrated design of platform and payload includes several aspects as follows: integrated configuration, integrated thermal control, integrated electronic system and integrated micro-vibration suppression.

3.1 Integrated Configuration Design

Integrated configuration is the foundation of the integrated design technology, which has a critical impact on the mechanical, electronic and thermal design. Integrated configuration design has the following features and advantages:

1. Light structural weight and excellent mechanical properties

The payload uses an embedded (Ikonos-2, Pleiades, WorldView-1) or fusion (EAROS-1A/B, GeoEye-1) installation, which can effectively lower the centre of gravity and improve the mechanical environment during launching stage. In addition, the main load-bearing structures of platform and payload are directly connected to each other, resulting in enhanced structure strength, excellent mechanical properties and lower structure weight. According to statistics, the structure weight of traditional remote sensing satellite is generally accounted for 15–20% of the whole satellite, while the data will drop below 7% using integrated design.

2. Small rational inertia

The rotational inertia of satellite is proportional to the square of the distance between the equipments and the rotating axis. The equipments of the integrated configuration satellite are arranged around the payload or even directly mounted on it, which are closer to the axis of the satellite, so it can significantly reduce the rational inertia and improve the agility of the satellite. Moreover, the satellite solar array usually adopts the rigid structure or parallel configuration to reduce the influence of flexible components on the attitude maneuver.

3. High image positioning accuracy

In the satellites with integrated configuration, the star trackers, gyros and other attitude sensors are generally installed directly on the payload (such as Ikonos-2, WorldView, Pleiades), which can maintain the stability of the high precision measurement datum and the optical axis of the remote sensor, avoid the adverse effects of the force and thermal deformations of the platform, and improve the positioning accuracy of the satellite image.

4. Saving development and launch costs

Remote sensing satellite with integrated configuration has more compact structure and smaller occupied space, which is more applicable to the flexible launch of low cost launch vehicles.

3.2 Integrated Thermal Control Design

With the increase of resolution, long focal length and large aperture remote sensor is more sensitive to platform and space environment temperature changes.

Therefore, the temperature field change has become an important factor affecting the quality of on-orbit imaging. The high performance remote sensing satellite with integrated design has the characteristics of high function density, multi imaging mode and long working time on orbit. The high function density causes limited heat radiation area and high temperature level. Multi imaging mode requires frequent satellite attitude maneuver, resulting in a large range of heat flux and temperature fluctuations. On-orbit long time imaging results in increased local high temperature and thermal noise. All these factors increase the difficulties of thermal control. Due to the heat transfer process of the platform and the payload are tightly coupled, we should give full play to the advantages of the whole satellite thermal control, and carry out the integrated thermal control design of the platform and the payload.

In order to meet the requirements of high performance imaging, a thermal control method based on integrated remote sensing satellite is proposed, which aims at the minimum occupancy rate of satellite resources [5]. To solve the problems of limited power and high power density and long working time of CCD, the following measures are taken: (1) Turning the waste of platform equipments into heating power of the payload. (2) Using phase change materials to increase the thermal inertia of CCD and extend the remote sensor time. The heat pipe and special radiating surface are connected to CCD to improve the radiating efficiency. Moreover, precision temperature control of radiating surface is adopted to reduce the temperature fluctuation.

3.3 Integrated Electronic System Design

With the development of remote sensing technology, the imaging modes of remote sensing satellites are becoming increasingly complex, requiring the satellite to have the capability of efficient and flexible task planning and self-management. In this context, the satellites adopting integrated design will be no longer only a collection of individual functions of devices, but an integrated system capable of unified scheduling and management of tasks, functions and resources.

Integrated electronic system is composed of hardware and software modules, which can complete the entire satellite management functions such as data management, management attitude and orbit control, satellite ground downlink data management, thermal control management, time management, power management, payload management, autonomous fault detection and security management of satellite. The essence of its function is the generation, processing, transmission and distribution of information flow. Integrated electronic system contains most of the electronic equipments of satellite, whose design standards determine the performance of the satellite to a large extent.

As shown in Fig. 4, WorldView satellites adopt advanced data system technology based on integrated electronics [6], which enable the satellites to have the

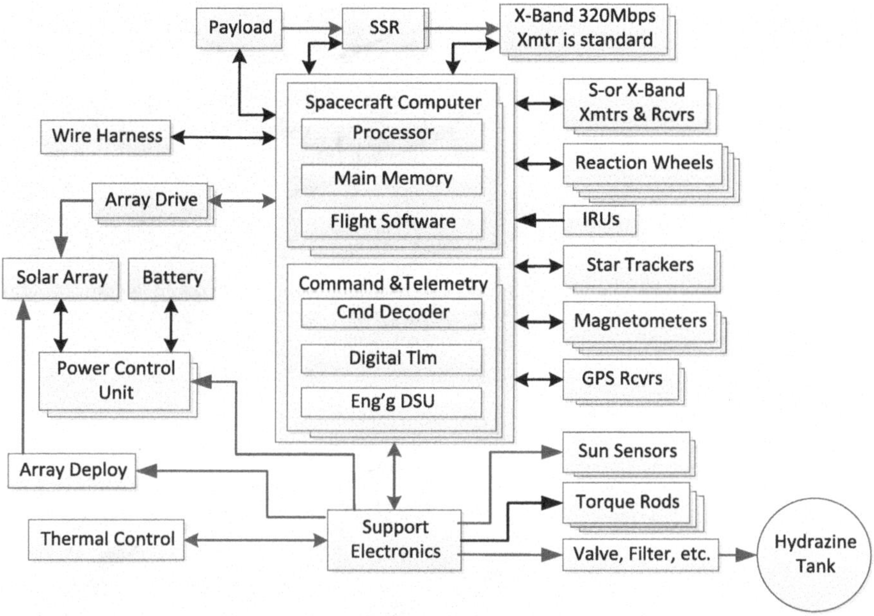

Fig. 4 Integrated electronic system of WorldView satellites

ability of task management, autonomous execution and autonomous fault processing, enhancing the ability of autonomous operation on orbit in full life cycle.

Pleiades satellites also use the integrated design concept of the whole satellite electronics system, taking the integrated management unit OBMU as the core [7], which can not only complete the routine satellite service management mission, but also fulfil the payload, attitude and orbit control, thermal control and power management as well as telemetry and telecontrol tasks.

3.4 *Integrated Micro-vibration Suppression*

As mentioned above, it is necessary to consider the influence of the micro-vibration in the integrated design of remote sensing satellite. For micro-vibration suppression, optimized layouts of vibration source and vibration isolation devices are mainly adopted to reduce the influence on payload imaging. First, the reaction wheel, CMG, solar array actuators and so on should be mounted as far away from the payload as possible, in order to increase the micro-vibration transfer path and attenuation (such as the CMG module mounted at the bottom of the platform in Fig. 2). Moreover, micro-vibration suppression devices are also frequently used,

Fig. 5 WorldView vibration isolator

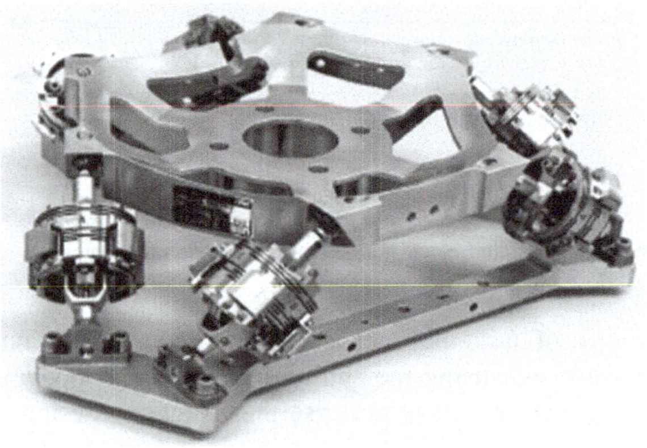

Fig. 6 Chandra X-ray vibration isolator

which consist of vibration source isolator, payload vibration isolator and flexible appendages damper, etc.

Vibration source isolators are installed between reaction wheels/CMGs and satellite structure to separate high frequency vibration transmitted to the satellite. For instance, an eight connecting rod isolation system was installed between the CMG mounting brackets and the main support beam of service module in WorldView; Chandra X-ray observer installed Stewart vibration isolation platform for each reaction wheel [8], as shown in Figs. 5 and 6.

The payload vibration isolators are installed between the imaging device and the satellite structure. Both WorldView-2 and GeoEye-1 used Bipod mechanisms [9] to realize the vibration isolation between payload and platform as well as optical components and remote sensor structure, as shown in Figs. 7 and 8.

Flexible accessory dampers are connected between the root of the flexible appendages and the structure of satellite to improve the damping parameter of the

Fig. 7 GeoEye-1 vibration isolator

Fig. 8 WorldView-2 vibration isolator

flexible appendages. As shown in Fig. 9, a damper was used on the root of the solar array of the Hubble telescope, which effectively reduces the amplitude of sloshing and improve the attitude stability [10].

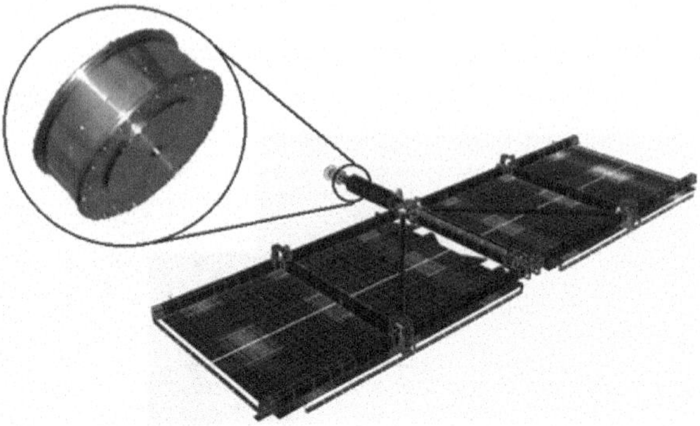

Fig. 9 HST solar array damper

4 Analysis of the Development Trends of Integrated Design Technology

4.1 Multidisciplinary Design Optimization of Remote Sensor Based on Structure-Thermal-Optics Performance

As the center of the integrated design of platform and payload, the space optical remote sensor has a complex coupling relationship with the external space environment and other satellite equipments. Therefore, it is necessary to carry out multidisciplinary design optimization considering optics, structure and thermotics, etc. For example, there are a number of connection points between the platform and payload structure under the integrated configuration. The deformation of the platform structure will directly affect the stability of payload. In addition, the heat radiation of the equipment of the platform will affect the thermal environment of the remote sensor, thereby affecting the imaging quality.

Traditional remote sensors are usually designed based on serial mode. After the optics, structure and electronics design is completed, the design thermal control system is carried out. The serial mode is easy to cause the design process repeatedly modified. Therefore, according to the platform and payload integrated design requirements, the development mode should be changed from traditional serial mode to STOP (Structural-Thermal-Optical Performance) multidisciplinary design optimization [11]. The flow of STOP is shown in Fig. 10. Firstly, thermal analysis model and structure analysis model are established based on the same remote sensor configuration, while the thermal analysis model contains the preliminary thermal control scheme. After solving the transient temperature field data of the remote sensor, the temperature field is mapped to the structural analysis model, and the calculation of the displacement field is completed. Then the optical component surface and position change data is extracted from the displacement field. After fitting the data into the optical analysis software, the assessment of the optical performance is completed. If the optical performance is not satisfied, the thermal

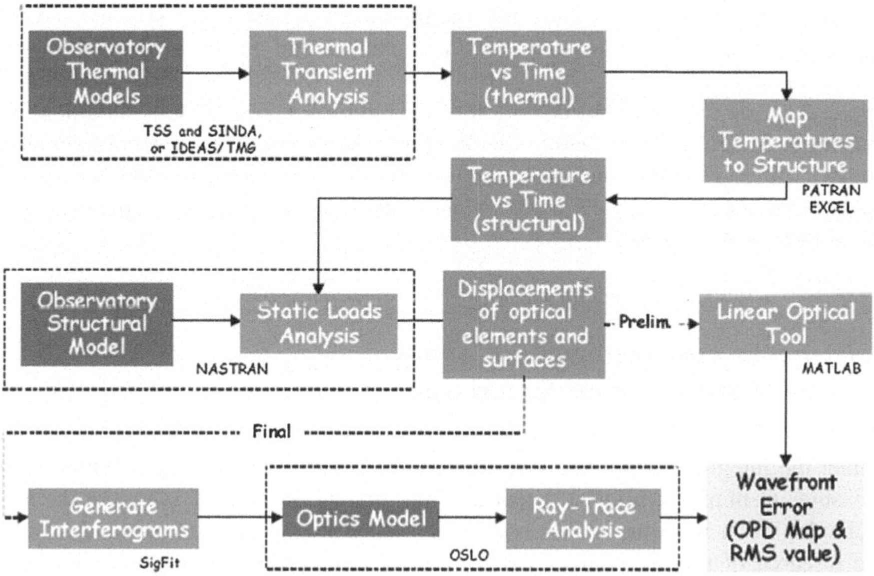

Fig. 10 STOP analysis process map

model (thermal control scheme) and structural model (structural design) should be improved iteratively until the final performance is satisfied.

The most striking feature of the STOP design mode is to achieve the multidisciplinary design optimization. Through a large number of data exchange based on temperature field, displacement field and optical surfaces, we can debug thermal, structure and optical design parameters repeatedly, which can fulfill real-time assessment for the optical performances of remote sensor and ultimately achieve overall optimization.

4.2 Development of Reconfigurable Integrated Electronic Technology

The integrated electronic system fulfils all the management functions of the satellite, which is the key to ensure the effectiveness of remote sensing satellite. The integration and miniaturization of the system have important influence on the integrated design of satellite. The spacecraft integrated electronic system has been developed to the fourth stage, the main development trends are from the special interface definition to the standard interface definition; from electronic equipments simple fusion to the top level system optimization and integration; from decentralized design to the top-down integrated design.

According to the above trends, the satellite integrated electronic system needs to be designed adopting top-down and integrated design methods. The hardware resources and software functions should be fully played to realize the unified application, deployment and operation of satellite resources. System should adopt the FDIR design and have strong autonomous fault detection and reconfiguration ability, which also has emergency hardware circuit and telemetry and telecontrol backup, in support of safe mode while fault occurs to ensure reliability and robustness of the satellite.

4.3 Focus on Improving the Ability of Micro-vibration Suppression

Under the integrated configuration, taking appropriate micro-vibration suppression measures to improve on-orbit dynamics environment of payload, is one of the key technologies in imaging quality assurance. Therefore, we should pay attention to the research of the mechanism and method of micro-vibration suppression. First, the disturbance source on-orbit should be determined. Perturbation amplitude and spectrum distribution must be identified by the vibration analysis and testing as a basis for micro vibration suppression design. Second, appropriate installation locations need to be selected for the vibration source and imaging equipments according to the mode of the whole satellite structure (avoiding the vibration antinodes, as close as possible to the node). Third, the micro-vibration transmitted to the satellite structure can be reduced by the micro-vibration source device optimization design (such as reducing static and dynamic unbalance amount of high speed rotor) and isolation (such as vibration isolator, flexible damper). Finally, it is one of the effective methods to solve the problem of micro-vibration by reducing the sensitivity of payload, which includes compensating the key sensitive components and taking the vibration reduction measures.

4.4 Innovation of Development Process

For traditional remote sensing satellite, the platform and the payload can be divided into separate modules for parallel development and production. After the platform structure and equipments are delivered, the AIT (Assembly Integration and Test) of the service module can be carried out first. Payload can be delivered at the late stage of the AIT of the platform. The integrated configuration satellite needs to use the payload structure as a support to complete the satellite equipments installation, so the payload delivery cycle needs to be advanced. Therefore, we should carry out AIT program research and process innovation for remote sensing satellite adopting integrated design to shorten the development and delivery cycle of satellite.

5 Conclusions

By investigating the typical applications of integrated design technology in remote sensing satellites, the technical features of the technology including integrated satellite configuration, integrated thermal control, integrated electronic system and integrated micro-vibration suppression are detailed summarized in this paper. Finally, four development trends of the platform and payload integrated design technology are proposed, which include multidisciplinary design optimization taking into account structural-thermal-optical performance, integrated electronic system based on reconfiguration modules, focus on improving the ability of micro-vibration suppression and innovation of satellite development process, which can provide a reference for the platform and payload integrated design of optical remote sensing satellite.

References

1. DigitalGlobe, "IKONOS data sheet", http://global.digitalglobe.com/sites/default/files/DG_IKONOS_DS.pdf
2. DigitalGlobe, "Introducing WorldView: DigitalGlobe's next generation system", http://www.auric-ht.com/Coasts/documents/DigitalGlobe_Satellite_Constellation_Presentation?WorldView%20Satellites.pdf
3. A. A. Barnes III, "The Ball Commercial Platform and their applications, QuickBird, QuikSCAT and ICE-SAT", *49th International Astronautical Congress, Melbourne*, pp. 1–8, September 1998
4. L. Perret, E. Boussarie, J.M. Lachiver, and P. Damilano, "The Pleiades System High Resolution Optical Satellite and Its Performances", *53rd International Astronautical Congress of the International Astronautical Federation (IAF), Houston*, pp. 10–19, October 2002
5. A. Baudoin, E. Boussarie, P. Damilano, G. Rum, and F. CALTAGRONE, "Pleiades: A Multi Mission and Multi Co-operative Program", *52nd International Astronautical Congress, Toulouse*, pp. 1–5, October 2001
6. WANG Dong, ZHANG Hong-sheng, "Study on Thermal Control Method for High Performance Optical Satellite", *Space Optical and Electromechanics Association Professional Committee of China Space Science Academic Annual Meeting 2013. Changchun: Chinese Space Science Association*, pp. 229–234, 2013
7. D. Dantes, C. Neveu, T. Gilbert, and J.M. Biffi, "Design and performances of a new generation of detection electronics for earth observation satellites", *Proceedings of SPIE-IS&T Electronic Imaging SPIE*, vol. 5017, pp. 205–216, 2003
8. A.J. Bronowicki, "Forensic Investigation of Reaction Wheel Nutation on Isolator", *49th AIAA/ASME/ASCE/AHS/ASC Structures, Structural Dynamics, and Materials Conference, Schaumburg*, pp. 1–11. April 2008
9. G. Matthews, K. Havey Jr, and R. Egerman, "A paradigm shift to enable more cost-effective space science telescope missions in the upcoming decades", *Modeling, Systems Engineering, and Project Management for Astronomy IV, Proc. of SPIE*, vol. 7738, 773824, pp. 1–7, 2010
10. J.R. Maly, B.B. Reed, M.J. Viens, B.H. Parker, and S.C. Pendleton, " Life cycle testing of viscoelastic materials for Hubble Space Telescope solar array 3 damper", *Smart Structures*

and Materials 2003: Damping and Isolation, Proceedings of SPIE, vol. 5052, pp 128–140, 2003
11. J.D. Johnston, J.M. Howard, G.E. Mosier, K.A. Parrish, M.A. McGinnis, and A.M. Bluth, et al, “Integrated Modeling Activities for the James Webb Space Telescope: Structural-Thermal-Optical Analysis”. *Optical, Infrared, and Millimeter Space Telescopes, Proceedings of SPIE, Bellingham*, vol. 5487, pp. 600–610, 2004

Precision Thermal Control Technology of Secondary Mirror Structure in Large F Number Korsch Space Camera

Shikui Luo, Dongjing Cao and Xinyang Song

Abstract In a large F number Korsch space camera with 10 m of focal length and 16 of F number, Temperature vibration of secondary mirror structure should be controlled within range of ±0.3 °C. According to traditional precision thermal control technology based on direct bonding of heater films and closed-loop control, temperature control precision is ±0.5 °C. Temperature vibration of secondary mirror structure cannot be controlled within required range, with traditional precision thermal control technology. Precision thermal control technology based on heat shield and radiation heat transfer is proposed. Compare with traditional thermal control technology, better temperature control precision, better temperature control uniformity and better suppression performance of ambient temperature disturbance can be achieved by adopting proposed technology. In experiments and applications, temperature vibration of secondary mirror structure can be controlled within range of ±0.3 °C, by adopting proposed technology. Problem about precision temperature control of secondary mirror structure in the large F number Korsch space camera is resolved. Precision temperature control technology based on heat shield and radiation heat transfer is also adopted on other parts and in other space cameras.

Keywords Heat shield · Radiation heat transfer · Thermal control · Korsch optical system · Large F number · Secondary mirror · Space camera

1 Introduction

In space camera, traditional precision thermal control technology is based on direct bonding of heater films and closed-loop control. Heater films and thermistors are bonded to structure directly, and multilayer insulation is wrapped outside. With precision Thermal Control technology Based on Direct Bonding of heater films and

---

S. Luo (✉) · D. Cao · X. Song
Beijing Institute of Space Mechanics and Electricity, Beijing 100094, China
e-mail: shk_1@sina.com

H.P. Urbach and G. Zhang (eds.), *3rd International Symposium of Space Optical Instruments and Applications*, Springer Proceedings in Physics 192,
DOI 10.1007/978-3-319-49184-4_3

Closed-loop Control (TCBDBCC), no better temperature control precision than ±0.5 °C can be achieved due to limitation of several factors.

In a large F number Korsch space camera, the focal length is 10 m and the F number is 16. To maintain good image quality, temperature vibration of secondary mirror structure should be controlled within range of ±0.3 °C.

If TCBDBCC is adopted, temperature vibration of secondary mirror structure cannot be controlled within required range. This will cause frequent defocusing of space camera.

Precision Thermal Control technology Based on Heat Shield and Radiation Heat transfer (TCBHSRH) is proposed by author due to accidental inspiration from engineering prototype stage thermal experiment of the space camera. In experiments and applications, by adopting TCBHSRH, temperature vibration of secondary mirror structure can be controlled within range of ±0.3 °C. Problem about precision temperature control of secondary mirror structure in the large F number Korsch space camera is resolved.

2 Secondary Mirror Structure in Korsch Optical System

In Korsch optical system [1], incident light, reflected by primary mirror, converges to secondary mirror, reflected by secondary mirror, forms imperfect image near primary mirror. Third mirror or even forth mirror is often adopted to make the image perfect. Folding mirror is often adopted to make light path layout more suitable for manufacturing. Typical Korsch optical system is shown in Fig. 1.

In order to reduce outer dimension and weight of optical system, Modulation Transfer Function (MTF) of optical system is controlled near 0.1 by means of large F number, and later image quality will be improved by means of Modulation Transfer Function Compensation (MTFC). In low MTF optical system, manufacturing tolerance of MTF is strict and all manufacturing tolerances are strict. For example, if MTF decrease of 10% is allowed, the decrease amount is 0.02 in optical

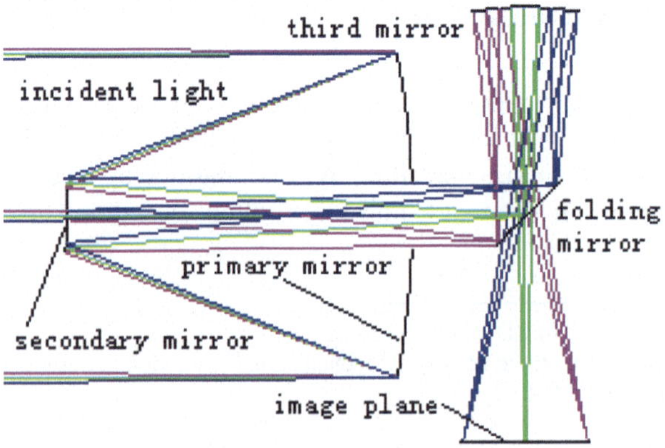

Fig. 1 Typical Korsch optical system

system with MTF of 0.2 but the decrease amount is only 0.01 in optical system with MTF of 0.1. Low MTF optical system is strict tolerance optical system which is more difficult to develop.

Secondary mirror is in the middle of incident light, support structure must pass through incident light to reach secondary mirror. MTF of optical system is decreased due to blocking of incident light by secondary mirror and its support structure. So in application, cross section area of secondary mirror structure must be reduced to minimize incident light blocking.

There are two types of secondary mirror support structures in application. In angular poles support structure, as shown in Fig. 2a, three angular poles extend from structure outside primary mirror to secondary mirror. In front barrel support structure, as shown in Fig. 2b, a barrel (front barrel) extends from structure outside primary mirror along axial direction of primary mirror to the height of secondary mirror, then three poles extend from top of front barrel to secondary mirror along radial direction of primary mirror. With the same support stiffness, incident light blocking area of angular poles support structure is bigger than that of front barrel support structure. MTF of optical system adopting angular poles support structure is smaller than MTF of optical system adopting front barrel support structure. Strict tolerance optical system often adopts front barrel structure to support secondary mirror.

In front barrel support structure of secondary mirror, three radial poles are in incident light path. Stray light forms when incident light reflected by poles goes into

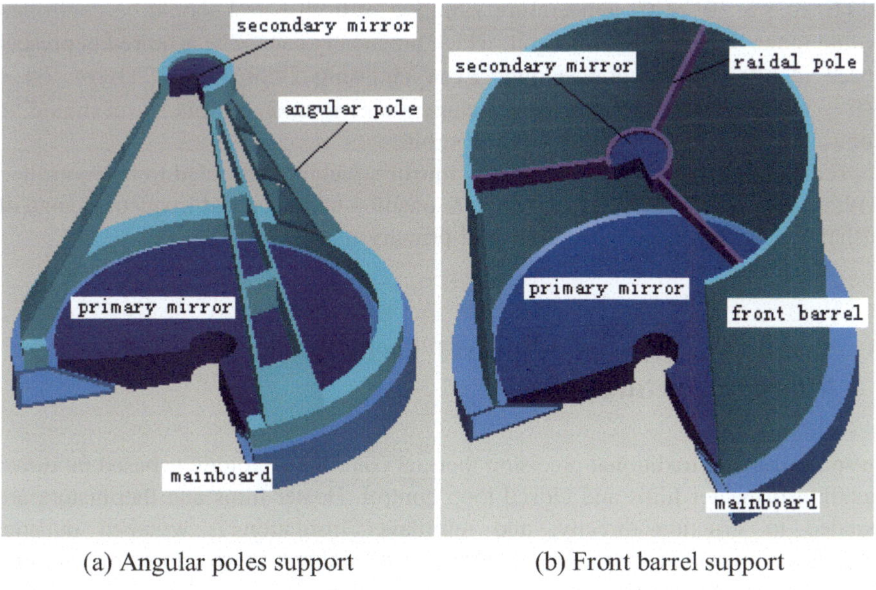

(a) Angular poles support (b) Front barrel support

Fig. 2 Typical support structure of secondary mirror in Korsch optical system image plane detector. Stray light elimination measure [2], such as matt black paint, should be taken to radial poles. Stray light forms when light reflected by inner surface of front barrel goes into image plane detector. Stray light elimination measure, such as vanes ore honeycombs, should be taken to inner surface of front barrel.

In Korsch optical system, tolerance of angle and position of secondary mirror according to primary mirror are strict [3]. In the space camera, radial translation of 5 μm on secondary mirror, or axial translation of 2 μm on secondary mirror, or rotation of 3" around radial direction on secondary mirror, each will cause decrease of 0.01 on MTF of optical system. Strict tolerance of secondary mirror requires that support structure must possess good stability and temperature of support structure must be stable and uniform.

In Korsch optical system, position of image plane changes while distance between secondary mirror and primary mirror changes [4]. In the space camera, half depth of focus is about 120 μm, but 2 μm change of distance between secondary mirror and primary mirror will cause about 300 μm position change of image plane, and will cause defocusing. To deal with low frequency defocusing, focusing mechanism is always adopted in space camera. But focusing mechanism should be controlled manually, so it cannot solve the problem of frequent defocusing. Frequent movement of image plane should be controlled within range of ± 20 μm. To avoid frequent defocusing, frequent change of distance between secondary mirror and primary mirror should be controlled within range of ± 0.2 μm.

In the space camera, front barrel support structure is adopted to support secondary mirror. If front barrel is made of Tc4 with Coefficient of Thermal Expansion (CTE) of $8 \times 10^{-6}/°C$, temperature vibration of front barrel should be controlled within range of ± 0.038 °C on orbit, which practically cannot be acquired in present space remote sensing engineering. Even choosing C/SiC with CTE of $7.5 \times 10^{-7}/°C$ to manufacture front barrel, temperature vibration of front barrel should be controlled within range of ± 0.4 °C on orbit.

Temperature vibration on secondary mirror structure is decided to be controlled within range of $\pm$ 0.3 °C, taking angle & position tolerance and frequent change of distance between secondary mirror and primary mirror into account.

3 Traditional Thermal Control Technology on Secondary Mirror Structure

In space camera, traditional precision thermal control technology is based on direct bonding of heater films and closed-loop control. Heater films and thermistors are bonded to structure directly, and multilayer insulation is wrapped outside. Multilayer insulation suppresses disturbance of ambient temperature [5], thermistors measure the temperature of structure, and heater films heat the structure according to measured temperature. Structure is controlled around certain temperature higher than ambient temperature via TCBDBCC.

In space camera, with traditional TCBDBCC, no better temperature control precision than ±0.5 °C can be achieved due to limited measurement accuracy of thermistor [6, 7], slow response speed of measurement system, on-off control mode or unneglectable overshoot of closed-loop control [8], large heat transfer nonuniformity of heater film [9] and unavoidable disturbance of ambient temperature [10].

In traditional TCBDBCC, heater films are bonded directly to structure. Heat transfer from heater films to direct bonding areas by means of heat conduction, strong thermal coupling exits between heater film and direct bonding area of structure. Heat transfer from heater films to none bonding areas by means of small angle coefficient heat radiation, weak thermal coupling exits between heater film and none bonding area of structure.

Overshooting temperature vibration occurs in direct bonding area of structure due to strong thermal coupling between heater film and direct bonding area. Due to weak thermal coupling between heater film and none bonding area, temperature of none bonding area is controlled by means of heat conduction between direct bonding area and none bonding area. So there must be difference in temperature exit between direct bonding area and none bonding area. Amplitude of difference in temperature between direct bonding area and none bonding area depends on thermal resistance between two areas and status of heat leakage between none bonding area and environment.

Controlled with TCBDBCC, large temperature nonuniformity exits, frequent temperature vibration exits, and temperature of structure is unstable due to disturbance of ambient temperature.

Adopting traditional TCBDBCC, heater films and thermistors are bonded to three radial poles, and matt black paint is coated, but none multilayer insulation can be wrapped outside due to stray light elimination requirement. Heater films and thermistors are bonded to outer surface of front barrel, multilayer insulation is wrapped outside, vanes or honeycombs are installed on inner surface of front barrel, but none multilayer insulation can be wrapped inside due to stray light elimination requirement. Temperature of secondary mirror structure is easily influenced by external heat flux due to direct exposing of radial poles and front barrel in external heat flux [11].

Via simulation we know that, adopting traditional TCBDBCC, temperature vibration of secondary mirror structure cannot be controlled within range of ±0.3 °C [12].

4 TCBHSRH in Secondary Mirror Structure

In the space camera with focal length of 10 m and F number of 16, via simulation we know that, adopting traditional TCBDBCC, temperature vibration of secondary mirror structure cannot be controlled within range of ±0.3 °C. Because no better thermal control technology is available, TCBDBCC is adopted in engineering prototype. TCBHSRH is proposed by author due to accidental inspiration from thermal experiment of the space camera in engineering prototype stage. In experiments and applications, by adopting TCBHSRH, temperature vibration of secondary mirror structure can be controlled within range of ±0.3 °C.

4.1 Inspiration from Thermal Experiment

Layout of thermal experiment is shown in Fig. 3. Space camera adopting TCBDBCC is wrapped with heat control multilayer outside. Thermal control bin provides simulated ambient temperature. Infrared cage provides simulated external heat flux [13]. Space camera, thermal control bin and infrared cage is installed in low temperature vacuum tank. Collimator is adopted to test MTF of space camera. Walls of thermal control bin are made of aluminum sheet with thickness of 1 mm. Bonded with Heater films and thermistors, walls of thermal control bin are wrapped with multilayer insulation outside. Closed-loop control is adopted to obtain good temperature performance of thermal control bin. Heat transfer between thermal control bin and space camera by means of heat radiation, weak thermal coupling exits between thermal control bin and space camera.

MTF and temperature control precision are to be tested in high temperature condition and low temperature condition. In high temperature condition, temperature of heat control bin is set to +40 °C. In low temperature condition, temperature of heat control bin is set to −30 °C.

Something unexpected happened when the experiment begins. Resistances of heater films on space camera do not match adopted temperature controller for

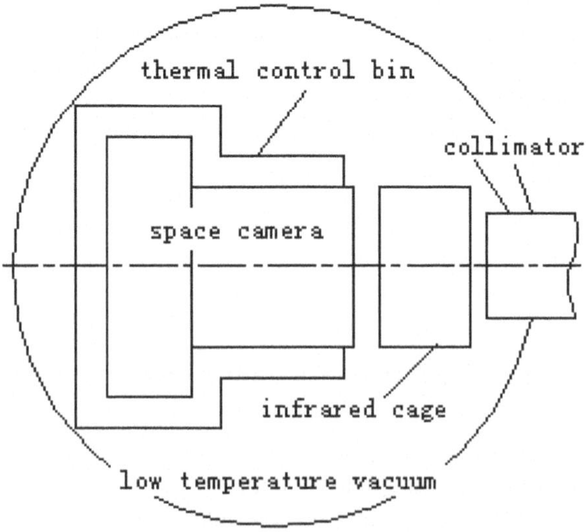

Fig. 3 Layout of thermal experiment

general use. Minimum unit adjustment in output voltage of temperature controller will induce several degrees centigrade of change in temperature on space camera. Equipped general use controller cannot fulfill the task of temperature control on space camera.

Testing of temperature is canceled. In order to finish MTF testing in both high temperature condition and low temperature condition, manual control of heater films on space camera is decided. Space camera is controlled near expected temperature via manual adjustment of output currents from power adaptors.

Under this condition, a phenomenon attracts the attention of author. Under thermal equilibrium state, keeping power supply of heater films on space camera unchanged, adjusting temperature of thermal control bin from +40 °C down to −30 °C, reaching thermal equilibrium state again, only no more than 20 °C of difference in temperature of space camera exits between the two thermal equilibrium states. Under passive condition, without thermal control, temperature of space camera changes no more than 20 °C, while temperature of thermal control bin changes 70 °C.

According to this phenomenon, if adding a "thermal control bin" to secondary mirror structure, and controlling temperature vibration of "thermal control bin" within range of ±1 °C, can we control temperature vibration of secondary mirror structure within range of ±0.3 °C?

4.2 *Proposing of TCBHSRH*

Controlling temperature of structure with "thermal control bin" outside, temperature of structure is controlled indirectly by means of controlling temperature of "thermal control bin". Heat transfer between "thermal control bin" and structure by means of heat radiation, weak thermal coupling exits between "thermal control bin" and structure.

With thermal control based on "thermal control bin", temperature vibration amplitude of structure is much smaller than that of "thermal control bin" outside, and they share the same vibration center. With same thermal control equipments, better temperature control precision and better temperature control uniformity can be obtained via "thermal control bin". For example, if temperature vibration of "thermal control bin" is controlled within range of ±1 °C, perhaps temperature vibration of structure can be controlled within range of ±0.3 °C; if temperature uniformity of "thermal control bin" is controlled within range of ±1 °C, perhaps temperature uniformity of structure can be controlled within range of ±0.3 °C.

With thermal control based on "thermal control bin", temperature of a point on wall of "thermal control bin" influences temperature of certain nearby area of structure, and temperature of a point of structure is influenced by temperature of certain nearby area on wall of "thermal control bin" [14]. This leads to better temperature uniformity on structure.

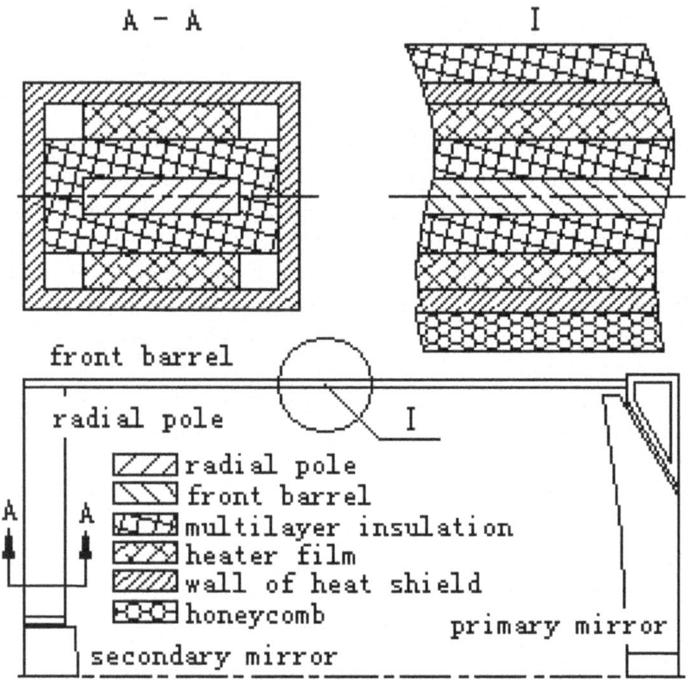

Fig. 4 Thermal control of secondary mirror adopting TCBHSRH

In order to add “thermal control bin” to three radial poles, walls are added outside each pole, as shown in Fig. 4, we name the walls as “heat shield”. Heater films and thermistors are bonded, and closed-loop control is taken to control the temperature of heat shield. Only thermistors are bonded to poles in order to monitor the temperature of poles. In order to obtain proper thermal resistance and proper heat radiation angle coefficient, certain distance is set between heat shield and pole, thermal control coating is coated to pole and inside surface of heat shield [15], and multilayer insulation is place between pole and heat shield. Matt black paint is coated to outside surface of heat shield to fulfill the requirement of stray light elimination. Adopting heat shield, temperature of radial poles is controlled via heat radiation, and direct disturbance of external heat flux on radial poles is prevented.

Similar measures are taken to add heat shield for front barrel, as shown in Fig. 4. Vanes or honeycombs are installed on inner surface of heat shield. Multilayer insulation is wrapped outside to suppress disturbance of ambient temperature.

Due to the adoption of heat shield to three radial poles, incident light blocking area is increased, and MTF of the space camera is decreased by about 0.003. None other influence but slight enlarge of diameter is done to front barrel by the adoption of heat shield to front barrel.

Thermal control of secondary mirror structure adopting TCBHSRH is submitted to project team. Supplement and refinement and detailed design are done by thermal control designer. Corresponding changes of secondary mirror structure are done by structure designer. Via simulation we know that, adopting TCBHSRH, temperature vibration of secondary mirror structure can be controlled within range of ±0.3 °C.

4.3 Temperature Control Precision of TCBHSRH in Experiments and Applications

TCBHSRH is adopted to control temperature of secondary mirror structure in identification stage of the space camera. In thermal experiment, temperature vibration of secondary mirror structure, including three radial poles and front barrel, is controlled within range of ±0.3 °C.

The space camera was launched more than a year ago. On orbit state, temperature vibration of secondary mirror structure is controlled within range of ±0.2 °C.

5 Extended Application

Within the space camera, besides secondary mirror, other components, such as primary mirror and main board, are controlled by TCBHSRH. Temperature control precisions of all components are satisfying.

TCBHSRH is already adopted by several following space camera, effects of thermal control precision are improved.

6 Conclusions

In Korsch optical system, secondary mirror is in the middle of incident light, support structure must pass through incident light to reach secondary mirror. None multilayer insulation can be wrapped outside due to stray light elimination requirement. Temperature of secondary mirror structure is easily influenced by external heat flux due to direct exposing.

In space camera, traditional precision thermal control technology is based on direct bonding of heater film and closed-loop control. Temperature control precision is ±0.5 °C.

In a Large F Number Korsch Space Camera, temperature vibration of secondary mirror structure is to be controlled within range of ±0.3 °C. Traditional TCBDBCC cannot fulfill the requirement.

TCBHSRH is proposed. In experiments and applications, by adopting TCBHSRH, temperature vibration of secondary mirror structure can be controlled within range of ±0.3 °C. Problem about precision temperature control of secondary mirror structure in the large F number Korsch space camera is resolved. TCBHSRH is also adopted on other parts and in other space cameras.

References

1. He Baokun, Lan Liyan, "Simulation Analysis on Distortion Stability of Coaxial Three-mirror System Used in Field Bias ", *SPACECRAFT RECOVERY & REMOTE SENSING*, 2012, 33 (6): 56–60.
2. Zhong Xing, Jia Jiqiang, "Stray light removing design and simulation of spaceborne camera", *Optics and Precision Engineering*, 2009, 17(3):621–625.
3. Ming Ming, Wang Jianli, Zhang Jingxu, "Error Budget and Analysis for Optical System in Large Telescope", *Optics and Precision Engineering*, 2009, 17(1): 104–108.
4. Wang Hong, Han Changyuan, "Study on the Thermal Effects of the Optical System on a Aerospace Camera", *Optical Technique*, 2003, 29(4):452–457.
5. Yang Ming, Wu Xiaodi, You Jun, "Analysis of Thermal Properties at Surface of Multilayer Insulation of Space Target", *IFRARED*, 2009, 30(5): 32–35.
6. Zhang Jiaxun, Wang Hong, Sun Jialin, "The Application Analysis of Thermistor in Spacecraft", *CHINESE SPACE SCIENCE AND TECHNOLOGY*, 2004, 6: 56–59.
7. Wu Jifeng, Xu Xiaodong, Xue Min, Liu Yaping, "Accuracy Analysis of On-Board Telemetry Acquirement", *SPACECRAFT ENGINEERING*, 2007, 16(5): 54–58.
8. Zhang Chunxi, Li Sen, Chen Guangjian, "Digital temperature control system design for FOG inertial measurement unit", *Journal of Chinese Inertial Technology*, 2008, 16(5): 544–547.
9. Fu Liangliang, He Xin, Wu Jin, Wang Zhongshan, "Design and Analysis of the Third Mirror Mantle Structure for a Space Camera", *Opto-Electronic Engineering*, 2013, 4(7): 44–49.
10. Yao Genhe, Wu Huahong, "Precise Aptitude Temperature Control Technology of Satellite", *The 5th Space Science Conference*, 2004: 59–69.
11. Liu Chen, "Calculation of transient space heat fluxes for space cameras working in sun-synchronous circle orbit ", *Chinese Optics*, 2012, 5(2): 148–153.
12. Song Xinyang, Gao Juan, Zhao Zhenming, Lu Pan, "Application of Indirect Thermal Control Technology for Constant Temperature Control of HR Optical Remote Sensor", *SPACECRAFT RECOVERY & REMOTE SENSING*, 2012, 33(6): 56–60.
13. Guan Fengwei, Liu Chen, Yu Shanmeng, Huang Yong, Cui Kang, "Space heat flux simulation and programmable load for thermal test of space optical remote sensor", *Chinese Optics*, 2014, 7(6): 983–988.
14. Robert Siegel, John R. Howell, *THERMAL REDIATION HEAT TRANSFER*, Washington: Hemisphere Publishing Corporation, 1980:233–275.
15. Shimazaki K, Ohnishi A, Nagasaka Y, "Development of spectral selective multilayer film for a variable emittance device and its radiation properties measurements", *International Journal of Thermophysics*, 2003, 24(3):757–769.

Space-Borne Integrated Design Analysis of Remote Sensing Camera on Geosynchronous Orbit

Yue Wang, Shiqi Li, Feng Yu, Wenpo Ma, Minlong Lian, Jie Dong and Nana Xu

Abstract This Paper presents a space-borne integrated design analysis of remote sensing camera on Geosynchronous orbit. Considering the engineering practice, a typical structure forming method is given out which is applied to the Geosynchronous remote sensor camera. The organic unity of the integral work of satellite platform and the payload provides the steady support of the equipment, in addition, the difficulties of camera structure design are reduced remarkably. Simultaneously, the full consideration is given on the integration design of camera's thermal control structure and the satellite structure. As a result, the integrated design makes the less technical difficulties of the design and installation of thermal control structure, higher efficiency of passive thermal dissipation and less on-orbit power waste. Integrated design result shows, the payload had outstanding characteristics of high rigidity and high stability. Consequently, it successfully passed the qualification level of vibration test and thermal vacuum test. This paper details an integrated design method of space-borne remote sensing camera for further reference.

Keywords Geosynchronous orbit · Remote sensing camera · Space-borne · Integrated design

1 Introduction

As is well known, the satellites working on the geosynchronous orbit (GEO) have the potential to achieve a stable orientation to the earth [1]. Thus the GEO remote sensors can provide wide and continuous coverage of ground observation [2]. Nations all over the world attach importance of GEO remote sensors because of its real-time, continuous observation and rapid response for tracking and staring

Y. Wang (✉) · F. Yu · W. Ma · M. Lian · J. Dong · N. Xu
Beijing Institute of Space Mechanics and Electricity, Beijing 100094, China
e-mail: 77792088@qq.com

S. Li
Huazhong University of Science and Technology, Wuhan 430074, China

H.P. Urbach and G. Zhang (eds.), *3rd International Symposium of Space Optical Instruments and Applications*, Springer Proceedings in Physics 192,
DOI 10.1007/978-3-319-49184-4_4

objects [3]. The GEO remote sensors have provided continuous coverage for US and international operational missions which respond to a broad range of environmental and military mission needs since 1970s [4]. For the environmental applications, the GEO remote sensors can provide real time observing services such as weather forecast, hurricane, thunderstorm, flood, and earthquake. The sensors can also enable monitoring of volcanic eruptions and forest fires using IR channel [5]. Environmental GEO programs have been implemented by nations around the world and include GOES (US), Meteosat (Europe), Himawari (Japan), GOMS (Russia), INSAT (India), and Feng-Yun (China).

Up to now, most of earth observing imagers on the meteorological satellites are scanning type imagers. Comparing with scanning type imager, the staring imager has advantages in aspects such as imaging frequency, observing efficiency, energy consumption, and instrument weight [6, 7].

Comparing with the LEO remote sensors, the GEO remote sensors have advantages such as real-time, continuous observation and rapid response [8]. But there are disadvantages in high level radiation intensity, severe thermal environment and high launch cost. Consequently, there are difficulties in camera structure design, anti-radiation design and on-orbit thermal design of the GEO remote sensor [9].

In recent years, the thermal and structure integrated design method is widely used in system design of space remote sensors to achieve their lighter mass, lower power consumption and better structure optimization [10–13].

In this paper, we discussed the function needs of the GEO staring remote sensing camera, and carried out the optimization design using the integrated design method in the key aspects of distribution of the camera's main structure, assembly method of the camera and the satellite, and the thermal design of the camera. The vibration test and the thermal vacuum test were carried into execution and the veracity and the rationality of the design were validated by the test results.

2 Design Background

Working on the GEO orbit, the GF-4 camera is a multi-spectral imager with visible and infrared spectrum. The camera can provide information that enables observation and forecasting of local weather, forest and natural disaster. Table 1 shows the key design parameters of the camera. A RC double reflection optical system with corrective lens was selected to provide high image quality for the camera (Fig. 1). The two optical channels are formed by the dichroic beam splitter.

Considering the technological maturity and the scheduling requirement of the project, the mature design method of the satellite bus was carried out. The bus structure is built on a octagon shape with a carbon-carbon central bearing tube to ensure a maximum stability [14]. The top of the central bearing tube is designed as the mounting interface of the camera main structure. The other payloads, such as the star tracker and the antenna system, are also mounted directly on the top panel of the satellite bus. The 3D model of the satellite bus is shown in Fig. 2.

Table 1 Key features of GF-4 camera

Optical design (μm)	Refractive with dichroicly channels: 0.45–0.9 and 3.5–4.1
Sensor bands	Panchromatic/multispectral/MWIR
Spatial resolution (km)	0.05VIS/0.4MWIR
Full disk coverage time (min)	30
Swath Width (km)	400 × 400
NEdT (k)	0.2
Sensor cooling	Active
Mass (kg)	≤400
Power (W)	≤800

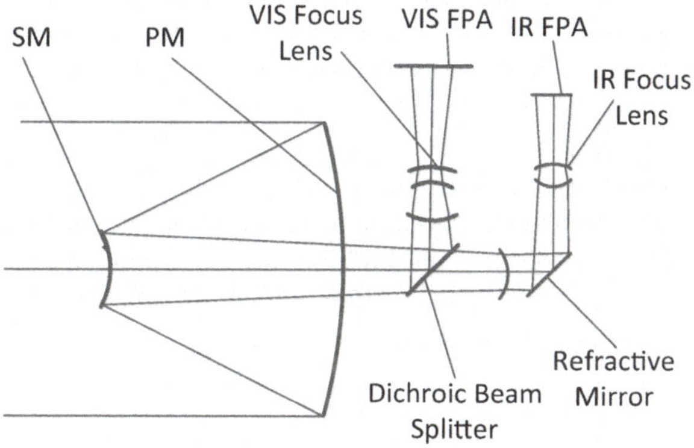

Fig. 1 A space remote sensing camera's optical system configuration

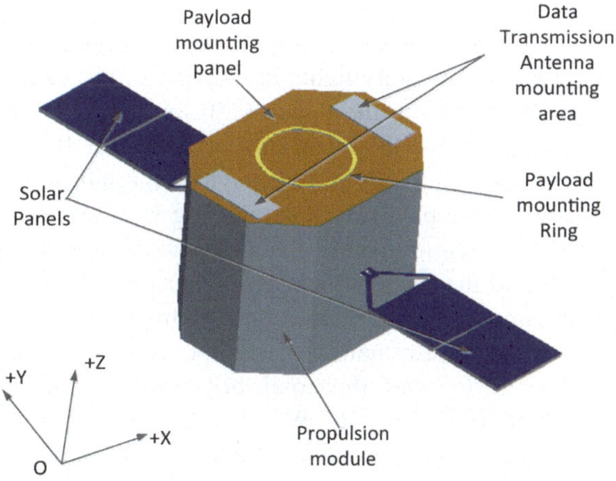

Fig. 2 The 3-D model of satellite

3 Integrated Design of the Camera

The main structure of the camera integrates the optical, mechanical and protective components into a complete instrument. This structure must afford high optical alignment stability on orbit and ensure the optical system safe through launching environment [15]. With the increase of the diameter of the remote sensing camera, the integrated design method of the camera possesses incomparable advantage than the separated design method [16]. The Pleiades-HR camera which launched by CNES in 2009 is integrated inside a satellite bus. The star trackers, gyroscope heads and the highly integrated detection unit are directly supported by the camera to avoid thermal distortion by the satellite bus [17, 18]. The high image location accuracy and the high resolution are achieved by minimizing the interface between the camera and the bus. However, a comprehensive consideration in aspects such as total mass, power consumption and payload stabilization should be noticed to carry out integrated design of the camera and the bus when the configuration of the bus is confirmed.

With the comparison of the truss configuration [19], the plate and frame configuration is more suitable for the typical optical system of the GF-4 camera. This paper will discuss the building and mounting mode of the baffle, the connection mode of camera and satellite and the thermal dissipating mode of the camera from the perspective of system integration.

3.1 Design Restrictions and Cases Discussion

Dimension restrictions: according to the optical system requirement, a baffle is necessary for the stray light suppression of the camera, besides, the aperture stops should be placed in the upper 1350 mm part of the baffle. The camera needs an adapter to achieve the connection with the satellite bus. The distance between the interface of the camera and the payload mounting panel of the bus is 700 mm. The diameter of the payload mounting ring is 1280 mm.

Stiffness requirements: the first modal frequency in lateral direction of the camera and the baffle should be higher than the satellite bus in a certain range to avoid camera damage caused by coupling resonance.

Thermal requirements: the temperature needs of the main optical system is 20 °C ± 2 °C, and the total power consumption restriction of the camera should be less than 712 W.

Weight restrictions: the total mass of the camera should be less than 400 kg.

Two integrated design case of the camera and the satellite is shown in Fig. 3. In case a, the camera baffle is directly mounted on the main structure of the camera, the camera and the bus are connected by the payload adapter. In case b, the camera baffle is divided into two parts. Separated from the main structure of the camera, the baffle is mounted directly on the payload adapter. The flexible support is designed

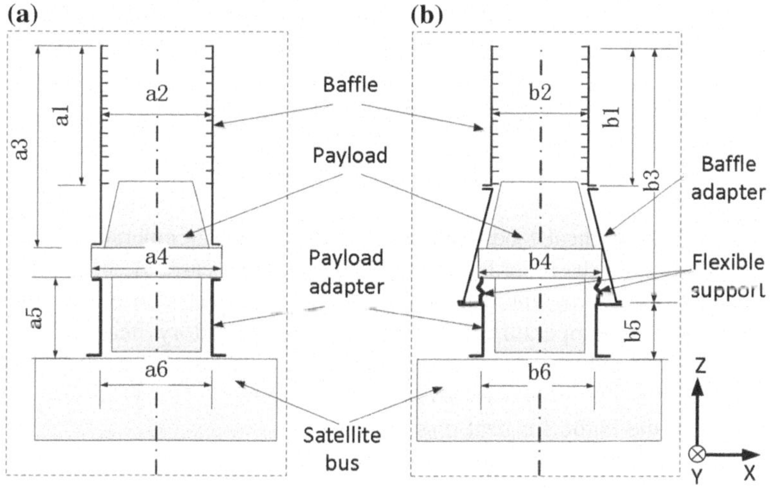

Fig. 3 Two cases of integrated design for payload and satellite

Table 2 Cases comparison of the integrated design of the camera and the satellite

	a	b
Main structure of the camera	Larger in size, higher stiffness requirement, greater mass	Smaller in size, lower stiffness requirement, lesser mass
The baffle	Larger diameter, difficult to thermal design	Smaller diameter, easy to thermal design
The payload adapter	Larger in size, greater mass	Smaller in size, lesser mass
Connection way of the camera	Higher accuracy requirement, without vibration isolation capability	Lower accuracy requirement, with vibration isolation capability

to realize the connection of the camera and the satellite bus. Table 2 shows the advantages and disadvantages between the two cases discussed before. Obviously, case b has tremendous advantages.

3.2 *Integrated Design of the Camera Baffle*

On the one hand, the baffle temperature should be controlled below 55 °C avoiding infrared unwanted ray at high temperature cases, so heat dissipation methods are needed; on the other hand, the temperature should fulfill the stability requirement and the camera main body should be controlled at 20 °C ± 2 °C at low temperature cases, so heat compensation methods are also needed. The best thermal design should consider both the high temperature cases and the low temperature cases. The balance equations according to the energy conservation law are as bellow:

$$Q_1 + Q_2 = \varepsilon \times \delta \times A \times (T_{high}^4 - 4^4) \quad (1)$$

$$Q_3 + Q_{com} = \varepsilon \times \delta \times A \times (T_{low}^4 - 4^4) \quad (2)$$

where Q_1 is absorbed heat load of the outer baffle at high temperature case, Q_2 is absorbed heat load of the inner baffle at high temperature case, T_{high} is temperature of the baffle at high temperature case, Q_3 is absorbed heat load of the outer and inner baffle at low temperature case, Q_{com} is compensatory heat load, T_{low} is temperature of the baffle at low temperature case, ε is infrared emissivity of the baffle surface, σ is the Stefan-Boltzmann constant($5.67 \times 10^{-8} W/m^2 \cdot K$), A is the surface area on the baffle for heat dissipation.

From Eqs. (1) and (2) we can deduce the following equation:

$$\frac{Q_1 + Q_2}{\varepsilon \times \delta \times \left(T_{high}^4 - 4^4\right)} \leq A \leq \frac{Q_3 + Q_{com}}{\varepsilon \times \delta \times (T_{low}^4 - 4^4)} \quad (3)$$

which gives the thermal design principle. In the equation, the area A should content the heat dissipation requirement at high temperature case, and Q_{com} should content heat compensation requirement at low temperature case. The design is optimal when "=" is content in Eq. (3).

Table 3 The two design plans of the baffle

	A	B
Baffle configuration	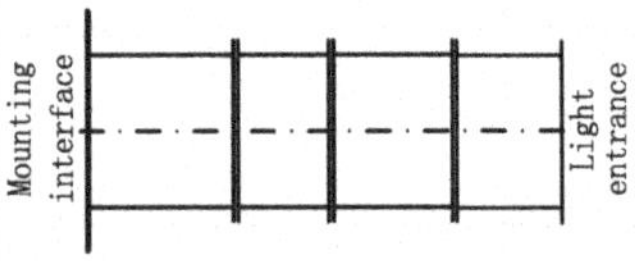	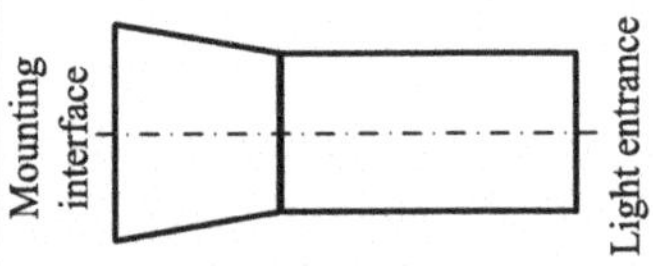
Manufacturing method	All four cylinder barrels: aluminum honeycomb structure	Conical barrel: carbon fiber composite structure Cylinder barrel: aluminum honeycomb structure
Total mass (kg)	56.4	66.2
First modal frequency (Hz)	75	77
Temperature control requirement	All four cylinder barrels: −70 ~ 55 °C	Conical barrel: −70 ~ 50 °C Cylinder barrel: −150 ~ 55 °C
Total power consumption (W)	242	78

Table 3 gives two plans of the baffle. In plan A, the baffle connects with the main body of the camera, so the baffle should avoid large scale temperature fluctuation for the high temperature requirement of the main body of the camera. Compared with plan B, T_{high} is lower and T_{low} is higher in plan A, so the area A and the compensation heat load Q_{com} is bigger. The design results of the two plans are stacked up in Table 3, we can see that plan B has big advantage on compensation heat load.

3.3 *Integrated Design of Flexible Supporting Structure*

The dynamic load during launch stage is the main cause of the damage or performance degradation for large aperture remote sensing camera. Besides, the bus disturbance source and the thermal distortion can also cause the reduction of the image quality. An effective way to solve this problem is to design a mechanics which can provide both functions such as vibration isolation during launch [20] and flexible unloading on orbit [21].

The forced vibration equation for the N-DOF system shown in Fig. 4 can be described as

$$[M]\{\ddot{q}(t)\} + [C]\{\dot{q}(t)\} + [K]\{q(t)\} = \{Q(t)\} \tag{4}$$

where M, C and K are matrices which present mass, damping and stiffness. $q(t)$ is the displacement response of the camera, $Q(t)$ is the excitation load. The system transmissibility TR can be described as follows by solving Eq. (5).

$$\mathrm{TR} = |q/Q| = \frac{\sqrt{1 + (2\xi\omega/\omega_n)^2}}{\sqrt{\left[1 - (\omega/\omega_n)^2\right]^2 + (2\xi\omega/\omega_n)^2}} \tag{5}$$

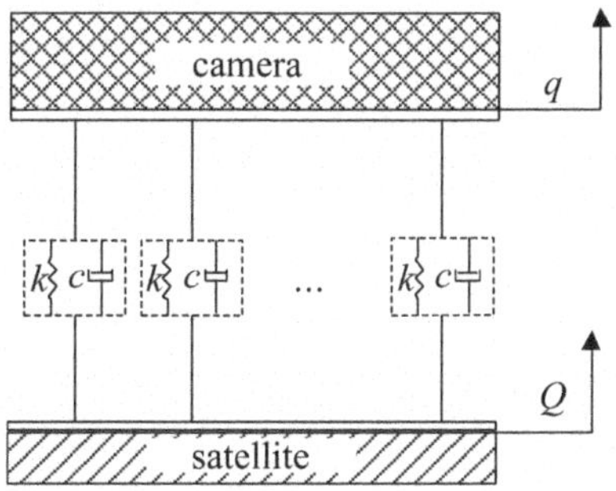

Fig. 4 Forced vibration of a N-DOF system

where $\omega_n = \sqrt{k/m}$ is the system natural frequency, $\xi = c/2\sqrt{km}$ is system damping ratio.

Based on the theory of passive vibration isolation, the displacement transmissibility $\mathrm{TR} = |q/Q|$ is determined for the different system damping ratio ξ. The camera is isolated when the normalized frequency $(\omega/\omega_n) > \sqrt{2}$. In this frequency range, the response of the camera q will always be less than exciting input Q, which means the $\mathrm{TR} < 1$ [22].

We have already known that the first two order lateral natural frequencies of the satellite bus are $f_{satL1} = 20$ Hz and $f_{satL2} = 56$ Hz, and the first order vertical natural frequency is $f_{satV1} = 50$ Hz. The most sensitive component of the camera is the SM component, and its first order vertical natural frequency is $f_{smV1} = 120$ Hz. In order to achieve the decoupling design of the camera, the design region of the camera frequencies in all three directions (f_x, f_y, f_z) must meet the following conditions:

$$\begin{cases} \sqrt{2} f_{satL1} < f_x = f_y < \frac{1}{\sqrt{2}} f_{satL2} \\ \sqrt{2} f_{satV1} < f_z < \frac{1}{\sqrt{2}} f_{smV1} \end{cases} \tag{6}$$

As the severe launching load usually happened in low frequency region, we define the frequency design area of the camera as (37–43)Hz in lateral and (75–90) Hz in vertical to reduce the resonance response of the camera.

To meet the design requirements, a modified vibration isolation truss structure (VITS) which derived from the bi-pod supporting structure is presented to connect the camera and the satellite bus through overall consideration of mounting space restrictions. Figure 5 shows the 3D model of the VITS. The system damping is provided by the material which is filled between the inner and the outer truss

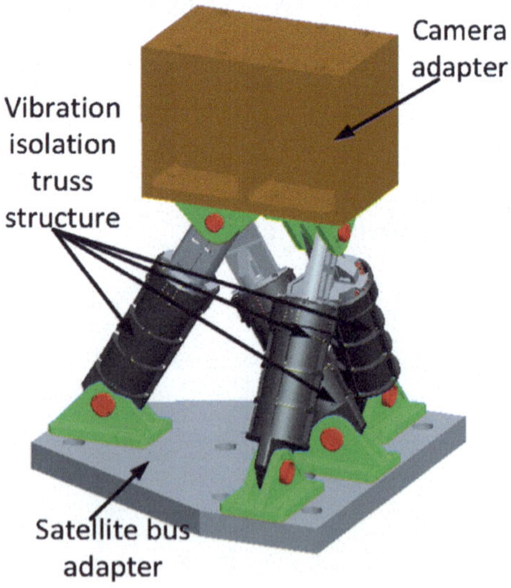

Fig. 5 The 3D model of the VITS

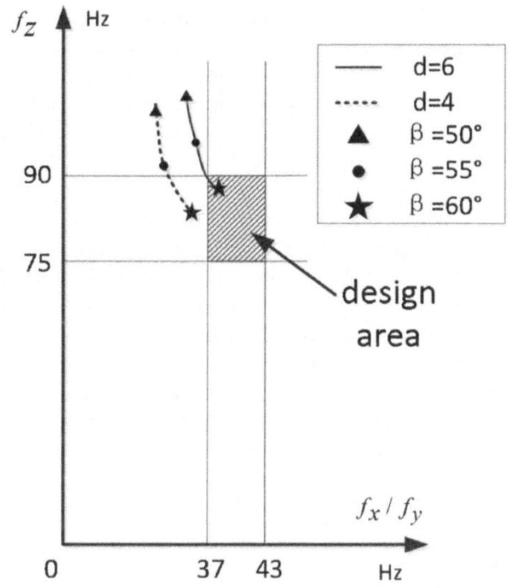

Fig. 6 The design area and the optimization process of VITS

Table 4 Modal frequency comparison between simulation and test

	Design requirement (Hz)	Simulation result (Hz)	Test result (Hz)
fx	37–43	40	38.3
fy	37–43	40	37.8
fz	75–90	88	84.1

cylinder through shearing movement of the truss, and the system stiffness is provided by the metallic truss and the rubber layers on the top of the truss. A dozen of VITS trusses which are divided into three groups are assembled to realize the connection of the camera and the satellite bus. The optimization process (Fig. 6) is achieved by adjusting the parameters of rubber layer d and V type truss angle β.

The modal frequency results meet all requirements after a series of optimization, and confirmed by test results in Table 4. The vibration test details are described in Sect. 4 of the paper.

3.4 *Integrated Design for Camera's Heat Dissipation with the Satellite*

This chapter gives the simulation and analysis results of solar heat radiation for geosynchronous orbit. In the simulation, +Z points to the earth, +X points to the fly direction of the satellite, +Y is defined according to the right-hand rule. The results of the solar radiation for the typical date are showed in Fig. 7.

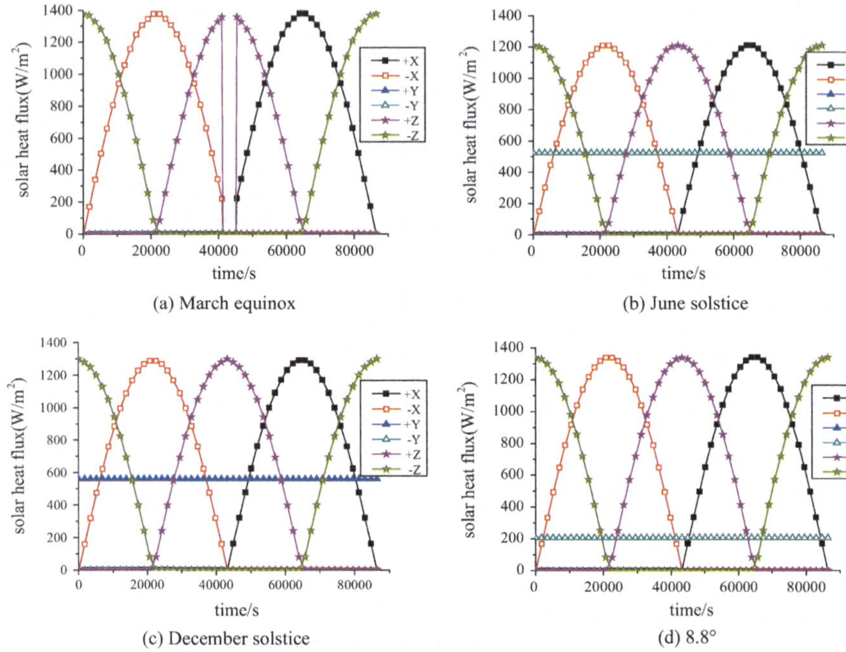

Fig. 7 Periodic heat of solar radiation

From Fig. 7, we can see that the space thermal environment of geosynchronous orbit is even worse compared with sun-synchronous orbit. The remote sensor will endure the cold black space's low temperature which is 4 K, and also will receive high heat flux of solar radiation which can reach to 1400 W/m^2. The hot and cold alternating time can up to several hours, so the outer temperature of the remote sensor will change from −150 to 150 °C, which is a big challenge for the high precision temperature control. Beside, all the surfaces will receive solar radiation and no shady side exists. ±Y are the best direction for heat dissipation due to the steady heat flux and only one direction will receive solar radiation at each time. Based on the analysis, we can connect the ±Y surfaces with heat pipe for heat

Fig. 8 The principle of the camera thermal dissipation

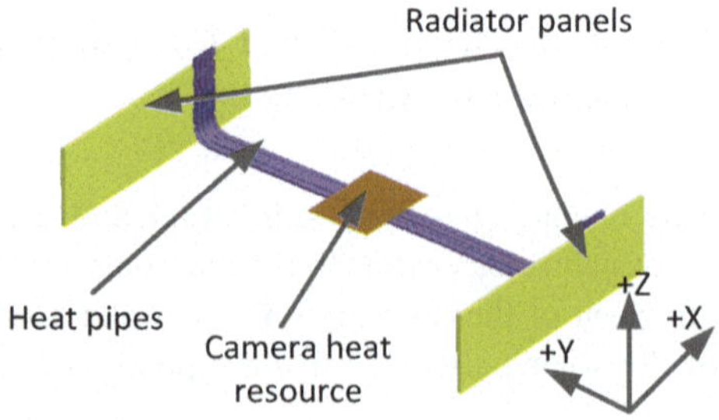

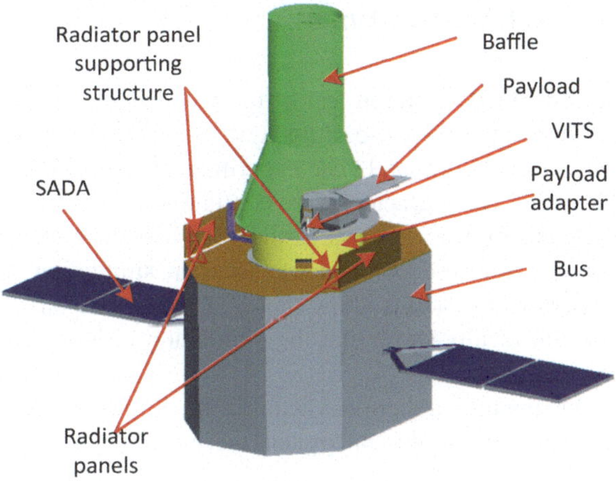

Fig. 9 The configuration of camera and satellite

dissipation, which is sketched in Fig. 8. Compared with using one direction for heat dissipation, the heat dissipation surface area can reduce 25% and the compensation heat load can reduce 25% too. Besides, the heat sources are collected to the bottom of the camera and the heat dissipation surfaces can install on the plain of the satellite directly, which reduced the design difficulty and mass of the supporting structures. In the final optimal design, the supporting structures weight only 5.6 kg. The integrated optimal design is show in Fig. 9.

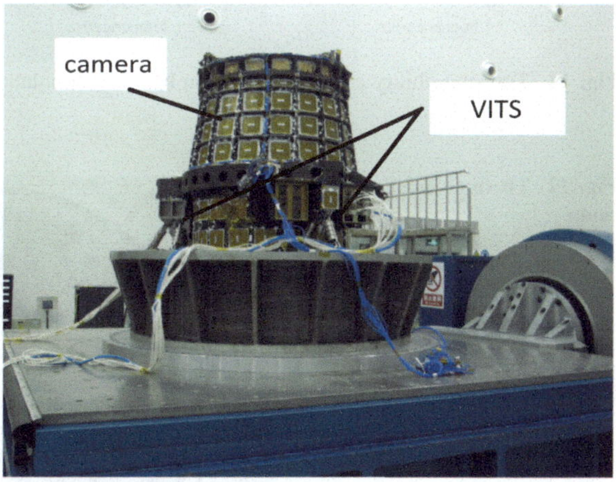

Fig. 10 Vibration test setup

4 Test Verifications

A Sinusoidal vibration test of the payload subsystem was completed to determine the frequency response of the system in a launch configuration. The camera system was a qualified model. The experimental setup for the qualification test is shown in Fig. 10. Accelerometers were configured to measure the lateral (X), lateral (Y), and vertical (Z) accelerations at the base and the top of the VITS. The resulting frequency response comparison between simulation and test is shown in Fig. 11. Experimental results show that VITS meet the structural requirements for launch, and the results show that the simulation curves agree well with that obtained by model tests.

Meanwhile, a thermal vacuum test setup (Fig. 12) was developed to demonstrate the veracity of the integrated thermal design using the large vacuum tank. In

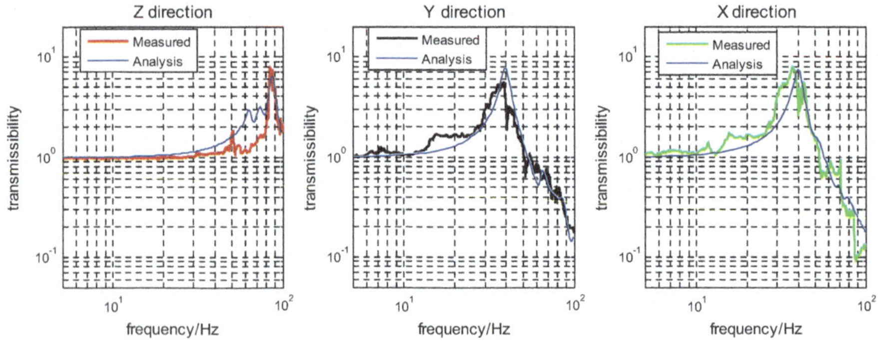

Fig. 11 Transmissibility curves comparison between simulation and vibration test

Fig. 12 Thermal vacuum test setup

Table 5 Temperature data comparison between thermal balance test and on orbit

Component	Transfer orbit		Working orbit	
	Test result (°C)	On orbit (°C)	Test result (°C)	On orbit (°C)
PM	16.2	16.2	18.4–20	17.4–19.8
SM	19.7	19.6	19.9–20.4	19.9–20.5
Support structures of SM	19.8–20.2	19.8–20.2	19.8–20.4	19.8–20.3
Optical lens	19.8–20.2	19.9–20.2	19.8–20.3	20.0–20.4
Main structures	19.7–20.3	19.6–20.4	19.8–20.5	19.6–20.6
FPA	7.8–7.9	7.8–7.9	7.9–12.6	7.7–10.8
Cryocooler	1.5–1.8	1.5–1.8	7.3–8.5	7.4–8.2

Table 6 Evaluations of the MTF test before and after the environment tests

Channel	Spectrum range (μm)	MTF requirement	MTF (before vibration test)	MTF (after vibration test)	MTF (after thermal vacuum test)
B1	0.45–0.90	≥0.14	0.143	0.145	0.148
B2	0.45–0.52	≥0.15	0.156	0.158	0.162
B3	0.52–0.60	≥0.15	0.155	0.157	0.162
B4	0.63–0.69	≥0.15	0.153	0.155	0.158
B5	0.76–0.90	≥0.12	0.137	0.139	0.140
B6	3.50–4.10	≥0.15	0.159	0.162	0.163

contrast with the thermal vacuum test results and on orbit test data (Table 5), we conclude the accuracy of the integrated thermal design.

The system MTF test results before and after environmental tests are shown in Table 6. The result data either before or after the two environmental tests meet all requirements of the six optical spectrum bands. The structure of the camera is tested stable, and the integrated design method is successfully proved reasonable and effective.

5 Conclusions

With full consideration of structural layout optimization, frequency decoupling optimization and power consumption optimization, this paper developed a space borne integrated design method for the GEO remote sensing camera. The vibration test and the thermal vacuum test were presented to illustrate the feasibility and validity of the integrated design method. Such developments can provide reference for integrated design of similar types of remote sensing camera.

References

1. Brian C. Busch. "Space-based solar power system architecture" [D]. Monterey, California: Naval Postgraduate School, 2012.
2. Jeffery J. Puschell, Lacy Cook, Yifal J. Shaham, Maciej D. Makowski, John F. Silny. "System engineering studies for advanced geosynchronous remote sensors: some initial thoughts on the 4th generation" [C]. *Proc. of SPIE*, 2008, Vol7087, 70870:G1-18.
3. Linghua Guo, Enyu Gao, Bo Liu, Dianjun Wang, Biru Wang. "Research on complicated imaging condition of GEO optical high-resolution remote sensing satellite" [C]. *Proc. of SPIE*, 2013, Vol 8889:88891X1-7.
4. Jason C. Eisenreich. "The All Seeing Eye: Space-Based Persistent Surveillance in 2030", ADA539694 [R], Alabama, Maxwell Air Force Base, 2009.
5. Timothy J. Schmit, Steven J. Goodman, Daniel T. Lindsey, et al. "Geostationary Operational Environmental Satellite (GOES)-14 super rapid scan operations to prepare for GOES-R"[J]. *Journal of Applied Remote Sensing*, 2013, Vol. 7, 073462:1-20.
6. S. Kalluri, J. Gurka, R. Race. "Improved Observations of Earth and Space Weather From GOES-R" [C]. *Proc. of SPIE*, 2009, Vol. 7456, 74560M:1–8.
7. CUT Dunjie. "A Staring Imager Concept for Geostationary orbit Meteorologic Satellite" [C]. *Proc. of SPIE*, 2013, Vol. 3501, 412-416.
8. Bernhard Haisch, Paul Robb, Keith Strong, Donald Shemansky. "Design for a low-cost geosynchronous 2.4-meter UV EUV solar system observatory" [C]. *Proc. of SPIE*, 1998, Vol. 3356, 678–688.
9. R.J. Koczor. "Technology needs for geostationary remote sensors" [C]. *Proc. of SPIE*, 1993, Vol. 1952, 134–140.
10. Paul Eccleston, Tom Bradshaw, John Coker , et al. "Mechanical and thermal architecture of an integrated payload instrument for the Exoplanet Characterisation Observatory" [C]. *Proc. of SPIE*, 2012, Vol. 8442, 8442U:1–15.
11. Matthew A. Greenhouse. "The JWST Science Instrument Payload: Mission Context and Status" [C]. *Proc. of SPIE*, 2015, Vol. 9143, 914307:1–12.
12. William R. Oegerle, Lee D. Feinberg, Lloyd R. Purves, et al. "ATLAST-9.2m a large-aperture deployable space telescope" [C]. *Proc. of SPIE*, 2012, Vol. 7731, 77312M:1–10.
13. Perret L, Boussarie E, Lachiver J M, et al. "The Pleiades System High Resolution Optical Satellite and Its Performances" [C]. 53rd International Astronautical Congress of the International Astronautical Federation, 2002.
14. Xie Xiaoguang, Yang Lin. "Spaceborne integration design of smart small earth observation satellite structure" [J]. *Infrared and Laser Engineering*, 2014, 43:53–58. (in Chinese)
15. G.E. Zurmehly, R.A. Hookman. "Thermal and structure design constraints for radiometers operating in geo orbit" [C]. *Proc. of SPIE*, 1992, Vol. 1693, 304–312.
16. XU Qinghe, FAN Lijia, GAO Hongtao, et al. "Integrated Configuration Design of Platform and Payload for Remote Sensing Satellite" [J]. *SPACECRAFT RECOVERY & REMOTE SENSING*, 2014, 35(4):9–16. (in Chinese)
17. Plaisant G, Le Goff R, Deswarte D, et al. "Design of the Focal Plane for the Pleiades Instrument" [C]. 53rd International Astronautical Congress of the International Astronautical Federation, 2002.
18. Baudoin A, Boussarie E, Damilano P, et al. "Pleiades-A Multi Mission and Multi Co-operative Program" [C]. 52nd International Astronautical Congress, 2001.
19. Shin Utsunomiya, Susumu Yasuda, Taihei Yano, et al. "Structure design of the telescope for Small-JASMINE program" [C]. *Proc. of SPIE*, 2015, Vol. 9143, 91430Z:1–6.
20. Allen J. Bronowicki. "Vibration Isolator for Large Space Telescopes" [J]. *JOURNAL OF SPACECRAFT AND ROCKETS*, 2006, Vol. 43, 45–53.

21. P. S. Wilke, T. A. Decker, L. C. Hale. “Highly damped exactly-constrained mounting of an x-ray telescope” [C]. *Proc. of SPIE*, 1995, Vol. 2445:2–13.
22. Julian R. Jarosh. “Active and Adaptive Control for Payload Launch Vibration Isolation” [D]. Ohio: AIR FORCE INSTITUTE OF TECHNOLOGY, 2000.

Dynamic Parameter Identification of Damping Reinforced Components and Its Application in Space Optical Instrument Stabilization

Shaohui Li, Guangyuan Wang and Guoqiang Ni

Abstract Damping reinforced components is an efficient way to improve the pointing stability of optical instrument. Traditionally, modeling methods of damping reinforced components are mostly based on simple load test data, such as time domain method or half power band width method. Modeling methods based on response under static load or dynamic load at a certain frequency cannot cover the dynamic characteristic in a wide frequency range. An experimental modeling method of damping reinforced components in wide frequency band was presented and its effect in optical structures was evaluated based on the model. First, a cantilever beam of damping reinforced component was excited in a sine sweep test to get its frequency response. A parametric viscous damping model was proposed and the frequency response was solved with Laplace transform. Then, the parameters were identified by solving an unconstrained optimization problem with pattern search method, which minimized the deviation between test result and calculated result by least square criteria. Finally, the pointing stability of a typical optical structure adopting damping reinforced components was analyzed based on the derived dynamic model, which indicated that the line of sight jitter was attenuated because of the absorption of vibration energy by damping reinforced components.

Keywords Dynamic · Damping · Optical structure · Parameter identification

1 Introduction

As the development of high performance space optical instrument, there is increasingly urgent demand of precise pointing. The optical payload becomes more and more sensitive to structure vibrations of satellite buses. Although these

S. Li · G. Ni
Beijing Institute of Technology, Beijing, China

S. Li · G. Wang (✉)
Beijing Institute of Spacecraft System Engineering, Beijing, China
e-mail: zhuichilun@126.com

H.P. Urbach and G. Zhang (eds.), *3rd International Symposium of Space Optical Instruments and Applications*, Springer Proceedings in Physics 192,
DOI 10.1007/978-3-319-49184-4_5

vibrations generally have extremely low magnitude, they can result in significant image quality degradation to an optical payload. The suggestion of using vibration isolators and dampers to isolate payload from the satellite bus has been presented in 1980s' [1]. Recently, WorldView-2 achieved its perfect image quality via using a set of low frequency isolators with fluid dampers [2]. Some of the optical payload manufacturers begin to provide vibration isolators as standard parts together with their main products. Damping reinforced components provided an efficient way to isolate the vibration transmitted to the optical payload onboard the satellite.

Traditionally, structure damping can be identified in both time domain and frequency domain. In time domain method, the damping parameter is calculated from the response to impulse excitation or the energy dissipation per cycle under sine excitation [3, 4]. Usually, the excitation frequency and amplitude cannot cover the load range of the actual working load, which will introduce apparent error in the identified damping model for load dependent components. Frequency domain methods are based on Fourier transform of infinite time domain signal. However, the length of time domain signal is limited in engineering application and numerical error would be introduced during signal processing [5, 6].

The limitation of dynamic models for damping reinforced components brings difficulties in structure design. The design result based on the models identified by traditional methods can only be applied in certain load condition because the working load is beyond the testing load. The dynamic model in the real circumstance varies from the design model, which may cause unexpected vibration. Therefore, it is necessary to model the damping reinforced components based on wide frequency band dynamic experiments.

2 Experiment Setup

The setup of the experiment system is shown in Fig. 1, where accelerometers a_1 and a_2 are arranged on the top of the clamp for control the shake table excitation and at the free end of the specimen for measuring the specimen vibration, respectively. The arrangement of the sensors can also be seen clearly in Fig. 2.

In the test, the specimen is fixed on the shake table by a specially designed clamp as shown in Fig. 2.

3 Modeling Method

3.1 Dynamic Model of the Specimen

For the sake of convenience in the further numerical analysis for a structure with damping reinforced components, the damping of the specimen is equalized as

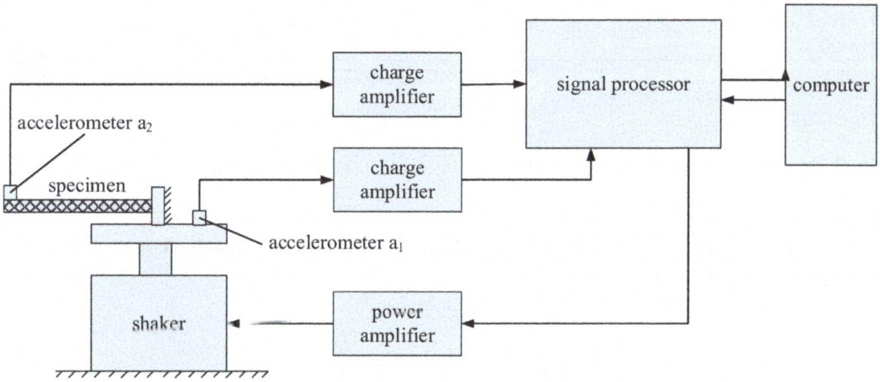

Fig. 1 Experiment setup

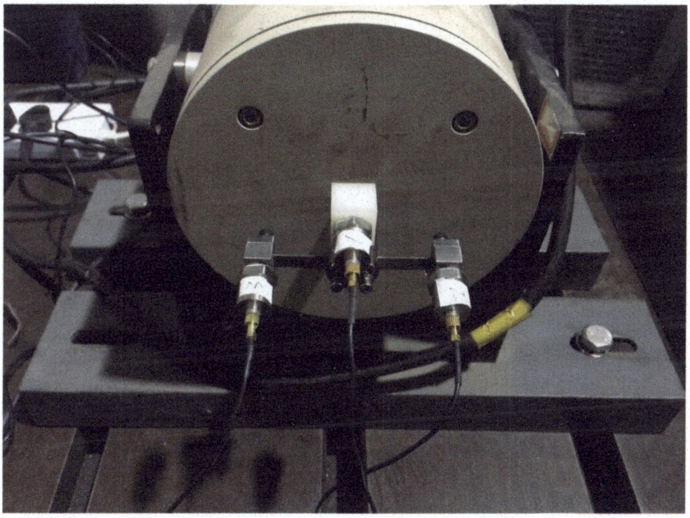

Fig. 2 Test rig and specimen

viscous damping. In the experiment, with a proper design of the clamp, the specimen can be taken as a cantilever beam with a lumped mass at its free end. Here the lumped mass is the accelerometer. The mechanic model of the specimen in the experiment is shown in Fig. 3. The equivalent flexibility rigid, the equivalent mass per unit length and the length of the specimen are EI, ρA, and l, respectively. The acceleration amplitude of the excitation is R and the excitation frequency is ω. The lumped mass is m. Let y be the transverse displacement of the beam, which is in the direction of perpendicular to the axis of the beam and a function of time t and

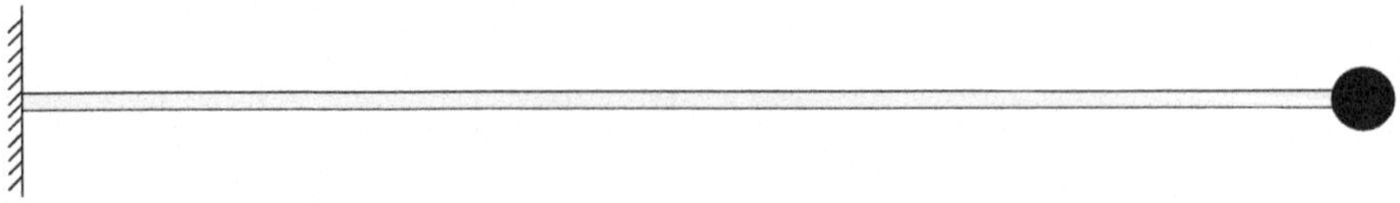

Fig. 3 Mechanic model of the specimen

dimensionless variable $x \in [0,1]$. The partial differential equation governing the motion y of the beam can be written as

$$EI\frac{\partial^4 y}{\partial (lx)^4} + \rho A\frac{\partial^2}{\partial t^2} + c\frac{\partial y}{\partial t} = 0 \tag{1}$$

with boundary conditions

$$\left.\frac{\partial^2 y}{\partial t^2}\right|_{x=0} = R\sin(\omega t)$$

$$\left.\frac{\partial y}{\partial (lx)}\right|_{x=0} = 0$$

$$\left.\frac{\partial^2 y}{\partial (lx)^2}\right|_{x=1} = 0$$

$$EI\left.\frac{\partial^3 y}{\partial (lx)^3}\right|_{x=1} + m\frac{\partial^2 y}{\partial t^2}|_{x=1} = 0$$

Applying the Laplace transform on Eq. (1), there is [7]

$$(\rho As^2 + cs + \frac{EI}{l^4}\frac{\partial^4}{\partial x^4})\bar{y}(x,s) = (\rho As + c)u_0(x) + \rho Av_0 = \bar{f}_{el}(x,s) \tag{2}$$

Let

$$\eta(x,s) = \left\{ \bar{y}(x,s) \quad \frac{\partial \bar{y}(x,s)}{\partial x} \quad \frac{\partial^2 \bar{y}(x,s)}{\partial x^2} \quad \frac{\partial^3 \bar{y}(x,s)}{\partial x^3} \right\}^T \in C^4$$

$$q(x,s) = \left\{ 0 \quad 0 \quad 0 \quad \frac{\bar{f}_{el}(x,s)}{EI} \right\}^T \in C^4$$

The dynamic equation of motion and the boundary conditions can be written as

$$\frac{\partial}{\partial x}\eta(x,s) = Q(s)\eta(x,s) + q(x,s) \tag{3}$$

$$M(s)\eta(0,s) + N(s)\eta(1,s) = \gamma(s) \tag{4}$$

In the above equations

$$Q(s) = \begin{bmatrix} 0 & 1 & 0 & 0 \\ 0 & 0 & 1 & 0 \\ 0 & 0 & 0 & 1 \\ -\frac{(\rho As^2 + cs)l^4}{EI} & 0 & 0 & 0 \end{bmatrix}$$

$$M(s) = \begin{bmatrix} 0 & 1 & 0 & 0 \\ 1 & 0 & 0 & 0 \\ 0 & 0 & 0 & 0 \\ 0 & 0 & 0 & 0 \end{bmatrix}$$

$$N(s) = \begin{bmatrix} 0 & 0 & 0 & 0 \\ 0 & 0 & 0 & 0 \\ 0 & 0 & 1 & 0 \\ -ms^2 & 0 & 0 & \frac{EI}{l^3} \end{bmatrix}$$

$$\gamma(x,s) = \left\{ 0 \quad \frac{f}{s^2+\omega_0^2} \quad 0 \quad ms\dot{y}(1,0) + my(1,0) \right\}^T \in C^4$$

In this way, a general solution of the equation can be written as

$$\eta(x,s) + \int_0^1 G(x,\xi,s)q(\xi,s)d\xi + H(x,s)\gamma(s) \tag{5}$$

where $G(x,\xi,s)$ is the Green function and it has the following form

$$G(x,\xi,s) = \begin{cases} e^{Q(s)x}(M(s)+N(s)e^{Q(s)})^{-1}M(s)e^{-Q(s)\xi}, & \xi < x \\ -e^{Q(s)x}(M(s)+N(s)e^{Q(s)})^{-1}N(s)e^{Q(s)(1-\xi)}, & \xi > x \end{cases} \tag{6}$$

$H(x,s)$ is the transform matrix between $\gamma(s)$ and $\eta(x,s)$

$$H(x,s) = e^{Q(s)x}(M(s)+N(s)e^{Q(s)})^{-1} \tag{7}$$

Thus, the acceleration transmissibility from the supported end to the free end is

$$h(s) = H_{12}(1,s) \tag{8}$$

3.2 Parameter Identification

In the experiment, when the excitation frequency is ω, the amplitude and the phase of the vibration at the beam's free end are $E(\omega)$ and $\varphi(\omega)$, respectively. They can be obtained from the experiment. With the least squared method, the identified equivalent Young's modulus E and equivalent c are those can minimize the value of the following equation

$$y_1(E,c) = \int_{\omega_l}^{\omega_h} (h(E,c,\omega) - g(\omega))^2 d\omega \tag{9}$$

In practical applications, instead of a continuous frequency response function, a sequence of frequency responses sampled at discrete frequencies with a certain interval will be obtained. Supposing there that N samples, Eq. (9) can be written as

$$y_1(E,c) = \sum_{i=1}^{n} (h(E,c,\omega_i) - g(\omega_i))^2 \tag{10}$$

The parameter identification problem can be considered as an optimization problem can be described as

$$\min y(x) \quad x = (E,c)^T \tag{11}$$

There is no constraint on the optimization and therefore there are many direct methods available for the solution. A method used here is the pattern search method. This method is simple and reliable for achieving the global minimization [8, 9].

3.3 Results and Discussion

In the experiment, the lumped mass is 0.064 kg. The frequency range of the sine sweep excitation is from 5 to 500 Hz with the amplitude of 0.5 g. For the convenience of comparing the identified results, a dimensionless damping ratio is introduced here, which is

$$\zeta = \frac{c}{2\sqrt{\rho AEI}} \tag{12}$$

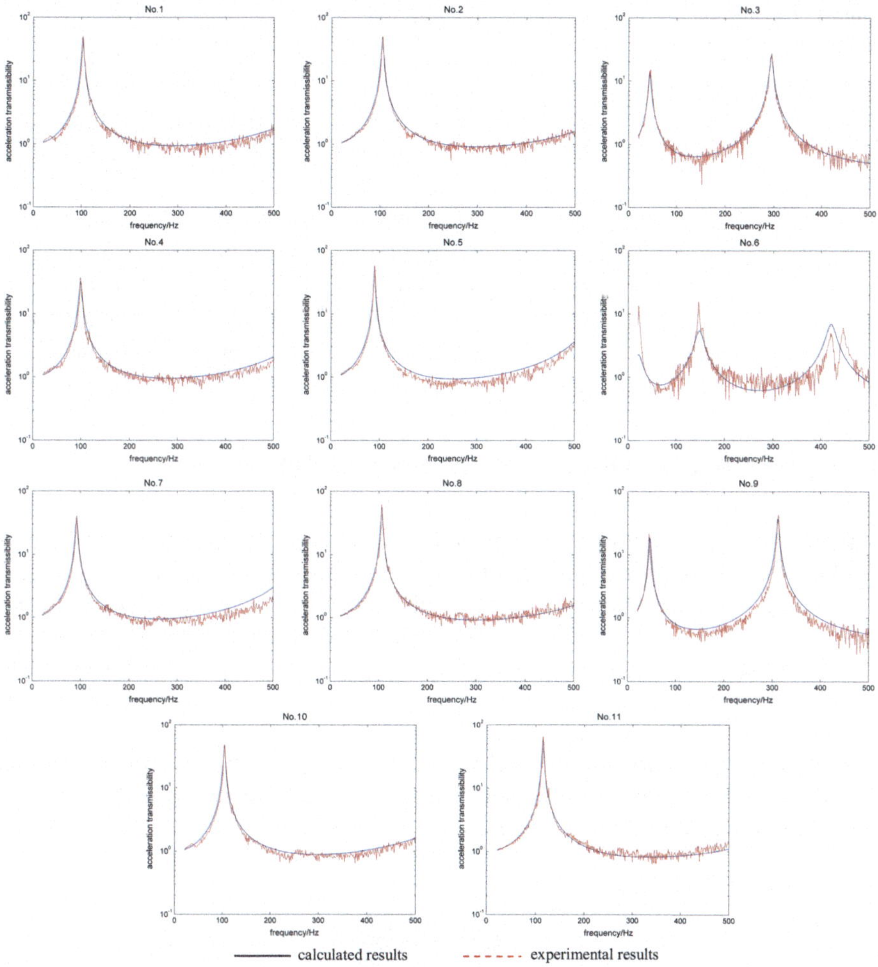

Fig. 4 Amplitude-frequency curves

To examine the accuracy of the identified results, we substitute the identified parameters back to the governing equation to calculate the amplitude-frequency curve and compare it with the experimental curve. Figure 4 is comparisons of the calculated curves with measured ones. It can be seen that they are very close to each other in most of the frequency range. Therefore, a conclusion can be drawn is that the viscous damping model can well describe the dynamic characteristics of the damping reinforced components used in the experiments (Table 1).

Table 1 Parameter identification results

Index	E/Pa	c	ζ
1	4.97E+10	46.38	0.9116
2	3.38E+10	30.58	0.5953
3	3.45E+09	18.53	0.5574
4	5.31E+10	78.28	1.5693
5	3.29E+10	30.45	0.6439
6	1.34E+09	89.44	3.8036
7	4.12E+10	55.01	1.1441
8	3.56E+10	26.50	0.5158
9	4.01E+09	14.51	0.4268
10	4.70E+10	43.60	0.8569
11	3.26E+10	21.26	0.3972

4 Application in Optical Instrument Stabilization

With the raise of resolution, optical payloads are becoming increasingly sensitive to satellite jitter. Inserting damping reinforced components between the payloads and the satellite is an effective way to block the transmission of disturbance. In the present study, an isolator using damping reinforced components has been developed for a multi-axis isolation system of optical payloads. The finite element model of the whole satellite with a damping reinforced isolator connecting the payload and satellite bus is as shown in Fig. 5.

The on-orbit vibration of the satellite structure has a very low magnitude, and its typical acceleration and displacement levels are usually several μm respectively. The vibration isolators should be specially designed to meet the new requirements

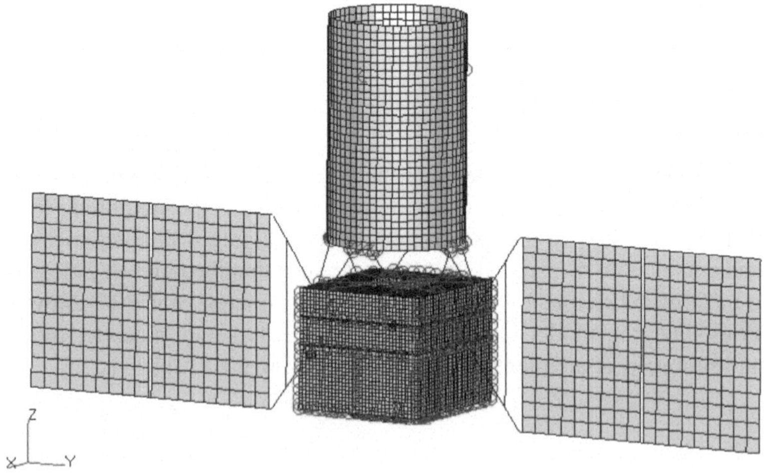

Fig. 5 Whole satellite finite element model

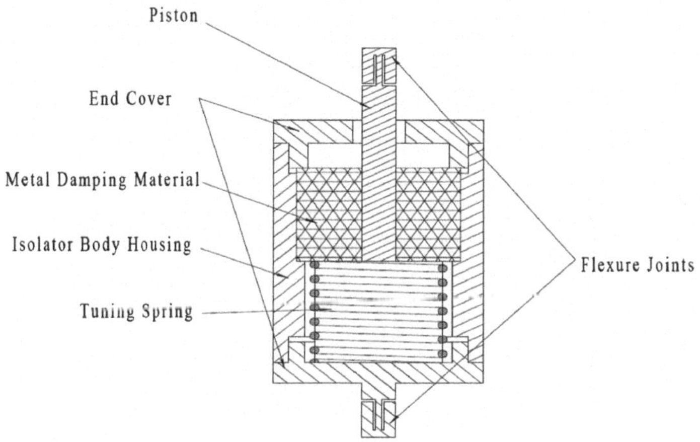

Fig. 6 Schematic of micro-vibration isolator

arisen from such extremely low amplitude. Dampers using the viscous fluid should be avoided because of some disadvantages such as difficulty to seal, insensitive to micro-vibration, and etc. A highly sensitive damper with excellent micro-vibration damping capabilities was developed using damping reinforced components, as shown schematically in Fig. 6. Energy is dissipated in the deformation of the damper as the piston is pumped by the vibrational displacements. Meanwhile, recovery force needed in the isolation is obtained through the deformation of the damper together with the tuning spring [10].

By applying flexure joints at both ends, desired transverse stiffness of the isolator can be achieved. Consequently, six isolators can serve as an assembly to implement vibration isolation in all six DOFs. The isolation frequencies are shown in Table 2.

A frequency response analysis was carried out for the purpose of estimating the reduction in vibration magnitude due to the isolation. The finite element model of the satellite integrated with the model of the isolation system was used to perform the analysis. Transmissibility from the disturbing source to the line of sight pointing error is plotted in Fig. 7, which shows that the amplitude of the response has been significantly attenuated. Compared to rigid connection, the amplitude of the pointing error response was reduced by more than 94% above 60 Hz.

Table 2 Isolation frequencies

Orders	Mode frequency (Hz)	Mode shape description
1	9.15	Pitch
2	10.41	Roll
3	19.76	Vertical translation
4	26.28	Horizontal translation 1
5	27.18	Horizontal translation 2
6	27.86	Yaw

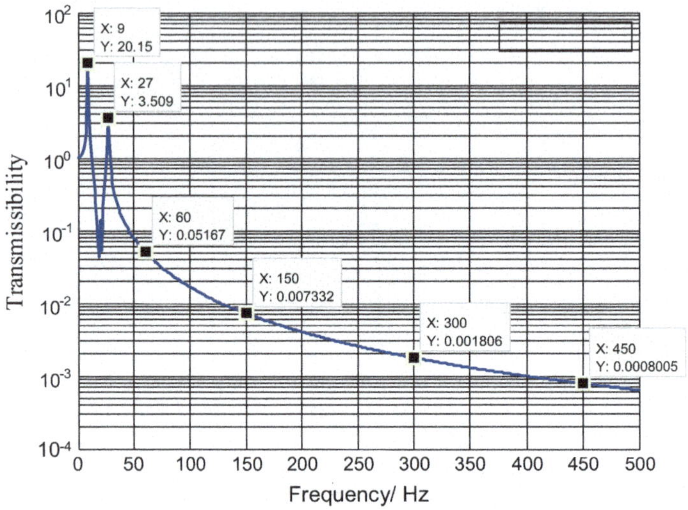

Fig. 7 Transmissibility form the disturbing source to pointing error

5 Conclusion

An experimental modeling method of damping reinforced components in wide frequency band was proposed and its effect in optical structures was evaluated in the present paper.

First, a cantilever beam of damping reinforced component was excited in a sine sweep test to get its frequency response. A parametric viscous damping model was proposed and the frequency response was solved with Laplace transform. Then, the parameters were identified by solving an unconstrained optimization problem with pattern search method, which minimized the deviation between test result and calculated result by least square criteria. Finally, the pointing stability of a typical optical structure adopting damping reinforced components was analyzed based on the derived dynamic model, which indicated that the line of sight jitter was attenuated because of the absorption of vibration energy by damping reinforced components.

References

1. R.A. Laskin, S.W. Sirlin. Future Payload Isolation and Pointing System Technology. J. Guidance, 1986, 9: 469–477.
2. WorldView-2 Offers Unsurpassed Imaging Capabilities. Earth Imaging Journal, Online Version.
3. Standard Test Memod for Measuring Vibration-Damping Properties of Materials, ASTM International E756-04.

4. Neal Granick, Jesse E. Stern. Material damping of aluminum by a resonant-dwell technique, NASA TN D-2893.
5. Benjamin J. Lazan. Damping of materials and members in structural mechanics, Pergamon Press, Oxford, London, 1968.
6. Stephen H. Crandall, On scaling laws for material damping, NASATN D-1467.
7. B. Yang, C. A. Tan. Transfer Functions of One-Dimensional Distributed Parameter Systems [J]. Journal of Applied Mechanics, 1992, 52: 1009–1014.
8. CHEN Bao-lin. Optimization Theory and Algorithm[M]. Beijing: Tsinghua University Press, 1989:397–403.
9. Tu Yong-qing, Wang Guang-yuan. Damper Parameter Identification and Study of Metal Foam Filled Steel Tube. Journal of Materials in Civil Engineering, 2010, 22(4): 397–402.
10. Guan Xin, Wang Guangyuan, Zheng Gangtie. Experimental Demonstration of 1.5 Hz Passive Isolation System for Precision Optical Payloads[C/OL]// International Conference on Space Optics. Rhodes Island, Greece, 2010 [2011-04-07]. http://congrex.nl/icso/Papers/Session%209a/guan.pdf.

Optical Design and Measurements of a Dynamic Target Monitoring Spectrometer for Potassium Spectra Detection in a Flame

Haiyan Luo, Wei Xiong, Shuang Li, Zhiwei Li and Jin Hong

Abstract We present the optical design, laboratory calibration and first field measurements of a dynamic target monitoring spectrometer (DTS) for detection of potassium emission lines at 766.491 and 769.897 nm in a flame under the two-dimensional orthogonal observation mode. DTS can provide high-spectral-resolution, high-spatial-resolution and high-throughput measurements by using imaging spatial heterodyne spectroscopy. Combined with a 0.035 nm spectral resolution and a 1.2 mrad spatial resolution in one dimension for each channel, 7.2° × 7°(vertical × horizontal) field of view is achieved in DTS. The high spectral resolution is important for identifying a very dim signal by a few spectra lines, and the high spatial resolution is a significant parameter for inferring accurate location of a flame per exposure and monitoring the trace of a dynamic flame varying from exposure to exposure. Finally, we demonstrate the ability to measure potassium spectra in a rapidly changing flame by DTS through a series of laboratory calibrations and field measurements.

Keywords Spatial heterodyne spectroscopy (SHS) · Fourier transforms · Emission line · Remote sensing and sensors

1 Introduction

The importance of the flame detection in remote and real-time monitoring of fire control and escape in has been recognized for 20 years [1, 2]. Nevertheless, the typical method of thermal infrared or ultraviolet video detection has resulted increasing the false warning rate on account of infrared and ultraviolet ambient background radiation [3–5]. In NIR spectral approach for flame detection,

---

H. Luo (✉) · W. Xiong · S. Li · Z. Li · J. Hong
Key Laboratory of Optical Calibration and Characterization,
Anhui Institute of Optics and Fine Mechanics, Chinese Academy of Sciences,
Hefei 230031, China
e-mail: luohaiyan@aiofm.ac.cn

H.P. Urbach and G. Zhang (eds.), *3rd International Symposium of Space Optical Instruments and Applications*, Springer Proceedings in Physics 192,
DOI 10.1007/978-3-319-49184-4_6

SHIM-Fire can observe the potassium lines using ISHS on breadboard stage in laboratory with moderate spectral resolution of 4 nm [6], and SHIM-Fire with one dimensional spatial resolution is not sufficient to identify the accurate location of flame or fire. Furthermore, the shape and width of potassium emission lines are related to the thermal background and density of the potassium. Fortunately, different background can be easily identified with a high-spectral-resolution instrument [7, 8].

The primary goal of DTS is to identify the accurate location and monitor varied trace of flame with the low false alarm rate and low cost in real-time, and to verify the availability of two dimensional orthogonal viewing for field measurements. Using ISHS and the specific orthogonal viewing mode, the DTS instrument in the near infrared offers high spectral resolution, low false alarm rate and high spatial resolution in two dimensions. ISHS was first developed in the early 2000s by NRL's Space Division for observing the vertical profile of OH density in the Earth middle atmosphere, which was applied in the ultraviolet spectral band with high spectral resolution of 0.03 nm and spatial resolution of 0.8 mrad. Environmental tests have verified that the design is suitable for formidable mechanics environments [9–11].

2 Optical Design and Instrument Integration

2.1 Properties of Imaging Spatial Heterodyne Spectroscopy

Spatial heterodyne spectroscopy(SHS) was originally presented in publication by J.M. Harlander [12, 14]. Compared to the SHS instrument, the ISHS instrument which containing several anamorphic lenses in the collimating or imaging system can image extended targets onto the detector in the spatial dimension, while illuminate the detector and completely defocus the extended targets in the spectral dimension. The basic ISHS configuration is presented in Fig. 1, and they have the same imaging system in both directions.

Compared to SHS, some important properties of ISHS are:

1. In contrast to SHS which produces the same interferogram in every row of the detector along the spatial dimension, ISHS produces the different interferogram from different slice in the field of view. This drives different requirements of the imaging quality in the horizontal and vertical directions, and if the field of view is binned into N slices with the M $\times$ M detector, the M $\times$ N sampling numbers of the interferogram can be recorded.
2. Compared to SHS, ISHS offers an additional advantage concerning scattered light. The radiation from one slice, that is scattered on the grating localization plane, is refocused on the detector and does not generally contaminate other slices in the field of view.

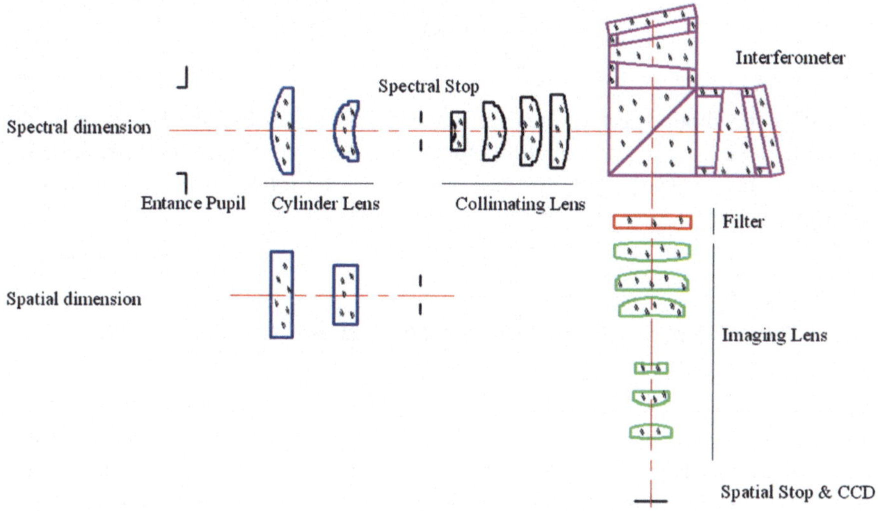

Fig. 1 ISHS configuration of the DTS instrument

3. Under two dimensional orthogonal viewing, the ISHS field of view projected onto the observed scene is a square area, and the observed scene is separately imaged by each channel of DTS. If the field of view is binned into N slices for each channel, the observed scene is finally binned into N × N elements. The number of N is depending on the signal-to-noise ratio of the spectrum information (SNR_{DTS}) and the restriction of the optical aberrations, and SNR_{DTS} at t rows binned into one slice is given by

$$SNR_{DTS} = SNR \cdot \sqrt{t} \tag{1}$$

Where, SNR in the spectrum is the signal-to-noise of the detector's each row. For a M × M detector, the t is given by Eq. 2.

$$M = t \cdot N \tag{2}$$

The relationship between SNR in the spectral domain and $SNR_{x=0}$ in the detector is [14]

$$SNR = \sqrt{\frac{2}{M}} \cdot \frac{B(\sigma)}{\overline{B_e}} \cdot SNR_{x=0} \tag{3}$$

where $B(\sigma)$ is the effective intensity at wavenumber σ, $\overline{B_e}$ is the mean signal in the spectral domain, and $SNR_{x=0}$ is the signal-to-noise ratio in the detector domain at zero path difference.

$$SNR_{x=0} = \sqrt{2 \cdot \frac{A \cdot \Omega \cdot \tau \cdot T \cdot \delta\sigma \cdot L(\sigma)}{M \times M}} \tag{4}$$

where A is the input aperture area, Ω is the solid angle of field of view, τ is the optical efficiency (including quantum efficiency), T is the integration time, $\delta\sigma$ is the spectral resolution, $L(\sigma)$ is the spectral radiance of the flame, and M is the number of samples.

2.2 Requirements of the Optical Design for DTS

The purpose of the design is a ultra spectral resolution, near infrared instrument that uses two array detectors for simultaneously detecting two dimensional spatial information under two-dimensional orthogonal viewing. Note that every slice in the field of view produces an interferogram along several rows of the detector as shown in Fig. 2. By the optic axis' registries for each channel of DTS, the extended target is binned into more than 35 slices in the spatial dimension (35 × 3 mrad = 6°) for each channel, and DTS finally yields at least 70 kinds of the interferograms per exposure.

For optimum performance and simplicity, a separate input fore-optical system is designed for each channel, rather than relying on dichroic beam splitter with its associated energy losses and polarization issues. This allows the design to be optimized for telecentric optics on the grating and the spectral stop surface in each channel, but DTS will then be co-aligned during assembly to the required co-registration levels.

From calculated spectrum of the potassium double emission lines at 766.491 and 769.897 nm with oxygen as depicted in Fig. 3, it is obvious that the potassium spectral lines lie within an absorption band of atmospheric oxygen. If the potassium line lies directly on top of an oxygen absorption line using the spectrometer with a moderate spectral resolution, it will result in reduced spectral intensity of the potassium line.

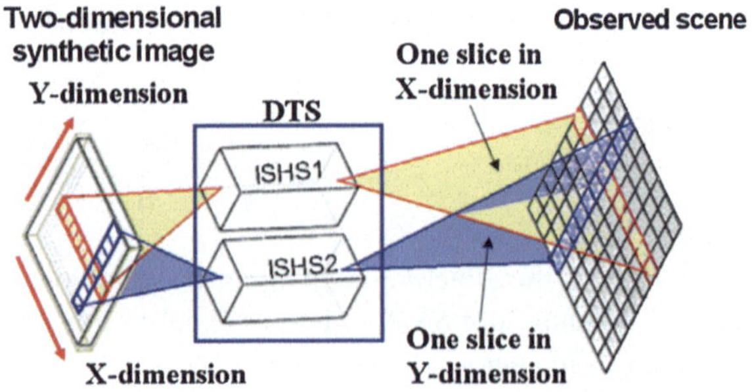

Fig. 2 Imaging concept of the DTS instrument

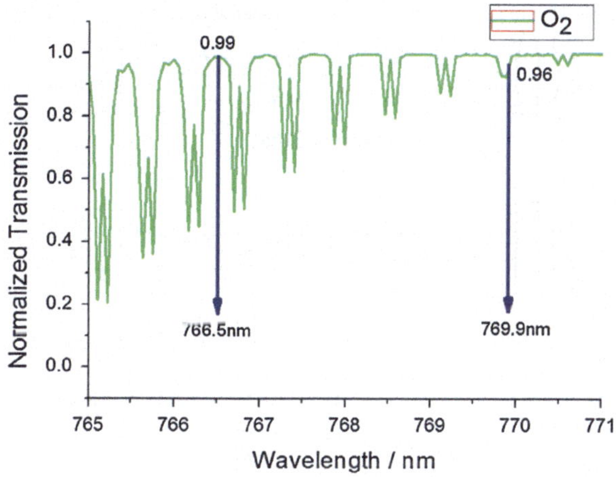

Fig. 3 Normalized atmosphere transmittance (*solid green line*) reduced by HITRAN

2.3 *Optical System Design for DTS*

It is composed of two functional subsystem: the cylinder system and collimating system. The field of slice scene is imaged in spatial dimension on the grating while the other dimension of the grating is used to record the spectral information in the form of the interferograms.

To ensure that spatial variation within one slice does not contaminate the other slice's interferogram, the cylinder system contains two cylindrical lenses, and the collimating system contains four spherical lenses as shown in Fig. 4. The four collimating lenses are made of BK7, and silica for the two cylinder lenses. A flare stop is placed between the cylinder lens and collimating lens. In the x-dimension, the four collimating lenses and two cylinder lenses images the entrance aperture stop on the grating (where the fringes are located), thus the interferogram is recorded along the x-dimension of the grating. In the y-dimension, the four collimating lenses images the observed scene on the grating, thus different interferogram is recorded along the y-dimension of the grating according to different angular height of the object.

The interferometer is consisted of beam splitter, two wedge spacers, two wedges, two plane spacers and two non-polarizing gratings as shown in Fig. 5. The cube square beam splitter is chosen to minimize the mass and dimension, which has a dielectric coating to eliminate polarization effects.

The main design parameters of the interferometer are shown in Table 1 and these associated equations are explained by J.M. Harlander et al. [14]. Note that the DTS interferometer has two plane spacers between the wedges and gratings [9]. By making this modification, our intention is to simplify the instrument and to make it

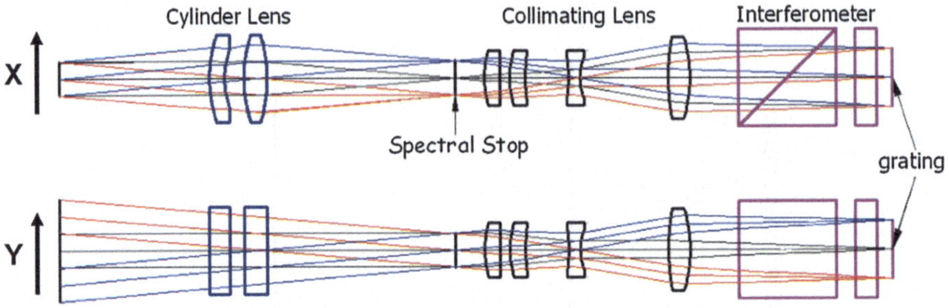

Fig. 4 The fore-optical system designed to identify the different scene of spatial information

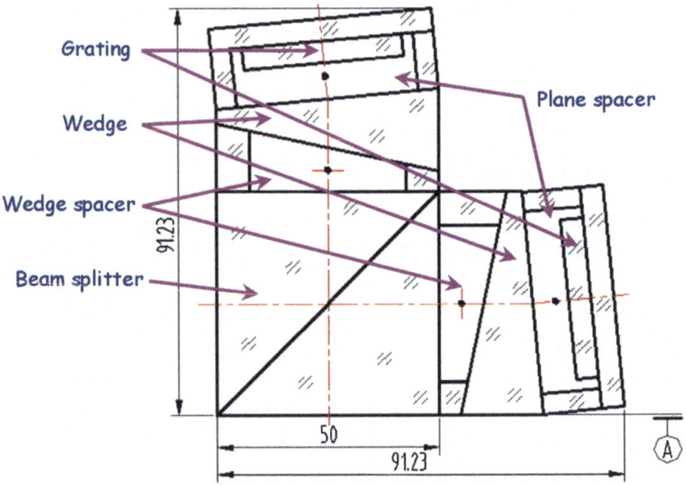

Fig. 5 The optical layout of the interferometer

Table 1 The interferometer design specification

Aspect	Attribute	Design
Gratings	Clear aperture	16 mm × 16 mm
	Groove density	600 gr/mm
	Littrow angle	13.174°
	Rotated angle	5.260°
Field widening prisms silica	Wedge angle	17.154°
	Incident angle	11.870°
Beam splitter silica	Clear aperture	26.2 mm × 26.2 mm
	Size	50 mm × 50 mm

Table 2 Aberration coefficients of image system

Item	Specifications	Design results
Axial spherical aberration/mm	0.09	0.028
Residual axial spherical aberration/mm	0.54	0.012
Position chromatic aberration/mm	0.09	0.058
Spherochromatic aberration/mm	0.36	0.051
Spectral chromatic aberration/mm	0.09	0.046
Sine aberration/mm	0.0025	3.78E-4
Relative distort	0.1%	0.06%

have the same high throughout through a pair of the slight different apex angle field-widening prisms.

The interferometer elements are made of silica, a classical material in optical application for thermal stability reasons, and the final adjustment of the interferometer is performed using the tunable laser. The introduction of a small tilt onto the gratings, which is produced by the field widening prism, rotates the image plane by an angle 5.323° [15].

$$\theta_{ima} = \frac{\alpha(n-1)}{n} \tag{5}$$

To minimize the aberrations introduced by insertion of the field widening prisms into the converging beam, the gratings are rotated in the same approximately 5.260° as the angle calculated to be 5.323° (as shown in Eq. 5), so that the imaging system is not introduced uncorrectable astigmatism and coma due to adding the field widening prisms.

For the case of the imaging system in which magnification is changed with adjustment of the focus plane, the magnification change can be overcome with a double telecentric optics. It re-images the fringes located very close to the gratings onto the focal plane detector, and each channel images the fringes onto its individual detector array. The detector array made by e2v technologies has 1024 × 1024 usable pixels with a 13 μm pixel pitch, so that the 1024 pixels are in the dispersion direction. The imaging system has a magnification of 1.592× and f/5.4 for each channel. The aberration properties of the imaging system are accurately modeled as shown in Table 2.

For deviations from telecentricity at the detector will result in different defocusing at the different place of the image plane and degrade the visibility of the interferogram modulation [14], we use the telecentric optics at both the grating and detector to correct deviations over the full field of view. The maximum deviation from 0° at the detector is 0.51°.

3 Calibration and Characterization

The fringe frequency of a specific wavenumber σ_i in Eqs. (6) and (7) is related to the Littrow wavenumber σ. Using the measured fringe frequency (f_i) of the tunable laser and their known spectral line position, we can retrieve the parameters of the least square fit function defined in Eq. (8) when series of monochromatic sources are observed. With spectrum tuning range from 0.38 to 1.1 μm, the tunable laser MBR110 for the spectral calibration manufactured by Coherent, Inc. has the spectrum line-width of 1 MHz. And an integrating sphere without speckle noise are used to form a laser-illuminated integrating sphere source.

$$f_i = 4(\sigma_i - \sigma)\tan\theta \tag{6}$$

$$f_i = \hat{k}\sigma_i + \hat{d} \tag{7}$$

$$\begin{cases} \hat{k} = \dfrac{\sum_{i=1}^{m} f_i \cdot \sigma_i - \frac{1}{m}\left(\sum_{i=1}^{m} f_i\right)\left(\sum_{i=1}^{m} \sigma_i\right)}{\left(\sum_{i=1}^{m} \sigma_i^2\right) - \frac{1}{m}\left(\sum_{i=1}^{m} \sigma_i\right)^2} \\ \hat{d} = \bar{f} - \hat{k}\bar{\sigma} \end{cases} \tag{8}$$

As described in Eq. 8, the regression coefficients $\hat{k}$ and $\hat{d}$ can produce a series of the corrected fringe frequency f_i for the wavenumber σ_i as a result of the symmetry of the cosine interferogram, where $\bar{f}$ and $\bar{\sigma}$ correspond to the average value of f_i and σ_i respectively. The Littrow wavelength is about 763.5 nm as shown in Fig. 6.

Length of the grating:

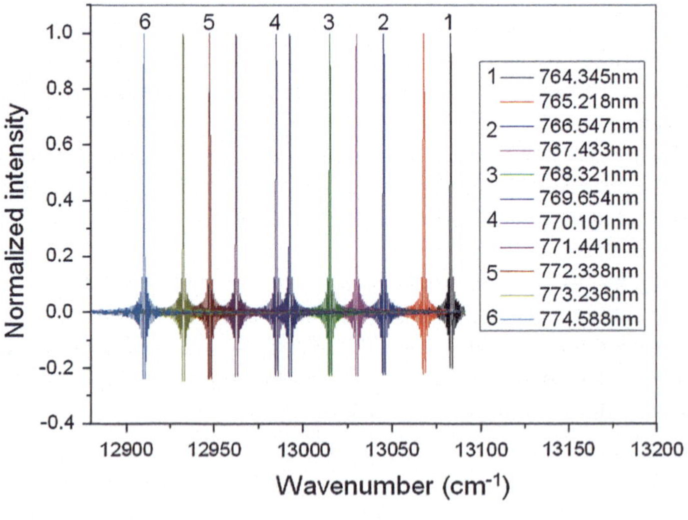

Fig. 6 Spectrum results by the tuneable laser

$$\left.\begin{array}{l}\delta\sigma = \frac{1.207}{2\cdot 4x_{\max}\tan\theta}\\ x_{\max} = \frac{1}{2}w_{grating}\cos\theta\end{array}\right\} \Rightarrow w_{grating} = \frac{1.207}{4\cdot\delta\sigma\,\sin\theta} \tag{9}$$

and the maximal path difference $4x_{\max}\tan\theta$ which determines the width of the instrument line shape (ILS).

As indicated in Table 1, grating angle θ is 13.174°, and the Littrow wavenumber is 13158 cm^{-1}. We can conclude that spectral resolution of the DTS instrument reaches around FWHM ~0.035 nm (0.59 cm^{-1}), and the error in the beam splitter assembling has greatly reduced the accuracy of the Littrow wavelength.

The FOV ($\arctan(d/f)$) can be determined by a definite source or a star test pattern with the diameter of d located on the telescope focal plane with a focal length f:

$$\frac{\arctan(d/f)}{N'} = \frac{1.592\times 13\,\mu\text{m}}{172.8\,\text{mm}} \tag{10}$$

Where 1.592 is the desired magnification of the imaging system, 13 μm is the pixel pitch of detector, and the 172.8 mm is the focal length of the collimating system in Eq. 10.

Figure 7 shows the schematic diagram of the FOV calibration with an incandescent source. With a star test pattern located on the image plane, the telescope provide a collimated beam for illumining one slice of the two channels. Using a 1.5 mm diameter star test pattern, the 1500 mm focal length telescope and the row number or column number N' of the illuminated FPA (the measured result is 8), the entire FOV of each channel can be calculated, and the measured entire FOV is 7.2° in the spatial direction.

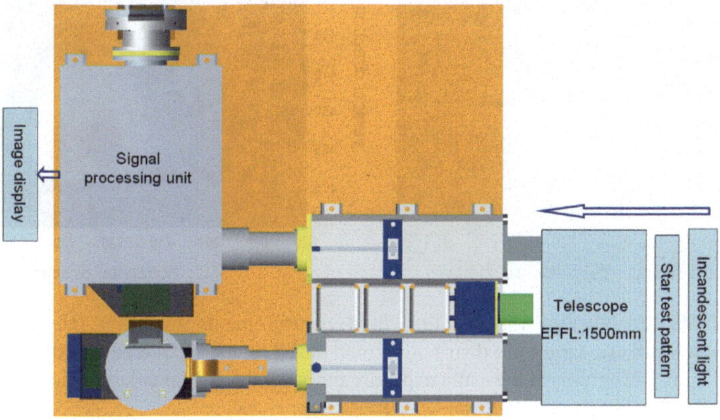

Fig. 7 The schematic diagram for FOV calibration

4 First Field Measurements

Because the field of view of the monitoring camera is relatively large, the geometrical information and auxiliary target recognition can be obtained. For the DTS instrument, the focal plane array (FPA) is binned into N slices, which yields 2 N interferograms per exposure. With a flame on a scale size of about 200 mm × 200 mm located on the distance of 50 m from the DTS entrance aperture, we obtain the potassium spectral lines from the partial enlarged drawing of Fig. 8a as shown in Fig. 8(c2).

Based on the size of the flame and the distance from the instrument for the field measurements, we can calculate the theoretical interferogram in 34 of the 1024 rows of the detector, and the measurement interferograms is in 46 rows because of the smoking diffusion by wind.

The preliminary results yield the potassium relative spectrum intensity of the flame from the first field measurements by DTS in good agreement with the standard model certified reference information from the National Institute of Standards and Technology (NIST) [15]. With the spectral intensity of two channel's potassium salt, the distribution map of the combustion source can be obtained as shown in Fig. 8 (c2). Furthermore, by a different fore-optical system that expands the FOV of the instrument, we can monitor the trace of the dynamic target of large scale.

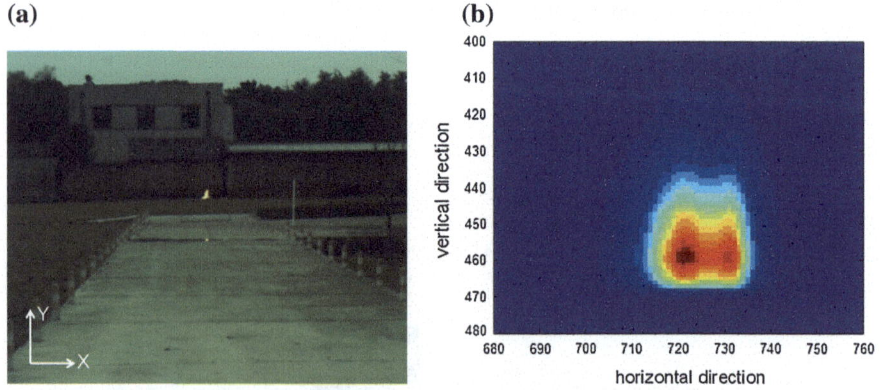

Fig. 8 **a** K_2SO_4 in alcohol for the first field measurements by DTS (Hefei, Anhui, for 15/11/2014). **b** Synthetic target/fire distribution map based on the recovered spectral intensity with two channel's interferograms at the same exposure time and the partial enlarged drawing of Fig. 8a

5 Conclusion

We successfully demonstrated a theoretical optical design, calibration and the first field measurements for DTS. As demonstrated by the calibration and subsequent field measurements, the interferograms of the different slices are recorded by DTS with a same exposure time, and the relative location of a flame can be inferred through the recovered spectrum from the two channel interferogram. System sensitivity and multi-target detection will be the focus of a future study. For the different thermal background through adding different combustion improvers in the measurements, the density of potassium salt as a function of the spectral intensity and spectral lines shape can be identified, as expected. Furthermore, the DTS concept allows the simultaneous detection with high spectral resolution and high spatial resolution, which is well suited for the observation of the trace elements for future space mission and aircraft mission.

References

1. A.C. Lloyd, Y.J. Zhu, L.K. Tseng, J.P. Gore, and Y.R. Sivanthanu, "Fire Detection Using reflected Near Infrared Radiation and Source Temperature Discrimination," NIST GCR 98: 747, (1998).
2. J.C. Owrutsky, and D.A. Steinhurst, "Fire Detection Method, "Patent Application Number 2005/0012626, January 20, (2005).
3. S.L. Rose-Pehrsson, J.C. Owrutsky, S.C. Wales, F.W. Williams, J.P. Farley, D.A. Steinhurst, C.P. Minor, J.A. Lynch, D.T. Gottuk, "Volume Sensor for Damage Assessment and Situational Awareness," AUBE 2004, Duisburg, Germany, September 14–16, (2004).
4. S.L. Rose-Pehrsson, J.C. Owrutsky, D.T. Gottuk, D.A. Steinhurst, C.P. Minor, J.P. Farley, F.W. Williams, "Volume Sensor for Shipboard Damage Control," 2004 NRL Review, p. 144, (2004).
5. D.A. Steinhurst, C.P. Minor, J.C. Owrutsky, S.L. Rose-Pehrsson, D.T. Gottuk, F.W. Williams, and J.P. Farley, "Long Wavelength Video-based Event Detection, Preliminary Results from the CVNX and VS 1 Test Series, ex-USS Shadwell, April 7–25, 2003," Naval Research Laboratory Memorandum Report NRL/MRMM/ 6110-03-8733, December 31, (2003).
6. C.R. Englert, J.T. Bays, J.C. Owrutsky, J.M. Harlander, "SHIM-Fire Breadboard Instrument Design, Integration, and First Measurements" Naval Research Laboratory Memorandum Report NRL/MR/ 7640-05-8926, November 23, (2005).
7. M.L. Parsons, W.J. Mccarthy, and J.D. Winefordner, "Approximate Half-Intensity Widths of a Number of Atomic Spectral Lines Used in Atomic-Emission and Atomic-Absorption," Applied Spectroscopy, Vol. 20, No. 4, 223–229, (1996).
8. H.A. Phillips, H.L. Lancaster, M.B. Denton, Karoly Rozsa and Pal Apai, "Self-Absorption in Copper Hollow Cathode Discharges: Effects on Spectral Line Shape and Absorption Sensitivity," Applied Spectroscopy, Vol. 42, No. 4, 572–576, (1988).
9. J.M. Harlander, F.L. Roesler, J.G. Cardon, C.R. Englert, and R.R. Conway, "SHIMMER: A Spatial Heterodyne Spectrometer for Remote Sensing of Earth's Middle Atmosphere," Applied Optics, 41, 1343–1352, (2002).

10. J.M. Harlander, F.L. Roesler, C.R. Englert, J.G. Cardon, R.R. Conway, C.M. Brown, J. Wimperis, "Robust monolithic ultraviolet interferometer for the SHIMMER instrument on STPSat-I," Applied Optics, 42, 2829–2834, (2003).
11. C.R. Englert, M.H. Stevens, D.E. Siskind, J.M. Harlander and F.L. Roesler. "Spatial Heterodyne Imager for Mesospheric Radicals on STPSat-1," J. Geophys. Res., 115, D20306, doi:10.1029/2010JD014398 (2010).
12. J. M. Harlander, R. J. Reynolds, and F. L. Roesler, "Spatial heterodyne spectroscopy for the exploration of diffuse interstellar emission lines at far-ultraviolet wavelengths," Astrophys. J. 396, 730–740 (1992).
13. James Howard, "Formulas for the coma and astigmatism of wedge prisms used in converging light", App. Optics, Vol. 24, No. 23, 4265–4268, (1985).
14. Alan Scott, S.H. Zheng, Stephen Brown, and Andrew Bell, "Spatial Heterodyne Spectrometer for FLEX", Proc SPIE, 6744, 67441 W, (2007).
15. W.C. Martin, J.R. Fuhr, D.E. Kelleher, A. Musgrove, L. Podobedova, J. Reader, E.B. Saloman, C.J. Sansonetti, W.L. Wiese, P.J. Mohr, and K. Olsen, "NIST Atomic Spectra Database (version 2.0, 1999)," http://physics.nist.gov/asd, 1999.

Optical Design of an Aperture-Divided MWIR Imaging Polarimeter

Xujie Huang, Yangming Jin, Zhicheng Zhao, Lin Han, Jiacheng Zhu and Weimin Shen

Abstract The polarization state of light emitted or reflected from scene contains massive important information about target, such as the geometry, surface properties, material characteristics and so on. Imaging polarimetry can measure the polarimetric information efficiently, and show significant advantages for target detection and discrimination in situations. In this paper, we design an aperture-divided optical system of MWIR imaging polarimeter, which consists of one co-aperture front Galilean telescope, four sub-aperture double-gauss objectives and one co-aperture Cooke triplet relay lens. Its focal length is 68 mm, F number F/2 and FOV 4° × 3.2°. The light emitted or reflected from target is collected by the co-aperture front telescope, and imaged simultaneously onto the four quadrants of the same FPA detector respectively by four decentered parallel subsystems, which divide the aperture into four parts. Each subsystem has different polarizing element, and the four different polarization images of the same target can be acquired simultaneously. The suggested optical system for imaging polarimeter employs relay lens, in order to implement 100% efficiency of its cold stop. It can be developed as payloads of reconnaissance aircraft, satellite or missile seeker etc.

Keywords Optical design · Aperture-divided · MWIR · Imaging polarimeter

1 Introduction

Imaging polarimetry can measure polarimetric information as well as the optical intensity of target. Compared with traditional optical imaging technics, imaging polarimetry can enhance contrast between target and background, and it also can

The original version of this chapter was revised: The spelling of the first author's name was corrected. The erratum to this chapter is available at doi: 10.1007/978-3-319-49184-4_50

X. Huang (✉) · Y. Jin · Z. Zhao · L. Han · J. Zhu · W. Shen
Institute of Modern Optical Technology, Soochow University, Suzhou 215006, China
e-mail: Jackhoang@126.com

H.P. Urbach and G. Zhang (eds.), *3rd International Symposium of Space Optical Instruments and Applications*, Springer Proceedings in Physics 192,
DOI 10.1007/978-3-319-49184-4_7

provide additional information about the target such as geometry, surface properties, material characteristics [1]. Imaging polarimetry can be widely used in the realms of astronomical observations, biomedical diagnostics, environmental monitoring, landmine detection, military reconnaissance etc. [2]. It is one of the most important technologies in the field of optical remote sensing.

Imaging polarimetry mainly includes sequential imaging polarimetry and simultaneous imaging polarimetry. Sequential imaging polarimetry measures several images sequentially in time [3–5], thus it is not appropriate for the dynamic targets, or moving platforms. Simultaneous imaging polarimetry can measure multiple images simultaneously, which is robust and has no moving parts. Simultaneous imaging polarimetry can be divided as amplitude-divided polarimetry [6], FPA-divided polarimetry [7] and aperture-divided polarimetry [8]. The aperture-divided polarimetry divides the aperture into several parts evenly by decentering mini-lens arrays in the optical path. There are different polarizers in each of the mini-lens arrays, thus it can acquire different polarization images onto a single FPA. The aperture-divided polarimeter is compact in size, and is convenient for data processing.

The worldwide scholars have designed aperture-divided polarimeters working in visible [9] and infrared bands [10]. The Polaris Sensor Technologies Incorporation has described a divided aperture MWIR imaging polarimeter, all of which field stop, relay optics and decentered subsystems are within a vacuum Dewar [10]. The vacuum Dewar of this imaging polarimeter has to be refitted to align the optical elements, which is much difficult and expensive.

However, all the optical elements of our designed aperture-divided MWIR imaging polarimeter are located outside of vacuum Dewar. There is no need to refit the IR detector's chilling equipment, thus this system is easier to be fabricated and aligned, and is compact in size, robust enough in practical use. The suggested optical system for imaging polarimeter employs relay lens, in order to implement 100% efficiency of its cold stop.

2 Design Considerations and Parameters

The layout of the described aperture-divided MWIR imaging polarimeter is shown in Fig. 1, including co-aperture front telescope, sub-aperture mini lens, co-aperture relay lens and IR detector. The classical Keplerian telescope or Galilean telescope can be used as the co-aperture front telescope. In this paper, we chose the Galilean telescope as the co-aperture front telescope for its advantages as compact in volume, less energy loss. The co-aperture front telescope in the system functions as limiting the FOV and compressing the light beam width.

The sub-aperture mini lens is composed of four decentered parallel subsystems, and the system stop is located in the optical path of subsystems. The decentered subsystems divide the aperture into four parts. There are different polarizers in each subsystem, so we get four different polarization images of a same target simultaneously onto the four quadrants of FPA. Considering that the four subsystems are all

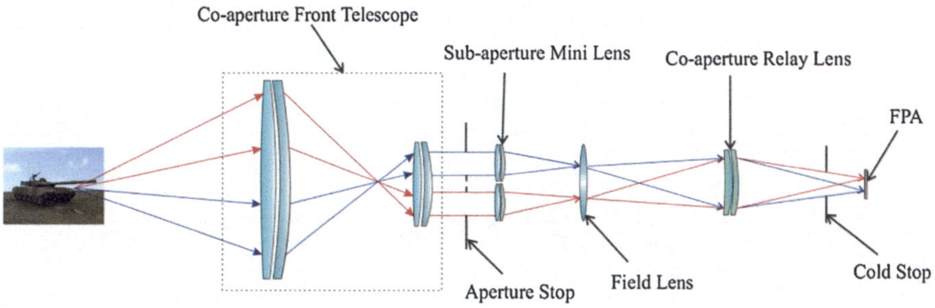

Fig. 1 Concept design for aperture-divided MWIR imaging polarimeter

decentered from the center axis, which leads to a larger FOV for each subsystem [9]. Moreover, there needs four groups of same optical lenses, which are relatively more complex in structure, more difficult for fabricating and aligning, therefore we chose the relatively simple double-gauss objective as the initial structure of the subsystem.

The rays makes a large angle after the field stop, there needs a field lens just after the field stop to turn the rays at large field angles in toward the optical axis. The field lens can reduce the diameter of the co-aperture relay lens, also it can reduce the off axis aberrations and distortion of the optical system. All the rays in the four subsystems are adjusted by the field lens, and then they are re-imaged by the same co-aperture relay lens onto the four quadrants of the FPA, which are four different polarization images of the same target. The co-aperture relay lens must satisfy the requirement that the F number equals to 2. In this paper, we chose the classical Cooke triplet as the initial structure of the co-aperture relay lens [12].

It is noteworthy that all four paths of rays in the subsystems are chilled by the same cold stop 200 mm away before the FPA. There is no need to refit the cold stop and vacuum Dewar, which is much easier for fabricating and aligning of the system. The system stop is re-imaged onto the cold stop by the co-aperture relay lens, which means they coincide with each other, leading to a 100% cold stop efficiency [11].

The aperture-divided MWIR imaging polarimeter described in this paper works in the 3–5 μm MWIR bands, focal length is 68 mm, F number F/2, and the FOV is $4^\circ \times 3.2^\circ$. The pixel size of the chosen PV staring IR detector is 15 μm × 15 μm, of which the effective pixel number is 640 × 512. The cold stop is 200 mm away from the FPA. The chosen polarizer is MWIR wire grid polarizer, whose extinction ratio is better than 1000:1.

3 Design Results and Analysis

Based on the system considerations listed above, by optimizing the optical system using ZEMAX software, finally we get the design result as is shown in Fig. 2. The main parameters of this system are listed in Table 1. The total size of the imaging polarimeter is about 78 mm(x) × 78 mm(y) × 356 mm(z).

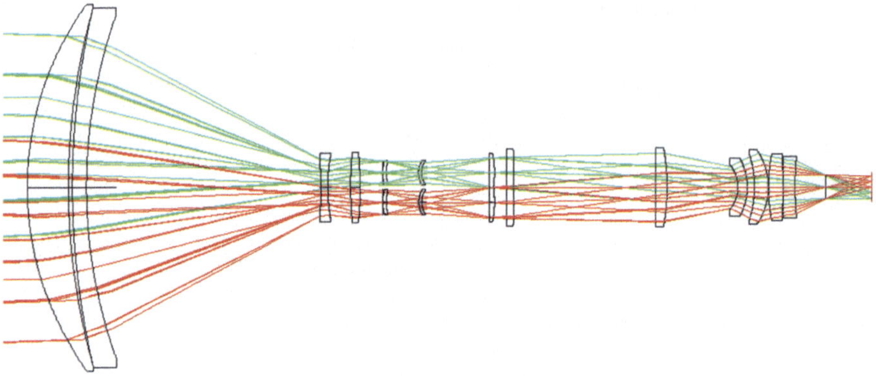

Fig. 2 Schematic of optical train for imaging polarimeter

Table 1 The main parameters of imaging polarimeter

Parameter	Value
Focal length	68 mm
F/#	2
FOV	4° × 3.2°
Wavelength	3–5 μm
Detector	640 × 512 IR detector
Pixel size	15 μm

We employ the modulation transfer function MTF, spot diagram and grid distortion diagram to evaluate the imaging quality of the optical system. The MTF curve is shown in Fig. 3, in which the black curve denotes the diffraction limit. The MTF is better than 0.5 at the Nyquist frequency of 33 lp/mm, which means the imaging quality is good enough to approach diffraction limit.

The spot diagram is shown in Fig. 4, every diagram is smaller than the Airy diameter, which means the system has the diffraction limited imaging characteristics. Figure 5 shows the encircled energy at 15 μm diameter is better than 73%.

Figure 6 shows the grid distortion diagram of the system, in which the crosses denote paraxial image locations, and solid lines denote the real image locations of each FOV. The figure shows that the main distortion of this system is pincushion distortion, and the max relative distortion is less than 1.13%, the max distortion is about 20 μm, which is slightly larger than a pixel. The distortion of this system could be corrected by image processing algorithm, after which the imaging polarimeter could satisfy the requirements of practical applications.

Finally we analyzed the tolerance requirements of the imaging polarimeter by Code V. The result shows that the most sensitive tolerance requirements are demanded by the co-aperture relay lens. The most sensitive tolerance requirements are listed in Table 2. The analysis result shows that the imaging polarimeter is much suitable for manufacturing and aligning.

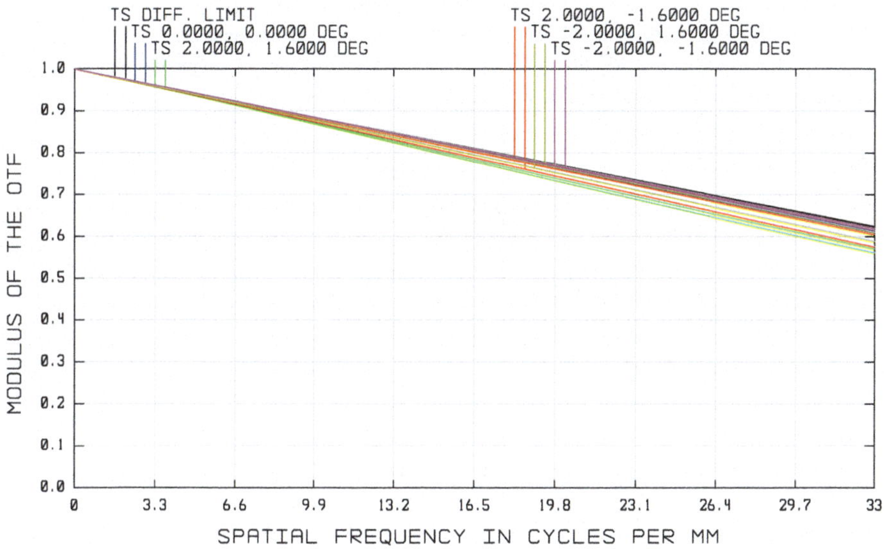

Fig. 3 MTF curves of the system

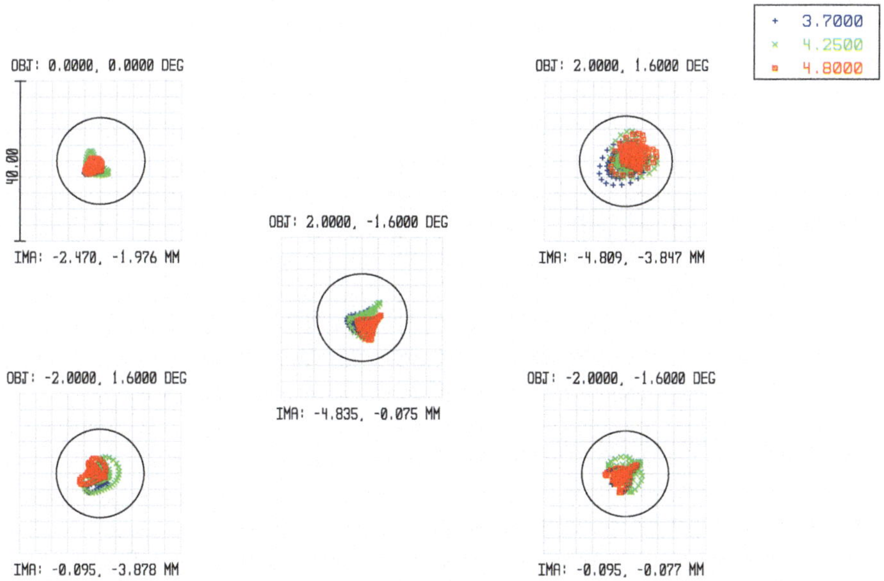

Fig. 4 Spot diagram of light tracing

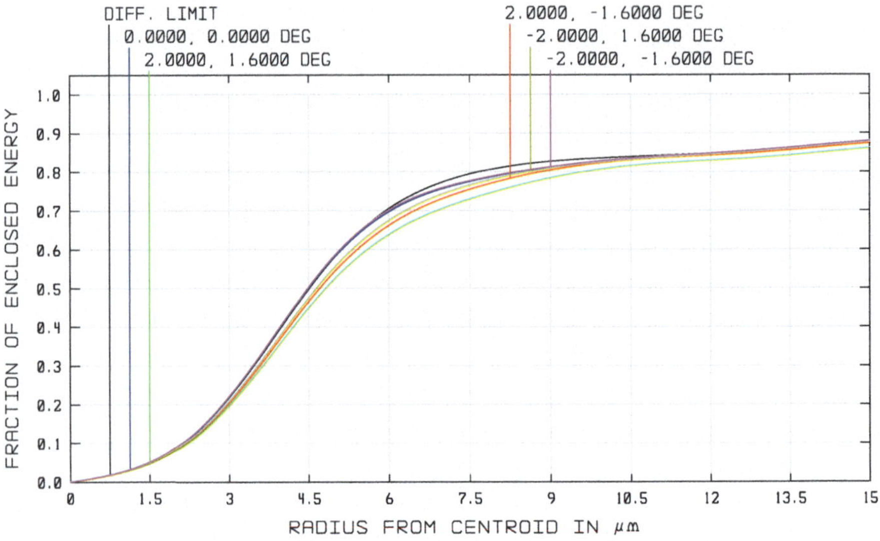

Fig. 5 Diffraction encircled energy curves

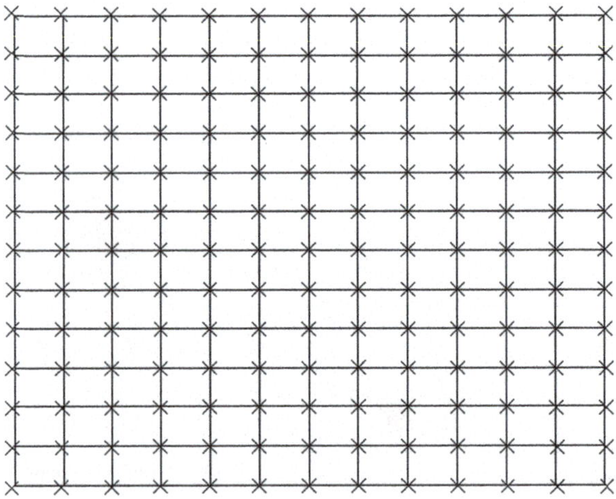

Fig. 6 Distortion grid of the system

Table 2 Most sensitive tolerance requirements of imaging polarimeter

Surface	Radius/mm	Thickness/mm	X Group displacement	Y Group displacement
23	0.02	0.01	0.01	0.01
24	0.01	0.01	0.01	0.01
25	0.02	0.01	0.01	0.01
28	0.02	0.01	0.01	0.01

4 Conclusion

An aperture-divided optical system of MWIR imaging polarimeter has been described. The system employs relay lens, leading to 100% efficiency of cold stop. There is no need to refit the vacuum Dewar or IR detector, hence it costs less and is easier to be manufactured and aligned. Finally we designed and optimized the imaging polarimeter consisting of co-aperture front Galilean telescope, sub-aperture double-gauss objective, polarizers, co-aperture Cooke triplet relay lens and IR detector. The imaging quality is evaluated by analyzing the MTF curves, spot diagrams, encircled energy and grid distortion diagram of the system, showing that it has a good imaging quality. The result of tolerance analysis shows that this imaging polarimeter is suitable to be manufactured. Moreover, if a rotating filter wheel is inserted inside the optical path of this imaging polarimeter, it can also detect spectrapolarimetric images of target, which realizes the spectral and polarimetric multi-dimensional information detection of target.

References

1. J.Q. Gong, H.G. Zhan, D.Z. Liu, "A Review on Polarization Information in the Remote Sensing Detection," *Spectroscopy and Spectral Analysis*, vol. 30, No. 4, pp. 1088–1095, April 2010.
2. F. Snik, J. Craven-Jones, M. Escuti, S. Fineschi, D. Harrington, A.D. Martino, et al, "An overview of polarimetric sensing techniques and technology with applications to different research fields," *Proceedings of SPIE—The International Society for Optical Engineering*, vol. 9099, pp. 90990B-90990B20, May 2014.
3. R. Walraven, "Polarization Imagery," *Proceedings of SPIE—Optical Polarimetry Instrumentation & Applications*, vol. 0112, pp. 14–18, October 1977.
4. M.H. Smith, P.D. Burke, A. Lompado, E.A. Tanner, L.W. Hillman. "Mueller matrix imaging polarimetry in dermatology," *BiOS 2000 The International Symposium on Biomedical Optics*, pp. 210–216, 2000.
5. D.A. Glenar, J.J. Hillman, B.N. Saif, J. Bergstrahl. "POLARIS-II: an acousto-optic imaging spectropolarimeter for ground-based astronomy," *Proceedings of SPIE—The International Society for Optical Engineering*, pp. 92–101, December 1992.
6. G.F.J. Garlick, G.A. Steigmann, W.E. Lamb, "Differential optical polarization detectors," *U.S. Patent*: 3992571, November 1976.
7. A.G. Andreou, Z.K. Kalayjian, "Polarization imaging: principles and integrated polarimeters," *Sensors Journal IEEE 2*, No.6, pp. 566–576, 2002.
8. J.S. Tyo, D.L. Goldstein, D.B. Chenault, J.A. Shaw, "Review of passive imaging polarimetry for remote sensing applications," *Applied Optics*, vol. 45, No. 22, pp. 5453–5469, August 2006.
9. H.C. He, Y.Q. Ji, J.K. Zhou, Z.C. Zhao, W.M Shen, "Optical Design of Decentered Aperture-Divided Polarization Imaging System," *ACTA OPTICA SINICA*, vol. 33, No.6, pp. 279–284, June 2013.
10. J.L. Pezzaniti, D.B. Chenault, "A division of aperture MWIR imaging polarimeter," *Proceedings of SPIE—The International Society for Optical Engineering*, vol. 5888, pp. 515–533, 2005.

11. R.E. Fischer, B.T. Galeb, P.R. Yoder, R. Galeb, *Optical System Design*, New York: McGraw Hill, pp. 223–226. 2000.
12. C.Y. Zhang, W.M. Shen, "Design of an athermalized MWIR and LWIR Dual-band Optical System," *Infrared and Laser Engineering*, vol.41, No.5, pp. 1323–1328, May 2012.

Analysis of High Resolution Two Line Array Camera Stereo Georeferencing Accuracy

Zhongqiu Xia, Qiaolin Huang, Hongyan He, Ruimin Fu and Chunyu Yue

Abstract The stereo images acquired by two line array camera of high resolution optical remote sensing satellite has been widely used in georeferencing of observed scene, and the georeferencing accuracy in plane and height without ground control points has significant importance for the geometric performance assessment, which can provide reference for design of the satellite platform and the remote sensor. According to the stereo georeferencing method, the theory of multivariate function error estimation and the assessment principle based the circular error probable, the stereo georeferencing accuracy algorithm of the two line array camera is derived and the influence of the stochastic error and the systematic error is analyzed with the parameters of the satellite platform and the remote sensor, which show that the plane accuracy is 25.53 m and the height accuracy is 29.01 m in condition that only the stochastic error is taken into account. After the systematic error is introduced, the plane accuracy is 26.75 m and the height accuracy is 43.98 m, which relate to the value of systematic error. This method that stochastic error and the systematic error are considered simultaneously is applicable to analyze the stereo georeferencing accuracy of high resolution two line array cameras.

Keywords High resolution · Two line array camera · Plane georeferencing accuracy · Height accuracy · Systematic error · Stochastic error

1 Introduction

Two line array camera set on the high resolution remote sensing satellite has been widely used in stereo georeferencing [1]. The analysis of georeferencing accuracy in plane and high without ground control point decided by the systematic error and the stochastic error has significant importance for the design, calibration [2] and application of the imagery system [3], which consists of the camera, GPS and star

Z. Xia (✉) · Q. Huang · H. He · R. Fu · C. Yue
Beijing Institute of Space Mechanics & Electricity, Beijing, China
e-mail: xiazhongqiu@hotmail.com

H.P. Urbach and G. Zhang (eds.), *3rd International Symposium of Space Optical Instruments and Applications*, Springer Proceedings in Physics 192,
DOI 10.1007/978-3-319-49184-4_8

tracker mainly [4]. To some extent the systematic error and the stochastic error reflect the measurement level and the system stability respectively.

In recent years, a number of calculate methods are introduced based on the different assessment systems, which have their own advantages and the disadvantages. The principal point error and the focal length error are regarded as the stochastic error in the error propagation investigation of interior orientation elements [5], which does not consider the exterior orientation elements error. This assessment system analyzed the influence of the interior orientation elements error briefly, but neglected the influence of the influence of the exterior orientation elements error. The position error, the photographic baseline error, the image point error and the focal length error are regarded as the stochastic error in the three line array camera mapping precision analysis [6]. This assessment system considers more comprehensive, but some parameters are not indexes in the satellite design directly (e.g. the photographic baseline error). In addition, these methods usually calculate the edge of the line only, which can not reflect the georeferencing performance of the whole image. And the system errors of various indexes are usually not introduced in the algorithm of the georeferencing accuracy.

This paper presents the georeferencing error distribution of the points tiled on the image using the circular error probability algorithm [7]. Considering the measure level and the stability [8], the influence on georeferencing accuracy of the systematic error and the stochastic error are analyzed. The assessment system in this paper is applicable to the georeferencing accuracy analysis of the high resolution optical remote sensing satellite which acquires the stereo images using the two line array CCD camera.

2 Model and Calculation Method

The stereo georeferencing method without the ground control point can be expressed as

$$\begin{cases} X = X_{S1} + N_1 X_1 \\ Y = Y_{S1} + (N_1 Y_1 + B_Y + N_2 Y_2)/2 \\ Z = Z_{S1} + N_1 Z_1 \end{cases} \tag{1}$$

where X, Y, Z are the coordinates of ground points, X_{S1}, Y_{S1}, Z_{S1} are the coordinates of position of projective center and photographic baseline B_X, B_Y, B_Z are calculated as

$$\begin{cases} B_X = X_{S2} - X_{S1} \\ B_Y = Y_{S2} - Y_{S1} \\ B_Z = Z_{S2} - Z_{S1} \end{cases} \tag{2}$$

The point projection coefficient $\mathbf{N_1}$, $\mathbf{N_2}$ are calculated as

$$\begin{cases} N_1 = \frac{B_X Z_2 - B_Z X_2}{X_1 Z_2 - X_2 Z_1} \\ N_2 = \frac{B_X Z_1 - B_Z X_1}{X_1 Z_2 - X_2 Z_1} \end{cases} \tag{3}$$

The auxiliary coordinate of the image space X_1, Y_1, Z_1 and X_2, Y_2, Z_2 are calculated respectively as

$$\begin{bmatrix} X_1 \\ Y_1 \\ Z_1 \end{bmatrix} = \begin{bmatrix} a_{11} & a_{12} & a_{13} \\ b_{11} & b_{12} & b_{13} \\ c_{11} & c_{12} & c_{13} \end{bmatrix} \begin{bmatrix} 0 \\ y_1 - y_{01} - D_1 \\ -f_1 \end{bmatrix} \tag{4}$$

$$\begin{bmatrix} X_2 \\ Y_2 \\ Z_2 \end{bmatrix} = \begin{bmatrix} a_{21} & a_{22} & a_{23} \\ b_{21} & b_{22} & b_{23} \\ c_{21} & c_{22} & c_{23} \end{bmatrix} \begin{bmatrix} 0 \\ y_2 - y_{02} - D_2 \\ -f_2 \end{bmatrix} \tag{5}$$

where y_{01}, y_{02} are the principal point, f_1, f_2 are the focal length, D_1, D_2 are the distortion, a_{11}, b_{11}, c_{11}, a_{12}, b_{12}, c_{12}, a_{13}, b_{13}, c_{13} are decided by φ_1, ω_1, κ_1 of front camera and a_{21}, b_{21}, c_{21}, a_{22}, b_{22}, c_{22}, a_{23}, b_{23}, c_{23} are decided by φ_2, ω_2, κ_2 of the rear camera.

Then based on the theory of multivariate function error estimation, the georeferencing accuracy in plane and high can be calculated as

$$\begin{cases} \sigma_X^2 = \left(\frac{\partial X}{\partial y_{01}}\right)^2 \sigma_{y_0}^2 + \left(\frac{\partial X}{\partial f_1}\right)^2 \sigma_f^2 + \left(\frac{\partial X}{\partial D_1}\right)^2 \sigma_D^2 + \left(\frac{\partial X}{\partial X_{S_1}}\right)^2 \sigma_{X_S}^2 + \left(\frac{\partial X}{\partial Y_{S_1}}\right)^2 \sigma_{Y_S}^2 \\ \quad + \left(\frac{\partial X}{\partial Z_{S_1}}\right)^2 \sigma_{Z_S}^2 + \left(\frac{\partial X}{\partial \varphi_1}\right)^2 \sigma_\varphi^2 + \left(\frac{\partial X}{\partial \omega_1}\right)^2 \sigma_\omega^2 + \left(\frac{\partial X}{\partial \kappa_1}\right)^2 \sigma_\kappa^2 + \left(\frac{\partial X}{\partial y_{01}} \varepsilon_{y_0} + \frac{\partial X}{\partial f_1} \varepsilon_f \right. \\ \quad \left. + \frac{\partial X}{\partial D_1} \varepsilon_D + \frac{\partial X}{\partial X_{S_1}} \varepsilon_{X_S} + \frac{\partial X}{\partial Y_{S_1}} \varepsilon_{Y_S} + \frac{\partial X}{\partial Z_{S_1}} \varepsilon_{Z_S} + \frac{\partial X}{\partial \varphi_1} \varepsilon_\varphi + \frac{\partial X}{\partial \omega_1} \varepsilon_\omega + \frac{\partial X}{\partial \kappa_1} \varepsilon_\kappa \right)^2 \\ \sigma_Y^2 = \left(\frac{\partial Y}{\partial y_{01}}\right)^2 \sigma_{Y_0}^2 + \left(\frac{\partial Y}{\partial f_1}\right)^2 \sigma_f^2 + \left(\frac{\partial Y}{\partial D_1}\right)^2 \sigma_D^2 + \left(\frac{\partial Y}{\partial X_{S_1}}\right)^2 \sigma_{X_S}^2 + \left(\frac{\partial Y}{\partial Y_{S_1}}\right)^2 \sigma_{Y_S}^2 \\ \quad + \left(\frac{\partial Y}{\partial Z_{S_1}}\right)^2 \sigma_{Z_S}^2 + \left(\frac{\partial Y}{\partial \varphi_1}\right)^2 \sigma_\varphi^2 + \left(\frac{\partial Y}{\partial \omega_1}\right)^2 \sigma_\omega^2 + \left(\frac{\partial Y}{\partial \kappa_1}\right)^2 \sigma_\kappa^2 + \left(\frac{\partial Y}{\partial y_{01}} \varepsilon_{y_0} + \frac{\partial Y}{\partial f_1} \varepsilon_f \right. \\ \quad \left. + \frac{\partial Y}{\partial D_1} \varepsilon_D + \frac{\partial Y}{\partial X_{S_1}} \varepsilon_{X_S} + \frac{\partial Y}{\partial Y_{S_1}} \varepsilon_{Y_S} + \frac{\partial Y}{\partial Z_{S_1}} \varepsilon_{Z_S} + \frac{\partial Y}{\partial \varphi_1} \varepsilon_\varphi + \frac{\partial Y}{\partial \omega_1} \varepsilon_\omega + \frac{\partial Y}{\partial \kappa_1} \varepsilon_\kappa \right)^2 \\ \sigma_Z^2 = \left(\frac{\partial Z}{\partial y_{01}}\right)^2 \sigma_{y_0}^2 + \left(\frac{\partial Z}{\partial f_1}\right)^2 \sigma_f^2 + \left(\frac{\partial Z}{\partial D_1}\right)^2 \sigma_D^2 + \left(\frac{\partial Z}{\partial X_{S_1}}\right)^2 \sigma_{X_S}^2 + \left(\frac{\partial Z}{\partial Y_{S_1}}\right)^2 \sigma_{Y_S}^2 \\ \quad + \left(\frac{\partial Z}{\partial Z_{S_1}}\right)^2 \sigma_{Z_S}^2 + \left(\frac{\partial Z}{\partial \varphi_1}\right)^2 \sigma_\varphi^2 + \left(\frac{\partial Z}{\partial \omega_1}\right)^2 \sigma_\omega^2 + \left(\frac{\partial Z}{\partial \kappa_1}\right)^2 \sigma_\kappa^2 + \left(\frac{\partial Z}{\partial y_{01}} \varepsilon_{y_0} + \frac{\partial Z}{\partial f_1} \varepsilon_f \right. \\ \quad \left. + \frac{\partial Z}{\partial D_1} \varepsilon_D + \frac{\partial Z}{\partial X_{S_1}} \varepsilon_{X_S} + \frac{\partial Z}{\partial Y_{S_1}} \varepsilon_{Y_S} + \frac{\partial Z}{\partial Z_{S_1}} \varepsilon_{Z_S} + \frac{\partial Z}{\partial \varphi_1} \varepsilon_\varphi + \frac{\partial Z}{\partial \omega_1} \varepsilon_\omega + \frac{\partial Z}{\partial \kappa_1} \varepsilon_\kappa \right)^2 \end{cases} \tag{6}$$

where σ_{y0}, σ_f, σ_D are the accuracy of principal point, focal length and distortion respectively, σ_{XS}, σ_{YS}, σ_{ZS} are the position accuracy of projective center, σ_φ, σ_ω, σ_κ are the attitude accuracy and ε_{y0}, ε_f, ε_D, ε_{XS}, ε_{YS}, ε_{ZS}, ε_φ, ε_ω, ε_κ are the systematic error correspondingly.

To reflect the georeferencing accuracy performance of the whole image, 11 points per line with the same interval are selected to calculate and this operation is repeated 10 times, which is equal to select 10×11 points in the image as the test points. The georeferencing error distribution and accuracy are calculated in the condition of stochastic error only. Then the systematic error and the stochastic error

Table 1 Values of parameters

Parameters	Value	
Altitude	$H = 500$ km	
Detectors per line	$n = 20{,}000$	
Detector size	$p = 10$ μm	
Focal length	$f = 15$ m	
Lens distortions error/accuracy	$\varepsilon_D = 5$ pixels	$\sigma_D = 1$ pixel
Principal point error/accuracy	$\varepsilon_{y0} = 5$ pixels	$\sigma_{y0} = 7$ pixels
Focal length error/accuracy	$\varepsilon_f = 0.03$ m	$\sigma_f = 0.02$ m
Position error/accuracy		$\sigma_{XS} = \sigma_{YS} = \sigma_{ZS} = 0.05$ m
Attitude error/accuracy		$\sigma_\varphi = \sigma_\omega = \sigma_\kappa = 9''$

are both considered, and the results are compared according to the assessment method of the circular error probable. The values of parameters are set as Table 1.

3 Results and Analysis

The georeferencing error distribution caused by systematic error only is shown in Fig. 1, which show that the systematic error leads to the georeferencing error along the line array direction obviously but not the flight direction and the value of the error increase from a position of the line array to the edge.

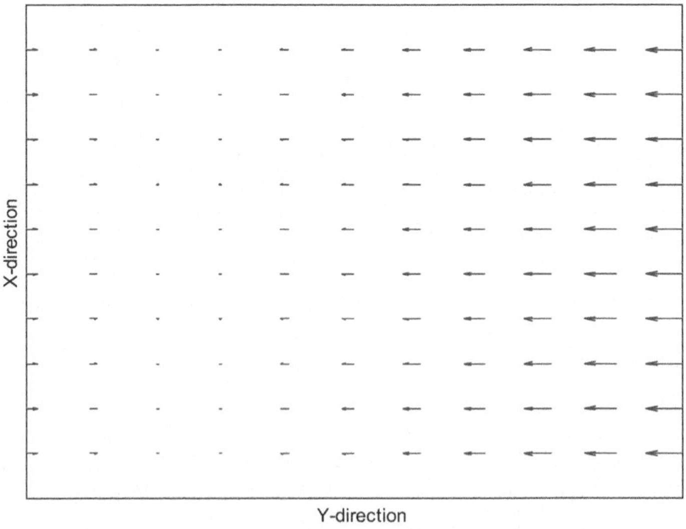

Fig. 1 Georeferencing error distribution caused by systematic error

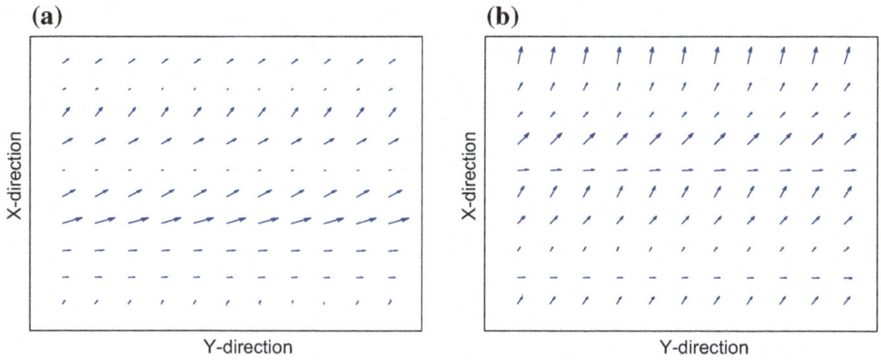

Fig. 2 Georeferencing error distribution caused by different error. **a** Stochastic error. **b** Systematic error and stochastic error

The georeferencing error distribution caused by stochastic error only is shown in Fig. 2a, which shows that the values of the georeferencing error in different position of the same line array along the line array direction or flight direction are nearly the same. In another words, the stability of the system make georeferencing error different in various line array corresponding to different imagery time. To estimate the georeferencing accuracy more precisionly, both the systematic error and the stochastic error should be considered. The georeferencing error distribution caused by systematic error and the stochastic error is shown in Fig. 2b, which is similar to the Fig. 2a and the georeferencing error in a certain position can not demonstrate the performance of the high resolution remote sensing satellite. So the georeferencing accuracy is calculated based on the circular error probable.

The circular error probable diagram of georeferencing error caused by stochastic error is shown in Fig. 3a. According to the circular error probable of 90%

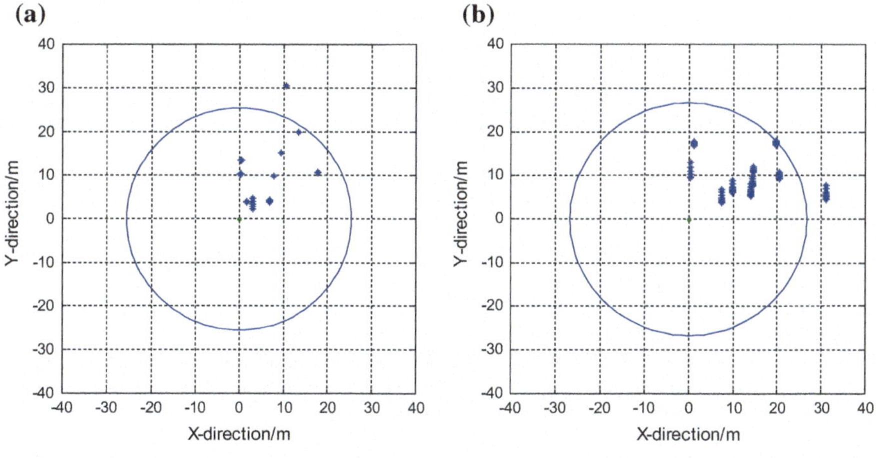

Fig. 3 Georeferencing error distribution based on the circular error probable. **a** Stochastic error. **b** Systematic error and stochastic error

algorithm, the radius is 25.53 m. After introducing the systematic error, the circular error probable diagram is shown in Fig. 3b and the radius increase to 26.75 m. For the high accuracy, the value increase from 29.01 to 43.98 m. To some extent this illustrates that the georeferencing accuracy in plane and high is decided by stochastic error mainly. During the design of the high resolution remote sensing satellite the systematic error should be minimized and eliminated as much as possible and the stochastic error should be restrained. That means the higher stability.

4 Conclusions

The factors on georeferencing accuracy consist of the systematic error and the stochastic error of principal point, focal length, distortion, position and attitude mainly. The method of considering both the systematic error and the stochastic error is more applicable to the analysis of the two line array camera stereo georeferencing accuracy. To some extent the systematic error represents the measurement level and the stochastic error represents the stability level of the high resolution remote sensing satellite geometric chain. The stochastic errors play a greater role in the plane georeferencing accuracy and the high accuracy, which means that the geometric chain should meet the higher requirement of the stability.

References

1. Q. L. Huang, and W. Jiang, "A Study of the New Direction of Space-borne Hi-resolution Optical Remote Sensor," *Spacecraft Recovery & Remote Sensing*, vol. 28, pp. 48–50, 2007.
2. Y. J. Zhang, M. T. Zheng, J. X. Xiong, Y. H. Lu, and X. D. Xiong, "On-Orbit Geometric Calibration of ZY-3 Three-Line Array Imagery With Multistrip Data Sets," *IEEE Transactions on Geoscience and Remote Sensing*, vol. 52. pp. 224–234, 2014.
3. R. D. Fiete, *Modeling the Imaging Chain of Digital Cameras*, Bellingham: SPIE Optical Engineering Press, pp, 5–12, 1997.
4. T. Jin, Z. Li, T. Li, and D. Yang, "System Design and Analysis for Improving Geometric Accuracy of High-Resolution Optical Remote Sensing Satellite Image," *Journal of Astronautics*, vol. 34, pp. 1160–1165, 2013.
5. J. P. Yu, and S. J. Sun, "Error Propagation of Interior Orientation Elements of Surveying Camera in Ground Positioning," *Spacecraft Recovery & Remote Sensing*, vol. 31, pp. 16–22, 2010.
6. X. Hu, and X. B. Cao, "Analysis on precision of stereo mapping microsatellite using three-line array CCD images," *Journal of Harbin Institute of technology*, vol. 40, pp. 695–699, 2008.
7. L. Zhang, W. Z. Li, Y. F. Ju, and H. L. Zhang, "Positioning Accuracy Evaluation Method Based on CEP," *Command Control & Simulation*, vol. 35, pp. 111–114, 2013.
8. H. Y. He, X. Y. Wang, and W. W. Qi, "The Relation between Imaging Parameter and In-orbit Imaging Quality of Remote Sensing Satellite TDICCD Camera," *The 23th China Space Exploration Academic Conference*, pp. 1–9, October 2010.

An Automatic Multiple-Slope Integration Algorithm for CMOS Image Sensor Based on Image Brightness

Tang Qi, Xie Jing, An Ran and Mu Yanna

Abstract In deep space exploration, illumination of environment changes tempestuously and it happens often that the contrast of target and background is large. Under this circumstance, an automatic multiple-slope integration algorithm is proposed for CMOS sensor based on the distribution characteristics of image brightness. First, the image is divided into a background region, some transitional regions and a target region according to its features of gray scale's distribution. The exposure slope and the proportion of multi-stage integral time length are obtained by calculating the percentage of number of pixels in total and average pixel brightness value of each region. Secondly, the weighted average brightness value of the whole image can be computed by assigning different weights to different regions. In the end, the total integral time length is obtained by comparing the weighted average brightness value with an expected brightness value to get an adjustment coefficient for integral time based on look up table. The algorithm extends dynamic range of CMOS sensor and enhances the image gray level. It can optimize the response curve of imaging system and improve the quality of image. Experiment results prove that this algorithm improves the dynamic range effectively and can be applied well not only in conventional light environment but also in space environment with greatly changing intensity of light.

Keywords Multiple-slope integration · CMOS image sensor · Brightness distribution · Dynamic range

1 Introduction

With fast development of the CMOS image sensor, it is increasingly challenging the conventional CCD image sensor in many application fields [1]. The CMOS image sensor with advantages of low cost, high resolution, high sensitivity and low

T. Qi (✉) · X. Jing · A. Ran · M. Yanna
Beijing Institute of Space Mechanics and Electricity, Beijing, China
e-mail: tangqi_qiqi@163.com

H.P. Urbach and G. Zhang (eds.), *3rd International Symposium of Space Optical Instruments and Applications*, Springer Proceedings in Physics 192,
DOI 10.1007/978-3-319-49184-4_9

noise is favorable for the small-size, and low-cost camera system [2]. Currently, camera systems using the CMOS sensor have been widely applied in industrial, medical, martial, aeronautic and astronautic fields [3]. However, the CMOS image sensor is usually based on active pixel structure and integrates many amplifiers and addressing circuits internally, which result in its small dynamic range and unfavorable shooting effect in conditions of large illumination range [4].

In order to improve the shooting effect in complex scenes, dynamic range expansion is one of the research hotspots in recent years. Most of the sensors adopt the technology of multiple-slope integration with conditional reset [5] at present. The technology extends the dynamic range by resetting pixels of the senor many times in one integration process. But the slope is generally set up to a fixed value such as double-slope, triple-slope or four-slope in current methods. This fixed setting may lead to the situation that the dynamic range is too large or too small because the variation of illumination is not well-considered. Therefore, the research of a method of multiple-slope integration which can adjust the dynamic range reasonably according to scenes has an active meaning.

An automatic multiple-slope integration algorithm is introduced in the paper. The algorithm adjusts relative configurations of multiple-slope integration according to the analysis of the distribution characteristics of image brightness and combines with a weighted auto-exposure method to get an appropriate dynamic range and improve the visual effect of the image. Experimental results show that the algorithm can extend the dynamic range reasonably and give consideration to the gray level at the same time. It can improve the visual quality of image effectively and have a good effect in conditions of high-contrast light.

2 Principle of the Multiple-Slope Integration

The dynamic range is generally defined as the ratio of the maximum signal to the minimum signal detected by the sensor [6]. It is an important index of the imaging quality. To an imaging system with linear response, the output voltage of the pixel is defined as (1).

$$V_{pout} = V_r - AIT_{\text{int}} \tag{1}$$

V_{pout} is the output voltage of the pixel, V_r is the reset voltage, A is the response coefficient, I is the light intensity and T_{int} is the integral time in (1).

Therefore, the output voltage of the pixel is determined by the incident light intensity when the reset voltage, response coefficient and integral time are the same to the pixel [7]. When the incident light intensity is too high or too low, the output signal will be saturated or below the noise and it cannot reflect the light intensity effectively [8]. But the adoption of multiple-slope integration can change the

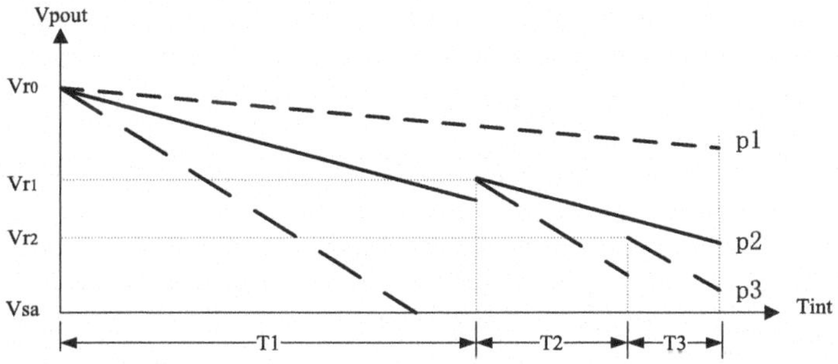

Fig. 1 The output curve in triple-slope mode

situation. This integration method adopts pixel structure with conditional reset and reset the pixel many times in one integration procedure, which makes the output signal can adapt larger range of change of the light intensity. Figure 1 shows the output curve of triple-slope mode. The output voltage of each pixel declines with time linearly when the integration starts. And the output voltage is reset to the value of the reset voltage in each reset operation if the output of the pixel is below the reset value. Then it declines linearly in the next integral time with the same slope. The value of the reset voltage is smaller than the last time and pixels of which the output voltage is larger than the reset value are not affected. The reset operation is equal to set inflexion points in the output curve. It makes the curve include more slopes and the dynamic range higher.

Multiple reset voltage can be set to get an output curve with the relative slope according to the different requirements and designs of the CMOS sensor. The more the inflexion point is set, the smoother the curve is because the curve is fitting of multiple straight lines by the effects of the inflexion point. The existing research proves that the expansion ratio of dynamic range increases progressively along with the increase of the ratio between the adjacent integral times [9].

3 The Automatic Multiple-Slope Integration Algorithm

The image brightness has been widely used in integral algorithms for it has a direct relation to the integral time. To get a better image effect, people always hope that the gray level of an image is uniformly distributed and the gray levels of central zone account for a larger proportion. However, images obtained in conditions of high-contrast light are usually have characteristic of local maximum, which means that the brightness distribution concentrates on the each end and are extremely unbalanced. The algorithm mentioned in the paper can configure corresponding parameters of the multiple-slope integration according to the brightness distribution to get an appropriate dynamic expansion effect and make the distribution uniformly. The algorithm also implements a good exposure control by calculating the weighted average brightness value and comparing it with an expected brightness value to get a suitable integral time.

At first, the image is divided into some regions according to the brightness distribution. In general, the regions can be distinguished as a dark background region, a target region, a bright background region and some transitional regions. Taking 8-bit quantization as example, the dark background region is the peak region of luminance histogram in which the brightness value is close to 0 while the bright background region is the peak region in which the brightness value is close to 255. These narrow peak regions include little gray levels and lower content complexity. In contrast, the target region is the central region of histogram includes more gray levels and higher content complexity. And transitional regions are in between. The brightness distribution can be judged preliminary by calculating the percentage of number of pixels in total of each region. In this case, the image is divided into five regions including a dark background region, a target region, a bright background region and two transitional regions. P1–P5 are used to represent the percentage of number of pixels in total of the above regions. If the sum of p2–p4 belongs to the preset interval (PL, PH), the brightness distribution is thought to be appropriate and the current integral mode can still be used. If the sum is less than PL, it indicates that the gray levels are concentrated in background areas and the dynamic range is low and the slopes should be changed according to the P1 and P5. If the sum is greater than PM, it indicates that the gray levels lack too much in background areas and the dynamic range is high and the slopes should be changed by further judgment of the percentage of the target region. After the number of slope is determined, the average value of pixel brightness of each region is calculated. A1–A5 are used to represent these values. The ratio between the adjacent integral times is determined by judging that whether A1–A5 belongs to the preset intervals (LMi, HMi) (i = 1, 2...5). If more average values of pixel brightness belong to the preset interval, it means the gray levels are good and the ratio can be set up to a small value. Similarly, the ratio should be set up to a corresponding value according to different situations. At last, the weighted average brightness value is obtained by assigning different weight coefficients to different regions. The data of the dark background should be ignored when the weight coefficient assigned if the A1 is too small because it would have a negative effect on the weighted average brightness value. Then the adjustment coefficient for integral time is obtained if the weighted average brightness value does not belong to the expected brightness interval. The adjustment coefficient is got in a look-up table by using the weighted average brightness value as index. The new integral time is calculated by current integral time multiple by the coefficient and the sectional integral times can also be calculated. Figure 2 shows the procedure of the algorithm.

Fig. 2 The procedure of the algorithm

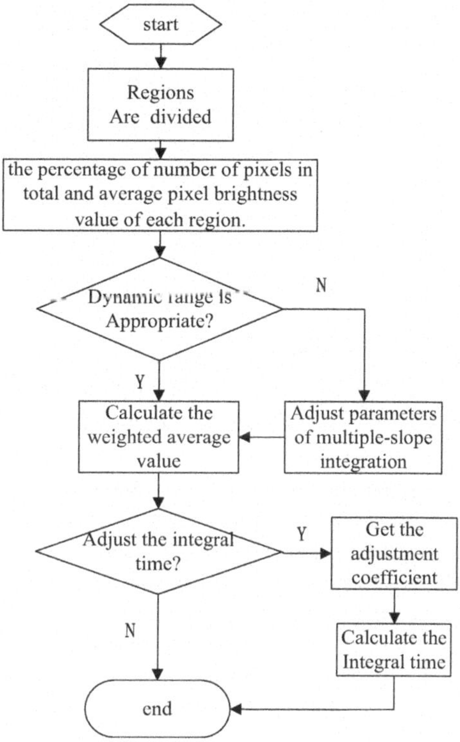

The preset values mentioned in above character can be changed on basis of the application of the camera and the analysis of relative experimental data. In general, the background regions should be narrow and the target region should be wide. The imaging effect should be well-considered when preset values are adjusted. For example, the region of interest (usually target region) should be given more weight than the background region. The form of the look-up table of adjustment coefficient for integral time can be fold line or smooth curve. Different average brightness values correspond to different coefficients. The curve can have various forms according to requirements to get an adaptive adjustment. Taking 8-bit quantization as example, the coefficients usually can be set up to 1 when the average is 128. Figure 3 shows one of the forms of the curve.

4 Experimental Results

The algorithm is applied in some type of CMOS camera and imaging test is taken to verify the improvement of the algorithm. Figure 4 shows the comparison result between the fixed slope setting and the algorithm. The camera is placed on the windowsill and takes photo of outside scene in the test. The test time is 4:00PM and

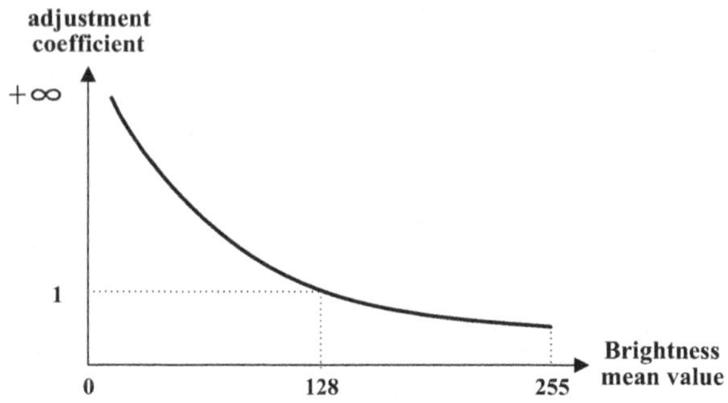

Fig. 3 The curve of the adjustment coefficient for integral time

the outside light intensity is a little great. Figure 4a shows the image in single-slope mode, Fig. 4b shows the image in four-slope mode, Fig. 4c shows the image applied the algorithm.

Figure 4a shows that the dynamic range is too small in single-slope mode and the image is extremely overexposed and the brightness distribution is too concentrated in the bright background area. Figure 4b shows that the brightness distribution is concentrated in the central zone and the gray levels is too less, which makes the visual effect of image weak. Figure 4c shows that the image in applied the algorithm is clear sharp and the buildings is exposed clearly in bright light, which indicate the dynamic range is extended reasonably. The gray levels of this image are distributed uniformed and the quality of image is good.

5 Conclusion

A new automatic multiple-slope integration algorithm for the CMOS image sensor is introduced in the paper. Unlike the conventional integral methods with fixed slope, the algorithm implements the automatic adjustment of multiple-slope on basis of the brightness distribution of image. The appropriate dynamic range is obtained by adjusting corresponding parameters. And the auto-exposure control is finished by applied the weighted exposure time calculation. The experimental results proves that the algorithm have a good effect on the reasonable expansion of dynamic range. The algorithm has low requirements of hardware and is easy to transplant. It can be widely used in conditions of tempestuous illumination variation and also have a good shooting effect in deep space environment with high-contrast light intensity.

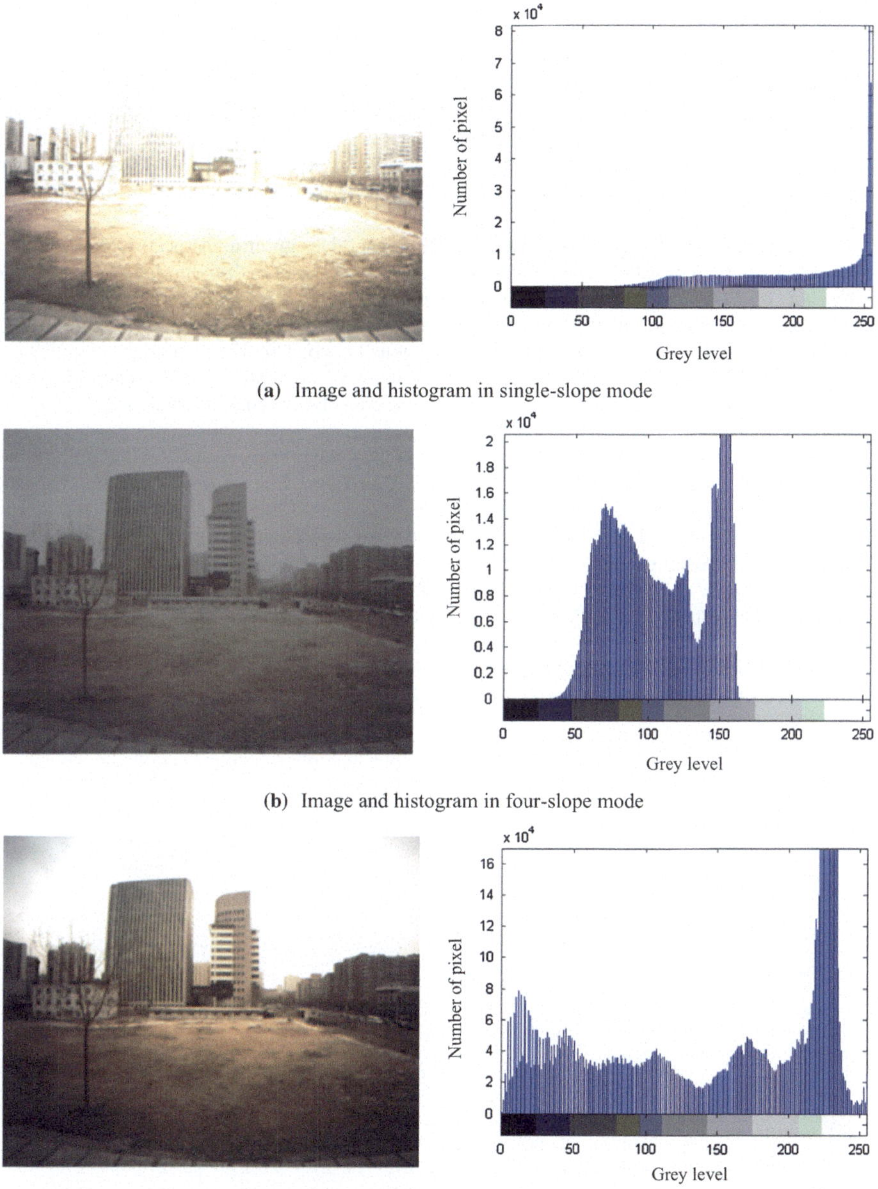

(a) Image and histogram in single-slope mode

(b) Image and histogram in four-slope mode

(c) Image and histogram using the algorithm

Fig. 4 Results of the imaging test

References

1. Nilsson M, Weerasinghe C, Lichman S, "Design and implementation of a CMOS sensor based video camera incorporating a combined AWB/AEC module," *International Conference on Acoustics. Hong Kong*, pp. 477–480, 2003.
2. Liu Meiying, Wang Hu, Wen Desheng, Liu Jie, Liu Yang, Xue Yaoke, et al. "Design of a large plane CMOS camera system FPGA," *Infrared and Laser Engineering. Tian Jin*, vol. 42, pp. 337–342, 2013.
3. Li Fangning, Wang Yanjie, Zhang Tao, Sun Honghai, "Design of high-speed high-resolution CMOS camera acquisition system based on AMV414 sensor," *Chinese Journal of Liquid Crystals and Displays. Chang Chun*, vol. 30, pp. 492–498, 2015.
4. Yu Shuai, Sun Dexin, "A Method of Enhancing Dynamic Range of CMOS Image Sensor," *Science Technology and Engineering. Bei Jing*, vol. 14, pp. 167–171, 2014.
5. Sasaki M, Mase M, Kawallito S, "A wide-dynamic-range CMOS image senor based on multiple short exposure-time read-out with multiple-resolution column-parallel ADC," *IEEE Sensors Journals.*vol. 7, pp. 151–158, 2013.
6. Jin Weiqi, Liu Bin, Fan Yongjie, Wang Xia, Xu Chao, Chen Yan, et al. "Review on infrared image detail enhancement techniques," *Infrared and Laser Engineering. Tian Jin*, vol. 40, pp. 2521–2527, 2011.
7. Gu Lin, Hu Xiaodong, Luo Changzhou, Xu Zhou, "Automatic adjustment of the exposure time of linear CCD based On cpld," *Acta Photonica Sinica. Xi An*, vol. 31, pp. 1533–1537, 2002.
8. Yang Wencai, Wen Desheng, Chen Shundan, Wang Hong, "Spatial Transient Light Detection Based on High-speed CMOS Image Sensor," *Acta Photonica Sinica. Xi An*, vol. 39, pp. 764–767, 2010.
9. An Ran, Chen Yan, Xie Jing, "Exposure algorithm for CMOS image sensor with adaptive dynamic range," *Infrared and Laser Engineering. Tian Jin*, vol. 42, pp. 88–92, 2013.

Influence of HY-2 Satellite Platform Vibration on Laser Communication Equipment: Analysis and On-Orbit Experiment

Qing-jun Zhang and Guang-yuan Wang

Abstract HY-2 satellite carried China's first laser communication equipment (LCE) into space. An on-orbit micro-vibration measurement unit (OMMU) was also onboard to provide assistant data for laser beam pointing prediction and correction. In the present study, firstly the pointing stability requirement of LCE was analyzed according to the payload parameters. Then the on-orbit data of OMMU was presented, based on which, the mechanical model of the satellite was modified to predict the angular jitter of the laser beam. Finally, the analyze results was verified by satellite-to-ground laser communication tests. It was concluded that the micro-vibration induced by moving parts onboard the satellite has little effect on satellite-to-ground laser communication.

Keywords HY-2 satellite · Micro-vibration · Laser communication · Pointing stability

1 Introduction

With the improvement of space technology, the data produced by spacecraft is expanding sharply. The traditional microwave communication equipment could not meet the requirement in the near future. Laser communication is becoming the most promising way to solve the problem for its wide band, high speed, small antenna size, and low power consumption.

Several laser communication terminals were tested in space in the past two decades, such as LCT onboard TSX-5 [1–3], PASTEL onboard SPOT-4 [4], OPALE onboard ARTEMIS [5–7], LUCE onboard OICETS [8–12], etc. China has sent laser communication equipment (LCE) into space onboard HY-2 satellite in 2011 and conducted a bunch of tests since then.

Q. Zhang (✉) · G. Wang
Beijing Institute of Spacecraft System Engineering, Beijing 100094, China
e-mail: zhuichilun@126.com

H.P. Urbach and G. Zhang (eds.), *3rd International Symposium of Space Optical Instruments and Applications*, Springer Proceedings in Physics 192,
DOI 10.1007/978-3-319-49184-4_10

Micro-vibration induced by the moving parts onboard the satellite may affect the pointing stability of the laser beam generated by the laser communication equipment. The communication terminal cannot connect if the pointing error exceeds the tracking view angle. Communication data would be error if the pointing jitter exceeds the laser beam divergence angle. Therefore, evaluation and attenuation of the satellite bus micro-vibration is essential for laser communication system.

An on-orbit micro-vibration measurement unit (OMMU) was installed on the HY-2 satellite to evaluate the micro-vibration level at the spacecraft structure and the laser communication equipment. The OMMU transmitted plenty of test data to the ground station, based on which, the pointing error of LCE was analyzed. The analysis results were provided for the laser catching and tracking experiment. The analysis method and results were presented in this paper.

In the present study, the pointing stability requirement of LCE was analyzed according to the payload parameters. Then the OMMU and its on-orbit data was presented, based on which, the mechanical model of the satellite was modified to predict the angular jitter of the laser beam. Finally, the laser communication test was described in detail.

2 Methodology

2.1 Affection Model of Micro-vibration on LCE [13, 14]

Accurate beam positioning is crucial for laser-based communication systems, capable of very high bit-rate data transfer. However, this technology is adversely affected by satellite jitter that is of little concern to common radio and microwave communication systems. Jitter is responsible for very small angular displacement of the source laser that corresponds to significant beam position errors when amplified by the hundreds of kilometers separating stations.

There are several types of errors associated with pointing, acquisition, and tracking (PAT) in laser communication systems. Pointing errors are associated with imprecise knowledge of the attitude and position of transmitting terminal and the initial position of the receiving satellite. The errors associated with tracking are boresight misalignments, satellite jitter, and point-ahead errors. They are due to the relative motion of the two platforms and the propagation delay caused by the large intervening distance. These errors are often combined in an rms (root mean square) fashion to provide an estimate of the systems PAT errors.

Assuming the signal light is Gaussian, when pointing error of the transmission terminal is ψ and the receiving terminal is ϕ, the receiving optical power is

$$P = \cos\phi \frac{4A_0 C_0^2}{\rho^2 \theta_b^2} \exp\left(-\frac{8\psi^2}{\theta_b^2}\right) \tag{1}$$

where A_0 is the area of the receiving antenna, θ_b is the divergence angle of signal light, C_0 is a constant. It can be seen that ψ and ϕ reduce the signal light power. Define $G(\psi)$ as the signal light power loss factor caused by pointing error of the transmission terminal and $G'(\phi)$ as the signal light power loss factor caused by pointing error of the receiving terminal. From Eq. 1 one can get

$$G(\psi) = \exp\left(-\frac{8\psi^2}{\theta_b^2}\right) \tag{2}$$

$$G'(\phi) = \cos\phi \tag{3}$$

The value of $G(\psi)$ and $G'(\phi)$ is between 0 and 1, where 1 stands for no power loss and 0 stands for maximum power loss. Equations 2 and 3 indicate that when the pointing error is in a small range like less than 1000 μrad, $G'(\phi)$ is quite close to 1 while $G(\psi)$ varies sharply. Enlarging θ_b can reduce the power loss effectively, while it is restricted by the laser generator hardware.

Optical power loss decreases the signal/noise ratio, raises the bit error rate, and affects the communication quality. Assuming the bit error rate in irrelative OOK modulation can be expressed as

$$\begin{aligned} E_{\mathrm{BER}} = & \int_0^\infty \int_0^\infty Q\left[\sqrt{N_{SNR}}\exp\left(-\frac{4qx^2}{\theta_b^2}\right)\right]\frac{xy}{\sigma_n^2\sigma_{ss}^2} \\ & \times \exp\left(-\frac{8qx^2}{\theta_b^2}\right)\exp\left[-\frac{y^2}{2\sigma_n^2}\exp\left(-\frac{8qx^2}{\theta_b^2}\right)-\frac{x^2}{2\sigma_{ss}^2}\right]\mathrm{d}x\mathrm{d}y \end{aligned} \tag{4}$$

where x is the amplitude of transmitting terminal jitter, y is the amplitude of receiving terminal jitter, N_{SNR} is the signal/noise ration with no tracking error, q is a detector constant, σ_n is the satellite jitter amplitude, σ_{ss} is the tracking error amplitude, $Q(x)$ is the area of Gauss function. From Eq. 3, one can conclude that the bit error rate decreases when N_{SNR} increases. Increasing the divergence angle can lower down the bit error rate.

2.2 *Dynamic Response Regeneration Based on Insufficient Sensors*

In engineering application, vibration sensors are often insufficient. For space industry, cost is quite sensitive to the complexity and weight of equipment. Therefore, the number of sensors for on-orbit micro-vibration measurement system is often limited to a reasonable level for financial consideration. On the other hand, angular transducers with high sensitivity are normally too large and heavy for a satellite to carry plenty of them. As a result, vibration data applicable from a on-orbit vibration measurement system is usually the acceleration at several points and angular at 1–2 points, which is far from enough to analyze the whole structure vibration [15, 16].

In this section, a procedure to regenerate the whole structure vibration based on limited measured data and analysis model is proposed.

The dynamic equation of a structure can be written as

$$k_S x + C_S \dot{x} + M_S \ddot{x} + K_J x + C_J \dot{x} + M_J \ddot{x} = f \quad (5)$$

where K_S, C_S, M_S stand for the stiffness, damping and mass matrix of the substructure separately. K_J, C_J, M_J stand for the connection stiffness, damping and mass matrix separately. Apparently, the dimension of $K_J\dot{x} + C_J\dot{x} + M_J\ddot{x}$ if force, which stands for the dynamic force that the connection applied on the substructures at the two ends. By assuming

$$\bar{f} = f - f_J. \quad (6)$$

$$f_J = K_J x + C_J \dot{x} + M_J \ddot{x} \quad (7)$$

the dynamic equation can be written as

$$K_S x + C_S \dot{x} + M_S \ddot{x} = \bar{f} \quad (8)$$

The right side of the equation can be seen as the summation of r + 1 harmonic excitation.

$$\bar{f} = \mathrm{Re}\left[\sum_{k=1}^{r} \bar{F}_k \exp(-2i\pi\omega_k t)\right] \quad (9)$$

Correspondingly, the response can also be seen as the summation of r + 1 harmonic response. For the kth harmonic response, the dynamic equation in frequency domain can be written as

$$K_S X_k + i\omega_k C_S X_k - \omega_k^2 M_S X_k = \bar{F}_k \quad (10)$$

Divide the structural degree of freedoms (DOFs) into 4 groups,

K measured inner DOF
F exciting DOF
E unmeasured inner DOF
J interface DOF

Assuming

$$D_k = K_S + i\omega_k C_S - \omega_k^2 M_S \tag{11}$$

Equations (4–6) can be expanded as

$$\begin{bmatrix} {}_{KK}D_k & {}_{KF}D_k & {}_{KE}D_k & {}_{KJ}D_k \\ {}_{FK}D_k & {}_{FF}D_k & {}_{FE}D_k & {}_{FJ}D_k \\ {}_{EK}D_k & {}_{EF}D_k & {}_{EE}D_k & {}_{EJ}D_k \\ {}_{JK}D_k & {}_{JF}D_k & {}_{JE}D_k & {}_{JJ}D_k \end{bmatrix} \begin{Bmatrix} {}_{K}X_k \\ {}_{F}X_k \\ {}_{E}X_k \\ {}_{J}X_k \end{Bmatrix} = \begin{Bmatrix} 0 \\ {}_{F}\bar{F}_k \\ 0 \\ {}_{J}\bar{F}_k \end{Bmatrix} \tag{12}$$

Since

$$\begin{Bmatrix} {}_{F}X_k \\ {}_{E}X_k \\ {}_{J}X_k \end{Bmatrix} = \begin{Bmatrix} {}_{F}\Phi_L \\ {}_{E}\Phi_L \\ {}_{J}\Phi_L \end{Bmatrix} q_k \tag{13}$$

Substitute Eq. (13) into Eq. (12), one can get

$$\begin{bmatrix} {}_{KK}D_k \\ {}_{FK}D_k \\ {}_{EK}D_k \\ {}_{JK}D_k \end{bmatrix} \{{}_{K}X_k\} + \begin{bmatrix} {}_{KF}D_{k_F}\Phi_L + {}_{KE}D_{k_E}\Phi_L + {}_{KJ}D_{k_E}\Phi_L \\ {}_{FF}D_{k_F}\Phi_L + {}_{FE}D_{k_E}\Phi_L + {}_{FJ}D_{k_E}\Phi_L \\ {}_{EF}D_{k_F}\Phi_L + {}_{EE}D_{k_E}\Phi_L + {}_{EJ}D_{k_E}\Phi_L \\ {}_{JF}D_{k_F}\Phi_L + {}_{JE}D_{k_E}\Phi_L + {}_{JJ}D_{k_E}\Phi_L \end{bmatrix} q_k = \begin{bmatrix} 0 \\ I \\ 0 \\ 0 \end{bmatrix} {}_{F}\bar{F}_k + \begin{bmatrix} 0 \\ 0 \\ 0 \\ I \end{bmatrix} {}_{J}\bar{F}_k \tag{14}$$

where Φ_L is the preserved low frequency modal vector. Expand the first 3 rows,

$$\begin{Bmatrix} {}_{KK}D_k \\ {}_{FK}D_k \\ {}_{EK}D_k \end{Bmatrix} \{{}_{K}X_k\} + \begin{bmatrix} {}_{KF}D_{kF}\Phi_L + {}_{KE}D_{kE}\Phi_L + {}_{KJ}D_{kE}\Phi_L \\ {}_{FF}D_{kF}\Phi_L + {}_{FE}D_{kE}\Phi_L + {}_{FJ}D_{kE}\Phi_L \\ {}_{EF}D_{kF}\Phi_L + {}_{EE}D_{kE}\Phi_L + {}_{EJ}D_{kE}\Phi_L \end{bmatrix} q_k = \begin{bmatrix} 0 \\ I \\ 0 \end{bmatrix} {}_{F}\bar{F}_k \tag{15}$$

The acceleration of group K (${}_{K}\ddot{x}$) and the excitation force applied on group F (${}_{F}\bar{f}$) are provided by on-orbit measurement and disturbing part ground test. Performing FFT on (${}_{K}\ddot{x}$) and ${}_{F}\bar{f}$, one can get the frequency domain form $-\omega^2{}_{K}X$ and ${}_{F}\bar{F}$. Substitute the kth harmonic $-\omega_k^2{}_{K}X_k$ and ${}_{F}\bar{F}_k$ into Eq. (15), one can get

$$\begin{bmatrix} {}_{KF}D_{kF}\Phi_L + {}_{KE}D_{kE}\Phi_L + {}_{KJ}D_{kE}\Phi_L \\ {}_{FF}D_{kF}\Phi_L + {}_{FE}D_{kE}\Phi_L + {}_{FJ}D_{kE}\Phi_L \\ {}_{EF}D_{kF}\Phi_L + {}_{EE}D_{kE}\Phi_L + {}_{EJ}D_{kE}\Phi_L \end{bmatrix} q_k = \begin{bmatrix} \frac{1}{\omega_k^2}{}_{KK}D_K & 0 \\ \frac{1}{\omega_k^2}{}_{FK}D_K & I \\ \frac{1}{\omega_k^2}{}_{EK}D_K & 0 \end{bmatrix} \begin{Bmatrix} -\omega^2{}_{K}X_k \\ {}_{F}\bar{F}_k \end{Bmatrix} \tag{16}$$

The row number of the equation is the DOF number of the whole structure, and the line number is the preserved mode number. Solve the equation and the response of the whole structure can be deducted

$$q_k = \begin{bmatrix} {}_{KF}D_{kF}\Phi_L + {}_{KE}D_{kE}\Phi_L + {}_{KJ}D_{kE}\Phi_L \\ {}_{FF}D_{kF}\Phi_L + {}_{FE}D_{kE}\Phi_L + {}_{FJ}D_{kE}\Phi_L \\ {}_{EF}D_{kF}\Phi_L + {}_{EE}D_{kE}\Phi_L + {}_{EJ}D_{kE}\Phi_L \end{bmatrix}^+ \begin{bmatrix} \frac{1}{\omega_k^2}\,{}_{KK}D_K 0 \\ \frac{1}{\omega_k^2}\,{}_{FK}D_K I \\ \frac{1}{\omega_k^2}\,{}_{EK}D_K 0 \end{bmatrix} \begin{Bmatrix} -\omega^2{}_K X_k \\ {}_F\bar{F}_k \end{Bmatrix} \quad (17)$$

where $[]^+$ stands for generalized inverse operator. Substituting Eq. (17) into Eq. (13), one can get the *k*th harmonic response of the unmeasured DOFs.

3 Micro-vibration Determination Requirement of HY-2 Satellite

3.1 LCE Pointing Stability Requirement Analysis

The laser communication equipment consists of six parts, which are the main body, terminal electrical control unit (TECU), high speed communication unit(HSCU), terminal data resource(TDR), thermal control unit(TCU), and DC-DC module. The main body is the key part of LCE, including a two dimensional pointing control mechanism (2DPCM), a fine pointing mechanism (FPM), laser generator, detector, and T/R optics, managing the transmission, reception, capture, and tracking of

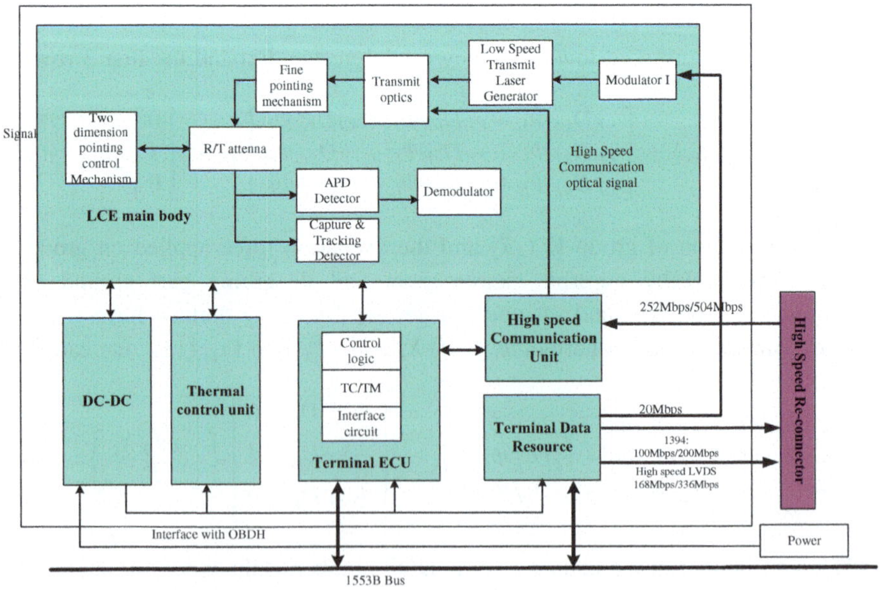

Fig. 1 LCE configuration sketch

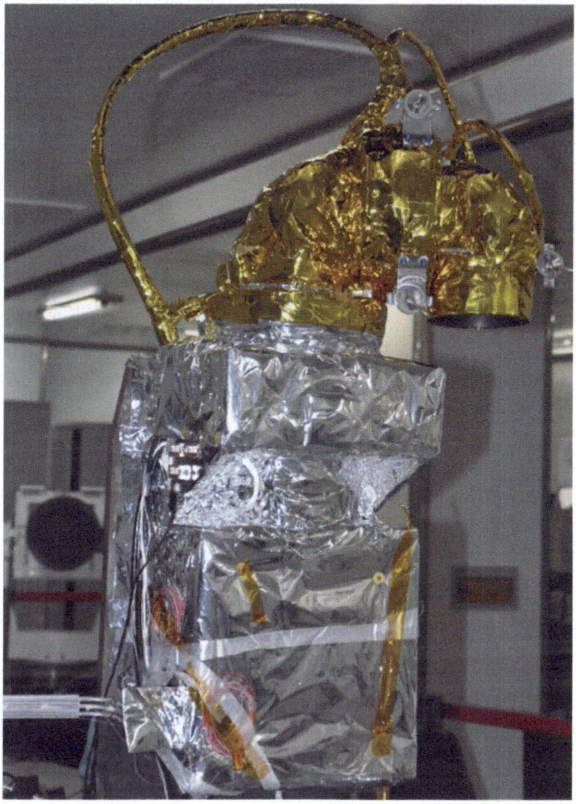

Fig. 2 LCE on satellite

optical signal. TECU controls the rough tracking mechanism, fine tracking mechanism, capture scanning, and communication with the satellite bus (Figs. 1 and 2).

The 2DPCM adjusts the angle of the LCE structure according to the orbit and attitude information of the satellite to ensure the incoming light get into the view of the pointing detector. Its function is compensating the rough angular error. The FPM adjust the optics to compensate the residual error of 2DPCM and the angular jitter of the laser beam.

Since the pointing mechanisms of LCE have the ability to track and compensate low frequency sway of the laser beam, laser communication is more sensitive to high frequency jitter. According to the specifications listed in Table 1, to ensure that there is enough optical energy was received, high frequency jitter should be less than 33 μrad, which is 1/3 of the transmitted signal light divergence angle.

Table 1 LCE specification

Item	Specification
Max capture time	35 s
Mean data error rate	7.8×10^{-8}
Rough tracking accuracy	±30 μrad (1σ, Single Axis)
Fine tracking accuracy	±1.5 μrad (1σ, Single Axis)
Data rate	20/252/504 Mbps
Signal light mean output power	71.1 mW
	1.01 W
Transmitted signal light divergence angle	98.5 μrad
	102.8 μrad
Capture view angle	Yaw 1601.2 μrad/pitch 2051.8 μrad
Tracking view angle	Yaw 260.73 μrad/pitch 235.13 μrad
Max power consumption	98.0 W
Weight	67.8 kg

3.2 *Disturbing Sources Onboard the Satellite*

There are many moving parts onboard the HY-2 satellite, including solar array driving mechanisms, antenna driving mechanisms, momentum wheels, scanning mechanisms, etc. Disturbing analysis of all the moving parts conducted in the design phase of the satellite showed that the laser beam jitter is mainly caused by three types of equipment, including the scatterometer, the radiometer, and the momentum wheels.

3.2.1 Scatterometer and Radiometer

The observing antenna of scatterometer scans conically while working, which means the rotation axis is vertical with the installation surface and the angle between the electrical axis and rotation axis is 35°(inner beam)/40.5°(outer beam). The rotation speed is 95°/s (default) or 105°/s with an accuracy of ±1°/s. The inertial of the rotation part is 9 kgm^2 with static imbalance of 400 gcm and dynamic imbalance of 4000 gcm^2.

The observing antenna of radiometer scans conically while working, which means the rotation axis is vertical with the installation surface and the angle between the electrical axis and rotation axis is 34°. The rotation speed is 95°/s (default) or 105°/s with an accuracy of ±1°/s. The inertial of the rotation part is 7.8 kgm^2 with static imbalance of 400 gcm and dynamic imbalance of 4000 gcm^2.

3.2.2 Momentum Wheels

Five momentum wheels were installed on the satellite in total, one of which was cold backup. Disturbance of momentum wheel comes from two mechanisms, as is shown in Fig. 3

(1) active disturbing force, which is produced by rotation with rotor imbalance, bearing imperfection, and motor instability;
(2) structural modulation, which changes the amplitude of active disturbing force while it transmits the wheel structure.

Because of the complexity of the rotating system, the spectrum of the disturbing force was composed by the rotation frequency and its fractional and integral frequencies. A typical waterfall of disturbing force against rotation speed and frequency is shown in the following figure. The disturbing force increases as the rotor accelerates and was amplified when the rotating frequency meets the structural natural frequency (Fig. 4).

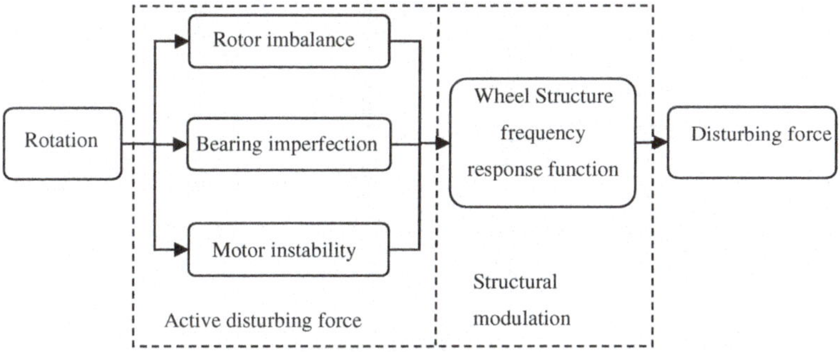

Fig. 3 Momentum wheel disturbing mechanism

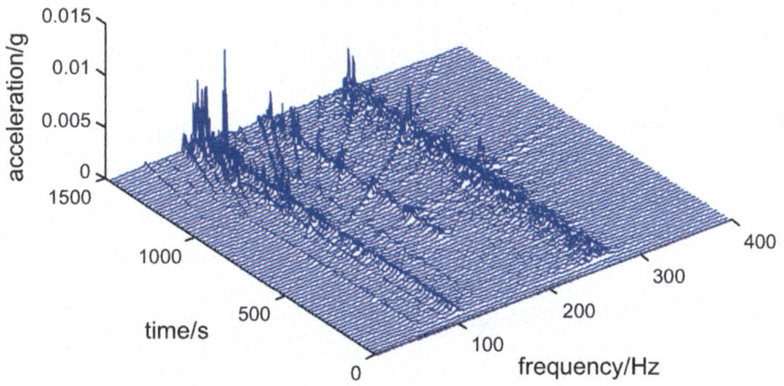

Fig. 4 Momentum wheel disturbing mechanism

4 Regeneration of LCE Angular Vibration Based on On-Orbit Data

4.1 On-Orbit Micro-vibration Measurement Hardware Configuration

OMMU consists of four parts, which are data acquisition module, control and storage module, power module, and sensors. Data acquisition module collects the signal from three dimensional acceleration transducers (3DAT) and generates data codes. Control and storage module records the data and call back the memory when necessary. Three sensors (3DAT) were installed at different positions on the satellite and connected with the OMMU by cables (Fig. 5).

Parameters of the data acquisition system is shown in Table 2.

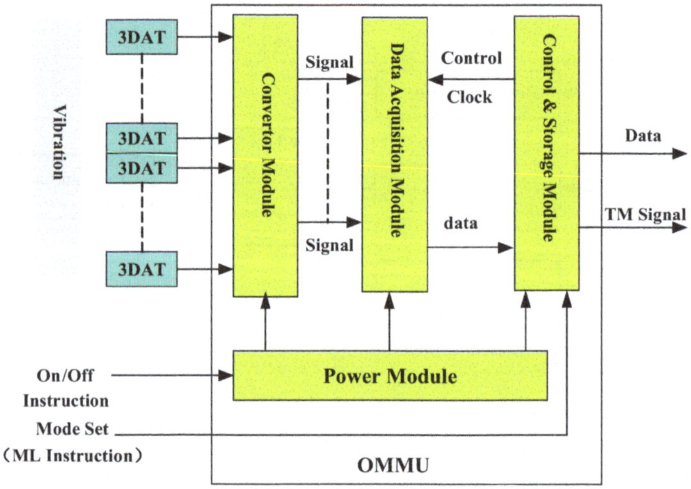

Fig. 5 OMMU configuration

Table 2 Data acquisition parameters

Item	Specification
Scale	0.001–0.1 g
Sensitivity	1000 mV/g
Accuracy	8%
Frequency band	0.1–150 Hz
Channels	3 × 3

4.2 On-Orbit Micro-vibration Measurement Data

4.2.1 Disturbing Source Vibration

According to the starting sequence of the momentum wheel, the scatterometer, and the radiometer, the acceleration of the installation surface of the moving parts was measured. Figure 6 showed the disturbance of one momentum wheel and the radiometer in time domain. The peak value of the disturbance caused by one momentum wheel is 11 mg, while caused by the radiometer is 38 mg.

The measured signal was processed by FFT method. Figure 7 showed the spectrum of momentum wheel disturbance. It can be seen from the figure that the disturbing energy was concentrated at the frequency band above 60 Hz. Figure 8 is the spectrum of radiometer disturbance, which showed that the disturbing frequency band is above 50 Hz.

4.2.2 LCE Vibration

In order to evaluate the vibration transmitted from the momentum wheel to the structure, the dynamic response of the installation surface of LCE and the radiometer was measured when the momentum wheel was working and the radiometer was off. Figure 9 showed that the vibration amplitude was 6 mg at one momentum wheel installation surface and 2 mg at the installation surface of LCE

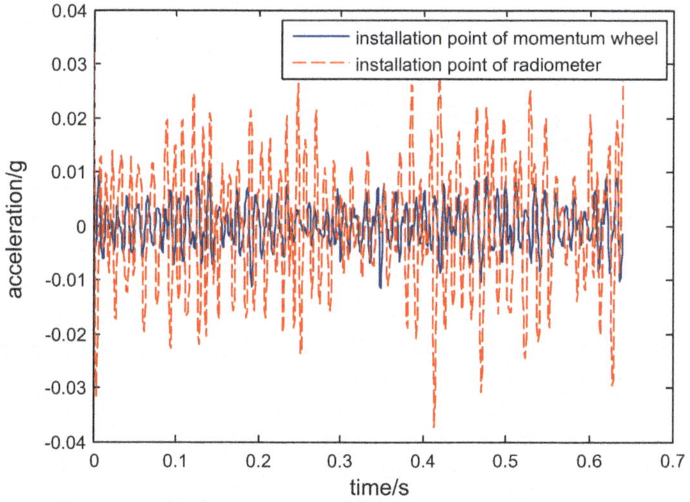

Fig. 6 Micro-vibration of the momentum wheel and the radiometer

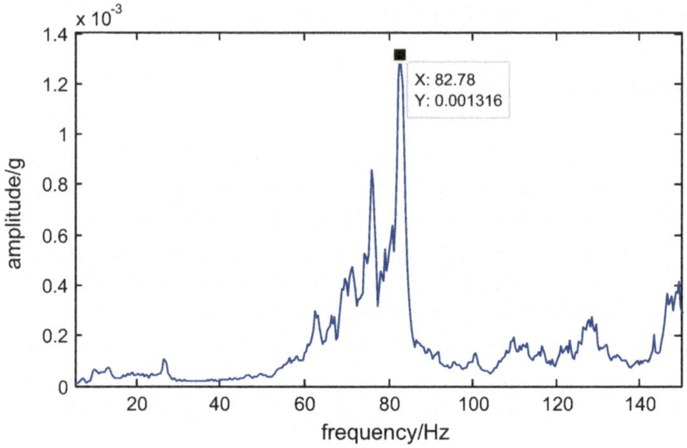

Fig. 7 Micro-vibration at the installation point of momentum wheel

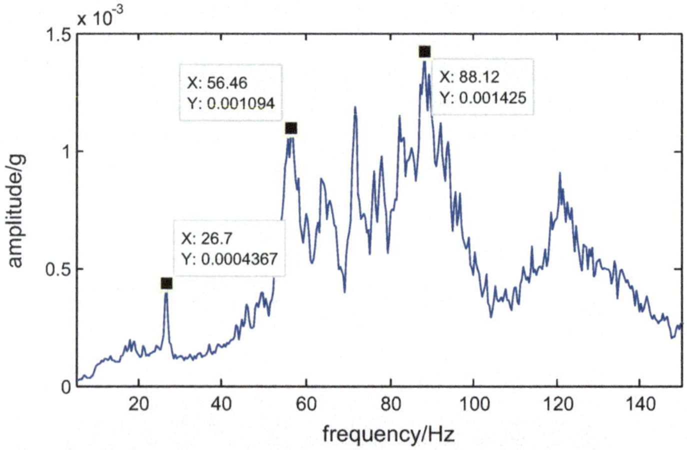

Fig. 8 Micro-vibration of the installation surface of radiometer

while four momentum wheels rotating. The disturbing energy was cut down to 34% through the transmission.

After the radiometer started, the vibration at LCE installation surface induced by all the moving part onboard the satellite, including four momentum wheels, radiometer, and scatterometer, was measured, as is shown in Fig. 10. The vibration level at LCE is about 20 mg, which shows that the LCE is more sensitive to radiometer and scatterometer.

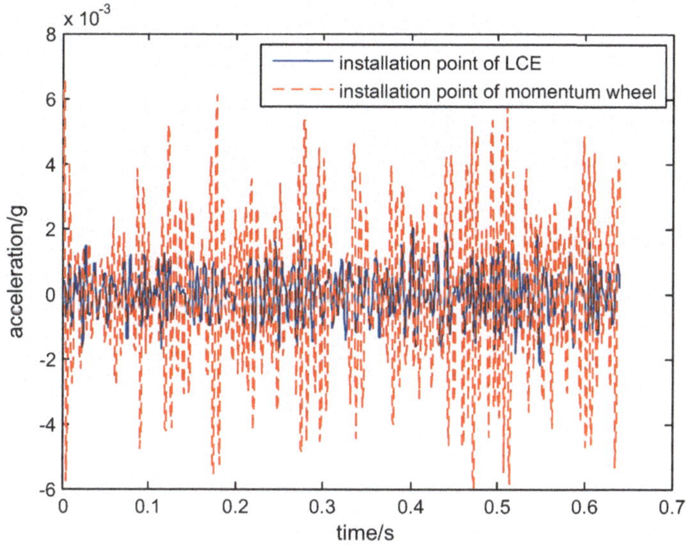

Fig. 9 Micro-vibration transmission from the momentum wheel to LCE

4.3 LCE Angular Vibration Analysis

The on-orbit vibration was predicted by the finite element model (FE model) of HY-2 Satellite pre-launch (Fig. 11). When the on-orbit data was applicable, the FE model was modified. Since the number of 3DAT was limited, the measured degrees of freedom (DOF) were much less than the FE model. The 3DAT can only provide the linear acceleration, while the angular displacement was the parameter directly correlated with the pointing jitter. Therefore, the following procedure was taken to analyze the LCE angular vibration,

(1) The FE model was condensed to limited DOFs, including all the 6 DOFs of each disturbing source installation surface and LCE installation surface;
(2) The stiffness matrix and damping matrix of the condensed model was modified to minimize the error between analyze results and on-orbit data at the measured DOFs;
(3) The angular vibration of LCE was calculated by applying the modified model.

Calculated angular vibration was shown in Fig. 12. It can be concluded that the laser beam jitter was far less than demanded, so that the laser communication would not be affected.

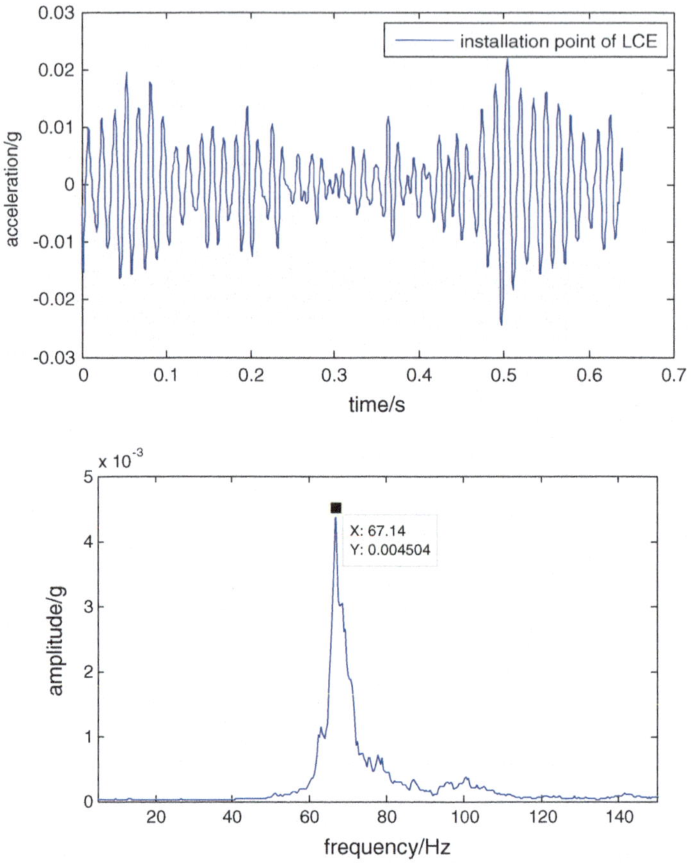

Fig. 10 Micro-vibration at LCE induced by all moving parts

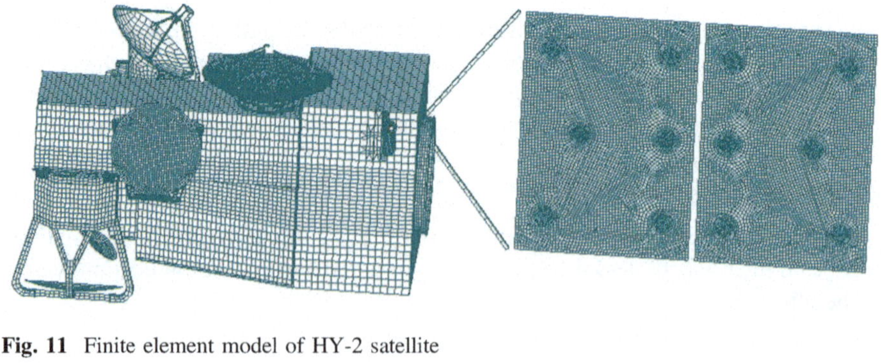

Fig. 11 Finite element model of HY-2 satellite

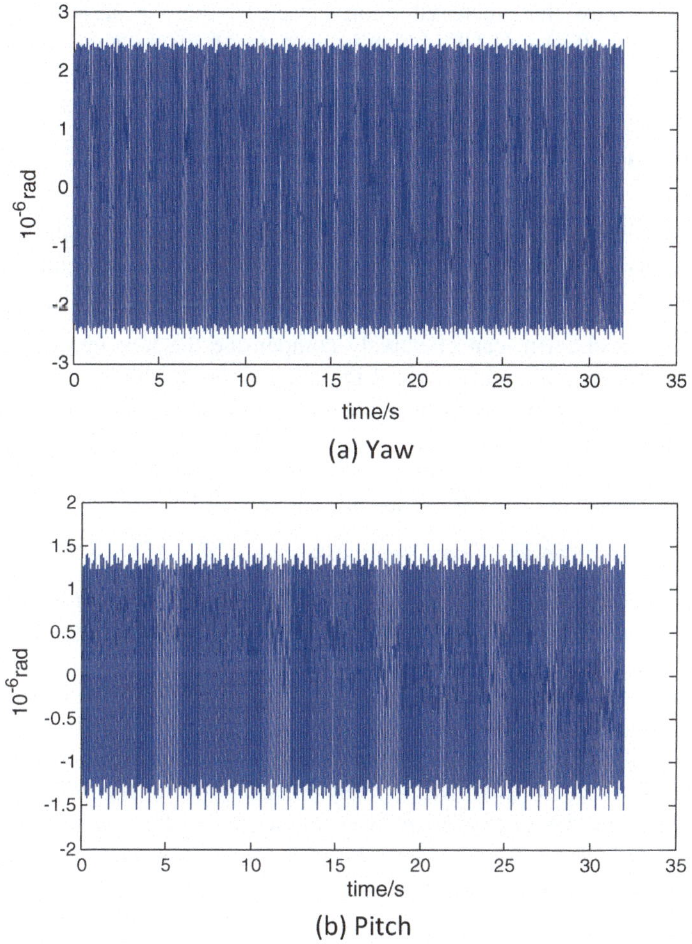

Fig. 12 LCE angular vibration

4.4 Satellite-to-Ground Laser Communication Tests

The LCE angular vibration analysis result was provide to the ground station to predict the laser beam jitter. From November 2011 to December 2011, 26 satellite-to-ground laser communication tests were conducted, 6 of which were invalid because of the weather. The low speed communication test was first succeeded at 10th November, and eight more low speed communication tests was succeeded in the month. High Speed communication test was first succeeded at 25th November and eleven more successful tests was finished in the following week. The data rate has achieved 504 Mbps.

Test results show that micro-vibration caused by moving parts onboard HY-2 satellite has little effect on LCE, which is consistent with the analysis result.

5 Conclusion

In the present study, the on-orbit micro-vibration measuring campaign of HY-2 satellite and its application in laser communication test was presented. It can be concluded that,

(1) Pointing mechanism can effectively compensate the low frequency sway of laser beam, which largely relaxed the requirement on high frequency jitter;
(2) On-orbit micro-vibration measurement can provide helpful data to predict the laser beam jitter;
(3) Satellite-to-ground laser communication test results showed that the moving parts onboard HY-2 satellite had little effect on LCE.

From HY-2, we can also learn some lessons,

(1) Three acceleration transducers are far from enough to provide ample information for FE model modification. More 3DAT channels are necessary in future missions.
(2) The frequency band of the 3DAT was so limited that high frequency vibration data above 150 Hz was not applicable. As LCE was sensitive to high frequency vibration, it will be helpful to expand the measuring frequency band.
(3) As the laser beam jitter was directly affected by the tilt of the supporting structure, angular displacement measurement would be more effective than linear acceleration measurement. Angular vibration sensors are suggested to be adopted in future missions.

References

1. J. L. Vanhove, C. Noldeke. In orbit demonstration of optical IOL/ISL-the SILEX project[J]. Intern. J. Satellite Communications, 1988, 6: 119–126.
2. E. Perez, M Bailly, J. M. Pairot. Pointing acquisition and tracking system for SILEX inter-satellite optical link. SPIE, 1989, 11: 277–298.
3. G. Oppenhauser, M. Witting. The European SILEX project: Concept, performance status and planning[C]. SPIE, 1990, 1218: 27–37.
4. M. Arnaud, A. Barumchereyk, E. Sein. An experimental optical link between an earth remote sensing satellite spot 4 and a European data relay satellite [J]. Intern. J. Satellite Communications, 1988, 6: 127–140.
5. A. Alonso, M. Reves, Z. Sodnik. Performance of satellite-to-ground communications link between ARTEMIS and the 0ptical Ground Station[C]. SPIE, 2004, 5572: 372–383.
6. M. Reyes, Z. Sodnik, P. Lopezdn Z. Preliminary resuhs of the in orbit test of ARTEMIS with the optical ground station [C]. SPIE, 2002, 4635: 38–49.

7. T. T. Nielsen, G. Oppenhiuser. In orbit test result of an operational intersatellite link between ARTEMIS and SPOT4, SILEX [C]. SPIE, 2002, 4635: 1–15.
8. M. Shikatani, M. T. Yoda. Ground system level development for the ETS-VI/LCE laser communications experiment[C]. SPIE, 1993, 1866: 21–29.
9. M. Toyoshima, K. Ta kizawa, T. Kuri, et al. Ground-to-OICETS 1aser communication experiments [C]. SPIE, 2006, 6304: 40B.
10. A. T. Nakamori. Present and future of optical inter-satellite communication research at the National Space Development Agency of Japan[C]. SPIE, 1994, 2123: 2–13.
11. T. Jono, Y. Takayama, N. Kuri, et al. OICETS on orbit laser communication experiments [C]. SPIE, 2006, 6105: 03.
12. K. Nakagasa, A. Yamamoto. Performance test result of LUCE (Laser Utilizing Communications Equipment) engineering model [C]. SPIE, 2000, 3932: 68–76.
13. I. Lopez J. M., Yong K. Acquisition tracking and fine pointing control of space-based laser communication equipment [J]. SPIE, 1990, 1218: 96–107.
14. Scheinfeild Michael, Kopcika Natan. Acquisition system for microsatellites laser communication in space[J]. SPIE, 2000, 3932: 166–175.
15. C P Pantelides, S R Tzan. Optimal design of dynamically constrained structures. Computers and Structures, 1997, 62(1): 141–150.
16. G D Cheng, Z Kang, G Wang. Dynamic optimization of a turbine foundation. Structural Optimization, 1997, 13: 244–249.

Microstructrue and Mechanical Properties of Selective Laser Melting AlSi10Mg

Weiyan Gong, Junfeng Qi, Zhe Wang, Yi Chen, Jiang Jiang, Zhen Wang and Yuanhao Qi

Abstract The probability of Selective Laser Melting of AlSi10Mg alloy applied to aerospace was investigated. The microstructure of AlSi10Mg alloy produced by SLM is compact, the grains are refined. In this work, mechanical properties like tensile strength, yield strength and elongation are investigated, and compared to conventionally cast AlSi10Mg parts. It is shown that the mechanical properties of AlSi10Mg parts produced by SLM are much higher than the alloy formed by conventional casting.

Keywords Selective laser melting · AlSi10Mg alloy · Microstructure · Mechanical properties

1 Introduction

Additive Manufacturing (AM) is a class of layer-by-layer manufacturing techniques that can fabricate highly complex components from a 3-dimensional CAD model [1, 2]. This model is sliced into 2D slices with a certain layer thickness, and subsequently each layer is built up by adding material. Among metal AM techniques, Selective Laser Melting (SLM) was originally generated as a precise additive process for fabricating complex components with similar properties of conventionally manufactured ones directly from metal powders. The SLM process can be applied in customized medical parts, tooling inserts with conformal cooling channels and functional components with high geometrical complexity such as porous and lattice constructs. Direct selective laser melting (SLM) is an additive freeform manufacturing process which creates 3-D parts through the application of laser energy to powder beds [3–5].

W. Gong (✉) · J. Qi · Z. Wang · Y. Chen · J. Jiang · Z. Wang
Beijing Spacecrafts, Beijing 100091, China
e-mail: gweiyan800@sina.com

Y. Qi
Xi'an Bright Laser Technologies Co., Ltd., Xi'an 710072, China

H.P. Urbach and G. Zhang (eds.), *3rd International Symposium of Space Optical Instruments and Applications*, Springer Proceedings in Physics 192,
DOI 10.1007/978-3-319-49184-4_11

With the development of resolution of space born remote sensor, the mirror lightweight design and manufacture have become the key technology in the large mirror engineering application. The technology to reduce the mass for space optical systems in the world has been summarized roundly in view of material choice, structural form and manufacture technology. Aerospace manufacturers are focusing on the SLM powder-bed technology for Ti-alloy and Ni-superalloy components [6] where the potential cost reduction, fewer steps in the production process and design-freedom are among the factors driving this technology. There has been an increasing number of reports on SLM of AlSi10Mg alloys recently, because of the demand from the industrial field for lightweight structures with complex geometries [7]. AlSi10Mg alloy has been widely used in the automotive and aerospace industry, because of its neat eutectic composition of Al and Si, it has good weldability. However, the aluminum alloy powder is very easy to be oxidized, and the oxide film is formed under the condition of laser action. This paper focuses on the microstructure and mechanical properties. Mechanical tests were performed on samples manufactured using optimized parameters. The influence of the build orientation (Vertically and horizontally build samples) on the tensile properties was investigated.

2 Experimental

The composition of AlSi10Mg powder is shown in Table 1, the size range was 20–70 μm, as measured using Coulter LS230 laser diffraction particle size analyses. Figure 2 shows a Scanning Electron Microscope (SEM) micrograph of the powder, it is obvious that the powder particles are not spherical. The particle shows a very irregular morphology, with many small irregular satellite particles attached to the big particles (Fig. 1).

All specimens were fabricated using EOS M290 system. All specimens were built using a Z-increment (Vertical) of 20–50 μm. The processing was carried out in an Argon atmosphere. After selective melting the islands, laser scans are carried out around the perimeter of the layer to improve the surface finish. For each subsequent layer, these islands are translated by 1 mm in the X and Y-directions, as illustrated in Fig. 2. The aim of the island deposition strategy is to balance the residual stresses in the build. Table 2 shows the process parameters in this study.

Rod-shape sample were fabricated using the optimized parameters. Samples were built vertically (Z-direction) and horizontally (XY-direction), as shown in Fig. 3. In the "vertical" samples, the long boundary of the sample is parallel to the

Table 1 Chemical composition of the investigated AlSi10Mg alloy (wt%)

Si	Mg	Fe	Cu	Pb	Ti	Zn	Mn	Sn	Ni	Al
9.0–11.0	0.2–0.45	≤0.55	≤0.05	≤0.05	≤0.15	≤0.10	≤0.45	≤0.15	≤0.05	Bal

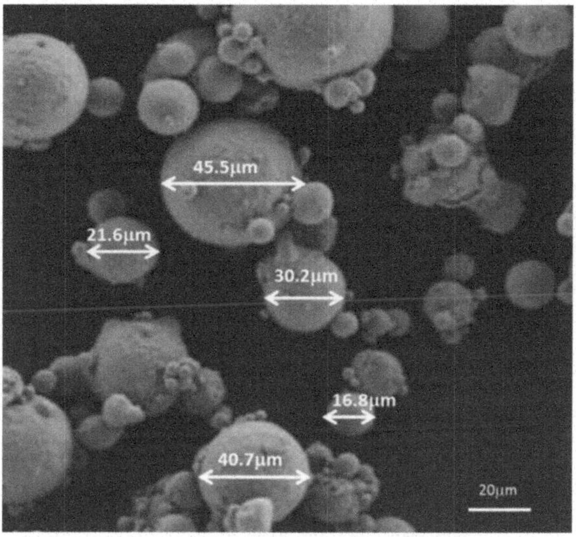

Fig. 1 The morphology of the AlSi10Mg powder

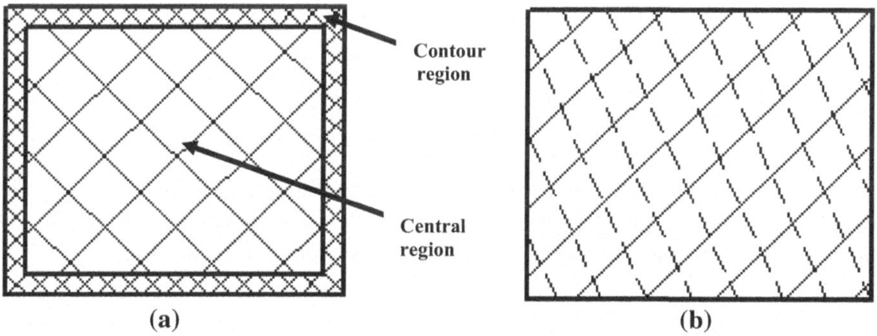

Fig. 2 Schematic illustration of the island scan strategy, showing **a** each layer is divided into Central region and Contour region, then **b** Rotation the scanning direction of 60° between the layers

Table 2 Process parameter in this study

Laser power	Central region	350–400 W
	Contour region	320–360 W
Scan speed	Central region	1200–1500 mm/s
	Contour region	1200–1350 mm/s
Scanning method		Rotation the scanning direction of 60° between the layers

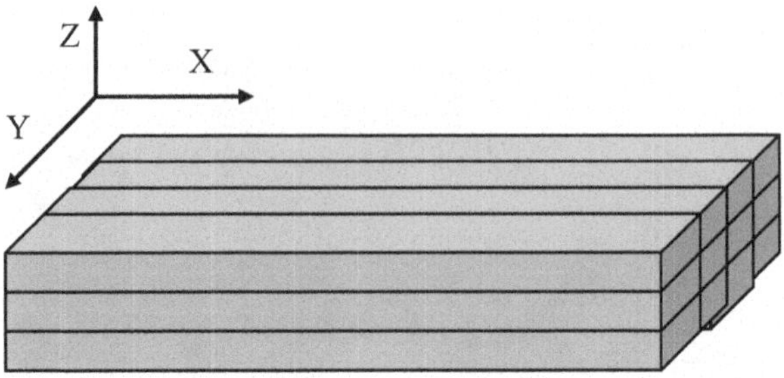

Fig. 3 Building direction of test samples

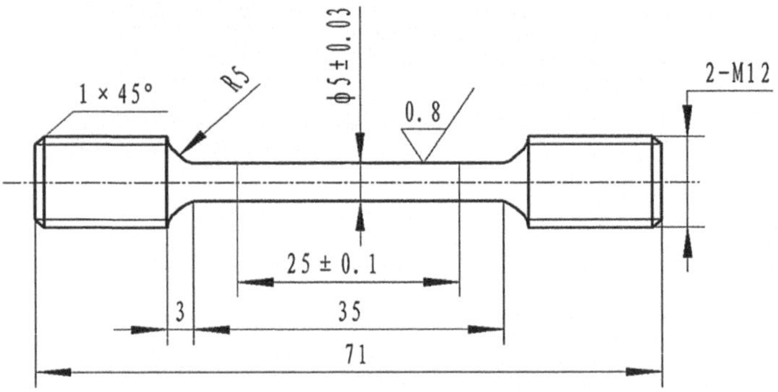

Fig. 4 Tensile sample of AlSi10Mg alloy

building direction, whereas the long boundary of the sample is perpendicular to the building direction in the "horizontal" samples. The tensile sample of AlSi10Mg is show in Fig. 4, all mechanical test results are the average of 3 samples.

3 Results and Discussion

3.1 Surface Features

The typical surface features are shown in Fig. 5, which shows the surface of the AlSi10Mg samples prepared by SLM in the as-processed. It is found that there exist a lot of bonded particles on the surface of the samples. The bonded particles on the surfaces are mainly caused by two reasons: (1) Phenomenon of partially melted raw metal particles on the boundary of each layer in the SLM process. During the SLM process, a new layer of metal particles is first scanned by the contour laser track, followed by the hatching laser track. Some aluminum alloy particles on the boundary are partially melted by the contour laser track. As a result, the partially

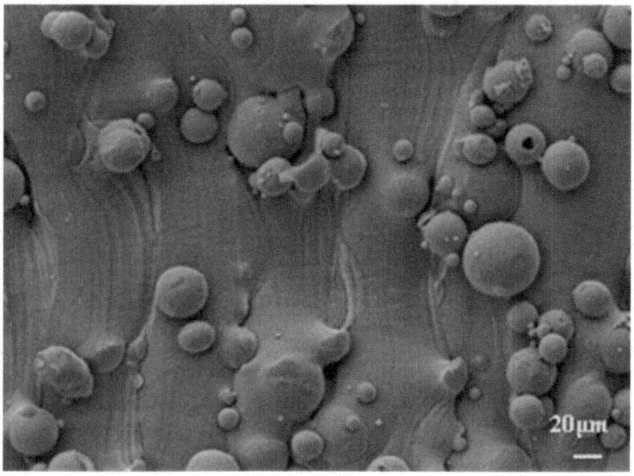

Fig. 5 The surface of the AlSi10Mg samples

melted particles are bonded to surfaces of the samples. (2) Angled struts are partially fabricated on the loose powder. Heat transport between the loose powder and built solid materials caused by a big difference in temperature leads to powder particles sticking to the strut surfaces [8, 9].

The surface roughness is one of key effects for the SLM precise additive manufacturing. Surface roughness measurements of the as-built SLM samples resulted maximum roughness of Ra 10 μm on both top and side surface and often not able to reach the user needs. In order to reduce the roughness value, the parts can be post-treated with some surface techniques. The surface roughness of the samples is Ra 3.13–5.70 after grinding and sand blasting.

3.2 Microstructure

The typical pore defect features of AlSi10Mg alloy fabricated by SLM are shown in Fig. 6. It can be seen that porosity is the main defect of SLM deposited AlSi10Mg alloy. First, the gas present between the powder particles may dissolve in the melting pool and may remain after solidification because of the high cooling rate. To resolve this problem, diminishing the volume of liquid metal through smaller layer thicknesses or lower laser powder is recommended. The other possibility is to reduce the gas pressure in the building. In addition, there also exit some non-fusion defects.

Figure 7 shows the micrograph of untreated AlSi10Mg specimens. The microstructure features of AlSi10Mg alloy fabricated by SLM in the XY and Z sectional directions are shown in Fig. 7. It shows the extension structure obviously through the Z sectional directions, the cladding well are well connected, the microstructure is homogeneous and there are no voids or defects; There are small cellular grain distribution in X-Y directions, the cladding line were cross of 60°,

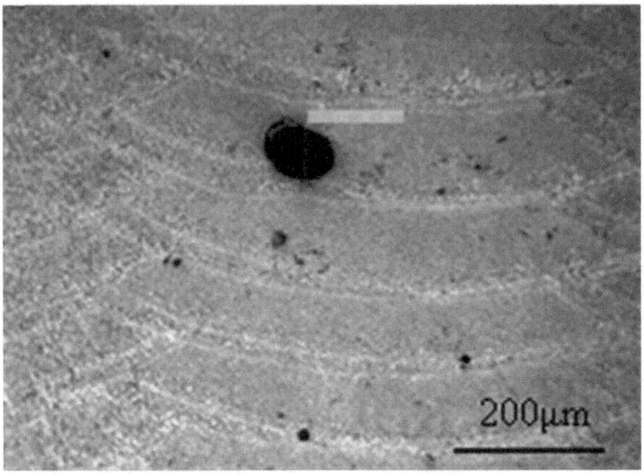

Fig. 6 Microstructure of the AlSi10Mg samples prepared by SLM (as build)

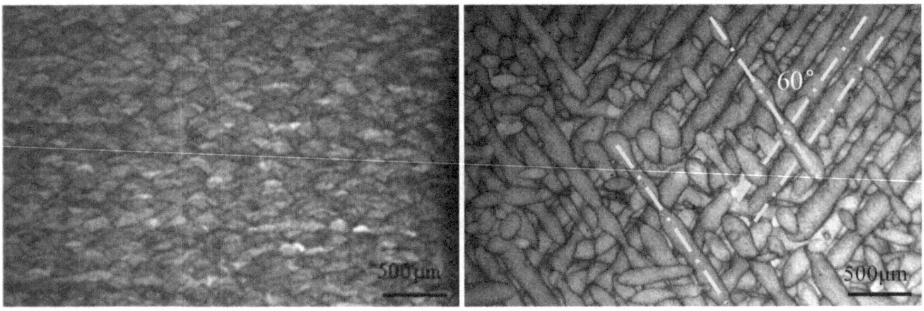

(a) Z Sectional directions (b) X-Y

Fig. 7 SEM micrographs of AlSi10Mg produced by SLM in different directions

there is no gap between the lading line. Interlayer melt boundaries can be observed clearly. The brighter areas are aluminum solid solutions and the darker are silicon. SLM samples show a finer microstructure than cast parts due to the faster solidification. Figure 7 shows the appearance of melt pools and layer development as a result of alternating (XY-direction) melt scans every other layer when viewed in the vertical reference plane (in the build direction) for a x-axis oriented cylinder. This melt pool layer appearance has been observed after the SLM fabrication of AlSi10Mg.

Table 3 Mechanical properties of specimens after annealing (300 °C/2 h)

Sample direction	Sample number	UTS/MPa	YS/MPa	El/%	Ra/%
Horizontal	XY-1	325	212	13.5	28
	XY-2	325	212	14	30
	XY-3	325	212	15	32
	Average	325	212	14.2	30
Vertical	Z-1	338	214	10	18
	Z-2	339	214	10	17
	Z-3	337	215	10	19
	Average	338	214	10	18

3.3 Mechanical Properties

Mechanical properties after annealing (300 °C/2 h) obtained from the tensile tests done on parts produced in 2 different directions (as depicted in Fig. 3), are shown in Table 3. The given values represent the mean values for 2 or 3 specimens with 95% confidence intervals. The results show that SLM AlSi10Mg parts have mechanical properties higher or at least comparable to the casted AlSi10Mg material. From these tensile test results, it can also be seen that the SLM samples show anisotropy in their properties. This is because of the optimal density scanning strategy which causes Z-oriented tensile samples to form more borderline porosity. A comparison between the stress-strain curves for the two directions show a different strengthening behavior. While tensile tests show a significant difference in ductility between parts produced in XY-direction and parts produced in Z-direction.

4 Conclusions

The major findings of this research are:

- The surface roughness of the samples is Ra 3.13–5.70 after post-treated; the as-built microstructure is characterized by cellular dendrites of aluminum solid solutions and interdentritic of Si-particles.
- The SLM AlSi10Mg parts have mechanical properties higher or at least comparable to the casted AlSi10Mg material, because of the very fine microstructure and fine distribution of the Si phase.
- SLM samples show some anisotropy in elongation at break. This is because of the optimal density scanning strategy which causes Z-oriented tensile samples to form more borderline porosity.

References

1. Casavola C, Campanelli SL, Pappalettere C, "Preliminary investigation on distribution of residual stress generated by the selective laser melting process", J Strain Anal Eng Des, pp. 93–104, 2009.
2. Osakada K, Shiomi M, "Flexible manufacturing of metallic products by selective laser melting of powder", Int J Mach Tools Manuf, pp. 88–93, 2006.
3. Kruth J, Froyen L, Van Vaerenbergh J. "Selective laser melting of iron-based powder", Journal of Materials Processing Technology, pp. 616–622, 2004.
4. Louvis, E., Fox, P. and Sutcliffe, J. "Selective laser melting of aluminium components," Journal of Materials Processing Technology, pp. 275–284, 2011.
5. Wong, M., Tsopanos, S., Sutcliffe, C.J. and Owen, I. "Selective laser melting of heat transfer devices," Rapid Prototyping Journal, pp. 291–297, 2007.
6. Gu D, Wang Z, Shen Y, Li Q, Li Y. "In-situ TiC particle reinforced Ti-Al matrix composites: powder preparation by mechanical alloying and selective laser melting behavior." Appl Surf Sci, pp. 9230–9240, 2009.
7. Dadbakhsh S, Hao L, "Effect of Al alloys on selective laser melting behaviour and microstructure and mechanical of in situ formed particle reinforced composites", J Alloy Compd, pp. 328–334, 2012.
8. Yan, C., Hao, L., Hussein, A., Raymont, D., "Evaluations of cellular lattice structures manufactured using selective laser melting." International Journal of Machine Tools & Manufacture, pp. 32–38, 2012.
9. Van Bael, S., Kerckhofs, G., Moesen, M., Pyka, G., Schrooten, "Micro-CT-based improvement of geometrical and mechanical controllability of selective laser melted Ti6Al4 V porous structures." Materials Science and Engineering A, pp. 7423–7431, 2011.

Calibration of Polarization Errors Introduced by Folding Mirror in Imaging Spectropolarimeters

Tingyu Yan, Chunmin Zhang, Qiwei Li and Yutong Wei

Abstract The folding mirror has been always avoided in Imaging Spectropolarimeters (ISPs) because of its polarization characteristics which can introduce errors to the polarization detecting system. However, it is a useful component in optic systems to fold the light path and save the volume of optic systems, especially in space remote-sensing field. In this paper, we proposed a calibration method based on Mueller matrix calculation to correct the polarization after a folding mirror for an ISP with a wide wavelength range. In the Mueller matrix of the folding mirror, the elements are expressed by reflection coefficients and change along with the incident angle and wavelength. Those two factors are considered in the calibration method. The first one is the incident angle. To a fixed folding mirror the incident angle changes very little because it depends on the little change of the field of view (FOV) angle of the ISPs. As a result, the influence of the change of incident angle is ignored. In our experiment, the reflect angle of the mirror was set to 45°, and the influence of the FOV angle was simulated. The second one is the wavelength, which should be considered because of its wide range. Therefore, in our experiment, a series of polarization before and after the reflecting mirror were acquired and the elements of the Mueller matrix of the folding mirror was calculated by solving the overdetermined equations. A 3D correction matrix was obtained, in which every page is a Mueller matrix for one wavelength. And the polarization of a wide wavelength range can be corrected by multiplying by the inverse matrix of it. The effectiveness of the method was proved by the experimental results. And the method offers a new way to save the volume of the IPSs and makes the optical system design of the IPSs more flexible.

Keywords Imaging spectropolarimeter · Calibration of polarization · Folding mirror

---

T. Yan (✉) · C. Zhang · Q. Li · Y. Wei
Institute of Space Optics, School of Science, Xi'an Jiaotong University,
Xi'an 710049, Shaanxi, China
e-mail: yantingyu2010@qq.com

H.P. Urbach and G. Zhang (eds.), *3rd International Symposium of Space Optical Instruments and Applications*, Springer Proceedings in Physics 192,
DOI 10.1007/978-3-319-49184-4_12

1 Introduction

Imaging spectropolarimetry is a novel multi-information detection technology which was developed based on imaging spectrometry and imaging polarimetry at the beginning of 21st-century. Imaging Spectropolarimeter (ISP) is a powerful tool to obtain the spatial, spectral, and polarization information of objects [1]. Image and spectrum can reveal the chemical composition of interesting targets while polarization provides a useful dimension of information that helps to characterize their shape and surface characteristics in optical imagery from the UV through the LWIR and beyond [2]. Therefore ISP has broad application prospects in remote sensing, geological exploration, environmental monitoring, food security, and military reconnaissance [1–3]. In general, folding mirrors have been always avoided in polarimetry systems, for its remarkable polarization characteristics which introduce changes for the state of polarization of incident light [4]. It is reasonable in some commercial compact polarimetry devices with relative small sizes, which only contain several optical components. However, in an ISP, when it is used for space remote sensing, there is a complex imaging spectrometry system besides the polarimetry system. So folding mirror is a necessary component to fold the light path and compress the length of the ISP, by which the volume of the IPSs can be saved. And that is preferred in aerospace field. As a result, to achieve this objective, calibration of polarization errors introduced by folding mirror should be implemented.

In this paper, we demonstrate the whole folding mirror calibration process for the ISP assembled in our laboratory. The polarization characteristic of folding mirror is expressed by a series of Mueller matrices, in which every page is a Mueller matrix for one wavelength. In the second part of this paper, the basic principles of folding mirror calibration are described, and the simulations of the influence of incident angle and wavelength are also presented. The experimental settings are demonstrated in the third part and the results are shown in the fourth part. The fifth part is the conclusion.

2 Principles

The principles of folding mirror calibration are based on polarization characteristics analysis. The most useful form of quantifying the polarization state is the Stokes parameter, which is suitable for incoherent illumination [4]. It is a description of the light in terms of intensity and can account for all possible states of polarization as well as the depolarized component usually present in scattered light. Hence, we use the Stokes parameter to describe the state of polarization of the incident light and the light reflected by the folding mirror. The definition of the Stokes parameter is described in terms of intensities as follow [4]:

$$S = \begin{bmatrix} S_0 \\ S_1 \\ S_2 \\ S_3 \end{bmatrix} = \begin{bmatrix} I_x + I_y \\ I_x - I_y \\ 2I_xI_y \cos\delta \\ 2I_xI_y \sin\delta \end{bmatrix} = \begin{bmatrix} I_H + I_V \\ I_H - I_V \\ I_{45} - I_{135} \\ I_R - I_L \end{bmatrix} \quad (1)$$

Here the four Stokes terms are described in terms of the intensities in the x–y-axis and the phase delay between the orthogonal waves δ, and in terms of intensities measured at specific angle such as H for horizontal, V for vertical, while R, L refer to the intensity in the right and left handed polarizations, respectively.

There are several ways to track changes in polarization such as Poincare sphere and Mueller calculus. Mueller calculus is a convenient method to describe the polarization transfer characteristics of a system, which can be implemented easily in the matrix calculus [4]. A polarization transfer function related to the polarization system such as a folding mirror is called the Mueller matrix. It maps the incident Stokes parameter to the reflected Stokes parameter. As is shown in (2), the Mueller matrix is a 4×4 transfer matrix, which is unique to a given folding mirror [4].

$$\begin{bmatrix} S_0^{out} \\ S_1^{out} \\ S_2^{out} \\ S_3^{out} \end{bmatrix} = \begin{bmatrix} M_{00} & M_{01} & M_{02} & M_{03} \\ M_{10} & M_{11} & M_{12} & M_{13} \\ M_{20} & M_{21} & M_{22} & M_{23} \\ M_{30} & M_{31} & M_{32} & M_{33} \end{bmatrix} \begin{bmatrix} S_0^{in} \\ S_1^{in} \\ S_2^{in} \\ S_3^{in} \end{bmatrix} \quad (2)$$

When we get the Mueller matrix of the folding mirror, we can get the correct incident Stokes parameter by left multiplying its inverse matrix on both sides of (2). In order to determine the Mueller matrix for a given folding mirror, it is necessary to illuminate with at least four known input Stokes states and measure the corresponding output Stokes states. Equation (3) shows the data taken to calculate the Mueller matrix.

$$\begin{bmatrix} S_{01}^{out} & S_{02}^{out} & S_{03}^{out} & S_{04}^{out} \\ S_{11}^{out} & S_{12}^{out} & S_{13}^{out} & S_{14}^{out} \\ S_{21}^{out} & S_{22}^{out} & S_{23}^{out} & S_{24}^{out} \\ S_{31}^{out} & S_{32}^{out} & S_{33}^{out} & S_{34}^{out} \end{bmatrix} = \mathbf{M_{mir}} \begin{bmatrix} S_{01}^{in} & S_{02}^{in} & S_{03}^{in} & S_{04}^{in} \\ S_{11}^{in} & S_{12}^{in} & S_{13}^{in} & S_{14}^{in} \\ S_{21}^{in} & S_{22}^{in} & S_{23}^{in} & S_{24}^{in} \\ S_{31}^{in} & S_{32}^{in} & S_{33}^{in} & S_{34}^{in} \end{bmatrix} \quad (3)$$

In above situation, $\mathbf{M_{mir}}$ is the Mueller matrix of the folding mirror, four known input Stokes states are used to calculate the Mueller matrix, which has a unique solution. In our experiments, we took eight groups different Stokes states to estimate the Mueller matrix, and an overdetermined equation group was solved using least squares method.

However, some other factors must be considered when we are estimating the Mueller matrix. The generic Mueller matrix for a mirror with the angle of incidence of the light given by i_{mir} is given as follow [5]:

$$\mathbf{M_{mir}}(i_{\mathrm{mir}}) = \begin{bmatrix} \frac{1}{2}(r_s^2 + r_p^2) & \frac{1}{2}(r_s^2 - r_p^2) & 0 & 0 \\ \frac{1}{2}(r_s^2 - r_p^2) & \frac{1}{2}(r_s^2 + r_p^2) & 0 & 0 \\ 0 & 0 & |r_p||r_s|\cos(\Delta) & |r_p||r_s|\sin(\Delta) \\ 0 & 0 & -|r_p||r_s|\sin(\Delta) & |r_p||r_s|\cos(\Delta) \end{bmatrix} \quad (4)$$

where r_p and r_s are the complex reflection coefficients for s- and p-polarized light, respectively, with p in the plane of reflection and s perpendicular to this plane, and Δ is the phase jump in this coordinate frame. All of them are given by Fresnel equations

$$r_p = \frac{n_2 \cos i_{\mathrm{mir}} - n_1\sqrt{1 - (n_1/n_2 \sin i_{\mathrm{mir}})^2}}{n_2 \cos i_{\mathrm{mir}} + n_1\sqrt{1 - (n_1/n_2 \sin i_{\mathrm{mir}})^2}} \quad (5)$$

$$r_s = \frac{n_1 \cos i_{\mathrm{mir}} - n_2\sqrt{1 - (n_1/n_2 \sin i_{\mathrm{mir}})^2}}{n_1 \cos i_{\mathrm{mir}} + n_2\sqrt{1 - (n_1/n_2 \sin i_{\mathrm{mir}})^2}} \quad (6)$$

$$\Delta = \arg(r_p) - \arg(r_s) \quad (7)$$

Here we can notice that the Mueller matrix of the folding mirror is a function of incident angle and the complex indices of refraction of the mirror material. Hence it is also a function of wavelength. It is difficult to use just one Mueller matrix to describe the folding mirror. We did some simulations to find the relationship between the complex coefficients and some parameters. Figures 1 and 2 show the Fresnel reflection coefficients and retardance as a function of incident angle and wavelength, respectively.

To the fixed folding mirror in our instrument, the incident angle changes very little because it depends on the little change of the field of view (FOV) angle of the ISP. In our experiment, the reflect angle of the mirror was set to 45°, and the half FOV angle is about 0.587°. According to the simulation, the relative error of r_p and r_s of the two

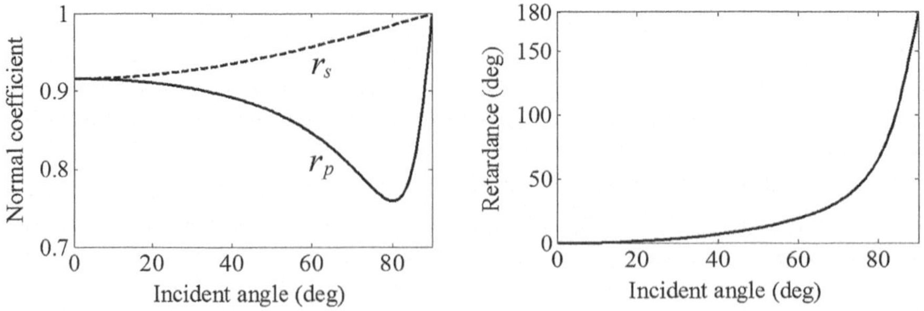

Fig. 1 Fresnel reflection coefficients (*left*) and retardance (*right*) as a function of incident angle for an air/aluminium interface at 500 nm

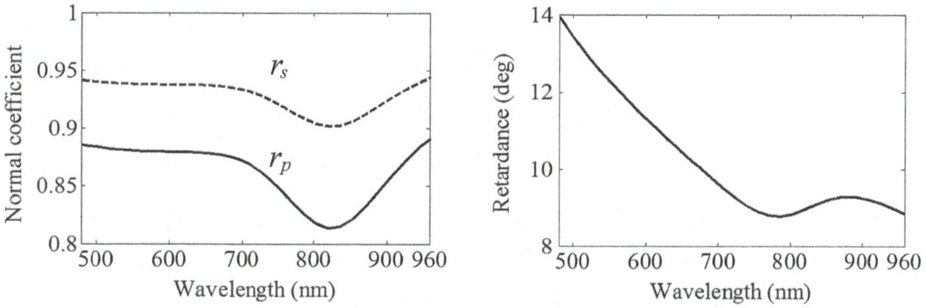

Fig. 2 Fresnel reflection coefficients (*left*) and retardance (*right*) as a function of wavelength for an air/aluminium interface at 45° incident angle

ends of the incident angle are 0.23 and 0.12%, respectively. As a result, the influence of the change of incident angle can be ignored. Because of the wide wavelength range of the ISP, which is 480–960 nm for our instrument, the Fresnel reflection coefficients and retardance change remarkably. For each wavelength there should be a Mueller matrix to describe its polarization characteristics. And a 3D matrix can be obtained, in which each page is a Mueller matrix for one wavelength.

3 Experimental Settings

A photograph of the core optics in ISP is shown in Fig. 3, those optical components are divided into three parts in this paper: fore optics, folding mirror and polarization spectrum detection module (PSDM). In our experiment, the PSDM had been calibrated using phase rearrangement calibration method [6]. And it can be used as a standard polarization detector to calibrate the folding mirror.

The whole procedure to calibrate the folding mirror are described here. An integrating sphere with polychromatic light source and a broadband wire grid

Fig. 3 A photograph of the core optics in ISP

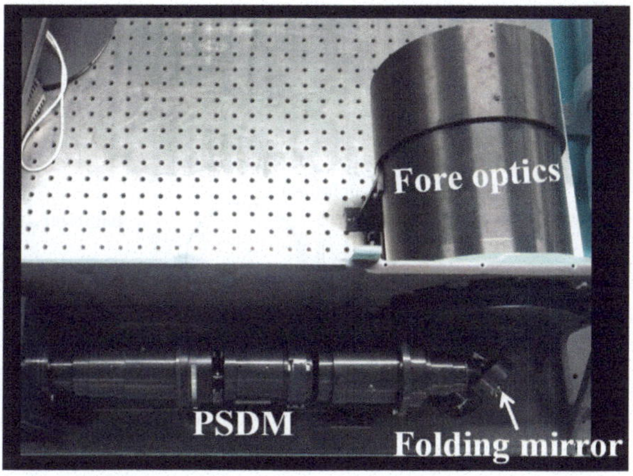

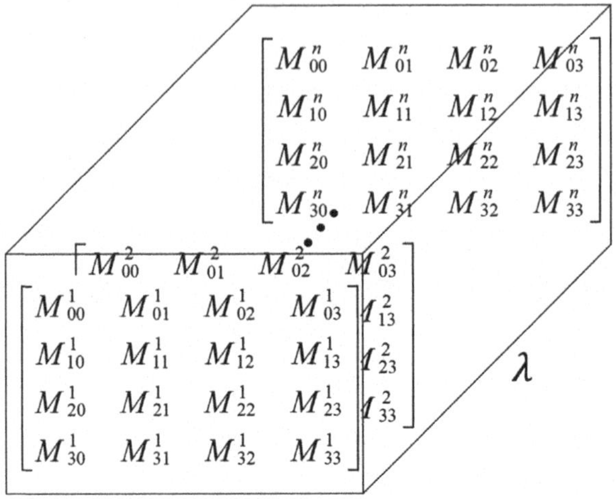

Fig. 4 The 3D Mueller matrix, each page is a Mueller matrix for one wavelength

polarizer were used in our experiment. Firstly, the integrating sphere and the polarizer were placed in front of the fore optics, and the polarization spectrum was acquired by PSDM. Secondly, the polarizer, which was set at the same polarization direction as the first step, was placed between the PSDM and the folding mirror. In order to get more accurate results, more than four groups of polarization spectrum in front of and behind the folding mirror were obtained. In our experiment, by rotating the polarization direction of the polarizer, eight groups of polarization spectrum data were obtained and used to calculate the Mueller matrix of the folding mirror. An overdetermined equation group was solved using least squares method. Then, as is shown in Fig. 4, the 3D correction matrix was obtained, in which every page is a Mueller matrix for one wavelength. And the polarization of a wide wavelength range can be corrected by multiplying by the inverse matrix of it.

Figure 5 shows the two steps of the calibration procedure. Because of the narrow gap between the PSDM and the folding mirror, it is difficult to generate a circular

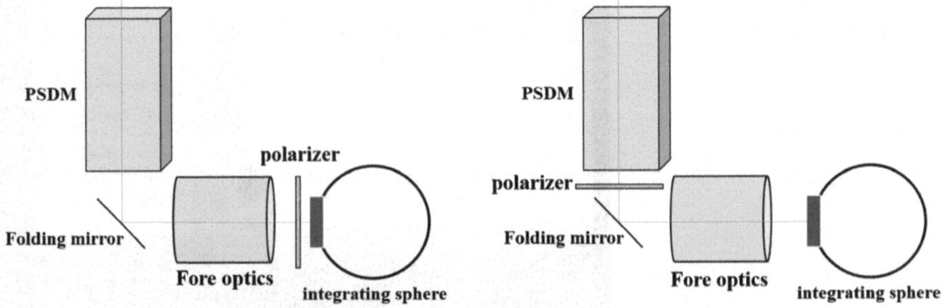

Fig. 5 The procedure to calibrate the folding mirror. *Left* is the first step and *right* is the second step, the polarization directions of the polarizer are same in the two steps

polarized light using a pair of wave plate and polarizer. Fortunately, the fourth column elements of the Mueller matrix can be get using its third column elements, according to (4). And the experimental results are shown in next part.

4 Results and Discussion

To demonstrate the effectiveness of the folding mirror calibration, the polarizer was set at 0° and 60° in front of the ISP, and the polarization spectrum of the halogen lamp was obtained using the channeled interferogram. Then the folding mirror calibration was implemented by multiplying the inverse matrix of the 3D Mueller matrix, and the corrected polarization spectrum was obtained. Figures 6 and 7 shows the 0° and 60° polarization spectrum without/with folding mirror calibration. In these figures, S0, S1, S2 and S3 represent the four Stokes spectrums of the halogen lamp, respectively.

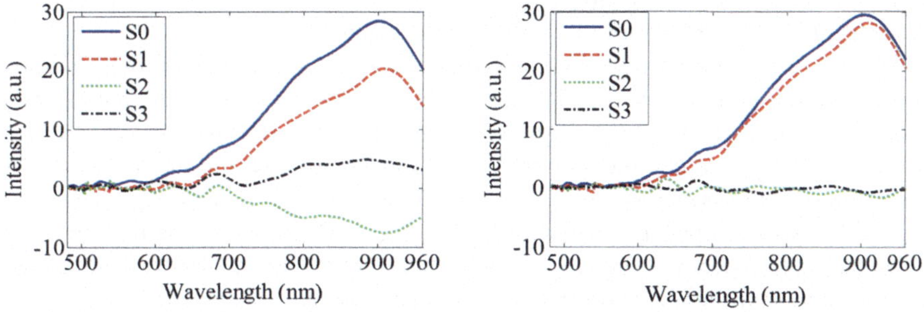

Fig. 6 The polarization spectrum reconstruction of 0° linear incident polarization states, the *left* one shows the result without folding mirror calibration while the *right* one shows the result using folding mirror calibration

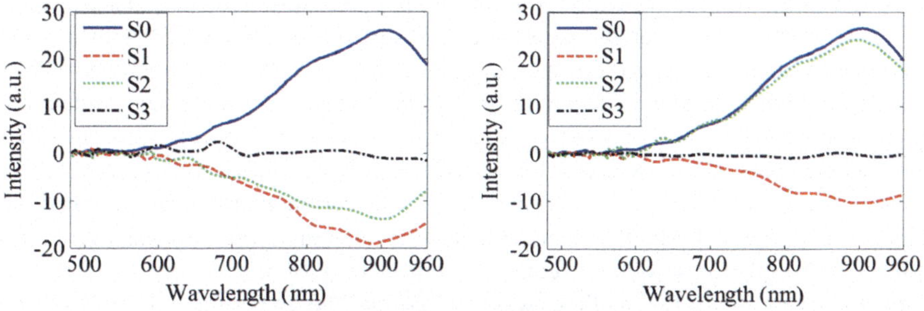

Fig. 7 The polarization spectrum reconstruction of 60° linear incident polarization states, the *left* one shows the result without folding mirror calibration while the *right* one shows the result using folding mirror calibration

Table 1 The comparison of relative errors of the results using folding mirror calibration or not

Stokes parameters	0°		60°	
	Without calibration (%)	After calibration (%)	Without calibration (%)	After calibration (%)
S1	30.84	6.91	45.53	3.03
S2	24.25	4.64	156.67	4.13
S3	16.62	1.69	3.29	1.02

Because of the effects of the folding mirror, some errors were introduced into the reconstruction polarization spectrum, significantly. To the 0° polarization spectrum, S0 and S1 should be same while S2 and S3 should be zero. However, before calibration, the intensity of S1 is much lower than S0 and S2 is not zero. We use the average relative errors to evaluate the deviation of the polarization states as follow:

$$E = \left| \frac{S_r - S_d}{S_r} \right| \times 100\% \tag{8}$$

where S_r is the theoretical normalized Stokes parameter, and S_d is the average normalized Stokes parameter of experimental results.

Table 1 shows the average relative errors of S1, S2 and S3 without/with folding mirror calibration. It is obvious that after folding mirror calibration, the polarization errors introduced by folding mirror has been corrected. Because of the effect of folding mirror, some circular polarization is introduced and the intensities of the linear polarization are changed. To the 60° linear incident polarization states, the sign of S2 is even opposite. After folding mirror calibration, the relative errors of the polarization spectrum reduced dramatically, and the residual errors are accord with the precision demand of the ISP in our laboratory.

5 Conclusion

In conclusion, we proposed a calibration method based on Mueller matrix calculation to correct the polarization after a folding mirror for an ISP with a wide wavelength range. In our experiment, the reflect angle of the mirror was set to 45°, and the influence of the FOV angle and wide wavelength range was simulated. Then a series of polarization before and after the reflecting mirror were acquired and the elements of the Mueller matrix of the folding mirror was calculated by solving the overdetermined equations. A 3D correction matrix was obtained, in which every page is a Mueller matrix for one wavelength. And the polarization of a wide wavelength range was corrected by multiplying by the inverse matrix of it. The effectiveness of the method was proved by the experimental results. The method offers a way to save the volume of the IPSs and makes the optical system design of the IPSs more flexible. The further research is about the calibration error analysis, and getting more accurate results by introducing circular polarization light.

References

1. Tyo, J. Scott, and T. S. Turner, "Variable-Retardance, Fourier-Transform Imaging Spectropolarimeters for Visible Spectrum Remote Sensing," *Applied Optics*, vol. 9, pp. 1450–1458, 2001.
2. Xin Meng, Jianxin Li, Defang Liu, and Rihong Zhu, "Fourier transform imaging spectropolarimeter using simultaneous polarization modulation," *Optics Letters*, vol. 38, pp. 778–780, 2013.
3. K Homma, H Shingu, H Yamamoto, and H Kurosaki, "Application of an imaging spectropolarimeter to agro-environmental sciences," *Proceedings of SPIE*, vol. 7, pp. 436–437, 2004.
4. Lukesh, Gordon W., S. M. Chandler, and J. K. Boger, "Laboratory experiment to demonstrate active nonimaging polarization distinction of materials," *International Symposium on Optical Science and Technology International Society for Optics and Photonics*, vol. 4481, 2002.
5. JM Krijger, R Snel, G Van Harten, JHH Rietjens, and I Aben, "Mirror contamination in space I: mirror modelling," *Atmospheric Measurement Techniques*, vol. 7, pp. 3387–3398, 2014.
6. Qiwei Li, Chunmin Zhang, Tingyu Yan, and Yutong Wei, "Polarization state demodulation of channeled imaging spectropolarimeter by phase rearrangement calibration method," in press.

Design of Nonpolarizing Narrow Band-Pass Filters with Wide Non-transmission Frequency Range

L. Wang, H.W. Dong, G. Wang, Y.L. Bai and P. Wang

Abstract Optical filters often exhibit polarization dependence properties at oblique incidence. We describe the designing results of nonpolarizing narrow band-pass filters with wide non-transmission frequency range using common coating materials. The designing analytical expression is derived from the theory of optical thin film characteristic matrix and simplex method is used in the optimization processes. Two kinds of non-polarizing film structure are shown at incident angles of 45° with low polarization sensitivity (PS), high transmittance (>80%) at central wavelength (550 nm), and low transmittance in the side band region (450–900 nm), which may have potential applications in nonpolarizing optical system or beam splitting system.

Keywords Nonpolarizing narrow band-pass filters · Oblique incidence · Simplex method

1 Introduction

Narrow band-pass filter has high transmittance in a narrow wavelength region and high rejection in all other wavelength regions, which is the key components of many optical apparatus and widely used in the field of detection. Fabry-Perot and induced-transmission filters are the typical narrow band-pass filters, which have received considerable attention for decades [1, 2]. With the deepening study of the electromagnetic waves and photonic crystals, many new ideas of designing narrow band-pass filters appear in recent years [3, 4]. By using photonic crystals, heterostructures and defect modes, multiple-channeled filters and narrow band-pass filters with wide non-transmission frequency range arise [5, 6]. Such common narrow band-pass filters, when used at nonnormal incidence, exhibit polarization

L. Wang (✉) · H.W. Dong · G. Wang · Y.L. Bai · P. Wang
Beijing Institute of Space Mechanics & Electricity, Beijing, China
e-mail: wangyleee@163.com

H.P. Urbach and G. Zhang (eds.), *3rd International Symposium of Space Optical Instruments and Applications*, Springer Proceedings in Physics 192,
DOI 10.1007/978-3-319-49184-4_13

dependent properties, owning to plane waves experiencing different reflectance R_S and R_P at oblique incidence. Such polarization-dependent properties are intolerable in many applications, and consequently many investigations have been carried out to design nonpolarizing coatings for beam-splitting mirrors, edged filters, omnidirectional reflectors, etc. [7–10].

To our knowledge, the design of a nonpolarizing band-pass filter has not been reported by using common dielectric or metal coating materials. An ideal nonpolarizing narrow band-pass filters would have high transmittance over the pass band and a wide non-transmission band, which would be independent of the polarization of the incident light. In the research described in this paper, the transmission and reflection characteristics of both s- and p-polarized light are studied by using transfer matrix method to minimize the linear polarization sensitivity and enlarge the non-transmission frequency range.

2 Test Principles

2.1 Optical Constants of Coating Materials

The optical constants of thin-film coating materials depend not only on the deposition process but also on the process parameters. In the design process of nonpolarizing narrow band-pass filters for production, it is important to use the optical constants corresponding to the actual deposition process. The refractive indices of silver, silicon dioxide (SiO_2), and titanium dioxide (TiO_2) coating materials used in the following calculations are shown in Fig. 1, which are referred from Ref. [11].

2.2 Numerical Design of Nonpolarizing Narrow Band-Pass Filters at 45°

The design of nonpolarizing narrow band-pass filters with wide non-transmission frequency range requires the choice of materials, incident angles, and layer thickness distributions to get the same optical spectrum characteristic of s-component and p-component. Correctly designed and fabricated multilayer can therefore be used as nonpolarizing narrow band-pass filters at angles close to 45°. The linear polarization sensitivity (PS) P is determined by the equation:

$$P = \frac{|Ts - Tp|}{Ts + Tp}, \tag{1}$$

where Ts and Tp are the transmissivities for s- and p-polarized incident light, respectively.

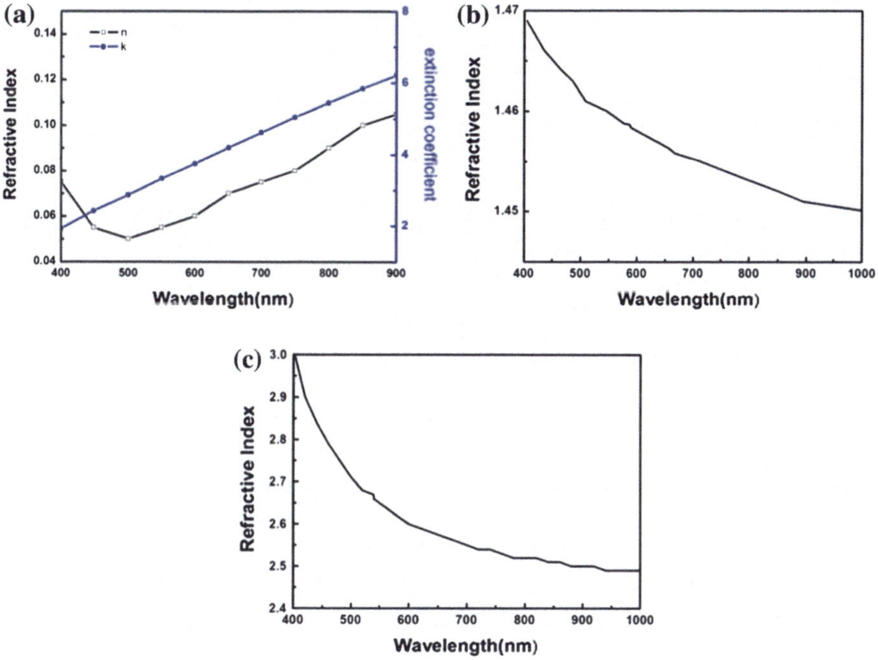

Fig. 1 **a** Refractive index and extinction coefficient of silver, **b** refractive index of SiO_2, **c** refractive index of TiO_2

The polarization dependence of a thin film used at nonormal incidence results from Fresnel's equations. The designed multilayer structure at an oblique angle of incidence can show the same optical performance for s- and p-polarized incident light as that of the normally incidence. For s-polarized light, each index of refraction in the design, including the massive medium indices, is replaced by effective index $\eta_s = n\cos\theta$, where n is the index of refraction and θ is the angle of incidence or refraction in that layer or medium. For p-polarized light, each index of refraction is replaced by effective index $\eta_p = n/\cos\theta$. The characteristic matrices of an isotropic thin film, Ms and Mp for s and p-polarized light, are represented by

$$\mathrm{Ms} = \begin{bmatrix} \cos\delta_j & (i/\eta_s)\sin\delta_j \\ i\eta_s \sin\delta_j & \cos\delta_j \end{bmatrix}, \tag{2}$$

$$\mathrm{Mp} = \begin{bmatrix} \cos\delta_j & (i/\eta_p)\sin\delta_j \\ i\eta_p \sin\delta_j & \cos\delta_j \end{bmatrix}, \tag{3}$$

respectively, where $\delta_j = 2\pi n_j d_j \cos\theta_j/\lambda$, n_j is the refractive index of the jth layer, d_j is physical thickness of the jth layer.

Using the matrix relationship and Fresnel's law yields reflectance R and transmittance T as follows

$$R = \left(\frac{\eta_0 B - C}{\eta_0 B + C}\right)\left(\frac{\eta_0 B - C}{\eta_0 B + C}\right)*, \tag{4}$$

$$T = \frac{4\eta_0 \eta_g}{(\eta_0 B + C)(\eta_0 B + C)*}, \tag{5}$$

where $\eta_0 = n\cos\theta$, $\eta_g = n_g\cos\theta_g$ for s polarization, and $\eta_0 = n/\cos\theta$, $\eta_g = n_g/\cos\theta_g$ for p polarization, B and C are given by

$$\begin{bmatrix} B \\ C \end{bmatrix} = \left\{\prod_j^n M_{nj}\right\}\begin{bmatrix} 1 \\ \eta_g \end{bmatrix}, \tag{6}$$

where η_g is optical admittance of substrate, M_n is Ms for s polarized light and Mp for p polarized light, respectively.

In the design process of nonpolarizing narrow band-pass filters, the transmissivity (or reflectivity) of s-polarized light should equal to that of p-polarized light, the degree of polarization sensitivity should be minimized. From Eq. (1), P tends to zero for equal reflectivity of s and p-polarized light, so the optimization results can be achieved by minimizing the merit function (MF),

$$MF = \frac{1}{m}\sum_{j=1}^{m} P(\lambda_j) \tag{7}$$

The layer thickness distribution is used as the independent variable and, during the recursive optimization, only randomly selected layer thickness changes, till the smaller MF can be got, finally leading to an optimized layer thickness distribution that provides a minimum value of MF. At each stage in the optimization the transmissivity (reflectivity) are calculated using the Eqs. (4), (5) and the illustrated optical constants shown in Fig. 1. The initial period distributions of the designs presented here were deduced using an analytic method, and the MF was then minimized using the standard simplex method.

Generally absorption is intolerable in some applications, we use silicon dioxide and titanium dioxide to design low loss nonpolarizing narrow band-pass filters. The optimized layer thicknesses are shown in Table 1. There are 65 layers in the final optimized stack. The transmittance of the final designed stack is shown in Fig. 2. During the optimization process, it is found that the value of MF would be smaller and the optical performance would be better with increasing the layer numbers, while the corresponding structure would be difficult to fabricate owning to the restriction of layer thickness monitoring technology. The production feasibility of the ultimate optimized results should be assessed.

As can be seen from Fig. 2, the transmittance spectra cut off from 450 to 900 nm covering the visible light wavelength range, except for the band-pass region. At incident angle of 45°, the transmissivity is higher than 90% at peak wavelength

Table 1 Layer thickness of designed nonpolarizing narrow band-pass filters at incident angle of 45° with all dielectric material

Layer no.	Materials	Layer thickness (nm)	Layer no.	Materials	Layer thickness (nm)	Layer no.	Materials	Layer thickness (nm)	Layer no.	Materials	Layer thickness (nm)
1	TiO_2	84.5	20	TiO_2	76.9	39	TiO_2	5.0	58	TiO_2	124.8
2	SiO_2	36.1	21	SiO_2	70.4	40	SiO_2	133.1	59	SiO_2	79.8
3	TiO_2	52.7	22	TiO_2	122.1	41	TiO_2	83.9	60	TiO_2	134.7
4	SiO_2	132.9	23	SiO_2	76.1	42	SiO_2	134.7	61	SiO_2	89.8
5	TiO_2	91.1	24	TiO_2	103.3	43	TiO_2	83.2	62	TiO_2	149.9
6	SiO_2	208.0	25	SiO_2	38.2	44	SiO_2	131.2	63	SiO_2	88.4
7	TiO_2	84.1	26	TiO_2	109.6	45	TiO_2	70.0	64	TiO_2	134.8
8	SiO_2	261.5	27	SiO_2	81.2	46	SiO_2	34.7	65	SiO_2	81.5
9	TiO_2	76.7	28	TiO_2	137.0	47	TiO_2	72.5			
10	SiO_2	120.5	29	SiO_2	87.0	48	SiO_2	129.4			
11	TiO_2	65.8	30	TiO_2	126.4	49	TiO_2	82.0			
12	SiO_2	68.4	31	SiO_2	2.8	50	SiO_2	134.9			
13	TiO_2	58.1	32	TiO_2	122.0	51	TiO_2	87.0			
14	SiO_2	116.8	33	SiO_2	85.2	52	SiO_2	147.5			
15	TiO_2	71.5	34	TiO_2	138.5	53	TiO_2	93.3			
16	SiO_2	48.7	35	SiO_2	124.3	54	SiO_2	138.2			
17	TiO_2	45.3	36	TiO_2	139.3	55	TiO_2	80.9			
18	SiO_2	111.9	37	SiO_2	84.8	56	SiO_2	126.0			
19	TiO_2	52.7	38	TiO_2	106.4	57	TiO_2	77.4			

Fig. 2 The transmission of designed nonpolarizing narrow band-pass filters at incident angle of 45° with all dielectric material, **a** for Ts and Tp, **b** for average light

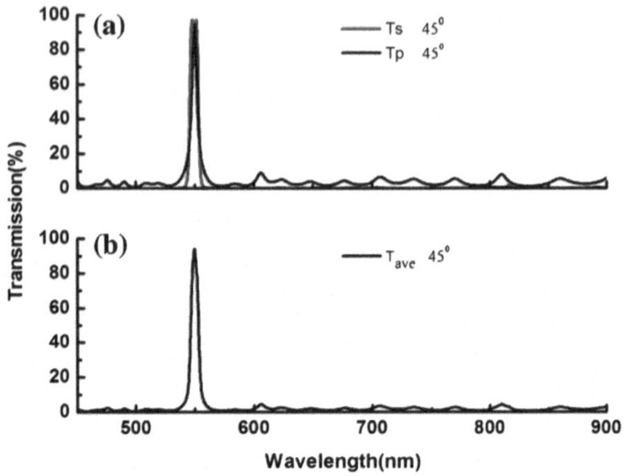

(550 nm), the average P is smaller than 0.91% in the wavelength range of 450–900 nm. There are some ripples for p-polarized light over the cut off range in the transmittance spectrum. The cause of the ripples is the lack of a proper impedance match or of admittance matching layers between the effective admittance of the stack and the admittances of the substrate and the medium for p-polarized light. These ripples can cause problems in some applications for which high reflection is required for the two polarized light. Because of the difference between the effective refractive indices for the two polarizations, good admittance impedance for p-polarized light does not ensure a good match for s-polarized light. New materials should be selected in designing better performance of nonpolarizing narrow band-pass filters with wide non-transmission frequency range.

As we all know, silver has large intrinsic absorption and often is used in designing narrow band-pass filters and absorbers [12]. In order to suppress the ripples shown in Fig. 2, we choose silver, silicon dioxide and titanium dioxide to design nonpolarizing narrow band-pass filters. The transmittance spectrum of the final new designed stack is shown in Fig. 3. The layer thicknesses are shown in

Fig. 3 The transmission of designed nonpolarizing narrow band-pass filters at incident angle of 45° with silver and dielectric material, **a** for Ts and Tp, **b** for average light

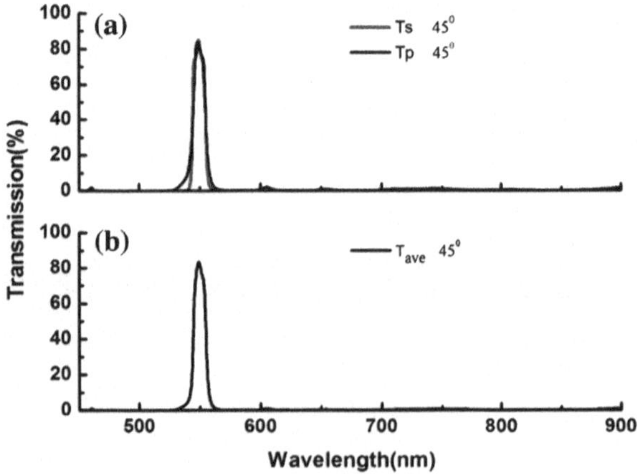

Table 2 Layer thickness of designed nonpolarizing narrow band-pass filters at incident angle of 45° with silver and dielectric material

Layer no.	Materials	Layer thickness (nm)	Layer no.	Materials	Layer thickness (nm)	Layer no.	Materials	Layer thickness (nm)	Layer no.	Materials	Layer thickness (nm)
1	TiO_2	71.613	13	TiO_2	72.776	25	TiO_2	78.7	37	TiO_2	10.117
2	SiO_2	115.914	14	SiO_2	115.285	26	SiO_2	64.174	38	SiO_2	77.661
3	TiO_2	71.008	15	TiO_2	71.186	27	Ag	25.731	39	TiO_2	61.124
4	SiO_2	106.265	16	SiO_2	113.05	28	SiO_2	121.083	40	SiO_2	116.871
5	TiO_2	56.911	17	TiO_2	66.59	29	Ag	1.924	41	TiO_2	77.68
6	SiO_2	80.57	18	SiO_2	93.256	30	SiO_2	121.501	42	SiO_2	58.33
7	TiO_2	48.876	19	TiO_2	55.41	31	Ag	24.432	43	Ag	13.44
8	SiO_2	70.975	20	SiO_2	106.811	32	SiO_2	62.517	44	SiO_2	78.466
9	TiO_2	48.661	21	TiO_2	100.465	33	TiO_2	79.783	45	TiO_2	124.153
10	SiO_2	100.706	22	SiO_2	0.624	34	SiO_2	128.882	46	SiO_2	13.933
11	TiO_2	70.434	23	TiO_2	43.75	35	TiO_2	135.136			
12	SiO_2	116.991	24	SiO_2	128.261	36	SiO_2	16.68			

Table 2. As we can see from the Fig. 3, the ripples are greatly suppressed combining silver and dielectric materials in stack designing, but the transmittance of s-polarized light and p-polarized light also decreases comparing to that of shown in Fig. 2. At incident angle of 45°, the average P is 0.71% in the wavelength range of 450–900 nm, the transmissivity is higher than 80% at peak wavelength (550 nm).

As an example, we gave two kinds of optimized multilayer system of nonpolarizing narrow band-pass filters at a special incident angle of 45°. With increasing of the incident angle, Fresnel's equation shows that the incident angle influences the performance of s and p-polarized light greatly and the transmission (or reflection) spectrum of s and p-polarized light will deviate. So the feasible optimized multilayer structure of nonpolarizing narrow band-pass filters will be difficult to get at large incident angle for common coating materials. In order to get good optical performance of nonpolarizing narrow band-pass filter used at large incident angle, new structures, such as two or three dimensional nano-structures may be used.

3 Conclusions

In this paper, two kinds of nonpolarizing narrow band-pass filters with wide non-transmission frequency range have been described. The all dielectric stack has higher peak transmittance, but with more side band ripples comparing to the structure containing silver layer. The designed filters all have good optical properties: high transmittance (>80%) at center wavelength (550 nm), low transmittance in the side band region (450–900 nm), low PS (<1%) in the target region, and can be used in the nonpolarizing optical or beam splitting system.

References

1. P. H. Berning and A. F. Turner, "Induced transmission in absorbing films applied to band pass filter design," J. Opt. Soc. Am., vol. 47, pp. 230–239, March 1957.
2. B. V. Landau and P. H. Lissiberger, "Theory of induced transmission filters in term of equivalent layers", J. Opt. Soc. Am., vol. 62, pp. 1258–1264, November 1972.
3. A. V. Kavokin, I. A. Shelykh, and G. Malpuech, "Lossless interface modes at the boundary between two periodic dielectric structures", Phys. Rev. B, vol. 72, pp. 233102–4, December 2005.
4. H. C. Zhou, G. Yang, K. Wang, H. Long, and P. X. Lu, "Multiple optical Tamm states at a metal–dielectric mirror interface", Opt. Lett., vol. 35, pp. 4112–4114, December 2010.
5. Q. Qin, H. Lu, S. N. Zhu, etc. "Resonance transmission modes in dual-periodical dielectric multilayer films", Appl. Phys. Lett., vol. 82, pp. 4654–4656, June 2003.
6. L. Wang, Z. S. Wang, Y. G. Wu, etc. "Enlargement of the non-transmission frequency range of multiple-channeled filters by the use of heterostructures", J. Appl. Phys., vol. 95, pp. 424–426, December 2004.
7. V. R. Costich, "Reduction of polarization effects in interference coatings", Appl. Opt., vol. 9, pp. 866–870, April 1970.

8. A. Thelen, "Nonpolarizing interference films inside a glass cube", Appl. Opt., vol. 15, pp. 2983–2985, December 1976.
9. Q. Y. Cai, H. H. Luo, Y. X. Zheng, and D. Q. Liu, "Design of non-polarizing cut-off filters based on dielectric-metal-dielectric stacks", Opt. Express, vol. 21, pp. 19163–19172, August 2013.
10. H. Qi, R. Hong, K. Yi, J. Shao, and Z. Fan, "Nonpolarizing and polarizing filter design," Appl. Opt., vol. 44, pp. 2343–2348, April 2005.
11. E. D. Palik, Handbook of Optical Constants of Solids, Academic press, Orlando, pp. 350–351, 1985.
12. G. Q. Du, H. T. Jiang, L. Wang, Z. S. Wang, and H. Chen, " Enhanced transmittance and fields of a thick metal sandwiched between two dielectric photonic crystals", Appl. Phys. Lett., vol. 108, pp. 103111-4, November 2010.

Design of Coaxial Four-Mirror System for Large-Scale Stereoscopic Mapping Camera

Tianjin Tang, Xiaoyong Wang, Bingxin Yang, Yun Su, Xiaolin Liu and Yingbo Li

Abstract In order to improve spatial resolution and geometric precision of the large-scale stereoscopic mapping camera, four-mirror coaxial optical system with ultra-low distortion and well symmetry is researched. The design method of the four-mirror system with ultra-low distortion based on primary aberration is explored. The basic structural parameters are calculated according to magnification and obscuration ratio of each mirror. The conic parameters are calculated according to primary aberration coefficient. The procedure for calculating initial structural parameters is programmed. Then a four-mirror system is designed with an focal length of 12 m, a field of view of $2° \times 0.1°$, and the wavelength range of 0.45–0.9 μm. The results indicate that the modulation transfer function (MTF) of the whole field is approached to the diffraction limit at Nyquist 71.51 p/mm, and the relative distortion is less than 0.01%. The design structure of four-mirror system can be widely used in the surveying and mapping cameras for high spatial resolution imaging.

Keywords Optical design · Four-mirror · Ultra-low distortion · Aberration correction

1 Introduction

As an important part of the aerospace-to-ground observation system, stereo mapping satellite is aimed to obtain the precise geographic information data of the targets all over the world. It can offer data resource for the construction, maintenance and update of national fundamental geographic information system. Meanwhile, it can satisfy the various respects needs, such as national fundamental surveying and mapping, national land resource survey and economic development.

T. Tang (✉) · X. Wang · B. Yang · Y. Su · X. Liu · Y. Li
Beijing Institute of Space Mechanics & Electricity, Beijing, China
e-mail: tianjin79526@aliyun.com

H.P. Urbach and G. Zhang (eds.), *3rd International Symposium of Space Optical Instruments and Applications*, Springer Proceedings in Physics 192,
DOI 10.1007/978-3-319-49184-4_14

In recent years, remote sensing satellites techniques have been developed rapidly in the domestic and overseas, and mapping accuracy has been improved increasingly. The topographic mapping scale has developed from 1:50,000 to 1:5000. Driven by the large-scale stereoscopic mapping application, the resolution of the stereo mapping satellite has been developed toward sub-meter. Optical system with long focal length and large aperture is needed to be designed for enough high resolution [1, 2]. Meanwhile, for meeting the improving requirement of land surveying and mapping accuracy, the absolute distortion must be controlled under the scope of micros level.

For the optical design of large-scale stereoscopic camera with high resolution, the advantage of reflective optics is more remarkable than refractive optics in the two aspects [3], chromatic aberration correct and the manufacture of large diameter element. The reflective optics mainly includes two forms: off-axis three mirror anastigmat (TMA) and on-axis TMA. The off-axis TMA optical system has no central obscuration, and large field is easy to be realized with small distortion. However, its volume and weight are both large, if it is applied in the stereoscopic camera with large aperture and long focal length. Correspondingly, the shooting cost of the whole satellite would increase. Although on-axis TMA optical system have central obscuration, degraded the imaging resolution [4]. It has many remarkable advantages, especially for the stereoscopic camera with large aperture and long focal length. These advantages include good symmetry, compact structure and small rotational inertia. In addition, temperature control with high precision and integrative design of camera and satellite are easier to be realized. In the overseas and domestic, many cameras using coaxial reflective optics have been on orbit successfully in the space high-resolution imaging area, such as the HR camera of pleiades series satellites developed by France, the WV110 camera of the WorldView series satellites developed by American and the PRISM stereoscopic mapping camera on ALOS satellite developed by Japan. However, the relative distortion of coaxial reflective optics is large, about one present, especially for the cameras with large swath width using long linear array CCD. The absolute distortion would cover multiple pixels, which would not satisfy the application requirements of the high accuracy mapping camera.

In the context of certain performance index, optical design using coaxial three-mirror reflective structure and off-axis three mirror reflective structure are performed and compared. Based on the coaxial three-mirror reflective structure, a spherical quaternary mirror with small focal power is inserted near the exit pupil plane. The absolute distortion can be controlled within the range of microns. The advantages including compact construction of coaxial three-mirror reflective optics and ultra-low distortion of off-axial three-mirror reflective optics can be achieved simultaneously. According to the primary aberration theory, design method of coaxial reflective optics with ultra-low distortion used for large-scale stereoscopic mapping camera is studied. The computation program for initial construction parameters is programed, then, the construction parameters are solved for four-mirror reflective optical systems. The optical system initial parameters are further optimized. Finally, a reasonable structure with the focal length of 12 m, the aperture of 1 m, the wavelength range of 450–900 nm, and field of view of 2° × 0.1° is obtained. Image quality is evaluated. A spherical quaternary mirror with small focal power is inserted near the exit pupil to correct the aberrations caused by primary mirror, and high image quality and low distortion can be achieved simultaneously. Four-mirror optics in this paper is also a feasible design scheme for high-resolution mapping camera.

2 Design Scheme

The design indexes of large-scale stereoscopic mapping camera are shown in Table 1.

As Table 1 shows that the optical system has two characteristics: long focal length and large aperture. Optical design based on the two structures, off-axis TMA and on-axis TMA, are implemented. The layouts of the two types of optical systems are shown in Fig. 1.

As Table 2 shows, the off-axis TMA has no central obscuration. The aperture stop is set on the secondary mirror, and the axis symmetry is good. The MTF can approach the diffractive limits, the relative distortion reach 0.01%. The on-axis TMA has central obscuration, so the MTF is lower than that of off-axis TMA about 0.04; the relative distortion is about one percent. However, at the same performance requirements, the volume of on-axis TMA is about one-third of the volume of off-axis TMA optical system.

The results above shows that the design results of the two optical systems cannot satisfy the design requirements. Therefore, a novel structural form must be

Table 1 Design indexes of large-scale stereoscopic mapping camera

Parameter	Value
Work spectrum (μm)	0.45–0.9
Field of view (FOV) (°)	2 × 0.1
Focal length (mm)	12,000
Entrance pupil diameter (mm)	1000
MTF	≥ 0.3@71.41 p/mm
Volume (mm)	≤ 1500 × 1500 × 1800

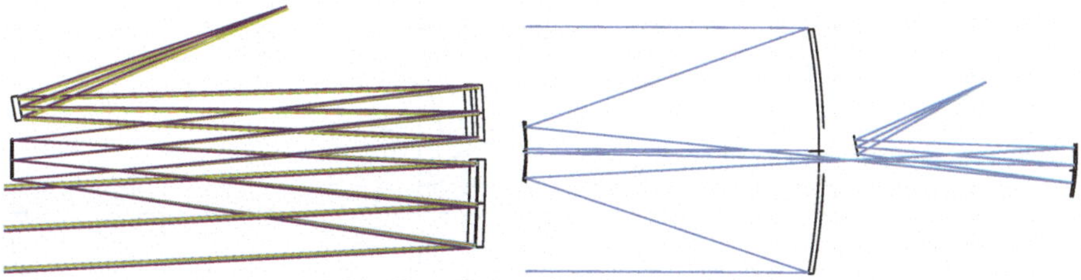

Fig. 1 The layouts of the optical system based on off-axis TMA and on-axis TMA

Table 2 Design results of the optical system

Parameter	On-axis TMA	Off-axis TMA
MTF	0.31	0.35
Relative distortion (%)	1	0.01
Volume (mm^3)	Ȼ1000 × 2300	3900 × 900

presented to further reduce the volume and satisfy the image quality requirements at the meantime.

3 Initial Parameters Solution

Based on the coaxial TMA optical system, a spherical mirror is added and set near the exit pupil. It would realize small distortion for large-scale stereoscopic mapping camera.

The initial structure of four-mirror optical system is shown in Fig. 2 [5]. It has twelve structure variables in total, including radius of curvature R_1, R_2, R_3, R_4, the conic coefficients e_1^2, e_2^2, e_3^2, e_4^2, the distance between the primary mirror and secondary mirror d_1, the distance between the secondary mirror and tertiary mirror d_2, the distance between the tertiary mirror and quaternary mirror d_3, and the distance between the quaternary mirror and the focal plane d_4. The symbol h_i represents the incident height to the mirror $(i = 1, 2, 3)$. L is the overall length of the optical system and d_s represents the back focal length.

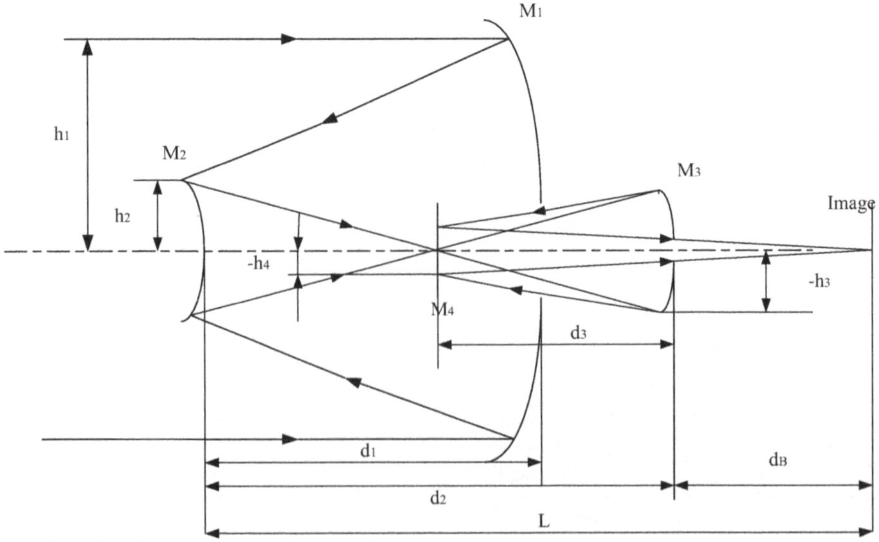

Fig. 2 The optical path of the four-mirror system

The primary mirror, secondary mirror, tertiary mirror and quaternary mirror share a common optical axis, and the aperture stop is set on the primary mirror. The primary mirror and the secondary mirror employ cassegrain structure. The light beam from infinite distance incident to the primary and then form intermediate image after the secondary mirror, and finally converge on the focal plane after the tertiary mirror and quaternary mirror. Generally, the intermediate image is not perfect. Field stop and inner light shield are often used to suppress the stray light out of FOV effectively. It would reduce the requirement of the outside light shield. The quaternary mirror with low focal power is set near the exit pupil to correct the aberration related to pupil, meanwhile, the manufacturing difficulty of primary mirror can be reduced.

For the four-mirror optical system, the aperture stop is set on the primary mirror. The object is located at infinite distance. The obscuration ratio α_i $(i = 1, 2, 3)$ and magnification β_i $(i = 1, 2, 3)$ are shown as follows [6]:

$$\begin{aligned} \alpha_1 &= \frac{l_2}{f_1'} \approx \frac{h_2}{h_1}, \alpha_2 = \frac{l_3}{l_2'} \approx \frac{h_3}{h_2} \\ \alpha_3 &= \frac{l_4}{l_3'} \approx \frac{h_4}{h_3}, \beta_i = \frac{l_i'}{l_i} \approx \frac{u_i}{u_i'} \end{aligned} \tag{1}$$

In Eq. (1), l_i and l_i' $(i = 1, 2, 3, 4)$ represent the object distance and image distance respectively; u_i and u_i' $(i = 1, 2, 3, 4)$ represent the incident angle and refracting angle of the chief ray on the mirrors respectively; f_1' represents the focal length of primary mirror. No matter the off-axis field or the paraxial field is used, the obscuration of primary caused by the secondary mirror is inevitable. If the paraxial field is used, the obscurations caused by the secondary mirror and quaternary mirror will further degrade the imaging quality.

The monochrome aberration mainly includes five types: spherical aberration, comma, astigmatism, field curve and distortion. According to the third-order aberration theory, the expression of primary aberration S_{I}, S_{II}, S_{III}, S_{IV}, S_{V} are shown as follows [7]:

$$\begin{aligned} S_{\mathrm{I}} &= \sum hP + \sum h^4 K \\ S_{\mathrm{II}} &= \sum yP - J\sum W + \sum h^3 yK \\ S_{\mathrm{III}} &= \sum \frac{y^2}{h}P - 2J\sum \frac{y}{h}W + J^2\sum \Phi + \sum h^2y^2K \\ S_{\mathrm{IV}} &= \sum \frac{\Pi}{h} \\ S_{\mathrm{V}} &= \sum \frac{y^3}{h^2}P - 3J\sum \frac{y^2}{h^2}W + J^2\sum \frac{y}{h}\left(3\Phi + \frac{\Pi}{h}\right) - J^3\sum \frac{1}{h^2}\Delta\frac{1}{n^2} \end{aligned} \tag{2}$$

and

$$P = \left(\frac{\Delta u}{\Delta 1/n}\right)^2 \Delta\frac{u}{n}, W = \left(\frac{\Delta u}{\Delta 1/n}\right)\Delta\frac{u}{n}, \\ \Pi = \frac{\Delta(nu)}{nu'}, \Phi = \frac{1}{h}\Delta\frac{u}{h}, K = -\frac{e^2}{R^3}\Delta n \tag{3}$$

If the object is located at the infinite distance, $l_1 = \infty$, $u_1 = 0$. If the entrance pupil is located on the primary mirror, $x_1 = y_1 = 0$. The quaternary mirror is located near the exit pupil, which is also near the intermediate image. Combining Expressions (1) and (2), we can obtain the relationships among the third-order aberration, the conic coefficients, obscuration ratio and magnification. The expressions are shown as follows:

$$\begin{aligned} S_{\rm I} &= \frac{1}{4}\begin{bmatrix} e_1^2\beta_1^3\beta_2^3\beta_3^3 - e_2^2\alpha_1\beta_2^3\beta_3^3(1+\beta_1)^3 + e_3^2\alpha_1\alpha_2\beta_3^3(1+\beta_2^3) - e_4^2\alpha_1\alpha_2\alpha_3(1+\beta_3)^3 \\ -\beta_1^3\beta_2^3\beta_3^3 + \alpha_1\beta_2^3\beta_3^3(1-\beta_1)^2(1+\beta_1) - \alpha_1\alpha_2\beta_3^3(1-\beta_2)^2(1-\beta_2) \\ +\alpha_1\alpha_2\alpha_3(1-\beta_3)^3(1+\beta_3) \end{bmatrix} \\ S_{\rm II} &= \frac{\beta_2^2\beta_3^2(\alpha_1-1)(1+\beta_1)}{4\beta_1}\left[(1-\beta_1)^2 e_2^2(1+\beta_1)^2\right] \\ &\quad + \frac{\beta_3^2(1+\beta_2)[\beta_1(1-\alpha_2)-\alpha_2(\alpha_1-1)]}{4\beta_1\beta_2}\left[e_3^2(1+\beta_2)^2-(1-\beta_2)^2\right] - \frac{1}{2} \\ S_{\rm III} &= \frac{\beta_2\beta_3(\alpha_1-1)^2(1+\beta_1)}{4\alpha_1\beta_1^2}\left[(1-\beta_1)^2-e_2^2(1+\beta_1)^2\right] + \frac{\beta_3(1+\beta_2)[\beta_3(1-\alpha_2)+\alpha_2(\alpha_1-1)]}{4\alpha_1\alpha_2\beta_1^2\beta_2^2} \\ &\quad \left[e_3^2(1+\beta_2)^2-(1-\beta_3)^2\right] - \frac{\beta_2\beta_3(\alpha_1-1)(1-\beta_1)^2}{\alpha_1\beta_1} - \frac{\beta_3(1-\beta_2)^2[\beta_1(1-\alpha_2)+\alpha_2(\alpha_1-1)]}{\alpha_1\alpha_2\beta_1\beta_2} \\ &\quad - \beta_1\beta_2\beta_3 + \frac{\beta_2\beta_3(1+\beta_1)}{\alpha_1} - \frac{\beta_3(1+\beta_2)}{\alpha_1\alpha_2} + \frac{1+\beta_3}{\alpha_1\alpha_2\alpha_3} \\ S_{\rm IV} &= -\beta_1\beta_2\beta_3 - \frac{\beta_2\beta_3(1+\beta_1)}{\alpha} + \frac{\beta_3(1+\beta_2)}{\alpha_1\alpha_2} - \frac{1+\beta_3}{\alpha_1\alpha_2\alpha_3} \end{aligned} \tag{4}$$

Equations (4) shows, the primary spherical aberration $S_{\rm I}$ is influenced by the conic coefficients of primary mirror and quaternary mirror namely e_1^2 and e_4^2; primary coma $S_{\rm II}$ and primary astigmatism $S_{\rm III}$ can be controlled by the conic coefficients of secondary mirror and tertiary mirror namely e_2^2 and e_3^2; the primary field curve S_{IV} is only related to the construction parameters and not related to the conic coefficients; The distortion S_V needs a significant consideration, which will distort the image, but will not influence the sharpness. The distortion aberration should be controlled through optimization. According to the paraxial formulas, the relational expressions between obscuration ratio α_i, magnification β_i and the construction parameters is shown in Eq. (4) as follows, the relational expression between the α_2, α_3, the overall length L and back image distance d_s are shown in Eq. (5) as follows [8].

$$\begin{aligned}\alpha_1 &= 1-\frac{2d_1}{r_1}, \beta_1 = -\frac{r_2}{r_2-r_1+2d_1}\\ \alpha_2 &= 1-\frac{2d_2(r_1-2d_1-r_2)}{r_2(r_1-2d_1)}, \beta_2 = -\frac{r_3}{\beta_1 r_1 - 2\beta_1 d_1 - 2d_1 - r_3}\end{aligned} \tag{5}$$

$$\begin{aligned}\alpha_3 &= 1-\frac{2d_3}{\alpha_1\alpha_2\beta_1\beta_2 r_1}, \beta_3 = -\frac{2f}{\beta_1\beta_2 r_1}\\ \alpha_2 &= \frac{\beta_2\beta_3(L-d_s)}{\alpha_1} - 1\\ \alpha_3 &= 1-\frac{\beta_2\beta_3 L+\alpha_1}{(L\beta_2\beta_3+\alpha_1)(1-\beta_3)+\beta_2\beta_3^2 d_s}\end{aligned} \tag{6}$$

$$\begin{aligned}r_1 &= \frac{-2}{\beta_1\beta_2\beta_3}, r_2 = \frac{-2\alpha_1}{\beta_2\beta_3(1+\beta_1)}\\ r_3 &= \frac{2}{1+\beta_2}\left[\left(L+\frac{\alpha_1}{\beta_2\beta_3}\right)\left(\frac{1}{\beta_3}-1\right)+d_s\right], r_4 = \frac{2}{1+\beta_3}\left(L+\frac{\alpha_1}{\beta_2\beta_3}\right)\end{aligned} \tag{7}$$

$$d_1 = \frac{\alpha_1-1}{\beta_1\beta_2\beta_3}, d_2 = L-d_s, d_3 = \frac{1}{\beta_2}\left[\left(L+\frac{\alpha_1}{\beta_2\beta_3}\right)\left(\frac{1}{\beta_3}-1\right)+d_s\right] \tag{8}$$

The target of the optical design is to eliminate the aberrations which are zero ideally. But twelve variables related to the construction parameters cannot calculated through four expressions. With the design experience, some variables should be set beforehand. Obscuration ratio and magnifications are set previously, and the relationship as follows should be obeyed.

$$\begin{aligned}&f_1' < 0, f_2' > 0, f_3' < 0, f_4' \approx \infty\\ &\text{On condition that } 0<\alpha_1<1, \beta_2>0, \beta_3>0; 0<\alpha_2<1, \alpha_2<-1\\ &\text{On condition that } \alpha_1>1, \beta_2<0, \beta_3<0; 0<\alpha_2<1, \alpha_2<-1\end{aligned} \tag{9}$$

If the obscuration ratio caused by the secondary mirror α_1 is too large, the amount of energy entering into the system will reduce, and the diffractive limits will reduce. If the obscuration ratio caused by the secondary mirror α_1 is too small, the difficulty to correct the aberration will be increased. Large field of view is different to achieve with long focal length. Considering the large aperture, α_1 is set to be 0.2–0.3, the area obscuration ratio will be less than 10%. The ratio of total length and focal length is expected to be small. The smaller the over length, the smaller the volume and weight will be, but the aberrations are more difficult to be corrected. In addition, for meeting the requirements of low distortion, the optical system should be good symmetry. By using MATLAB software, multiple initial solutions are given. Then focal power optimization is done using ZEMAX software. The flow chart for solving the initial parameters is shown in the Fig. 3.

Fig. 3 Flow chart for solving the initial parameters

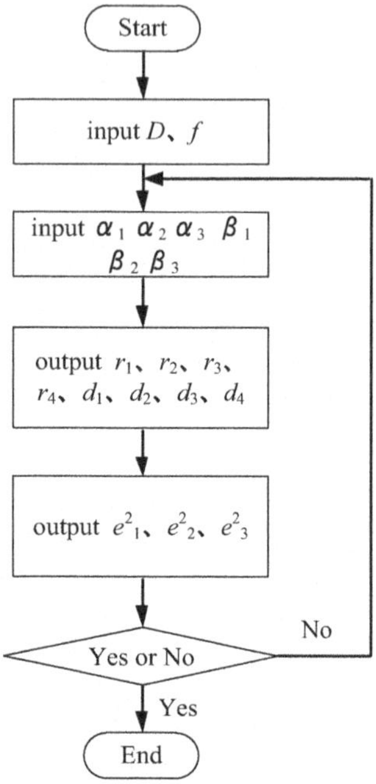

4 Optical Design

According to the design method above, the program acquiring initial structure is built. An appropriate initial structure is selected from multiple solutions of initial structure. The selected initial structure data is listed in Table 3. A coaxial optical system is obtained as shown in Fig. 4. The aperture stop is set on the primary mirror. For decreasing secondary obscuration, the tertiary mirror is placed near the focal plane, and the quaternary mirror is designed on the exit pupil position, which is near the first immediate image plane.

MTF is the core assessment indicator for imaging optical system. Figure 5 shows the MTF curves of the designed coaxial optical system. The MTF value at

Table 3 Parameter of the initial structure

Surface name	Radius (mm)	Surface type	Airspace (mm)
Primary	3080	Hyperboloid	1180
Secondary	−705	Ellipsoid	2120
Tertiary	−1095	Hyperboloid	750
Quaternary	4200	–	810

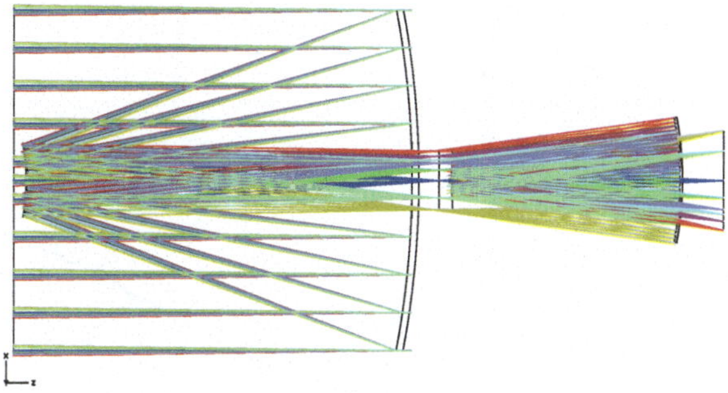

Fig. 4 Structure of the coaxial optical system

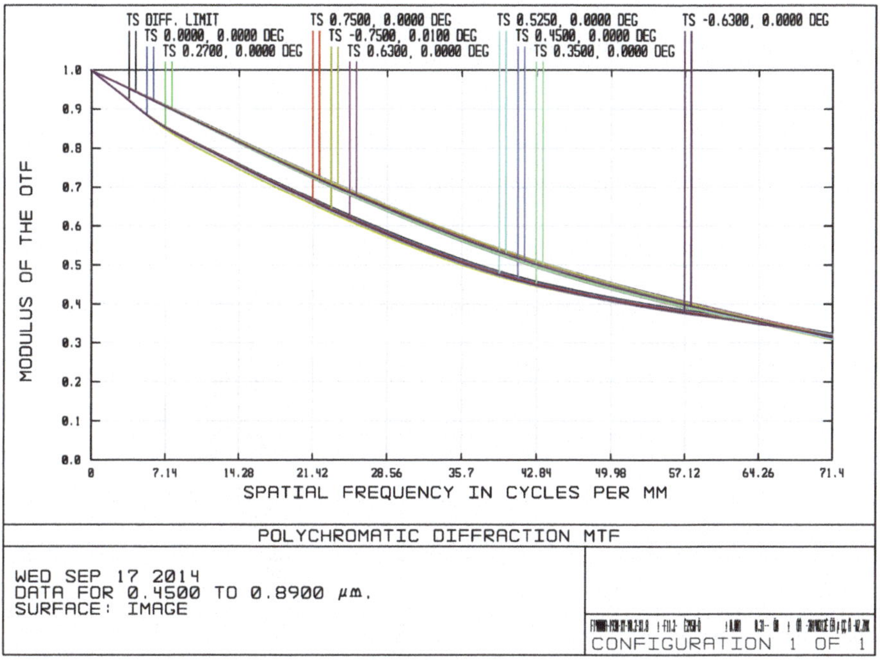

Fig. 5 MTF curves of the coaxial optical system

71.51 p/mm, which is Nyquist frequency for the CCD pixel of 7 μm, can reach 0.3. The relative distortion is shown in Fig. 6. The horizontal coordinate specifies the angle of half field of view. The relative distortion is less than 10^{-4} across the entire field.

According to geometric optics theory, for on-axis field, the tertiary mirror and quaternary mirror need to open holes on the central area for reducing secondary

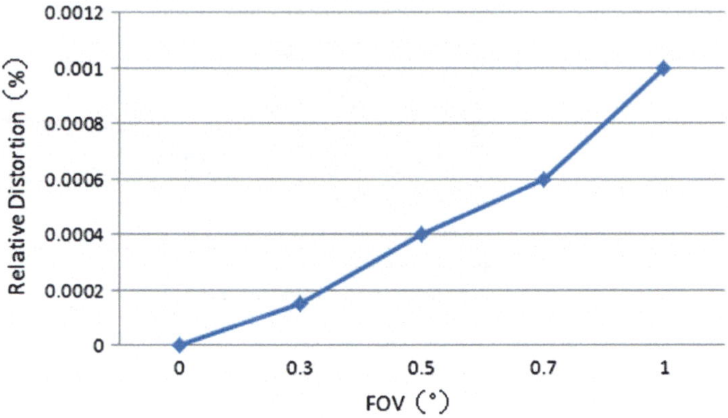

Fig. 6 Distortion of the coaxial optical system

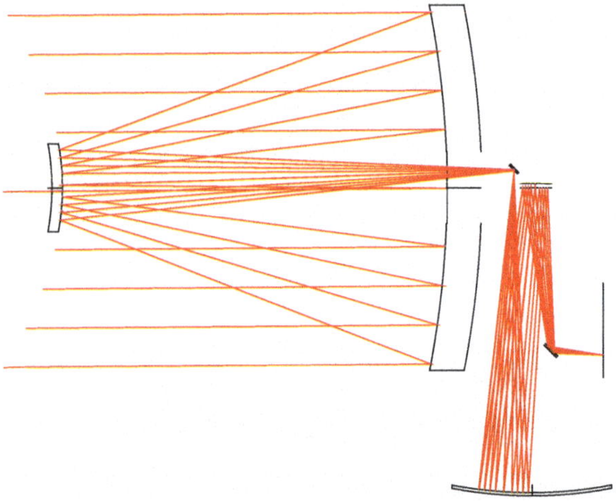

Fig. 7 Structure of the optical system after optimization

obscuration. Therefore, the MTF perform the bifurcation (astigmatism) on low and medium frequency parts. To improve imaging performance, off-axis field design is applied in this optical system. Figure 7 shows the layout of optical system with off-axis field design. Two fold mirrors are used in the light path for more compact structure design. The boundary dimension of this system is about 1800 mm × 1000 mm × 1500 mm. Figure 8 shows the MTF curves of the designed coaxial optical system. The relative distortion is shown in Fig. 9. The relative distortion is less than 10^{-4} across entire field.

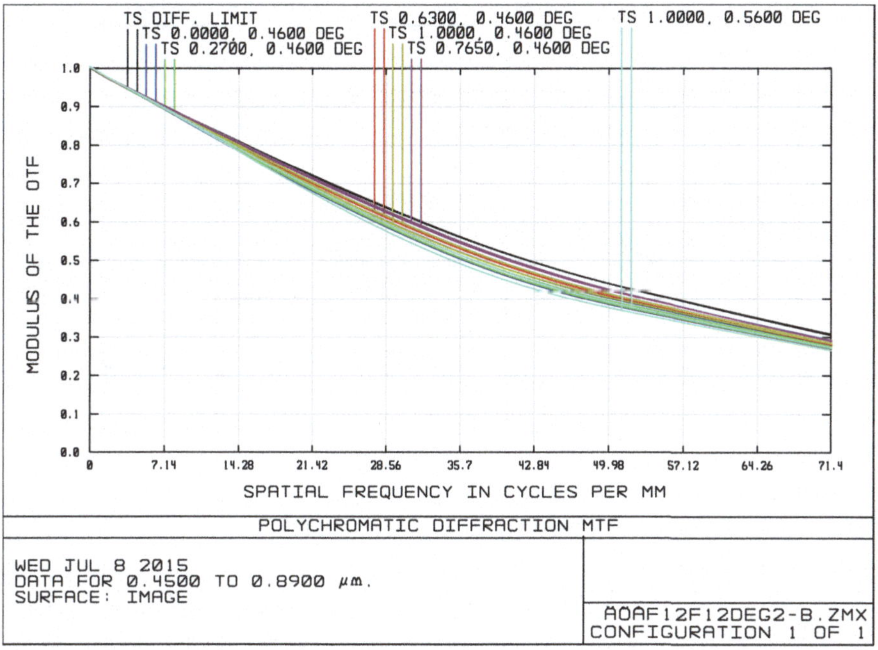

Fig. 8 MTF curves of the four-mirror optical system

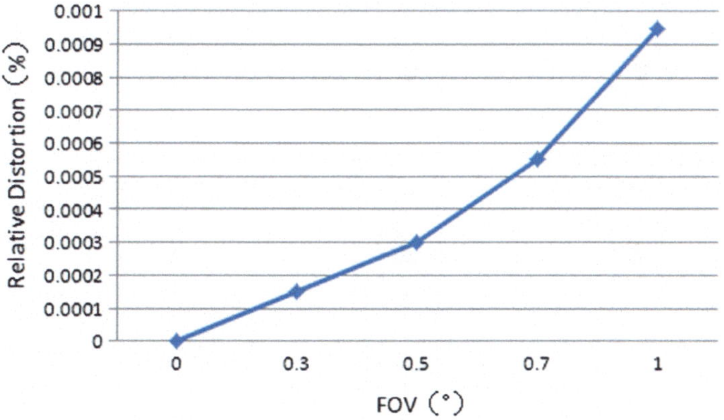

Fig. 9 Distortion of the four-mirror optical system

The transverse ray aberration is shown in the Fig. 10. The residual aberration is small in the whole field. The astigmatism is corrected well. Certain coma is existed on off-axis field, and few filed curvature is also existed. The spot diagram is shown in Fig. 11. The circle specifies the boundary of airy disk of t the field point. In the whole field, the maximum of root-mean-square (RMS) radius is 11 μm, less than the radius of airy disk.

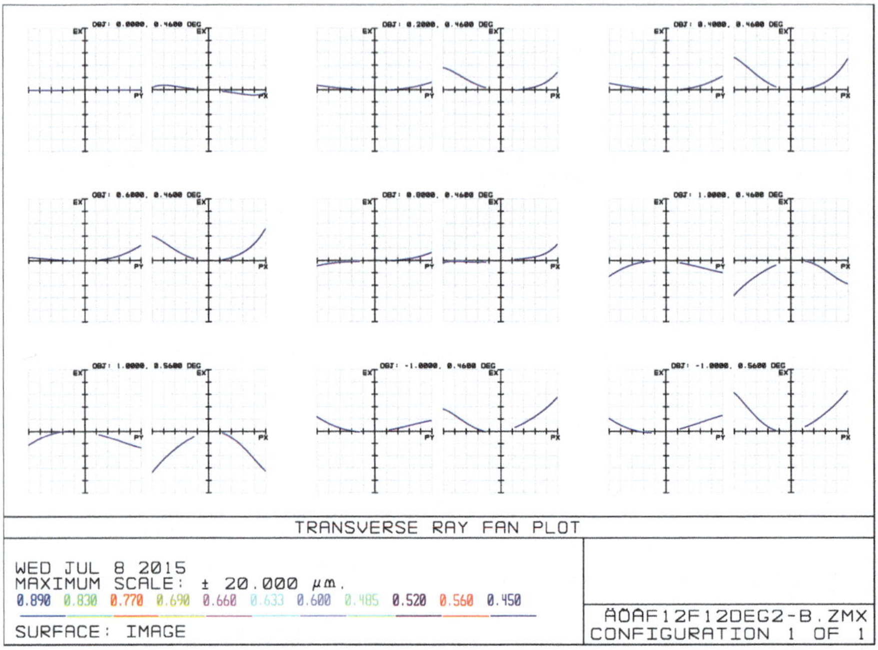

Fig. 10 Transverse ray aberration of the optical system

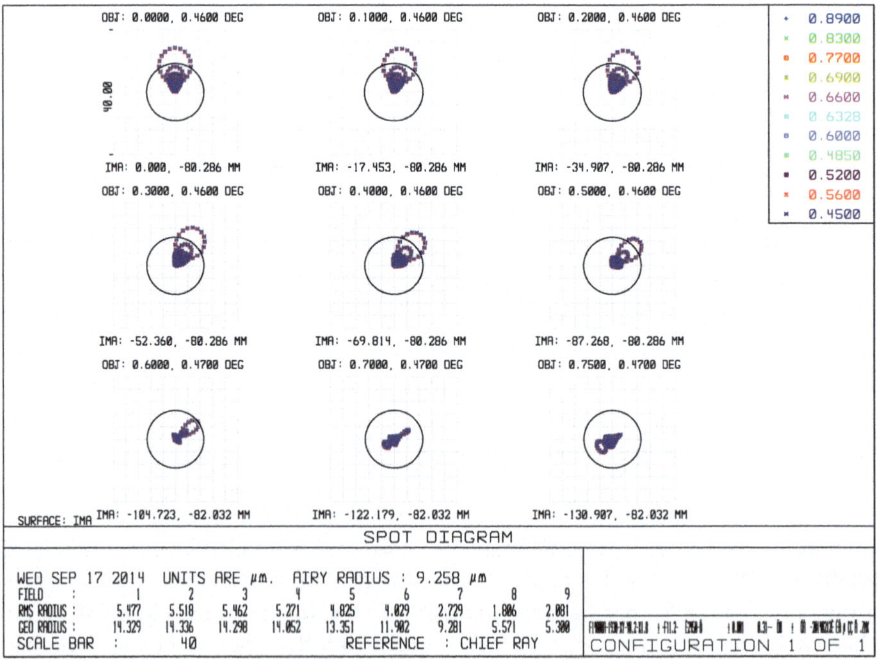

Fig. 11 Spot diagram of the optical system

5 Conclusion

Taking the large-scale stereoscopic camera as application background, the traditional off-axis TMA and on-axis TMA are design respectively and contrastive analyzed, a new structure, on-axis four-mirror optical system with low distortion, is proposed, which is complicated design of off-axis TMA. According to the primary aberration theory, the initial structure design method of the on-axis four-mirror optical system is expounded, and the structure parameter and conic coefficient are solved using our program. Finally, through further design and optimization, a reasonable structure with the focal length of 12 m, the aperture of 1 m and field of view of $2° \times 0.1°$ is obtained. The MTF approaches diffraction limit and the distortion is less than 10^{-4} across entire field. This kind of optical system has the characteristics of both compact structure and small distortion. The design method of on-axis four-mirror optical system with small distortion has extensive application prospect for high resolution stereoscopic camera, and can be expanded to multi-channel optical system.

References

1. Conrad W, Matthew C, "Opto-mechanical integration and alignment verification of the James Webb Space Telescope (JWST) optical telescope element," *SPIE*, vol. 7433, pp. 743303, 2009.
2. Qu Hongsong, Jin Guang, Zhang Ye, "Nextview program and progress in optical remote sensing Satellites," *Chinese Journal of Optics and Applied Optics*, vol. 2(6), pp. 467–476, 2009.
3. Chen Wei, Xue Chuang, "Design of Wide Field-of-view Off-axis Three-mirror Telescope for Imaging Spectrometer," *Acta Photonica Sinica*, vol.42(8), pp. 950–955, 2013.
4. Liu Zhaojun, Zhou Feng, "Study on Coaxis Three-mirror-anastigmat Space Camera with Long Focal Length," *Spacecraft Recovery & Remote Sensing*, vol. 6(32), pp. 46–52, 2011.
5. Yao Gang, Huang Ying, and Fu Danying, "An easily fabricated unobscured three-mirror reflective system with large Fov," *Spacecraft Recovery & Remote Sensing*, vol. 5(31), pp. 44–48, 2010.
6. Ding Xuezhuan, Wang Xin, Lan Weihua, and Liu Yinnian, "Design of four-mirror reflective anastigmat optic system," *Infrared and Laser Engineering*, vol. 37(2), pp. 319–321, 2008.
7. Zhang Yimo, Optical Application. Beijing: Publishing House of Electronics Industry, 2008. (in Chinese).
8. Chen Jianfa, Wang Helong, and Liuxin, "Design of compact four-mirror optical system," *Electronics optics & Contral*, vol.19(12), pp. 76–79, 2012.

Thermal Optical Analysis of Optical Window Glass

Ruoyan Wang, Zhishan Gao and Qun Yuan

Abstract Optical window is a multifunction window. It is not only a window that let the light come through, it is also a window, which isolate the vacuum environment inside the bottle from the outside atmosphere environment. At the same time, it is as well as a temperature-controlling window that can separate the heat sink inside the bottle away from the outside normal temperature. The thickness of the glass should be large enough to guarantee the strength and reliability of the window. But if the window glass is too thick, it would lead to the temperature change and, furthermore, would cause to the change of refractive index gradient and also, the thermal deformation will be more apparent. It may even be able to meet the requirements of the optical indicator when the situation gets worse. Based on the thermal optical analysis, with one side in normal pressure, and the other in 1×10^{-3} Pa vacuum, 100 K degrees, the thickness of the silica glass and the germanium optical window is optimized. The thermal environment of optical window is analyzed, including system average temperature shift as well as circumferential, radial and axial temperature differences. By calculate the result of steady state temperature field and mapping it to structural model, the deformation of optical window under the force-thermal coupling condition could be figured out. The thermal optical analysis could also be carried out after combining refractive index gradient analysis and glass surface change of optical window. With the differences of glass thickness, the optical path in root mean square error (RMS) of silica glass optical window with diameter of 230 mm and the germanium optical window with diameter of 150 mm is calculated. The results show that the optical window can not only meet the strength and reliability requirements, but also meet the optic index in the effective optical aperture. These results provide the guidance for the optical window design.

Keywords Optical window · Thermal optical analysis · Force thermal coupling

R. Wang (✉) · Z. Gao · Q. Yuan
School of Electrical Engineering & Photoelectric Technology, Nanjing University of Science & Technology, Nanjing, People's Republic of China
e-mail: 348521544@qq.com

H.P. Urbach and G. Zhang (eds.), *3rd International Symposium of Space Optical Instruments and Applications*, Springer Proceedings in Physics 192,
DOI 10.1007/978-3-319-49184-4_15

Table 1 The performance parameter of silica and germanium

Material	Density (kg/m^3)	Elastic modulus (Pa)	Poisson's ratio
Silica	2.21×10^3	7.27×10^{10}	0.16
Germanium	5.33×10^3	10.27×10^{10}	0.28

1 Introduction

As part of the solar simulator, optical window is the connector between the light source of solar simulator and the main optical system of the internal vacuum tank. Optical window is not only the window of light source; it is also the pressure isolated airtight window, and the temperature isolated window. Its optical performance directly related to the size and the uniformity of the solar simulator's flare radiation intensity of illumination. At the same time its air tightness and strength also related to the security problems of solar simulator system.

Window materials of solar simulator, on the one hand, should satisfy the requirement of the visible light and infrared transmission. Respectively, on the other hand, the designer should consider the mechanical and optical properties, the machining performance, stability and security and other factors of the materials. Material mechanics performance mainly requires that the deformation caused by weight of primary mirror be as small as possible. This requires that the specific stiffness of the window material (stiffness is defined as the ratio of elastic modulus and density, the E/ρ), which requires high elastic modulus, low density of the material. Material stability and security mainly requires mirror window have longer service life. Therefore, in real working environment, the material of mirror should not be easily deformed, or having performance change by environmental effect. As consider the cost of the material as well, the best choice would be Silica glass and germanium tablet. Its performance parameter is shown on the following Table 1.

According to the requirements of the optical aperture solar simulator system, the silica window diameter would be 230 mm, and the germanium window diameter would be 150 mm.

2 The Strength Analysis of the Optical Window

Silica glass and germanium plate is brittle material, under the influence of pressure, will not only cause error of the light wave which came through the window, but also will damage the window glass. When the window glass does not have enough plastic deformation, the surface will appear crack or break. So it is necessary to carry out strengthen analyze on the window glass, in which to make sure that the strength would meet the requirement of the minimum thickness of window glass. To brittle material, the maximum tensile stress caused by pressure difference should not exceed [1]

Table 2 Changes of silica glass surfaces under different pressure

Thickness (mm)	5	10	15	20	25	30	35
Center deformation (nm)	0.3240	0.0413	0.0126	0.0056	0.0030	0.0019	0.0001

$$\sigma_{\max} = \frac{\sigma_F}{F_S} \tag{1}$$

In the formula, the σ_{max} is the permitted maximum tensile stress, σ_F is the fracture stress of window material, F_s is safety factor. Optical mechanical engineering usually adopt the safety factor of 4. The minimum ratio between the thickness of the simply supported circular window and its diameter should be [2]

$$\frac{d}{h} = 2\left[\frac{8\sigma_F}{3\Delta p(3+u)F_S}\right]^{\frac{1}{2}} \tag{2}$$

In the formula, the aspect ratio of the window is d/h, Δp is the pressure difference between inside and outside the window surface, U for window material Poisson's ratio.

Silica fracture stress is 100 MPa, the fracture stress of germanium is 92.7 MPa. When the window has an atmospheric pressure difference on both sides, the window of the minimum thickness of 7.89 and 5.46 mm respectively.

When the pressure difference exists between the inside and outside surface, the pressure difference will cause the shape change to the window glass surface, resulting in the optical path, which goes through the window, change. Tables 2 and 3 shows the flat surface deformation situation of different thickness windows, under a standard atmospheric pressure difference.

As show above, with the glass getting thicker, the change caused by pressure is getting smaller.

3 Thermal Optical Analysis of Optical Window

One side of the optical window is normal temperature environment, while the other side is the vacuum tank at low temperature environment. Optical window heat exchange with the surrounding environment may lead to the temperature change of the window, and also the radial temperature difference, the circumferential temperature difference and axial temperature difference temperature, gradient refractive index, and even the thermal deformation, which may again, make optical window does not meet its optical indicator.

Circumferential temperature difference refers to the uneven temperature distribution in the window along the circumferential direction. If there is temperature difference in the window edge, it may cause Equal thickness difference, and that

Table 3 Changes of germanium tablet surfaces under different pressure

Thickness (mm)	5	10	15	20	25	30	35
Center deformation (nm)	0.04233	0.00556	0.00174	0.00090	0.00049	0.00032	0.00024

may further make the window in the presence of asymmetric deformations, and cause Spherical aberration, Coma and astigmatism optical errors.

Window material thermal conductivity is very low. When the edge of the center or temperature changes, this change can't transfer to the other end of the radial immediately. So that will form the radial temperature difference distribution from center to the edge, inside the lens. According to different temperature, the temperature distribution may performance in two types, one is that the center temperature goes higher than the edges and the other is on the contrary. The temperature difference on the radial can cause uneven change of the radius and the thickness of the curvature, and cause distortion.

The axial temperature difference refers to the window material temperature difference between two surfaces. As shown in Fig. 1.

The diameter of the circular plate was D, thickness of d, when the two flat surface have temperature difference ΔT, inside the flat would form the axial temperature difference, and may occur the bending deformation as shown in the picture. Then, the diameter difference of the two flat surfaces would be [3]

$$\Delta D = D \cdot \alpha \cdot \Delta T \tag{3}$$

In the formula, α is the flat glass linear expansion coefficient.

The radius of the circular plate glass after bending curvature is:

$$R = \frac{d}{\alpha \cdot \Delta T} \tag{4}$$

Fig. 1 The lens distortion under the axial temperature difference

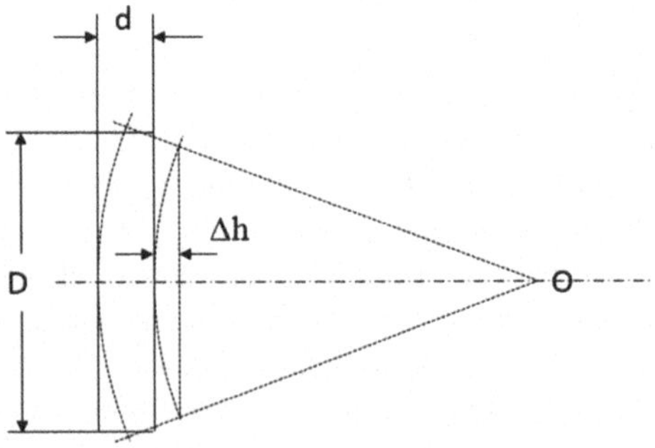

Increment of vector height

$$\Delta h = \frac{D^2 \cdot \alpha \cdot \Delta T}{8d} \quad (5)$$

When the lens its own have axial temperature gradient, the radius of curvature and focal length of the mirror will be changed.

3.1 The Influence of the Diameter Temperature Difference to the Shape of Window Surface

Do two hypotheses in this case, one kind is the edge temperature gets higher; the other is a higher temperature in the middle. The following Fig. 2 shows the thermal elastic deformation of silica glass when the edge temperature gets 5 °C higher.

When the edge temperature is 20 °C, and the center temperature is 15 °C, we analyze that how different thickness flat surface is affected by temperature. As shown in Tables 4 and 5, with the increase of the thickness of the window, the change of the type flat surface increases as well. It shows that the radial temperature difference would have a bigger effect to the thicker window type flat surface.

We also studied the case that the edge center temperature is 5 °C higher than the middle. We found that with the same radial temperature difference, the deflection of the mirror surface shape is similar, no matter the temperature of the middle is higher or lower than the edge one.

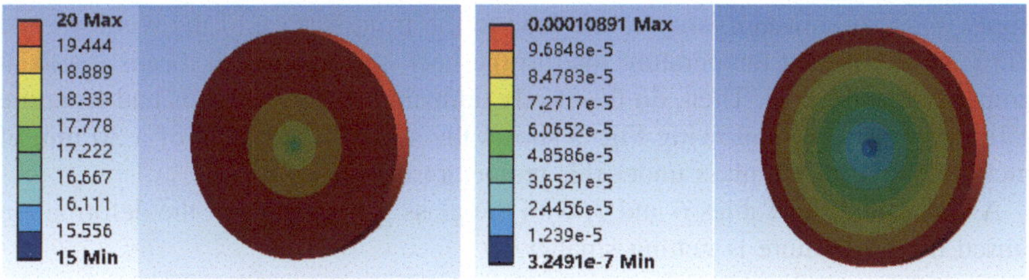

Fig. 2 The radial temperature distribution of the window and the deformation nephogram

Table 4 Changes of silica glass surfaces under caused by different radial temperature

Thickness (nm)	5	10	15	20	25	30	35
Center deformation (nm)	107.9	108.2	108.4	108.7	109.0	109.5	109.8

Table 5 Changes of germanium tablet surfaces caused by different radial temperature

Thickness (mm)	5	10	15	20	25	30	35
Center deformation (nm)	537.1	539.5	542.1	544.7	547.2	550	552.8

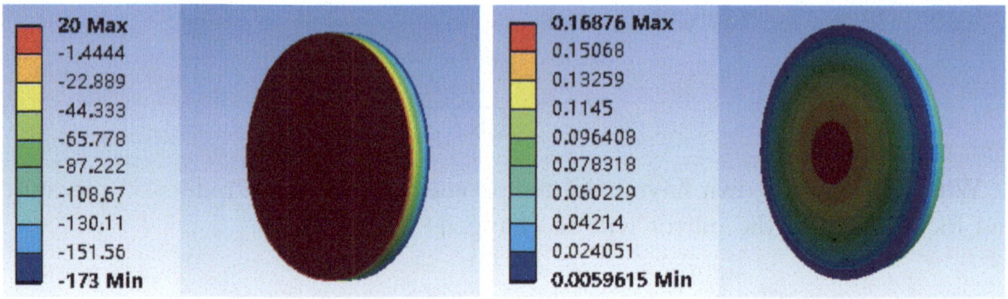

Fig. 3 The radial temperature distribution of the window and the deformation nephogram

Table 6 Changes of silica glass surfaces under caused by different axial temperature

Thickness (mm)	10	15	20	25	30	35
Center deformation (mm)	0.0455	0.0343	0.0259	0.0209	0.0176	0.0159

Table 7 Changes of germanium tablet surfaces caused by different axial temperature

Thickness (mm)	10	15	20	25	30	35
Center deformation (mm)	0.328	0.222	0.169	0.138	0.117	0.101

3.2 The Influence of the Radial Temperature Difference to the Shape of Window Surface

Apply 20 °C of constant temperature load to the front surface of the window, and −173 °C of constant temperature load to the back surface, which means the axial temperature is 193 °C. Then, do the steady temperature field analysis and structure statics analysis. The following Fig. 3 shows the surface's situation of a 30 mm of thick microcrystalline glass under this temperature field [4].

As shown on the Tables 6 and 7, with the glass getting thicker, the deformation caused by temperature is getting smaller.

4 Analysis and Optimization

As shown on all the simulation results, with the glass getting thicker, the deformation caused by pressure and radial temperature difference is getting smaller. On the contrary, with the increase of the thickness of the window, the change of the type flat surface increases as well by diameter temperature difference. compared with pressure and diameter temperature difference, the deformation caused by radial temperature difference is more significant, because the temperature difference is

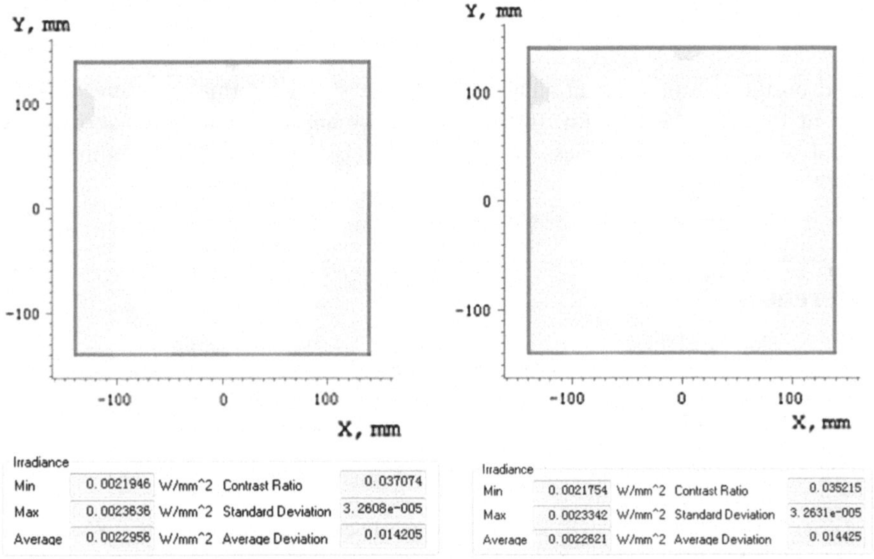

Fig. 4 The intensity and uniformity of light through the window without or with deformation

193 °C. In consideration of the deformation caused by radial temperature difference, with the difference of the thickness, the change of the center deformation keeps stable when the thickness of silica glass and germanium tablet is more than 20 mm. Combined with the reliability of window material and the price factor the thickness of the silica glass is chosen to be 30 mm, the thickness of the germanium plate chooses to 20 mm.

The deformation may affect the image quality in solar simulator.

The germanium optical window is the connector between the blackbody radiation source and the main optical system of the internal vacuum tank. Blackbody radiation optical system model is constructed in ZEMAX [5]. The optical system model contains convergent lens and field lens. The germanium plate's curvature radius of its back and front is calculated with the data simulated by Ansys. Take it as a consideration, with the output wave aberration and focal length as the optimization goal, the design parameter of convergent lens and field lens is received.

The silica glass is a window that let the light of xenon lamp come through. Silica glass is added into optical tunnel system. The intensity and uniformity of light is simulated by lighttools [6]. Figure 4 shows the intensity and uniformity of light through the window without or with deformation.

There is almost no influence when silica glass is deformated though the intensity and uniformity is better with the window in normal environment. So it is feasible that the thickness of the silica glass is chosen to 30 mm.

5 Conclusion

Based on the thermal optical analysis, with one side in normal pressure, and the other in 1×10^{-3} Pa vacuum, 100 K degrees, we analyzed the deformation of the optical window. The thickness of the window is determined and optimized by ZEMAX and lighttools.

References

1. Ming L, Qingwen W, Fei Y. Optimization of optical window glass thickness based on the thermal optical analysis [J]. Acta Optica Sinica, 2010, 30(1): 210–213.
2. Barnes W P. Some effects of aerospace thermal environments on high-acuity optical systems [J]. Applied Optics, 1966, 5(5): 701–711.
3. Wu F. The effects of temperature on quality of large optical mirrors [J]. Opto-Electronic Engineering, 1986, 4: 24–30.
4. DING Y, HAN S, LI J. Analysis for Thermo-Optical Sensitivity of Space Optical Window [J]. Opto-electronic Engineering, 2002, 5: 004.
5. Qingsheng X, Jie X, Linhua Y. Surface deformation analysis of off-axial paraboloidal mirror under space simulation conditions [J]. Optical Technique, 2011, 37(2): 153–157.
6. XIANG Y, ZHANG R, HUANG B. Irradiation Uniformity Simulation of KFTA Solar Simulator [J]. Spacecraft Environment Engineering, 2006, 5: 010.

Design and Analysis of Integration Structure of Space-Borne Fourier Transform Spectrometer

Caiqin Wang, Bin Tu and Pengmei Xu

Abstract As main supporting structure of conventional space-borne remote sensor, single layer panel design causes that the space of payload is occupied by large numbers of erose optical-mechanical assemblies and electrical instruments, and the size of payload is increased. The satellite space is seized, the number of satellite payload is decreased, and the function of satellite is limited. Combining current development situation of space optical remote sensor, and basing on the requirements of small size and decreased weight of payload, a kind of double layer integration structure is introduced into space-borne Fourier transform spectrometer, which takes full advantage of remainder space of height direction. The layout of payload is optimized, and the size is diminished and meets the requirement by using of this supporting structure. Firstly, the design method and advantage of this structural format are elaborated. As the most critical assembly of spectrometer, the carbon fiber aluminum honeycomb composite panel bears the weight of topper and lower assemblies, which takes great effect to the mechanical stability of spectrometer. The detailed design of carbon fiber aluminum honeycomb composite panel is illustrated in this paper, which includes structural format, the choice of reinforced fiber and resin matrix, and design of the reinforced frame. Secondly, the finite element model of carbon fiber aluminum honeycomb composite panel and

C. Wang (✉) · P. Xu
Beijing Institute of Space Mechanics and Electricity,
No. 99 Zhongguancundong Rd., Beijing, China
e-mail: wangcaiqin1980@126.com

P. Xu
e-mail: 13651065845@163.com

B. Tu
Beijing Institute of Space Mechanics and Electricity,
No. 104 Youyi Rd., Beijing, China
e-mail: 13520204741@139.com

H.P. Urbach and G. Zhang (eds.), *3rd International Symposium of Space Optical Instruments and Applications*, Springer Proceedings in Physics 192,
DOI 10.1007/978-3-319-49184-4_16

whole spectrometer are built, mechanical analysis are conducted, including the displacement of optical assemblies under the action of gravity and 20 ± 3 °C temperature range, and the model analysis of spectrometer. The displacement of carbon fiber aluminum honeycomb composite panel meets the design requirements by the analysis, which has proved that the carbon fiber aluminum honeycomb composite panel could bear the weight of topper and lower assemblies and guarantee the mechanical stability of spectrometer. The reliability of double-layer integration supporting structure of space-borne Fourier transform spectrometer has been validated.

Keywords Spectrometer · Integration structure · Carbon fiber aluminum honeycomb composite material · Design · Analysis

1 Introduction

The property requirement of optical remote sensors has been getting higher and higher, and function of them was gradual diversity, which along with the development of technology. The size of payloads has been larger and larger, whose weight has been increased and structural format has been complicated. Conventional main structural support form of space remote sensor is commonly that unilaminar panel fixed onto the satellite deck, all the opto-mechanical structure, electrical instruments and other assemblies (such as enclose and lens hood) are installed on this panel. Because the shape of most optical-mechanical structure and electrical instruments are anomalistic, the size of payload is getting larger and larger, and occupying the space of satellite payload bay. Conventional composition of satellite is that the payload lay on the panel which was towards Earth. If the payload gets larger, the amount of satellite will be decreased, and the function of satellite be astricted. In order to decrease the weight of satellite, and take more payloads, the demands of miniaturization and lightweight are increasingly urgent.

Because the distribution of its optical-mechanical structure of Fourier transform spectrometer is dispersive, if we choose conventional unilaminar panel supporting method, the horizontal size of spectrometer is large, which could not meet the requirement of satellite. Based on the above reasons, a kind of double layer Integration Structure is introduced into the spectrometer, which takes full advantage of remainder space in the height direction. Design method of the double-layer integration structure is introduced in this paper. Especially, the detailed design of carbon fiber aluminum honeycomb composite panel is also elaborated, which takes great effect to mechanical stability.

The finite model has been built in this paper, and static stiffness of composite panel and dynamic stiffness of spectrometer have been calculated, this provides theoretical basis for the correctness and reliability of double-layer integration supporting structure.

2 Operational Principle and Supporting Structure of Space-Borne Fourier Transform Spectrometer

2.1 Operational Principle of Spectrometer

The space-borne Fourier transform spectrometer is used to observe the spectral of atmosphere and get the atmospheric composition and concentration distribution by solar occultation from the satellite, which has high spectral resolution, high signal noise ratio and large spectral coverage, as in [1], the operational principle is shown in Fig. 1.

The spectrometer is pointed to Sun firstly, the solar radiance without atmosphere is measured out of atmosphere, which is illustrated in light 1 in Fig. 1, at this time Sun is a calibration source, a set of exo-atmospheric spectra and images are also taken as reference measurements. When the satellite runs to a certain location, the solar radiance is partially attenuated by chemical assemblies present in the atmosphere at sunrises and sunsets, which is showed in lights 2–4 in Fig. 1. By taking the ratio of these two readings, scientists can identify the specific portions of the solar energy that has been absorbed by Earth's atmosphere, as in [1]. And the data will be used to calculate the chemical constitution of atmosphere.

The constitution is showed in Fig. 2. The solar light is introduced into the primary mirror by the tracking mirror, which is driven by sun-tracking mechanism.

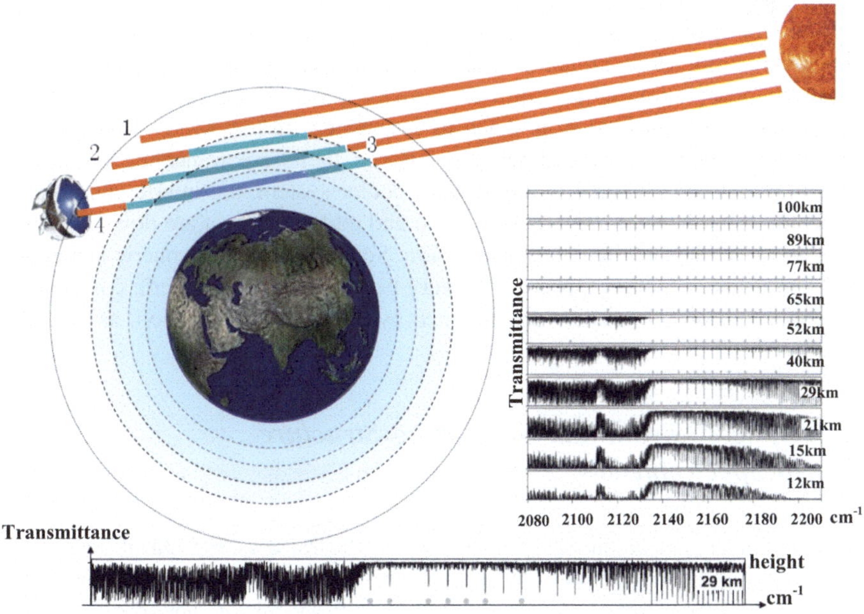

Fig. 1 Operational principle of space-borne fourier transform spectrometer

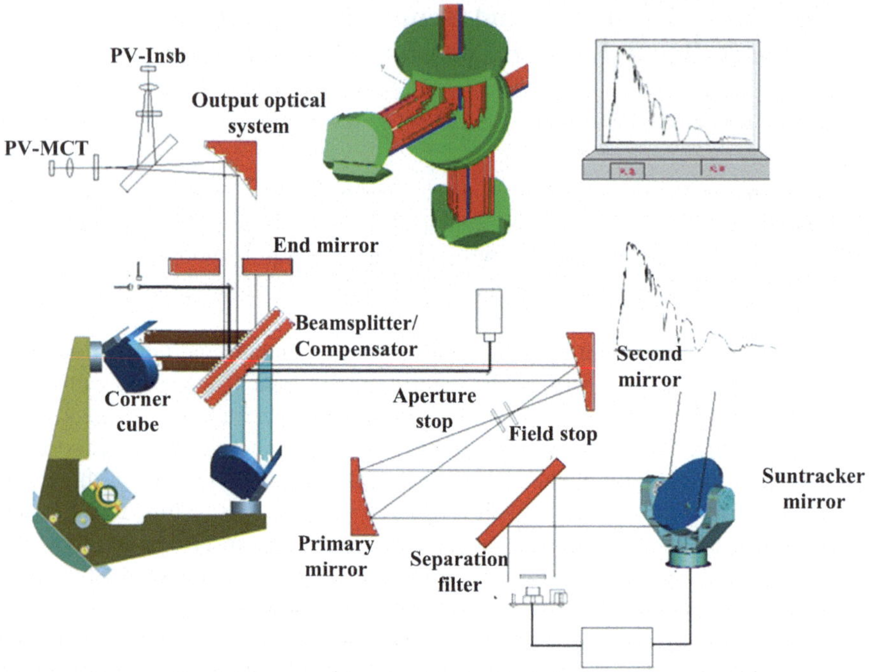

Fig. 2 Constitution of space-borne fourier transform spectrometer

The tracking mechanism is composed of two rotating shafts in the horizontal and vertical direction, and the function of it is to track solar light accurately and steadily during the observation of spectrometer. Primary mirror then transmits the lights to the aperture stop and field stop, reach to the second mirror, which collimates the beam to the interferometer. When the infrared lights are transmitted to output optical system and received to the detector-cooler system, the interference signals are acquired by detector. Both science data and additional data are down-linked to the ground by the satellite data transmission. Scientists get atmospheric absorption spectrum data products by spectral recovery process, and atmospheric composition and concentration information are obtained by atmospheric inversion, as in [2].

2.2 Optical Layout and Supporting Structure of Spectrometer

The original optical layout of spectrometer is illustrated in Fig. 3a, it consists of sun-tracking optical system (including tracking mirror, tracking camera and separation filter), input optical system (including primary mirror, aperture stop, field stop, and second mirror), interference optical system (including beam splitter,

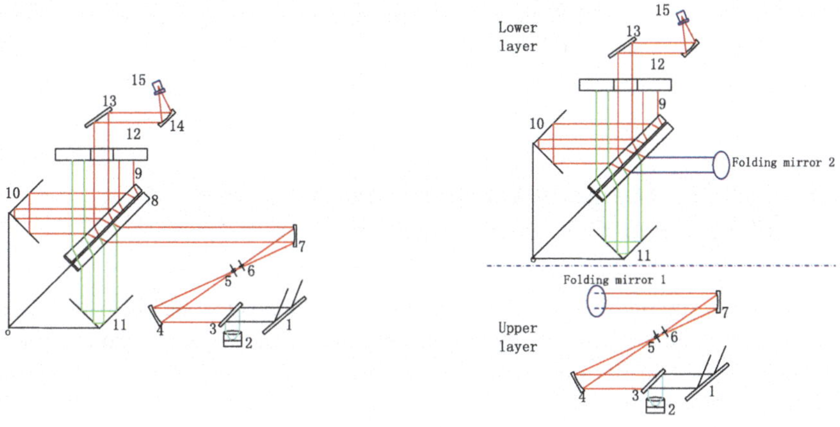

(a) Single layer optical layout (b) Double layer optical layout

1- tracking mirror; 2- tracking camera; 3- separation filter; 4- primary mirror; 5- aperture stop; 6- field stop; 7- second mirror; 8- beamsplitter; 9- compensator; 10, 11- corner cube; 12- end mirror; 13- output folding mirror; 14- output parabolic mirror; 15- detector

Fig. 3 Optical layout of spectrometer

compensator, corner cube and end mirror), output optical system (including output folding mirror and parabolic mirror) and detector-cooling system, as in [3]. If conventional unilaminar panel supporting method is introduced, the size of spectrometer will be larger than 1100 mm × 1000 mm × 400 mm, the size of horizontal directions did not meet the requirements of satellite.

Combining the original optical design, the size requirement of satellite and structural shape of optical assemblies of the spectrometer, two folding mirrors were added between the input optical system and interferometric optical system, which has changed the original single layer panel layout into two layers of up and down. The optimized optical layout is illustrated in Fig. 3b. The sun-tracking optical system, the primary mirror, aperture stop, field stop, second mirror, the first folding mirror and their supporting structure are installed on the upper layer, and the second folding mirror, interferometric module, output optical system, detector cooling system and signal processor are fixed on the opposite side. A carbon fiber aluminum honeycomb composite panel is in the middle and support upper and lower assemblies, which is linked with satellite mounting surface by a titanium alloy frame. A direct result of this optimization is the size of spectrometer has been reduced to 767 mm × 675 mm × 600 mm, which satisfies the design requirement.

With an increase of two folding mirrors, the transmittance of the whole optical system has been decreased 3.5%, and Signal Noise Ratio of spectrometer has been affected too, but they still meet the optical design requirement. At the same time, the weight of two folding mirrors and their supporting structure is 0.2 kg, that has influenced the weight of whole spectrometer weakly. But a consequence of the optical layout change is that the optical performance of the spectrometer is affected seriously by mechanical stability of the carbon fiber aluminum honeycomb composite panel.

2.3 Detailed Design of Carbon Fiber Aluminum Honeycomb Composite Panel

Conventional carbon fiber aluminum honeycomb composite panel is loaded on one single side, while the panel of spectrometer is loaded on two sides and has not any supporting point in the middle of the down side, so it must has excellent static stiffness and strength.

The design of carbon fiber composite panel and aluminum honeycomb sandwich structure is adopted because of its high stiffness, high strength, low weight, low thermal expansion, and excellent processing techniques. The thickness of whole composite panel is 40 mm, the thickness of carbon fiber is 1 mm, and aluminum honeycomb is 36 mm thickness. Integral framework is embedded between the aluminum honeycomb, which is helpful for improving structural stiffness and joint strength. The mounting points of optical assemblies and electric instruments are designed in the form of metal embedded parts, which are adhibited with aluminum honeycomb and framework.

As the requirement of high stiffness and strength, the carbon fiber of high modulus must satisfy the demand of highest stiffness in minimum quality. Most frequently used carbon fibers of high modulus are showed in Table 1. Combining the parameters in Table 1, M55J is selected because of its high tensile modulus, high strength and thermal conductivity.

The dimensional stability requirement of composite materials is more difficult to guarantee than other types of materials, it includes thermal dimensional stability due to temperature variation and the dimensional stability caused by the moisture absorption for composite material's structure of resin matrix. These characteristics of the composite material are determined by the resin matrix. Isocyanate resin is a

Table 1 Several frequently used carbon fibers of high modulus and their performance parameters

Brand	Tensile strength (MPa)	Tensile modulus (GPa)	Coefficient of thermal expansion (ppm/°C)	Thermal conductivity (W/m K)	Elongation rate (%)	Density (g/cm^3)
M40J	4410	377	−0.83	68.66	1.2	1.77
M46J	4210	436	−0.9	84.57	1.0	1.84
M50J	4120	475	−1.0	97.97	0.8	1.88
M55J	4020	540	−1.1	155.75	0.8	1.91
M60J	3920	588	−1.1	151.98	0.7	1.93

new type of high performance thermosetting resin, which includes two or more isocyanate functional groups. Excellent electrical insulation, low moisture absorption, high heat resistance, excellent dimensional stability, good mechanical properties and good forming process of isocyanate resin are endowed by its structural characteristics, as in [4]. Synthesizing the properties above, isocyanate resin is selected for the resin matrix of the carbon fiber composite panel.

In order to improve the stiffness and strength of carbon fiber aluminum honeycomb panel, strengthening framework is ensconced between the upper and lower carbon fiber panel. The section shape of strengthening framework is groove; its position is adjusted appropriately according to the embedded parts of the optical-mechanical assemblies installation.

3 Finite Element Analysis

3.1 Static Stiffness Analysis of Composite Panel

The finite element model of the carbon fiber aluminum honeycomb composite panel is built by MSC/Patran, honeycomb sandwich structure is simulated by 3 nodes composite shell element, lumped mass point is selected to simulated each assembly of the spectrometer, which is connected with the composite panel by RBE2. The finite element model is indicated in Fig. 4.

The essential of carbon fiber aluminum honeycomb composite panel's static stiffness is the positional stability of optical-mechanical assemblies under the action of gravity and 20 ± 3 °C temperature range. The mounting surface of composite panel has been constrained, and displacements of the optical-mechanical assemblies have been calculated in the working condition above, which are listed in Table 2. It can be seen that displacement of each optical-mechanical assembly on the composite has satisfied the design requirement from Table 2.

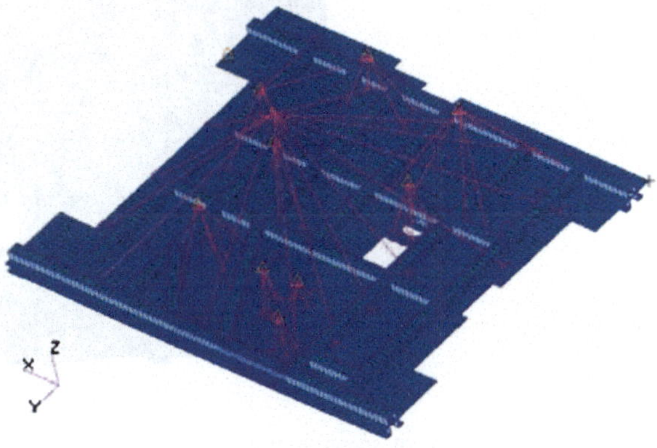

Fig. 4 Finite element model of carbon fiber aluminum honeycomb composite panel

Table 2 Displacements of optical-mechanical assemblies (—represent no requirement)

Name of optical-mechanical assemblies	Requirements (mm)	Displacements (mm)
Primary mirror assembly	Benchmark	Benchmark
Aperture stop and field stop	±0.003	0.001
Second mirror assembly	±0.005	0.002
The first folding mirror assembly	—	0.001
The second folding mirror assembly	—	0.001
Interferometer	±0.008	0.003
Output folding mirror assembly	—	0.002
Output parabolic folding mirror assembly	±0.005	0.002
Detector	±0.005	0.001
Sun-tracking mechanism	—	0.003
Separation filter	—	0.001
Tracking camera	0.008	0.001

3.2 *Dynamic Stiffness Analysis of Spectrometer*

The finite element model of the spectrometer has been built too, and the model analytical result is that its fundamental frequency is 94.4 Hz, which has met the requirement of the satellite. The first mode is showed in Fig. 5, it is the bending of the mounting surface of the interferometer assembly of spectrometer.

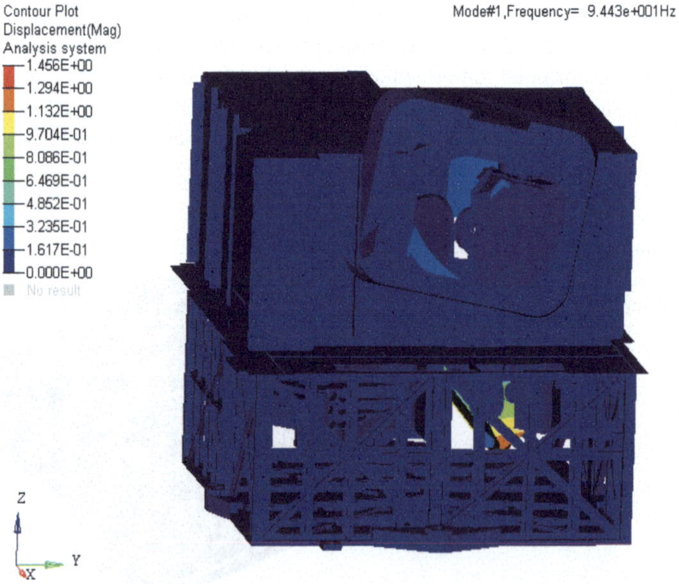

Fig. 5 The first mode of spectrometer

4 Conclusion

The optical layout of space-borne Fourier transform spectrometer is optimized, original single layer panel structure is changed into a kind of double layer integrated structure, which has achieved the purpose of reducing the size of the horizontal direction of spectrometer.

The composite panel could bear the assemblies of double layer through the finite element analysis. The correctness and reliability of the design of double layer integration structure of spectrometer have been validated, which is helpful for structural layout of other remote sensors.

References

1. FAN Dongdong and BAI Shaojun, "Influence of Micro-vibration on Space-borne Fourier Transform Spectrometer," *Spacecraft Recovery & Remote Sensing*, 4th ed., vol. 34, pp. 59–65, Aug 2013.
2. Burkert P, F Fergg, H Fischer. "A compact High-resolution Michelson Interferometer for Passive Atmospheric Sounding". *IEEE Transactions on Geoscience and Remote Sensing*, 3rd ed., vol. 21, pp. 345–349, 1983.
3. QI Weihong, Wei Haoyun and YI Lina, "Analysis on Infrared Spectrometer System Specification for Atmospheric Composition Detecting," *Spacecraft Recovery & Remote Sensing*, 5th ed., vol. 34, pp. 36–45, Aug 2013.
4. GUO Jiang, SHAO Mingdong, WANG Guoliang and SUN Jiming, "Design of optical-mechanical Structure Made of CFC in Sspace Remote Sensing Camera," *Optics and Precision Engineering*, 3rd ed., vol. 20, pp. 571–578, Mar 2012.

The Design and Assembly of Infrared Zoom Lens with Replicate Structure

Yang Huang, Tingcheng Zhang, Cong Wang, Chunyu Wang and Jiyou Zhang

Abstract With the rapid development of optical remote sensing technology, the micro or small remote sensors are widely used in aerospace and other military technical fields. Due to the structure space limitation of micro or small remote sensor and the needs of performance indicators, an infrared lens with replicate structure which is containing two mirrors is designed. The infrared lens can realize the zooming function by optical path automatic switching system, driving by the rotating motor. It not only can provide a large field of view for the pilot in the cruise, but also can switch to a small field of view with high resolution when the target is found. The lens has the characteristics of multi zoom, high image quality, small size and so on. But, compared with the traditional lens, the infrared zoom lens with replicate structure is more complex in structure. Therefore, it puts forward higher requirements on assembly. So, based on the traditional assembling for infrared lens, one method that contains computer-aided alignment and center error of the replicate structure is brought in practical engineering. Firstly, through the eccentric sensitivity for each lens is simulated and analyzed, the alignment priority is decided. Secondly, after the mirrors were fixed, the Opti-Centric IR which is a center error measuring instrument for infrared lens is used to complete the high precision assembling of one optical path which contains all of the lenses. Finally, when the infrared zoom lens is assembled, the imaging quality and the effective focal length of all optical path of the lens are measured in an infrared Modulation Transfer Function (MTF) measurement instrument. The result shows that the design of the infrared zoom lens with replicate structure is reasonable and feasible, and the method is successful and effective to assemble the infrared zoom lens with replicate structure. So the lens with this kind of structure can be widely used in the micro or small remote sensors.

Keywords Infrared zoom lens · Replicate structure · Design · Assembly

---

Y. Huang (✉) · T. Zhang · C. Wang · C. Wang · J. Zhang
Beijing Institute of Space Mechanics and Electricity, Beijing, China
e-mail: huangyang600@126.com

H.P. Urbach and G. Zhang (eds.), *3rd International Symposium of Space Optical Instruments and Applications*, Springer Proceedings in Physics 192,
DOI 10.1007/978-3-319-49184-4_17

1 Introduction

In recent years, with the rapid development of micro sensors, especially in the field of unmanned aerial vehicle, the miniaturization and high image quality is a very important criterion for the optical systems [1, 2]. Due to the limitation of the structure size, the physical space of the optical lens is limited, and the traditional lens structure and alignment method can't meet the requirements. Therefore, it is significant to study on the design and assembly of the optical systems for the micro or small remote sensors.

According to the research task, an infrared lens with replicate structure which is containing two mirrors is designed in present paper. The size of the optical system is 172 × 151 mm, and the focal length can be zoomed in 45 and 135 mm stably. Because of the lens has a folding structure, and the optical axis is always consistent to ensure that the image plane without jitter, it puts forward higher requirements on assembly.

2 Optical System Design

2.1 Design Parameters

Size: <200 mm × 200 mm; total length: <280 mm; aperture: 70 mm; wavelength: 3.7–4.8 μm; focal length: 45/135 mm (switch-zoom); back focal length: 20 mm; field of view: $2\omega \geq 15.28°/5.21°$; Modulation Transfer Function (@33lp/mm): ≥ 0.3; the angle between line of sight (LOS) when switching field of view: ≤ 0.35 mrad; field of view switching time: ≤ 0.5 s.

2.2 Consideration of Structural Type

According to the compensation method of image, the zoom system can be divided into continuous type (e.g. mechanical or optical compensation type) and cut in/out type. Their zoom principles and features are shown in Fig. 1 and Table 1.

According to the design parameters and the characteristics of the three kinds of zoom mode, the cut in/out type zoom system is selected in lens design.

2.3 Optical and Structural Design

Different from general optical system design, the infrared zoom lens must be sufficiently considered for the system size (e.g. size: <200 mm × 200 mm, total

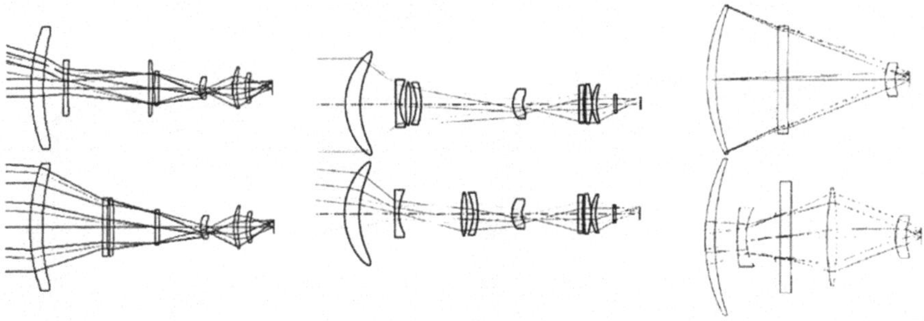

(a) Mechanical compensation type **(b)** Optical compensation type **(c)** Cut in/out typett

Fig. 1 Principles of three kinds of zoom mode

Table 1 Features of three kinds of zoom mode

Zoom mode	Advantages	Disadvantages
Mechanical compensation	① Zoom continuously	① Through the cam rotation to complete the zoom process, the requirements for the design and manufacture are highest
	② The field of view switching time is shorter	
	③ Maximum to avoid loss of target	
Optical compensation	① Through the lens linear motion to achieve zoom, the location accuracy is easy to achieve	① Zoom discontinuously
		② The field of view switching time is longer
		③ Target is easy to lose
Cut in/out	① The field of view switching time is short	① Zoom discontinuously
	② Target is not easy to lose	
	③ The principle is simple and easy to implement	② Only can switch two or three focal lengths

length: <280 mm, et.) and structure in the design process. From the Gauss optics, combined with the cut in/out type zoom mode and the principle of optical focal length distribution, the specific structural parameters of each element and the initial structure are calculated by using the theory of primary aberration. The results of optical design are shown in Figs. 2 and 3.

The optical elements can realize optical function through the mechanical structure. According to the optical structure design of infrared zoom lens, the cut in/out type zoom system is to use the way of rotation, that the two lenses of the system cut in the short focal length by the way of rotating around the axis. Mechanical structure design is shown in Fig. 4.

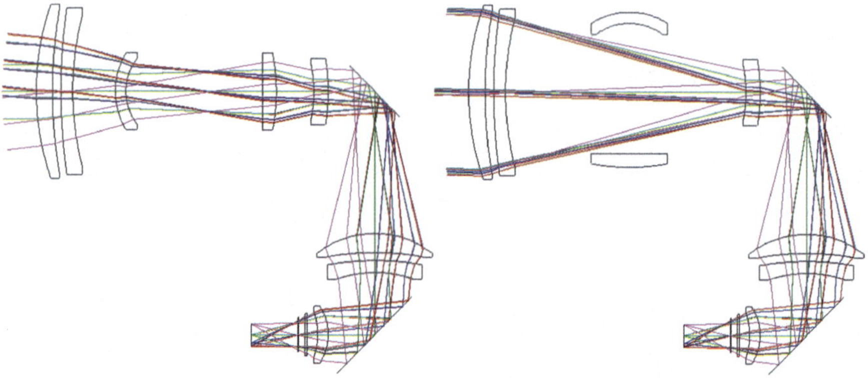

(a) Short focal length path **(b)** Long focal length path

Fig. 2 Optical system structure

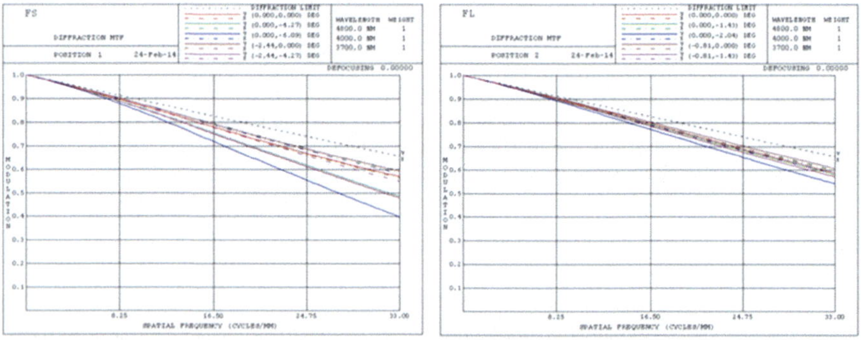

(a) MTF curve at short focal length path **(b)** MTFcurve at long focal length path

Fig. 3 Modulation transfer function (MTF) *curve*

3 High Precision Assembly

3.1 Computer-Aided Alignment

Computer-aided alignment is a combination of optical alignment and computer simulation [3–5]. Because the contribution of each optical element to the aberration of the optical system is different, in order to realize the high precision assembly, the misalignment sensitivity of each lens must be calculated firstly, and the alignment priority will be decided. Take the optical path of short focal length for example, the effect of the shift, tilt and distance of each lens on the image quality of the center field of view is shown in Fig. 5.

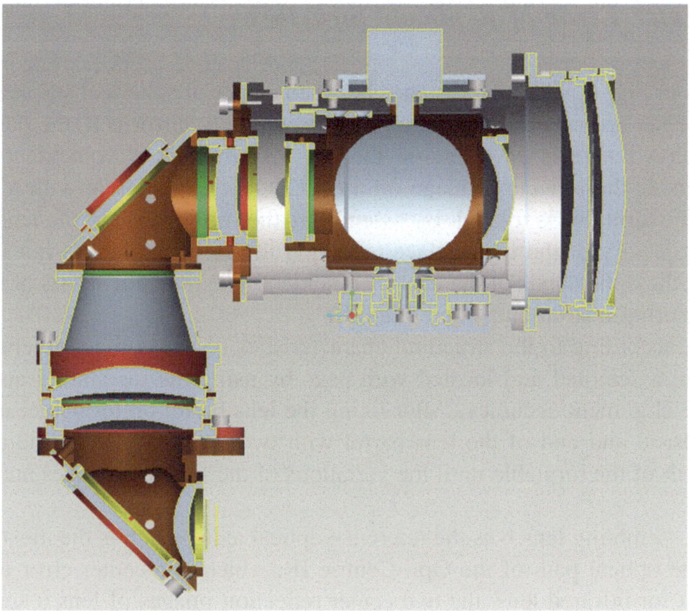

Fig. 4 Mechanical structure

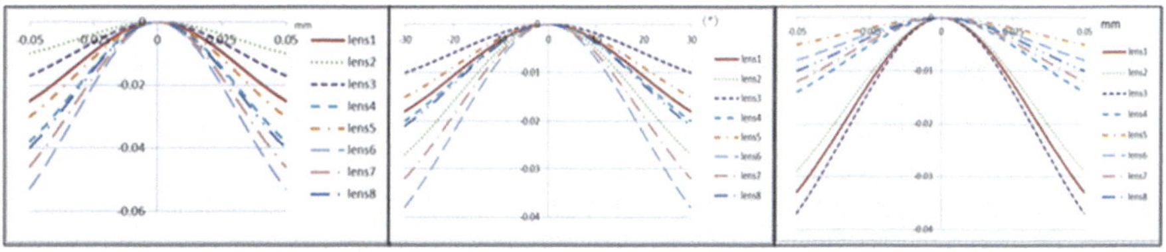

(a) The shift effect on MTF **(b)** The tilt effect on MTF **(c)** The distance effect on MTF

Fig. 5 The effect on the image quality (MTF@33lp/mm) from the misalignment of each lens

From Fig. 5, we can see that the shift misalignment of the lens 4–8 and the tilt misalignment of the lens 2, 6–7 have a bad effect on the image quality, and the distance misalignment of the lens 1–3 has a significant influence on the image quality of the system. Therefore, considering from the impact of the misalignment factors, we should focus on the control of these sensitive factors in assembly process to improve the image quality of the optical system effectively.

3.2 Center Error of Replicate Structure

As shown in Fig. 6, when the distance between center of spherical surface and axis *L* of the center error measuring instrument through the mirror is *d*, the deviation of the reflection image is 2*d*. Because the axis of turntable is coincident with the optical axis *L*, the reflection image on the CCD plane is a circle with a radius of 2*d* when the turntable is rotated. By assembling the shift and tilt of the lens until the rotation trajectory of the two surface reflection image of the lens is not moved, at this point, the optical axis of the lens is coincident with the optical axis of the center error measuring instrument (Figs. 7 and 8).

Firstly, according to the structural characteristics of infrared zoom lens, the two mirrors are assembled and located with pegs by using the machining accuracy to ensure the alignment accuracy. After fixing the lens barrel on turntable, measuring the cylindrical and end of the lens barrel with two dial gauges, and adjusting the shift and tilt of the turntable until the variations of the two dial gauges are less than 0.01 mm.

Then, assembling lens 6 as the reference optical axis, which is the most sensitive lens. In the optical path of the Opti-Centric IR, which is a center error measuring instrument for infrared lens, the two center reflection images of lens 6 surfaces are detected, and the deviations of the two center reflection images are measured when turning the turntable. Adjusting the shift of the lens by four screws in the lens

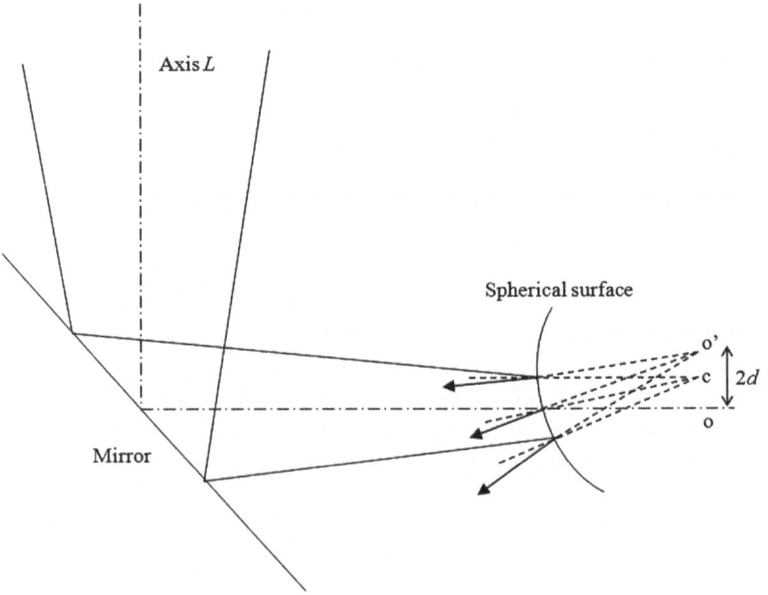

Fig. 6 Principle of center error measure for replicate structure

Fig. 7 Infrared zoom lens with replicate structure

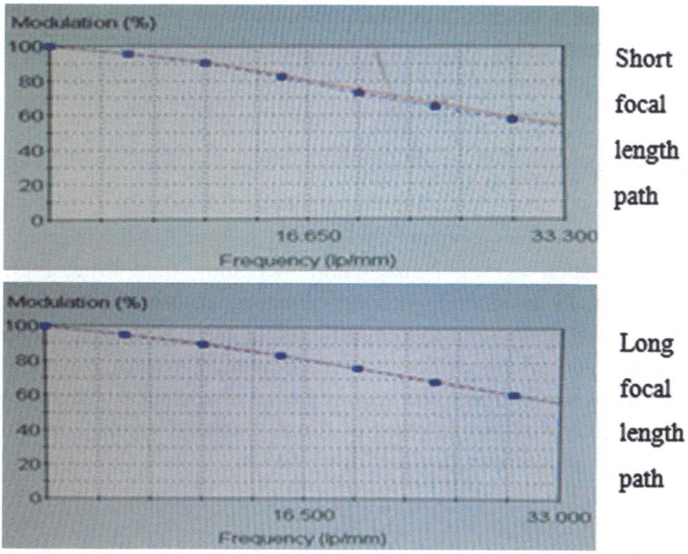

Fig. 8 The results of the optical system's MTF

barrel, and controlling the tilt of the lens by gaskets between the contact surfaces, until the deviations of the two center reflection images are less than 5″.

Lastly, in order of lens 7–8 and 5–1, completing the high precision assembling of the optical path of short focal length, according to the above methods to adjust

Table 2 The performance of the infrared zoom lens with replicate structure

Item	Parameters	Design value	Measured value
Size	<200 mm × 200 mm	172 × 151 mm	172 × 151 mm
Total length (mm)	<280	271	271
Aperture (mm)	70	69	68
Wavelength (μm)	3.7–4.8	3.7–4.8	3.7–4.8
Focal length (mm)	45/135	45/135	44.98/134.99
Back focal length (mm)	20	20	19.99
Field of view	$2\omega \geq 15.28°/5.21°$	15.4°/5.4°	15.4°/5.4°
MTF@33lp/mm	≥0.3	≥0.40/0.53	0.47/0.52 (average)
The angle between line of sight (LOS) when switching (mrad)	≤0.35	≤0.2	≤0.2
Field of view switching time (s)	≤0.5	≤0.5	0.3

the lens shift and tilt. On the bases of analyzing the effects on the image quality from the misalignment of each lens, the shift misalignment of the lens 4–8, the tilt misalignment of the lens 2, 6–7 and the distance misalignment of the lens 1–3 must be strictly controlled in assembly process, and others are in the assembly tolerance. The performance test results of the lens are shown in Table 2.

4 Conclusions

An infrared zoom lens with replicate structure is designed in this paper, which has many advantages such as small size, short switching field time, high imaging quality, and so on. In assembling process, dynamic control is realized combined with computer-aided alignment and center error of the replicate structure, and the alignment quality and efficiency for this type of optical system are improved. The works of this paper fully meet the requirements of technical specifications, and provide experience and technical reserves for the infrared zoom technology in the aerospace, military reconnaissance and other fields.

References

1. R.M. Fu, and L.Q. Yue. "Present state and perspectives of alignment technology for space optical remote sensor," *Spacecraft Recovery and Remote Sensing. China*, vol. 32, pp. 30–35, June 2011.
2. T. Yue, B. Li, X.L. Chen, and F. Zhou. "The current and future development of space optics," *Spacecraft Recovery and Remote Sensing, China*, vol. 32, pp. 1–9, October 2011.

3. Y.F. Huang, and L. Li. "Computer-aided alignment for space telescope optical system," *SPIE*, vol. 6149, pp. 0P1-6, 2006.
4. X.T. Duan, J.G. Bai, and R.K. Zhou. "An experimental study on the image quality loss of an objective caused by assemblage stress," *SPIE*, vol. 6149, pp. 3D1-5, 2006.
5. Z.B. Liao, C.Y. Wang, and M.J. Li. "Research on computer alignment based on refract optical system," *Infrared and Laser Engineering*, vol. 42, pp. 2453–2456, September 2013.

Aberration Analysis for the Computing Optical Design Method

Xiaopeng Shao, Jiaoyang Wang, Jie Xu and Jietao Liu

Abstract Traditional imaging methods often separate the two technical procedures of optical system design and imaging processing, where the lens subsystem is firstly optimized using an optical measure of performance and the image processing subsystem is optimized subsequently, resulting in the failures in efficient cooperation between the optical and digital elements and missing of the optimal solution. Therefore, an innovative approach is presented to combine optical system design and image processing together to obtain high resolution images from the final system. In this paper, the theoretical foundations for this computing optical design method is analyzed by considering the modulation transfer functions (MTF) and aberrations. According to the wave aberration theory and Zernicke formula, the relationship between the point spread function (PSF) and a certain monochromatic aberration can be obtained. Fourier transformation is taken on these PSFs and then modulus value calculation is adopted to get the MTFs. By analyzing the MTFs of idealized optical systems, the influence of each monochromatic aberration on the image can be concluded. Note that some of the MTFs values have zeros and others do not. Considering that imaging processing algorithm cannot recover the information lost at such zeros, thus some aberrations can be easily corrected through imaging processing algorithm, while others not. Aimed at a single monochromatic aberrations respectively, the optical system with single lens is designed in ZEMAX software and the relevant imaging processing algorithm is adopted. From the comparison among the final images, the theoretical foundations for the computing optical design method considering the MTF and aberrations can be verified. The compensation for aberration of imaging processing can be utilized to reduce the pressure of optical system design. The computing optical design method provides a new approach to simplify structure and reduce cost, as well as to gain high

X. Shao (✉) · J. Wang · J. Xu · J. Liu
School of Physics and Optoelectronic Engineering,
Xidian University, Xi'an, Shaanxi, China
e-mail: xpshao@xidian.edu.cn

H.P. Urbach and G. Zhang (eds.), *3rd International Symposium of Space Optical Instruments and Applications*, Springer Proceedings in Physics 192,
DOI 10.1007/978-3-319-49184-4_18

resolution images simultaneously, especially for designing certain complex optical system, which shows promising perspective of industrial application.

Keywords Computing optical design · Aberration theory · MTF · Imaging processing

1 Introduction

The traditional methods for designing optical imaging systems generally involve two stages: first, the optical system is designed to produce the highest quality optical image, second, the image processing is designed to correct some remaining defects in the images. As a result, the traditional optical imaging methods depend heavily on the performance of the optical system. With the development of the information technologies, high resolution cameras with low cost are required in more aspects of industry and life. However, the demand of higher image quality always leads to more complicated structure, larger volume, heavier weight and higher cost. Therefore, the traditional methods have to make a tradeoff between the image resolution and the cost. That is to say, the traditional methods can't meet the requirement of high image resolution and low cost simultaneously.

To balance the contradiction between the cost and the resolution, it is necessary to develop a novel type of optical design method. With the blossom of emerging computing imaging technology, new types of optical design method emerge recently, among which Wave Front Coding (WFC) [1–3] method and information-based method [4, 5] are very representative. To achieve the optimal performance, WFC method takes the integral optimization of optical system and image processing. However, it ignores information about the object source and relies on nonstandard optical elements such as cubic phase plates [1, 2]. In comparison, information-based method makes a source-channel-detector model of an imaging system and can be applied to both non-traditional and traditional optical elements [5], such as spherical lenses and mirrors. At present, information-based method is a promising novel optical design method.

In this paper, the theoretical foundations for this computing optical design method is analyzed by considering MTF and aberrations. The paper is organized as follows. In Sect. 2, according to the wave aberration theory and Zernicke formula [6], the relationship between MTF and a certain monochromatic aberration can be obtained. In Sect. 3, for the idealized optical systems, the property of MTFs that are due to equivalent severity among different optical aberrations is analyzed to verdict whether this aberration can be corrected through image processing. In Sect. 4, a singlet lens is designed with different aberrations weight in ZEMAX optical design software. Through the contrast between the final images, the aberration analysis for the computing optical design method can be proved. Finally, in Sect. 5, we conclude with some future research directions.

2 Wave Aberration Theory

A system with aberrations has a wave-front phase surface that deviates from the ideal spherical wave. Aberrations are found in most practical imaging systems, and their effect reduces image quality. In this section, we will study the relationship between MTF and a certain monochromatic aberration.

2.1 *Seidel Wave Aberration*

Seidel polynomials are often used to describe monochromatic aberrations for rotationally symmetric optical systems, such as most lenses and mirrors. A common form that is applied in conventional imaging systems is described by [6]

$$W(u_0;\rho,\theta)=\sum_{j,m,n}W_{klm}u_0^k\rho^l\cos^m\theta \tag{1}$$

where $k=2j+m$, $l=2n+m$, ρ is a normalized radial distance in the exit pupil, θ is defined relative to the x axis in a counter-clockwise direction. u_0 is a fractional image height, or normalized image height, defined along the u axis in the imaging plane. W_{klm} is the wave-front aberration coefficients.

The five primary Seidel aberrations can be expressed as:

$$\begin{aligned}W(u_0;x,y)&=W_d\left(x^2+y^2\right)+W_{040}\left(x^2+y^2\right)^2\\&=W_{131}u_0\left(x^2+y^2\right)x+W_{222}u_0^2x^2\\&=W_{220}u_0^2\left(x^2+y^2\right)+W_{131}u_0^3x\end{aligned} \tag{2}$$

The first term in this series is not one of the five primary aberrations, but is a defocus term. The other terms are known as spherical aberration, coma, astigmatism, field curvature, and distortion. The reference coordinate transfers from polar coordinates to rectangular coordinates.

2.2 *Modulation Transfer Function*

When considering aberrations, the pupil function for the diffraction-limited imaging system is corrected for generalized pupil function. $W(u_0,v_0;\,x,y)$ is defined as optical path difference between the actual wave-front and ideal wave-front in the exit pupil. If the pupil function is assumed as a standard unit circular and then the generalized pupil function can be expressed as [7, 8]:

$$P(u_0, v_0; x, y) = circ(\sqrt{x^2 + y^2}) \exp[ikW(u_0, v_0; x, y)] \quad (3)$$

Benefited from the diffraction limited system research, the optical transfer function (OTF) of incoherent imaging system including aberrations can be expressed as [7, 8]:

$$\mathrm{OTF}(u_0, v_0; \xi, \eta) = P(u_0, v_0; \lambda d_i \xi, \lambda d_i \eta) \otimes P(u_0, v_0; \lambda d_i \xi, \lambda d_i \eta) \quad (4)$$

where symbol $\otimes$ represents correlation operation. As following, OTF consists of amplitude and phase

$$\mathrm{OTF}(u_0, v_0; \xi, \eta) = \mathrm{MTF}(u_0, v_0; \xi, \eta) \cdot \exp[i\varphi(u_0, v_0; \xi, \eta)] \quad (5)$$

Since the autocorrelation can be expressed in Fourier transform, the relationship between MTF and a certain monochromatic aberration can be obtained.

$$\mathrm{MTF} = abs\{FT^{-1}[FT(P) \cdot conj(FT(P))]\} \quad (6)$$

3 MTF Analyze Among Different Aberrations

For the idealized optical systems and the pupil function assumed as a standard unit circular, the wavelength λ is chosen as 550 nm and the image distance d is 100 mm. Then the cut-off frequency of OTF can be calculated by the equation below.

$$\xi_0 = \eta_0 = 2/(\lambda d) \quad (7)$$

To study the influence of each single aberration, the wave aberration function $W(u_0, v_0; x, y)$ can respectively represent the five primary Seidel aberrations, spherical aberration, coma, astigmatism, field curvature, and distortion. Among different optical aberrations, equivalent severity is supposed that the maximum optical path difference equal to 1.5λ. Each aberrations coefficient W can be calculated from the maximum optical path difference.

The maximum field ($u_0 = 1$) is elected as a reference point. The MTFs that are due to equivalent severity among different aberrations is shown below.

From Fig. 1, the property of MTFs that are due to equivalent severity among different optical aberrations can be analyzed to verdict whether this aberration can be corrected through image processing. We can note that some of the MTFs have zeros and others do not. No amount of linear digital filtering can recover the information lost at such zeros. Thus some aberrations can be easily corrected through digital filtering, others cannot. Also the faster the MTFs decline, the lower the relative cut-off frequency is. So the optical system cannot transmit more information. According to the numbers of zeros and descent velocity of the MTFs, we can analyze the difficulty of aberration correction for image processing.

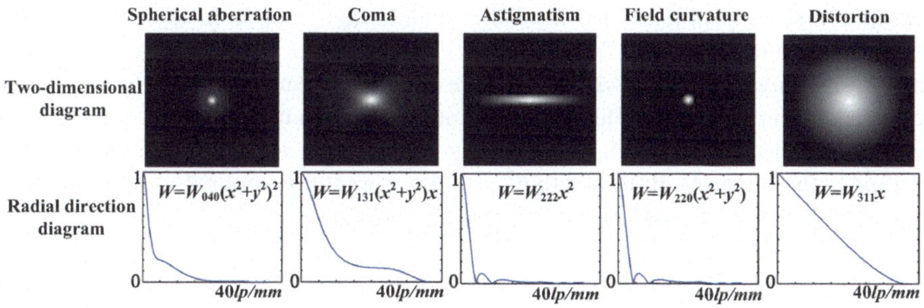

Fig. 1 MTF for different aberrations

We predict that the effective cooperation between appropriate aberration coefficient distribution and image processing algorithm can lead to the overall optimal solution.

4 Simulation Verification

In order to optimize the global performance of the whole system, we draw a general block diagram for the computing design methods as is shown in Fig. 2. We firstly utilize the optimization function of ZEMAX software to accomplish traditional optimal design [9]. Then we take the traditional optimal structural parameters as a

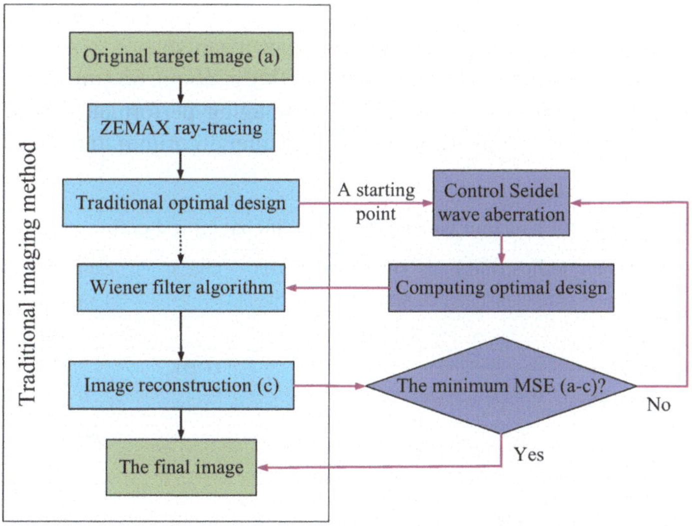

Fig. 2 A general block diagram for the computing design methods

reasonable starting point and control the Seidel wave aberration to search for the computing optimal design. Then Wiener filter algorithm is adopted to process the image simulation from ZEMAX software and the mean square error (MSE) is regarded as an evaluation criterion [10]. Finally in the comparison between the final images produced by the digital imaging system for the two different design methods, the theoretical foundations for the computing optical design method considering the MTF and aberrations can be verified.

4.1 Traditional Optical Design

We choose to start with designing simple rear meniscus lens. The singlet lens has the aperture stop 11 mm in front of the singlet. The thickness of the lens is 10 mm and the material is H-BK7 glass. A single wavelength of λ = 550 nm is used in this evaluation. The field of view is 30° and the object distance is 500 mm. The focal length of the optical system is 70 mm. The maximum image height equals to 18.8 mm. This is the point on the upper part of the image plane (y axis) to which the off-axis ray bundle is converging. The normalized image coordinate for this point is $u_0 = 0, v_0 = 1$. The maximum image height is associated with the maximum field of view for the system.

In order to utilize the automatic optimization feature provided by ZEMAX [11], the back focal length and curvature radius of the two surface is taken as the variable parameter and the default merit function is employed as optimization criteria. Through the automatic optimization, the traditional optical design structural parameters can be obtained as is shown in Table 1. The primary Seidel aberration coefficients as calculated by ZEMAX for this lens arrangement are listed in Table 2.

In the traditional optimal design, the root-mean-square size of the spot (SS-RMS) for the maximum field of view (FOV) is about 60 μm as is shown in Fig. 3. Considering singlet lens imaging, the imaging performance is acceptable. Then we input the image simulation into Wiener filter algorithm to realize the image

Table 1 The optimal lens structure parameters for traditional optical design

Surface	Curvature radius (mm)	Thickness (mm)	Glass	Semi-diameter (mm)
Object	Infinity	500.000		133.975
Aperture	Infinity	11.000		4.374
2	**−81.911**	10.000	H-BK7	7.327
3	−26.196	81.668		9.158
Image	Infinity			20.214

Table 2 Seidel coefficient values

Coefficient	W_{040}	W_{131}	W_{222}	W_{220} (M)	W_{311}
Value	4.3090λ	−0.1479λ	−1.5877λ	3.9481λ	34.0124λ

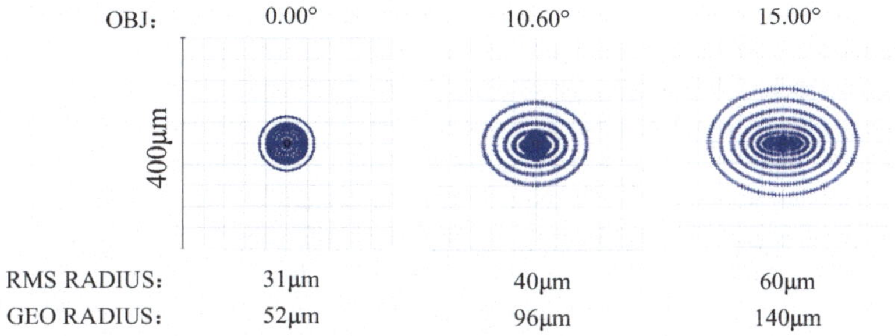

Fig. 3 SS-RMS diagram for traditional optimal design

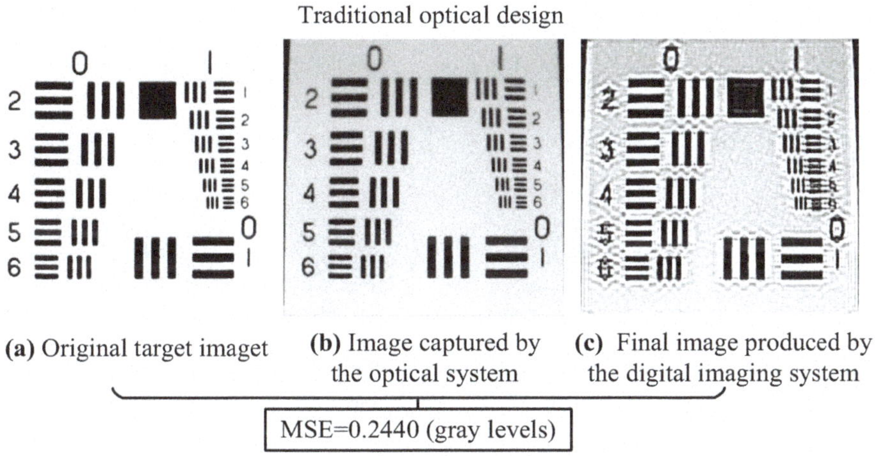

Fig. 4 The final image from traditional optical design

reconstruction. Through calculating the MSE between the final image produced by the digital imaging system and the original target image, the quality of the image can be measured and evaluated.

As shown in Fig. 4, we note that the final image from traditional optical design for the maximum FOV has higher contrast comparing with the image captured by the optical system, however, it also shows obvious ringing effect [12] that decreases image quality. In other words, the aberration coefficient distribution shown in Table 2 is not beneficial to digital filtering, resulting in information loss.

4.2 Integral Optical Design

In the integral optical design, we take the traditional optimal structural parameters as a reasonable starting point. In other words, we search for a better solution near the traditional optimal solution. We optimize the curvature radius of the front surface from −81.911 to −82.6 mm to search for the optimal aberration coefficient distribution (see Table 3).

In the integral optimal design, the SS-RMS for the maximum FOV is about 70 μm as is shown in Fig. 5, indicating that the optical imaging performance of integral optical design is worse than traditional optical design. However, the final image from integral optical design not only has higher contrast but also avoids the ringing effect, which has better image quality than the final image in Fig. 4. This shows that appropriate aberration coefficient distribution can reduce the pressure of image processing (Fig. 6).

Table 3 Seidel coefficient values

Coefficient	W_{040}	W_{131}	W_{222}	W_{220} (M)	W_{311}
Value	4.2644λ	-0.1473λ	-1.5597λ	3.9623λ	33.9847λ

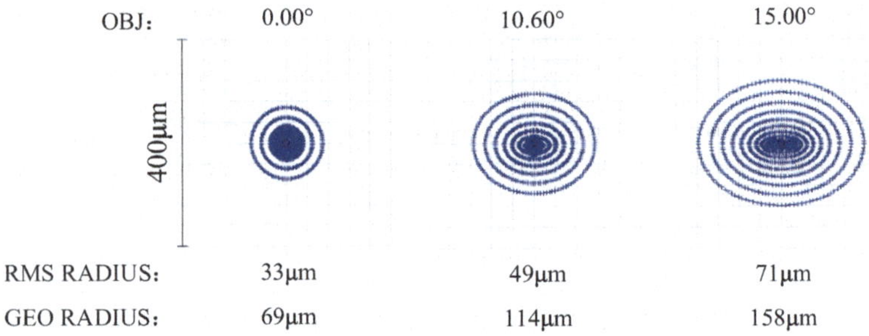

Fig. 5 SS-RMS diagram for traditional optimal design

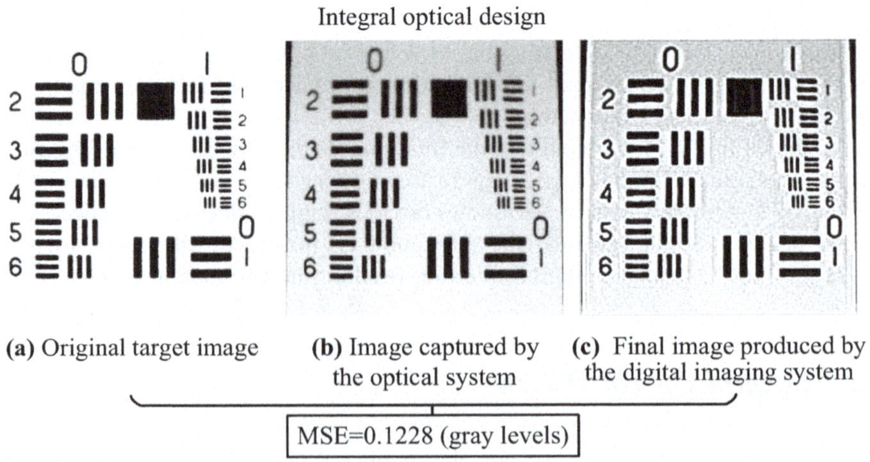

Fig. 6 The final image from integral optical design

5 Conclusion

In this paper, the theoretical foundations for computing optical design method is analyzed by considering MTF and aberrations. The property of MTFs that are due to equivalent severity among different optical aberrations is analyzed and we conclude that the effective cooperation between appropriate aberration coefficient distribution and image processing algorithm can lead to the overall optimal solution. The optical system with single lens is designed in ZEMAX to achieve simulation validation. The results have shown that an integral approach to optical system design can yield designs superior to the traditional methods. In the future, we will consider the influence of the target and detector to set up the whole information transmission model. We believe that this design method will provide a benefit for simple (and thus cheap) imaging systems and have a promising perspective of industrial application.

References

1. Cathey W T, Dowski E R. New paradigm for imaging systems [J]. Applied optics, 41(29): 6080–6092, (2002).
2. Johnson G E, Macon A K, Rauker G M. Computational imaging design tools and methods [C]. Optical Science and Technology, the SPIE 49th Annual Meeting. International Society for Optics and Photonics, 284–294, (2004).
3. Harvey A R, Vettenburg T, Demenikov M, et al. Digital image processing as an integral component of optical design [C]. Optical Engineering + Applications. International Society for Optics and Photonics, 706104-706104-11, (2008).
4. Robinson M D, Stork D G. Joint design of lens systems and digital image processing [C]. Contract Proceedings 2006. International Society for Optics and Photonics, 63421G-63421G-10, (2007).
5. Stork D G, Robinson M D. Theoretical foundations for joint digital-optical analysis of electro-optical imaging systems [J]. Applied Optics, 47(10): B64–B75, (2008).
6. David Voelz, Computational fourier optics [M]. SPIE press, 2011.
7. Liu Jifang, Modern optics [M]. Xidian university press, 2004.
8. Goodman J W. Introduction to Fourier optics [M]. Roberts and Company Publishers, (2005).
9. Fischer R E, Tadic-Galeb B, Yoder P R, et al. Optical system design [M]. New York: McGraw Hill, (2000).
10. Annadurai S. Fundamentals of digital image processing [M]. Pearson Education India, (2007).
11. Shixian L, Lenian Z. Optical Design Manual [M]. Beijing Institute of technology Publishers, 274, (1990).
12. Nasonov A V, Krylov A S. Scale-space method of image ringing estimation [C]. ICIP, 2793–2796, (2009).

Free-Form Surface Profilometry Based on Subtracting CMM Date from Enveloping Surface

Jianfeng Ren, Xiaojun Tang, Gang Wang, Qitai Huang, Yi Wang and Yin Ni

Abstract Freeform optical components (FOCs) can effectively simplify optical system configuration and improve optical preforms, they can be applied on imaging system, illumination system and phase compensation system. Scattered sagittal points reported by coordinate measuring machine (CMM) often exist uncertainty and makes contour error inaccurate. A new model which subtracts CMM date from enveloping surface calculated by theoretical equation includes tilt, rotation and shift. Experiments result shows this method introduced by this article could efficiently analysis scattered CMM data and PV (Peak-to-valley) less than 2 μm which matched accuracy of CMM (1.8+ L/300 μm) and accuracy of fabrication error of SPDT.

Keywords Freeform optical components · Enveloping surface · CMM · Optical profilometry

1 Introduction

Freeform optical components (FOCs) have great potential in the areas of military and spectacles markets [1–4]. However, the testing techniques for FOCs are relatively weak compared to their fabrications and thus restrict the production rates.

J. Ren (✉) · Q. Huang · Y. Wang · Y. Ni
College of Physics, Optoelectronics and Energy, Soochow University, Suzhou 215006, China
e-mail: jf.ren@suda.edu.cn

J. Ren · Q. Huang · Y. Wang · Y. Ni
Key Lab of Advanced Optical Manufacturing Technologies of Jiangsu Province, Soochow University, Suzhou 215006, China

J. Ren · Q. Huang · Y. Wang · Y. Ni
Key Lab of Modern Optical Technologies of Education Ministry of China, Soochow University, Suzhou 215006, China

X. Tang · G. Wang
Science and Technology on Solid-State Laser Laboratory, Beijing 100015, China

H.P. Urbach and G. Zhang (eds.), *3rd International Symposium of Space Optical Instruments and Applications*, Springer Proceedings in Physics 192,
DOI 10.1007/978-3-319-49184-4_19

The testing techniques, depending on the operations, are classified as contact and noncontact approaches [1]. Contact inspection methods are dominated by the Three Coordinate Measuring (CMM) techniques [5–9], while Moiré fringe projection [10], Shark-Hartman wave-front sensing technique [11]; Ronchi testing and interference methods play important roles on the noncontact measurements [12, 13]. The advantages and disadvantages of these techniques are well described by different authors [1–20]. Quicker, cheaper and reliable testing method is very important for manufacture and designer.

A new method was proposed, testing the FOCs with traditional CMM [1, 16, 18, 19]. The key technique is to cut down influence caused by uncertainty of CMM. To solve this problem, we get enveloping surface with sample point got from theoretical equation first and then compare testing point serious with enveloping surface includes tilt, rotation and shift. This is a quick and easy way to get precision testing result.

To ensure this method works and can get precision sagittal error distribution; we test a FOC part made by single-point diamond turning (SPTD) machine. Testing result shows we can get 1.5 μm PV (peak-to valley) error distribution matching to SPTD accuracy [1, 19]. This approach would be very useful for freeform optical components manufacturing and testing.

2 Principles of FOC Measuring

CMM is a device which can report position of probe [21]. Measurements are defined by a probe attached to moving axis. Figure 1 shows typical CMM machine including three moving axis. Each axis has a scale system that indicates the location of that axis [22]. The machine will read the position of touching probe, as directed by the operator or programmer. The machine then use the x y z coordinates of each of these points to determine size and position with micrometer precision. In fact, probe position stands for enveloping surface of the measured sample such as FOCs.

Figure 2 shows calculation process of sagittal error distribution of FOCs. Enveloping surface is calculated based on theoretical equation and then compare CMM testing result with enveloping surface involves tilt, rotation and coordinate shift.

2.1 Equation of Freeform Surface

Freeform surface means the surface do not rotational symmetry and can only described with complex formula. Extended polynomials (1) describes sagittal distribution of freeform surface [23]

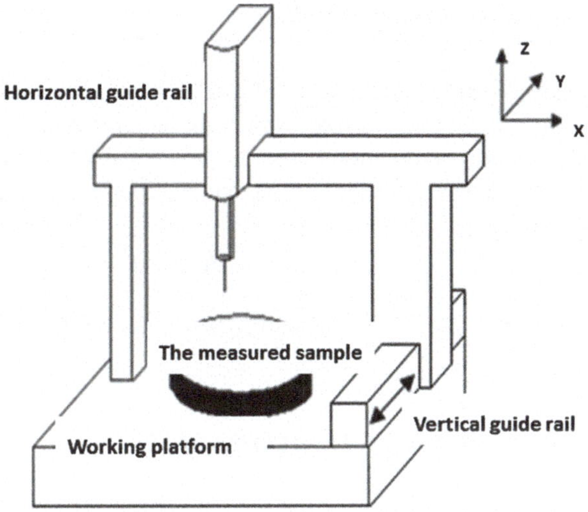

Fig. 1 Typical coordinate measuring machine

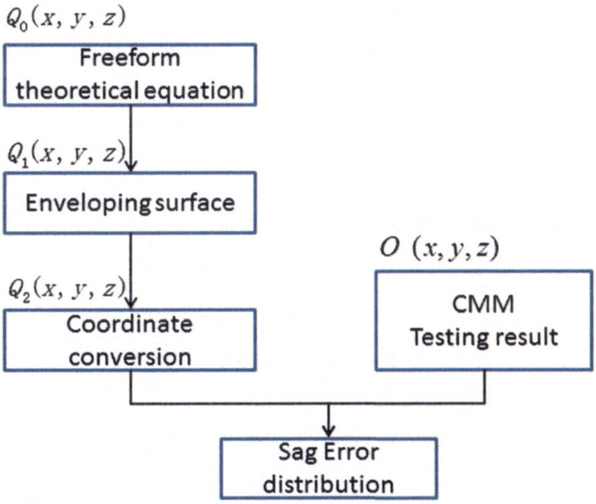

Fig. 2 Process of date model

$$z = \frac{cr^2}{1 + \sqrt{1 - (1+k)c^2r^2}} + \sum_{i=1}^{N} A_i E_i(x, y) \quad (1)$$

where Z is sagittal of point (x, y), c is curvature of the surface and k is conic of aspheric surface, N is the number of polynomial coefficients in the series and A_i is the coefficient on the ith extended polynomial term.

2.2 Enveloping Surface

Figure 3 shows relationship of touching point and probe center. Probe center (Q_1) on the enveloping surface. Any point on enveloping surface can be calculated with (2) and all computation based on theoretical equation of freeform surface (1). R is probe diameter and (a, b, c) is vector of Q_0.

$$\overrightarrow{Q_1 - Q_0} = a\vec{i} + b\vec{j} + c\vec{k} \tag{2}$$

where $R = \sqrt{a^2 + b^2 + c^2}$ and (a, b, c) can be derived from slope of Q_0.

2.3 Coordinate Transformations

Actually, we can't put FOCs on right position but we can get rid of position and rotation deviation during analysis process. Figure 4 shows how tilt, rotation and shift works on FOCs.

Enveloping surface includes rotation and shift can be described as

$$Q_2 = R_x R_y R_z Q_1 + (-\delta x, -\delta y, -\delta z) \tag{3}$$

where

$$R_x = \begin{bmatrix} 1 & 0 & 0 \\ 0 & \cos(\alpha) & -\sin(\alpha) \\ 0 & \sin(\alpha) & \cos(\alpha) \end{bmatrix} R_y = \begin{bmatrix} \cos(\beta) & 0 & \sin(\beta) \\ 0 & 1 & 0 \\ -\sin(\beta) & 0 & \cos(\beta) \end{bmatrix}$$

$$R_z = \begin{bmatrix} \cos(\gamma) & -\sin(\gamma) & 0 \\ \sin(\gamma) & \cos(\gamma) & 0 \\ 0 & 0 & 1 \end{bmatrix}$$

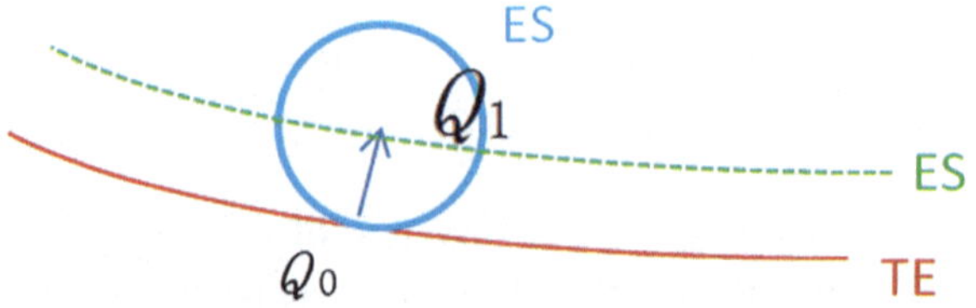

Fig. 3 Enveloping surface. TE stands for theoretical freeform surface, ES stands for enveloping surface, Q_0 is the touching point on TE and Q_1 is probe position on ES

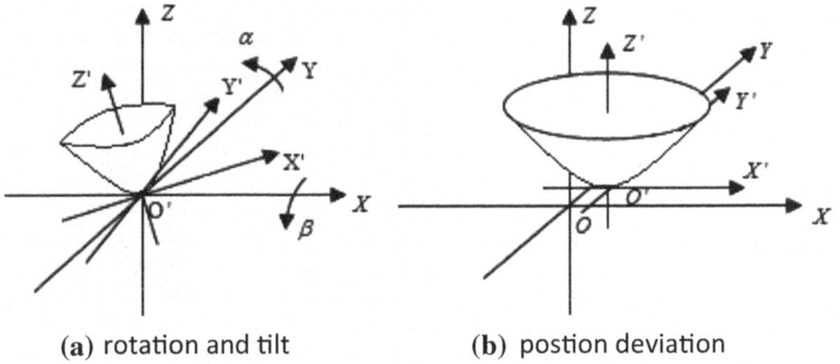

Fig. 4 Coordinate transformation. α is tilt angle of X axis, β is tilt angle of Y axis, γ is rotation angle of FOCs, $(\delta x, \delta y, \delta z)$ stands for shift of FOCs

2.4 Data Processing Model

Normally we need to fix $(\alpha, \beta, \gamma, \delta x, \delta y, \delta z)$ first and then calculate sagittal error distribution of FOCs. Equation (4) shows how to get PV (Peak-to-Valley), which describe sagittal error distribution information, from probe position and enveloping surface.

$$PV = \max(O_{(x,y,z)} - Q_{2(x,y,z)}) - \min(O_{(x,y,z)} - Q_{2(x,y,z)}) \tag{4}$$

where "max" stands for maximum value algorithm and "min" stands for Minimum value algorithm. Rotation angle and position shift $(\alpha, \beta, \gamma, \delta x, \delta y, \delta z)$ can't be measured directly but they can be determined by nonlinear Eq. (5)

$$MF = \sum (Q_{2i} - o_i)^2 \tag{5}$$

where Q_2 stands for enveloping surface calculated with (1)–(3), O stands for position of probe, i is normally over all points.

We optimize (5) with Matlab to get minimum MF and then calculate contour error distribution with (4) after getting coordinate rotation and shift $(\alpha, \beta, \gamma, \delta x, \delta y, \delta z)$.

3 Experiments

A FOC, made by 250UPL (SPDT, Single Point diamond Turning machine, made by Moor Nano technology System), is measured By CONTURA G2 (ZEISS, 1.8+ L/300 μm). Surface formula of freeform surface is $z = ax^3 + ay^3$, a = 1e − 5 and FOC diameter less than 15 mm.

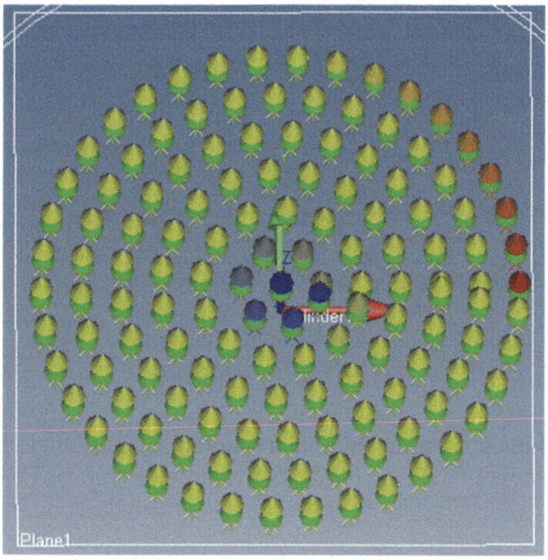

Fig. 5 Testing point array distribution

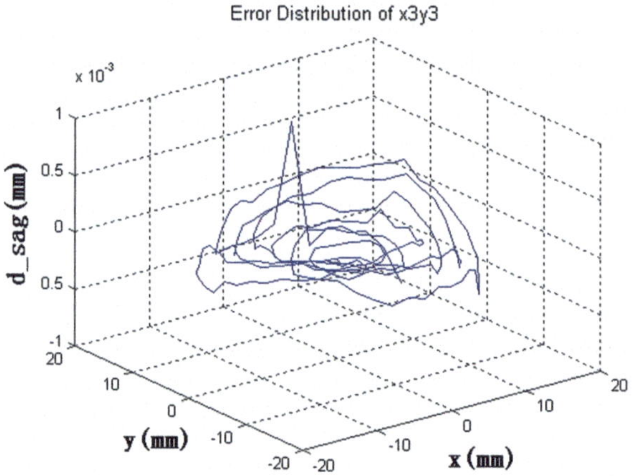

Fig. 6 Error distribution of FOC and its PV less than 2 μm

We plan testing path as Fig. 5 shows and less testing point helps to improve testing efficiency. We get rotation angle and position shift $(\alpha, \beta, \gamma, \delta x, \delta y, \delta z)$ with (5) and then calculate PV of FOCs with (4). Analysis result shows the PV of FOCs is 1.4 μm and Contour error distribution as Fig. 6 shows.

4 Conclusions

In summary, we propose a new method for surface profilometry based on CMM. Our method is general and only use testing value once which helps avoid brings uncertainty of measurements. Experiment result shows that PV is 1.4 μm and less than accuracy of CMM (1.8+L/300 μm) while matching accuracy of fabrication error of SPDT. Our studies might provide an easy solution to inspections of FOCs quickly.

Acknowledgements This work was supported by Science and Technology on Solid-State Laser Laboratory, China (Grant No. 9140C040103150C04018).

This work was also supported by the Project of Priority Academic Program Developments (PAPD) of Jiangsu Higher Education Institutions.

References

1. G. Qiu and C. Cui, "Fermat principle based reflector design for fast and contactless freeform optical component inspections," Opt. Lett. 38(18), 3510–3513 (2013).
2. W. Plummer, J. Baker, and J. Tassell, "Photographic optical systems with nonrotational aspheric surfaces," Appl. Opt. 38(16), 3572–3592 (1999).
3. X. Sun, Z. Zheng, and X. Liu, "Design of ultra-thin projection system with curved screen based on Zernike free-form surfaces," J. Zhejiang Uni. 43(8), 1428–1432 (2009).
4. O. Cakmakci et al., "Design of a free-from single-element head-worn display," Proc. SPIE 7618, 761803 (2010).
5. C. Cheung et al., "Measuring ultra-precision freeform surfaces using a robust form characterization method," Meas. Sci. Technol. 17(3), 488 (2006).
6. X. Jiang, X. Zhang, and P. Scott, "Template matching of freeform surfaces based on orthogonal distance fitting for precision metrology," Meas. Sci. Technol. 21(4), 045101 (2010).
7. C. Cheung et al., "Measuring ultra-precision freeform surfaces using a hybrid fitting and matching method," Meas. Sci. Technol. 20(10), 105103 (2009).
8. L. Kong et al., "Measuring optical freeform surfaces using a coupled reference data method," Meas. Sci. Technol. 18(7), 2252 (2007).
9. C. Cheung, L. Kong, and M. Ren, "Measurement and characterization of ultra-precision freeform surfaces using an intrinsic surface feature based method," Meas. Sci. Technol. 21 (11), 115109 (2010).
10. A. Arimoto et al., "Laser scanning system using a rotationally asymmetric aspheric surface," Appl. Opt. 30(6), 699–704 (1991).
11. W. Guo et al., "Adaptive centroid finding algorithm for freeform surface measurements," Appl. Opt. 52(10), D75–D83 (2013).
12. K. Cordero-Davila et al., "Appling Ronchi test to evaluate local and global surface errors without both approximations and integration," presented at Imaging and Applied Optics Technical Digest, Vol. 13, Paper JTu5A (2012).
13. L. Zhao et al., "Reference-free Shack–Hartmann wave-front sensor," Opt. Lett. 36(15), 2752–2754 (2011).
14. H. Shen et al., "Design and fabrication of computer-generated holograms for testing optical freeform surfaces," China. Opt. Lett. 11(3), 032201 (2013).
15. J. Martinez-Anton et al., "Enhancement of surface inspection by Moire interferometry using flexible reference gratings," Opt. Express 8(12), 649–654 (2001).
16. S. Li, S. Chen, and Y. Dai, "Inspection of free-form optics," Nanotechnology. Précis. Eng. 3 (2), 126–136 (2005).
17. D. Malacara, Optical Shop Testing, 3rd ed., Wiley, Dordrecht, Netherlands (2007).
18. Wencai Zhou et al., "Application of computed graphic holograph in testing the integrated wave-front coding unit," Proc. ICOEE, 2016.
19. Jianfeng Ren et al., "Study on four-step computer-generated hologram with same diffraction efficiency of zeroth and first order," Optical Engineering 50(8), 085801(August 2011).

20. Jianfeng Ren et al., " Design of original structure of illumination system in off-axis convex aspherical lens testing system with computer-generated hologram," ACTA OPTICA SINICA, 32(2), 0222005 (2012).
21. https://en.wikipedia.org/wiki/Coordinate-measuring_machine.
22. Jianming Wang et al., "Data processing of off-axis aspheric surface measurement by coordinate contour measuring machine", Optical Technique, 2013(4):291–296.
23. "Optical design program," Zemax Manual, June 9, 2009, 289.

On-Board Spectral Calibration for Chinese Medium Resolution Imaging Spectrometer

R.M. Fu, Y.X. Liu and M. Li

Abstract Spectral calibration is the basis for space-borne spectral imager to obtain accurately quantified spectral imaging data. The random vibrations during the launch process and the change of space environment will influence the veracity of ground spectral calibration. So the on-board spectral calibration is necessary to a space-borne spectral imager. Considering the specific mission requirements including the spectral range and 1 nm wavelength accuracy of Chinese Medium Resolution Imaging Spectrometer (CMERIS), the on-board spectral calibration is achieved by rare earth doped diffuser plate combining the Fraunhofer absorption lines and oxygen A-band. And the ground simulation was introduced in Sect. 3, which employs well-defined equations to calculate spectral calibration through data processing including unitary calculation, peak searching and regression analysis. Finally, the results of the simulation were verified by the standard lines of mercury lamp and the filter wavelengths. These results can satisfy the spectral calibration requirement of 1 nm wavelength accuracy.

Keywords Chinese medium resolution imaging spectrometer · On-board spectral calibration · Ground simulation

1 Introduction

Spectral imager is a kind of the optic remote sensing instrument, which can be used to obtain images and continuous spectral information at the same time [1].

R.M. Fu (✉) · Y.X. Liu · M. Li
Beijing Institute of Space Mechanics & Electricity, Beijing, China
e-mail: fu_ruimin@hotmail.com

Y.X. Liu
e-mail: 1225666088@qq.com

M. Li
e-mail: 508liming@sina.cn

H.P. Urbach and G. Zhang (eds.), *3rd International Symposium of Space Optical Instruments and Applications*, Springer Proceedings in Physics 192,
DOI 10.1007/978-3-319-49184-4_20

The accuracy calibration is important for spectral imagers because veracity and reliable quantified data is the basis for spectral imager's application. After launch and on-board activities, some characteristics of accessories of spectral imager will be changed due to the rough vibration and extreme outer-space environment. To obtain accurate spectral imaging data, on-board spectral calibration is considerable necessary for spectral imagers to revise the results of ground spectral calibration [2].

There are several on-board spectral calibration methods for a space-borne spectral imager. For example, MERIS [3] uses rare earth doped diffuser plate, Fraunhofer lines, oxygen A-band, Hyperion [4] uses Fraunhofer lines and atmospheric absorption lines, CHRIS [5] uses oxygen A-band and atmospheric absorption lines, OMI [6] uses atmospheric absorption lines, MOS-B [7] and MODIS [8] use filters and HISUI [9] uses filters and Mylar film. Space-borne spectral imagers usually use several on-board spectral calibration methods to improve the calibration accuracy by mutual complementation and verification.

CMERIS (Chinese Medium Resolution Imaging Spectrometer) has a spectral range from 400 to 950 nm with 1.25 nm spectral resolution and 1 nm wavelength accuracy. The spectral data is achieved by imaging via a dispersing grating onto a 2-D CCD array. Considering the mission requirements and characteristics of CMERIS, three on-board spectral calibration methods are chosen in Sect. 2 and the ground simulation is accomplished in Sect. 3. From ground simulation, the spectral calibration equation and uncertainty analysis are well modeled and conducted. Finally, to verify the precision of ground simulation results, the standard lines of mercury lamp test and the filter wavelengths test are introduced, the conclusion can be seen in Sect. 4.

2 The On-Board Spectral Calibration

2.1 System of On-Board Spectral Calibration

On-board calibration system of CMERIS consists of calibration disk, transmission mechanism, electromotor, resolver, light fender, lens hoods, earth window glass and other structures, which located at the front of the optical system. Calibration disk has 5 work positions: Earth diaphragm, radiometric calibration diffuser plate (main), radiometric calibration diffuser plate (backup), spectral calibration diffuser plate and shade position (see Fig. 1).

Radiometric calibration diffuser plate is made of Teflon, and spectral calibration diffuser plate is made of rare earth doped Teflon. From 400 to 950 nm, the reflectance of radiometric calibration diffuser plate is steady and high reflectance (95%). And the spectral calibration diffuser plate has 6 absorption peaks: 407.7, 451.9, 489.3, 521.4, 653.7 and 798.6 nm (see Fig. 2).

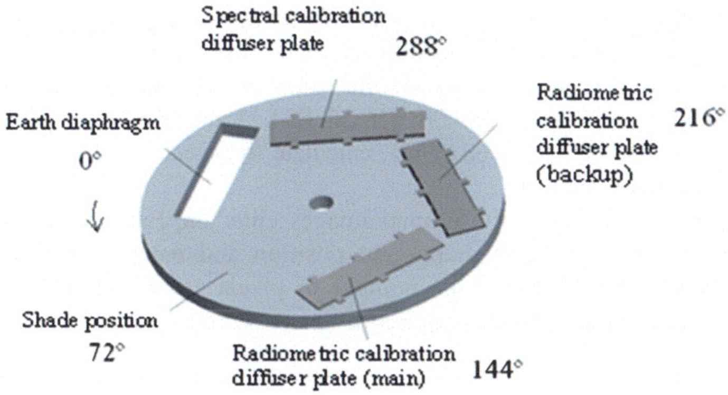

Fig. 1 Five positions of the calibration disk

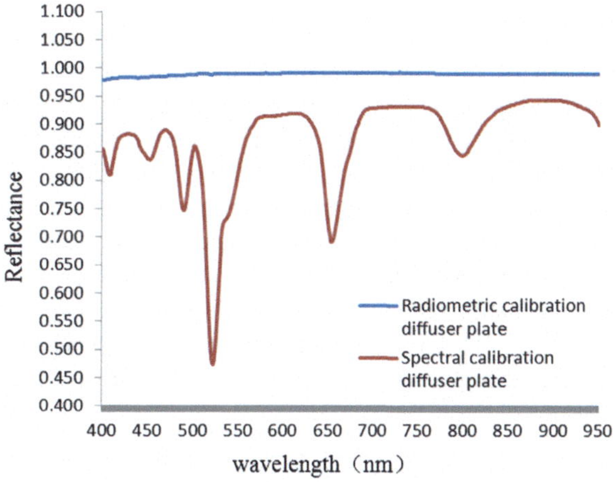

Fig. 2 Reflectance of radiometric calibration diffuser plate and spectral calibration diffuser plate

2.2 *Procedure of On-Board Spectral Calibration*

Three methods, rare earth doped diffuser plate, Fraunhofer lines and oxygen A-band, are chosen to achieve on-board spectral calibration by CMERIS.

(a) Rare earth doped diffuser plate
When sunlight enters the solar window, turns the calibration disk to radiometric calibration diffuser plate position and gets the first frame of data. Then, turn the calibration disk to spectral calibration diffuser plate position and get the second frame of data. The corresponding relation between the number of pixels and absorption peaks by contrasting the two frames of data near absorption peaks of spectral calibration diffuser plate can be perceived. The last step, calculate the spectral calibration equation using regression analysis.

(b) Fraunhofer lines When sunlight enters the solar window, turns the calibration disk to radiometric calibration diffuser plate position and acquires data. The corresponding relation between the number of pixels and Fraunhofer absorption lines (486.13, 589.00, 656.28, 854.21 and 866.22 nm) through data processing can be perceived. Then, calculate the spectral calibration equation with regression analysis.
(c) Oxygen A-band When the ground images enter the Earth window turn the calibration disk to Earth diaphragm position and receive data. The corresponding relation between the number of pixels and Oxygen A-band (near 760 nm) can be accurately acquired by peak searching.

2.3 *Uncertainty Analysis of On-Board Spectral Calibration*

The influence factors of calibration precision can be found through analysing the process of on-board spectral calibration. The influence factors include: (a) the uncertainty of absorption peak positions got by ground calibration as σ_a; (b) the uncertainty of wavelength positions which induced by the temperature changing on board as σ_b; (c) the shift of wavelength positions on focal plane which induced by the temperature changing on board as σ_c; (d) the uncertainty of peak searching as σ_d; (e) the uncertainty of regression analysis as σ_e.

Table 1 shows the estimate of every influencing factor. The uncertainty σ of on-board spectral calibration can be calculated by Eq. (1).

By Eq. (1), the uncertainty of on-board spectral calibration is ±0.60 nm, which satisfies the spectral calibration requirement of 1 nm wavelength accuracy.

$$\sigma = \sqrt{\sigma_a^2 + \sigma_b^2 + \sigma_c^2 + \sigma_d^2 + \sigma_e^2} \quad (1)$$

3 The Ground Simulation

The ground simulation of rare earth doped diffuser plate is been conducted during our experiment. The light of solar simulator enters the solar window with the same angle of on-board spectral calibration (see Fig. 3). First, turn the calibration disk to shade position and record the data of every pixel on CCD as D_{ij}, which is dark current. Second, turn the calibration disk to spectral calibration diffuser plate position and record the data A_{ij}. Then, turn the calibration disk to radiometric

Table 1 Uncertainty of on-board spectral calibration

Influence factor	σ_a	σ_b	σ_c	σ_d	σ_e
Uncertainty (nm)	0.45	0.10	0.04	0.33	0.19

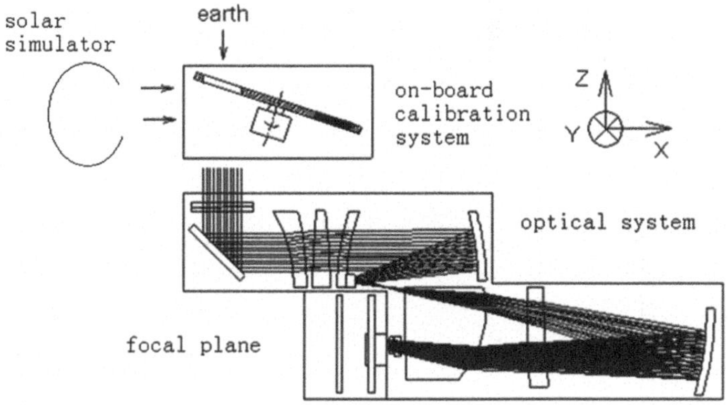

Fig. 3 Ground simulation of rare earth doped diffuser plate

calibration diffuser plate position and record the data B_{ij}. Processing the data as Eq. (2), the ratio C_{ij} of spectral calibration diffuser plate and radiometric calibration diffuser plate without noise can be calculated.

$$C_{ij} = \frac{A_{ij} - D_{ij}}{B_{ij} - D_{ij}} \tag{2}$$

i is the number of row meaning the number of pixels in spectral dimension; *j* is the number of array meaning the number of pixels in spatial dimension.

Make $j = 500$, the corresponding relation of C_{ij} and the number of pixels in spectral dimension at central field (see Fig. 4) can be perceived, which is demonstrated in Fig. 4, the 6 absorption peaks of spectral calibration diffuser plate is obvious.

For every spatial dimension ($1 < j < 1024$), repeatedly matching the peak wavelengths to the number of pixels by peak searching at the 6 absorption peak positions and using the regression analysis to the data groups of the peak wavelengths and the number of pixels by least squares method, the spectral calibration equation of CMERIS can be calculated.

Gaussian fitting is chosen to search the peaks. Equation (3) is Gaussian function.

$$S_j(X) = K \exp\left[\frac{1}{2}\left(\frac{X - X_j}{\sigma_j}\right)^2\right] \tag{3}$$

X is the number of continuous pixels near one absorption peak at array *j*; S_j is the unitary result of *X*; *K*, X_j and σ_j are solved-for parameters expressing as peak value, peak position and peak width of Gaussian curve.

We can get the data groups of peak positions and peak wavelengths [X_{ij}, Y_{ij}] of the absorption peaks S_j by peak searching. In the data groups, X_{ij} is the pixel number of absorption peak, and Y_{ij} is the peak wavelength of absorption peak.

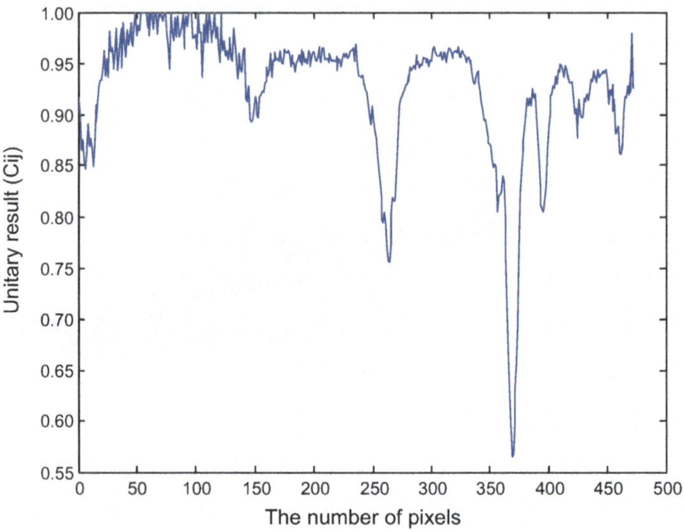

Fig. 4 Unitary result at central field

Table 2 Absorption peaks of spectral calibration diffuser plate and the number of pixels at central field

Peak wavelengths (nm)	407.7	451.9	489.3	521.4	653.7	798.6
Number of pixel	460.641	425.517	395.216	369.610	262.659	147.291

Make $j = 500$, to get the absorption peaks of spectral calibration diffuser plate and the number of pixels at central field (see Table 2).

Due to the low sample points towards the longer wavelength, the precision of spectral calibration equation may not meet the requirement standard at longer wavelengths. Thus, the ground simulation of Oxygen A-band and Fraunhofer lines is introduced as supplements.

At the ground simulation of Oxygen A-band and Fraunhofer lines experiments, the data is acquired when sunlight enter the solar window with the specific radiometric calibration diffuser plate position being conducted for the calibration disk. The corresponding relationship between the number of pixels and Oxygen A-band (760.0 nm) and Fraunhofer absorption lines (486.13, 589.00, 656.28, 854.21 and 866.22 nm) through data processing can be perceived (see Table 3). Since the bandwidth of Fraunhofer lines is narrow that the precision can only reach half pixel.

The model of spectral calibration equation is built by polynomial fitting as Eq. (4). Least squares method is chosen to make regression analysis to the data groups of Tables 2 and 3 as Eq. (5).

Table 3 Oxygen A-band, Fraunhofer lines and the number of pixels at central field

Peak wavelengths (nm)	760.0	486.13	589.00	656.28	854.21	866.22
Number of pixel	177.729	398.0	315.5	261.0	102.0	92.5

$$Y_{ij} = a_{0j} + a_{1j}X_{ij} + a_{2j}X_{ij}^2 + \cdots + a_{nj}X_{ij}^n \tag{4}$$

$$\begin{pmatrix} N & \sum_{i=1}^{N} X_{ij} & \cdots & \sum_{i=1}^{N} X_{ij}^k & \cdots & \sum_{i=1}^{N} X_{ij}^n \\ \sum_{i=1}^{N} X_{ij} & \sum_{i=1}^{N} X_{ij}^2 & \cdots & \sum_{i=1}^{N} X_{ij}^{k+1} & \cdots & \sum_{i=1}^{N} X_{ij}^{n+1} \\ \vdots & \vdots & \ddots & \vdots & \ddots & \vdots \\ \sum_{i=1}^{N} X_{ij}^k & \sum_{i=1}^{N} X_{ij}^{k+1} & \cdots & \sum_{i=1}^{N} X_{ij}^{2k} & \cdots & \sum_{i=1}^{N} X_{ij}^{n+k} \\ \vdots & \vdots & \ddots & \vdots & \ddots & \vdots \\ \sum_{i=1}^{N} X_{ij}^n & \sum_{i=1}^{N} X_{ij}^{n+1} & \cdots & \sum_{i=1}^{N} X_{ij}^{n+k} & \cdots & \sum_{i=1}^{N} X_{ij}^{2n} \end{pmatrix} \begin{pmatrix} a_{0j} \\ a_{1j} \\ \vdots \\ a_{kj} \\ \vdots \\ a_{nj} \end{pmatrix} = \begin{pmatrix} \sum_{i=1}^{N} Y_{ij} \\ \sum_{i=1}^{N} X_{ij}Y_{ij} \\ \vdots \\ \sum_{i=1}^{N} X_{ij}^k Y_{ij} \\ \vdots \\ \sum_{i=1}^{N} X_{ij}^n Y_{ij} \end{pmatrix} \tag{5}$$

a_{ij} (i = 0, 1, ..., n) are the coefficients of spectral calibration equation; N is the number of absorption peaks and N = 12 in this equation.

Table 4 shows spectral wavelengths of different polynomial fitting at the same pixel number. The difference of results can be neglected when the degree of polynomial reaches 3. Make n = 3, we can get the spectral calibration equation at central field as Eq. (6) and Fig. 5.

$$Y_{500} = 982.216 - 1.257X_{500} + 4.740 \times 10^{-5}X_{500}^2 - 5.526 \times 10^{-8}X_{500}^3 \tag{6}$$

Y is the value of wavelength; X is the row number of pixel, from 1 to 472; subscript 500 expresses the 500th array of spatial dimension corresponding to central field.

Table 4 Spectral wavelengths of different polynomial fit at central field

Degree of polynomial	Number of pixel (nm)						
	150	200	250	300	350	400	450
1	794.634	732.396	670.157	607.919	545.680	483.442	421.204
2	794.631	732.375	670.128	607.891	545.663	483.444	421.235
3	794.557	732.285	670.084	607.913	545.730	483.495	421.164
4	794.527	732.264	670.069	607.924	545.701	483.451	421.148

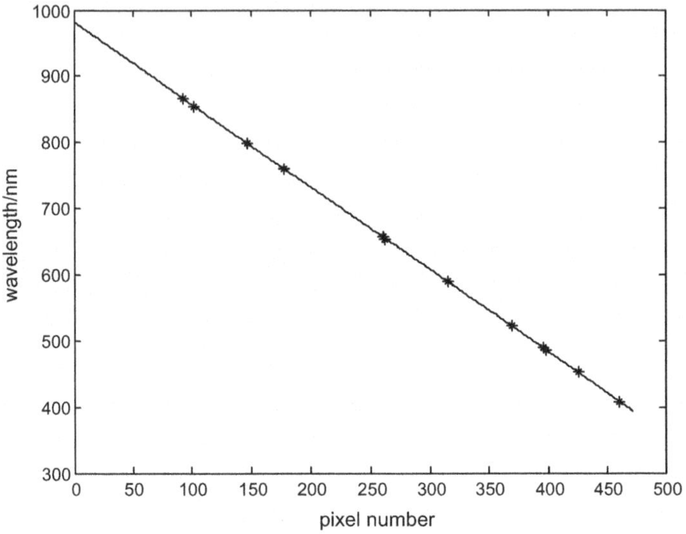

Fig. 5 Spectral calibration result at central field

Table 5 Comparison result of the mercury lamp standard lines at central field

Number of pixel	Standard lines of mercury lamp (nm)	Calibration wavelength (nm)	Error (nm)
463.325	404.66	404.532	0.128
437.927	435.84	436.225	0.385
349.420	546.07	546.452	0.382
324.387	576.96	577.488	0.528

Table 6 Comparison result of the filter wavelengths at central field

Number of pixel	Filter wavelengths (nm)	Calibration wavelength (nm)	Error (nm)
430.131	445.6	445.947	0.347
357.448	536.0	536.464	0.464
318.263	535.4	535.204	0.196
246.967	673.6	673.356	0.256
194.011	739.5	739.740	0.240
139.767	807.4	807.315	0.085
88.996	870.6	870.691	0.091

4 Verification of Precision

At the wavelength range from 400 to 950 nm, 4 standard lines of mercury lamp and 7 filter wavelengths are chosen to verify the precision of spectral calibration equation. The most error is less than 0.6 nm at central field (see Tables 5 and 6).

5 Conclusion

Considering the mission requirements and characteristics of CMERIS, rare earth doped diffuser plate combining the Fraunhofer absorption lines and oxygen A-band are chosen to achieve the on-board spectral calibration. From the ground simulation, the equation of spectral calibration was calculated. And, the precision of ground simulation was verified by the standard lines of mercury lamp and the filter wavelengths experiments, which has proved that this method satisfies the spectral calibration requirement of 1 nm wavelength accuracy. Furthermore, this ground simulation also validated the feasibility of rare earth doped diffuser plate in on-board spectral calibration, which can offer technical reference for follow-up applications.

References

1. 崔敦杰. 成像光谱仪的定标[J]. 遥感技术与应用, 1996, 11(3): 56–64.CUI Dunjie. "Calibration of Imaging Spectrometer"[J]. *Remote Sensing Technology and Application*, 1996, 11(3): 56–64. (in Chinese).
2. 李晓辉, 颜昌翔. 成像光谱仪星上定标技术[J]. 中国光学与应用光学, 2009, 2(4): 309–315. LI Xiaohui, YAN Changxiang. "Onboard Calibration Technologies for Hyper-spectral Imager"[J]. *Chinese Journal of Optics and Applied Optics*, 2009, 2(4): 309–315. (in Chinese).
3. S. Delwart, R. Preusker, L. Bourg, R. Santer, D. Ramon, J. Fischer. "MERIS In-flight Spectral Calibration'[J]. *International Journal of Remote Sensing*, 2007, 28: 49–496.
4. Barry P S, Shepanski J, Segal C. "Hyperion On-orbit Validation of Spectral Calibration Using Atmospheric Lines and an On-board System"[J]. Proc. of SPIE, 2002, 4480: 231–235.
5. M. A. Cutter, D. R. Lobb, T. L. Williams, R. E. Renton. "Integration & Testing of the Compact High Resolution Imaging Spectrometer (CHRIS)"[J]. Proc. of SPIE, 1999, 3753: 180–191.
6. Dobber M R, Dirksen R J, Levelt P F, et al. Ozone Monitoring Instrument Calibration[J]. IEEE, 2006, 44(5): 1209–1238.
7. G. Zimmermann, A. Neumann, H. Siimnich, H. Schwarzer. 'MOS/PRIROD-An Imaging VIS/NIR Spectrometer for Ocean Remote Sensing'[J]. Proc. of SPIE, 1993, 1937: 201–206.
8. X. Xiong, N. Che, Y. Xie, et al. "Four-years On-orbit Spectral Characterization Results for Aqua MODIS Reflective Solar Bands"[J]. Proc. of SPIE, 2006, 6361: 63–69.
9. Kenji Tatsumi, Nagamitsu Ohgi, Hisashi Harada, et al. "Onboard Spectral Calibration for the Japanese Hyper-spectral Sensor"[J]. Proc. of SPIE, 2010, 7826: 782625.

Calibration Method of High Spectral Infrared Atmospheric Sounder Onboard FY-3D Satellite

Chengli Qi, Mingjian Gu, Chunqiang Wu and Xiuqing Hu

Abstract High spectral infrared sounders possess distinct advantages that were exhibited in application of atmospheric temperature and humidity profiles retrieval, numerical weather prediction (NWP), climate change study and trace gas measurement due to high spectral resolution, low radiometric noise and high spectral and radiometric accuracy. A high spectral infrared atmospheric sounder (HIRAS) that was designed and manufactured completely by China will be carried onboard FY-3D and it is a sign of new age of infrared high spectral vertical sounding system in China. HIRAS is a Fourier interferometer and high radiometric precision is required for better application of data and also because radiometric noise and calibration errors are magnified in the inversion process to derive atmospheric parameters. Noise equivalent-temperature errors and calibration precision of the order of 0.1 °C and absolute errors of less than 0.3 °C are required to obtain temperature profile with rms errors of less than 1 °C. To develop calibration algorithm software and validate instrument data, complex calibration method is applied which using Cold Screen of 17 K and Internal Calibration Target of 313 K as calibration reference source and based on engineering model instrument vacuum calibration experiment data. Preliminary calibration results revealed that a strong absorbing peak at 800 cm^{-1} and need deep investigation. For FOV1, 2 and 4 the bias are less than 1 K for wide spectral range and are about 1–2 K just in location of less than 700 cm^{-1}. FOV3 reflects bias of 2–4 K in whole LWIR band and need more data to verify it. Future work will be focused on non-linearity and spectral correction and the radiometric bias can be expected reach less than 0.7 K.

Keywords FY-3D · HIRAS · Infrared calibration

C. Qi (✉) · C. Wu · X. Hu
National Satellite Meteorological Center, Beijing, China
e-mail: qicl@cma.gov.cn

M. Gu
Shanghai Institute of Technical Physics (SITP),
Chinese Academy of Sciences (CAS), Shanghai, China

H.P. Urbach and G. Zhang (eds.), *3rd International Symposium of Space Optical Instruments and Applications*, Springer Proceedings in Physics 192,
DOI 10.1007/978-3-319-49184-4_21

1 Introduction

High spectral Infrared Atmospheric Sounder (HIRAS) is a new infrared sounding instrument which will be carried onboard FY-3D satellite to replace IRAS and be launched in end of 2016. HIRAS was designed and manufactured completely by Shanghai Institute of Technical Physics (SITP), Chinese Academy of Sciences (CAS) and it is a sign a new age of vertical sounding system due to its distinct advantages of high spectral resolution, low radiometric noise and high spectral and radiometric accuracy. HIRAS is also a Fourier transform spectrometer (FTS) and is the third FTS for operational meteorological application that follows the Infrared Atmospheric Sounding Interferometer (IASI) on the European Metop-A/B satellite and the Cross-Track Infrared Sounder (CrIS) on board the Suomi-NPP operational environmental satellite [1]. HIRAS provides measurements of radiance spectra of 1370 channels in the three spectral bands: the long wave IR (LWIR) band from 650 to 1136 cm^{-1}, middle wave IR 1(MWIR1) band from 1210 to 1750 cm^{-1}, and middle wave IR 2 (MWIR2) band from 2155 to 2550 cm^{-1}, with spectral resolution of 0.625, 1.25, 2.5 cm^{-1}, respectively. The HIRAS instrument measures interferograms which then converted to radiance spectra by the ground pre-processing system. HIRAS includes a beamsplitter, a porch swing moving mirror, a stationary mirror, and a metrology system. The interferogram measurements are double sided with optical path difference (OPD) varying from a negative maximum path difference (MPD) to a positive MPD through the zero path difference (ZPD). A 852 nm laser metrology system is used for accurately sampling an interferogram at equal intervals in OPD. The laser metrology injects a laser beam into the center of the interferometer optical path and metrology detector convert the modulated laser fringes into electrical pulses used as interferogram sampling signals. Radiance is modulated through interferometer and then converged by the telescope onto the LWIR, MWIR1, MWIR2 detector focal planes through the Aft optics. Four detectors on each focal planes are arranged into a 2×2 grid which define the field of regard (FOR). CrIS digitized interferograms are further processed by a complex finite impulse response (FIR) digital band-pass filter to reject out-band signals and noise [2], also the complex interferogram is compressed with a bit trim encoding scheme. Due to lack of feasible onboard data decimation technique so HIRAS transfers all the original inteferogram samples to ground, interferogram data also using a bit trim compressing method.

HIRAS measurement scan sequence consists of 33 interferogram sweeps, including 29 Earth View (ES), 2 Deep Space (DS), 2 Internal Calibration Target (ICT) measurements. One complete scan sequence takes 10 s. View angle of each detector (Field of View, FOV) is 1.1° corresponding the nadir spatial resolution of 16 km (Fig. 1). To compensate the in-track spacecraft motion, the scan mirror rotates slightly along the in-track axis so that a FOV footprint on the earth surface is fixed during the measurement. Detailed instrument characteristics and spectral band channels and sensitivity are listed in Tables 1 and 2 [3].

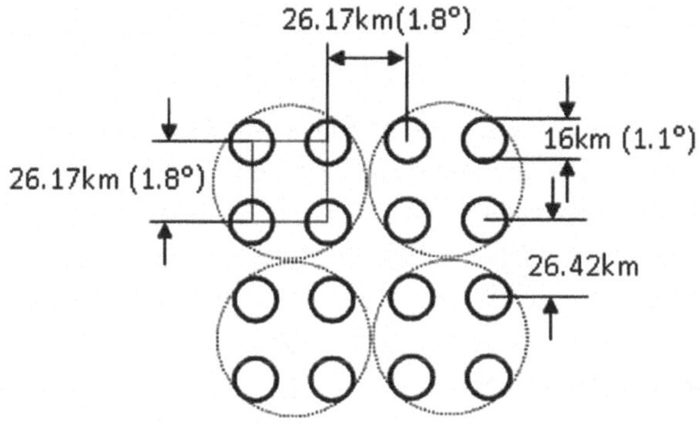

Fig. 1 Ground footprints of the FOVs arrays

Table 1 FY-3D/HIRAS characteristics

Parameters	Specification
Scan period	10 s
View angle	1.1°
Pixels per scan line	58
Scan angle	±50.4°
Radiative calibration accuracy	0.7 K
Spectral calibration accuracy	7 ppm
Direction pointing bias	<±0.25°

Table 2 FY-3D/HIRAS spectral characteristics

Band	Spectral range (cm^{-1})	Spectral resolution (cm^{-1})	Sensitivity (NEΔT@250 K)	No of channels
LWIR	650*–1136 (15.38–8.8 μm)	0.625	0.15–0.4 K	778
MWIR1	1210–1750 (8.26–5.71 μm)	1.25	0.1–0.7 K	433
MWIR2	2155–2550 (4.64–3.92 μm)	2.5	0.3–1.2 K	159

* Spectrum range from 650 cm^{-1} but the specification evaluation range from667 cm^{-1}

Radiometric calibration is the process of assigning absolute radiance to the spectral axis. The instrument should be linear in energy. Therefore a non linearity correction step is performed on the interferograms before the raw spectral are computed. Then two radiation calibration sources of Internal Calibration Target (ICT), a high-precision blackbody and Deep Space (DS), a source with negligible infrared radiance are used for radiometric calibration. A classic radiometric calibration method for high spectral infrared interferometer is complex calibration described in [4] and applied in current two in orbit high spectral sounders of

NPP/CrIS and Metop/IASI. Interferogram is not mirror-symmetrical about the ZPD point which will getting a non-zero imaginary component out of the Fourier transfer and combined a complex spectra. The asymmetry originates from two different sources, either extrinsic or intrinsic. Extrinsic sources may arise from sampling errors. Intrinsic asymmetry may be due to wavenumber-dependent phase delays of either the optics or the electronics. Instrument phase, sometimes called "phase error" is more of a characteristic than an error as such.

2 Radiotric Calibration

A. *Calibration principle and method*

The recorded interferogram is the sum of the interferogram of the viewed scene (the first port) and the interferogram corresponding to the self-emission (the second port) of the instrument. The detected signal is the sum of the contribution from the two ports. The second optical port of the interferometer is 180° out of phase with respect to the first port which also calls a dual phase response. It is important to take into account other-than-scene contributors to the measured signal, mainly the instrument self-emission, also called self-radiance that is emitted principally by the beamsplitter, which undergoes phase shift different from the emission from the scene which goes through the whole optical chain of the interferometer. For a perfectly compensated interferometer with a second port at 0 K, the observed signal would correspond to the pure scene, but in practice, the term corresponding to the instrument self-emission is not null. This explains IGM reversal when scene becomes colder than the terminator of the second port (complex calibration takes care of this effect). Figure 2 shows typical interferograms of blackbody view, deep space view and earth view, IGM of cold space view is reversal of the other two object views.

The basic mathematical relation between interferograms and un-calibrated spectra is the Fourier transform. As the measured interferogram is not symmetrical (because of dispersions effects in the beamsplitter and electronics, among other

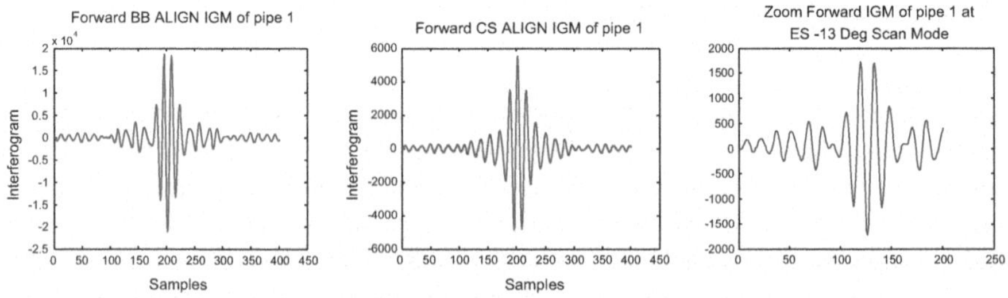

Fig. 2 Typical interferograms of blackbody view (*left*), deep space view (*central*), earth view (*right*)

contributors), the resulting spectrum will be complex. Prior to perform Fourier transform process, 30 ICT and DS IGMs that sweeps in the same direction should be collected and averaged.

The basic approach for determining absolute radiance measured by a FTIR spectrometer is the same as that used for filter radiometers and has been used successfully for other interferometric applications. The detectors and electronics are designed to yield an output which is linear with respect to the incident radiance for all wavenumbers in the optical bands of the instrument, and two reference sources are viewed to determine the slope and offset which define the linear instrument response at each wavenumber. The basic mathematical relation between interferograms and spectra is the Fourier transform. The general relationship between an interferogram and its equivalent spectrum over each IR band can be expressed as [4, 5]:

$$C'_v = r_v\{L_v + L_v^0 \exp[i\varphi^0(v)]\}\exp[i\varphi(v)] \tag{1}$$

where C' denotes the un-calibrated spectra and r_v is the overall spectral responsivity of the instrument referred to as slope which is complex. L_v is the true incident spectral radiance from the scene and L_v^0 is the instrument emission that does not come from the scene. $\varphi(v)$ is the phase of scene and $\varphi^0(v)$ is the phase difference of self-emission with scene. Applied equation to ICT and DS can get:

$$C'_{hv} = r_v\{B_v(T)_h + L_v^0 \exp[i\varphi^0(v)]\}\exp[i\varphi(v)] \tag{2}$$

$$C'_{cv} = r_v\{B_v(T_c) + L_v^0 \exp[i\varphi^0(v)]\}\exp[i\varphi(v)] \tag{3}$$

where $B_v(T_h)$ is blackbody Planck radiance, ICT temperature is the average of six temperature sensor of platinum resistance thermometer (PRT), $B_v(T_c)$ is radiance that sensed by detector when observing object is DS, the radiance is zero when instrument is in status of on-orbit operation, while it is calculated from temperature of He screen when in ground test lab. According Eqs. (2) and (3) can get slope and offset:

$$r_v = |C'_{hv} - C'_{cv}|/[B_v(T_h) - B_v(T_c)] \tag{4}$$

$$L_v^0 \exp[i\varphi^0(v)] = \frac{C'_{hv}\exp[-i\varphi^0(v)]}{r_v} - B_v(T_h) \tag{5}$$

When applied to observing object, the earth view scene radiometric calibration equation is:

$$L_v = \mathrm{Re}[(C'_v - C'_{cv})/(C'_{hv} - C'_{cv})][[B_v(T_h) - B_v(T_c)]] + B_v(T_c) \tag{6}$$

The calibration algorithm removes phase dispersion of the sensor over each IR band. Once removed, the complex spectral data is thus remapped into a cosine transform where the imaginary portion of the spectra can be discarded for a 1.41 noise factor improvement. Only the cosine transform data (real part of the calibrated spectrum) is delivered to the user production application algorithm. The sin transform data (imaginary part of the calibrated spectrum) can also be used to estimate noise [6].

B. *Vacumm testing data*

HIRAS engineering model was manufactured and integrated during 2015 and performed thermal vacumm (TVAC) testing in Jan 2016, the primary objective of this test system was to establish transfer relation of detected counts and radiance and characterize key performance parameters of instrument under simulated nominal and extreme mission conditions and so can identify any shortcomings or issues in the design. This allowed any issues that were found in engineering model to be addressed prior to finalizing the HIRAS flight model design, greatly reducing overall program risk. Another objective of the test program was to develop and check out the test support equipment needed to test the HIRAS flight unit.

In a vacumm infrared calibration test, the instrument viewed an External Calibration Target (ECT) whose temperature changed from 190 to 340 K. HIRAS operated in focus and scan models in tests. Focus model observing data was used mainly to assess instrument noise (NEdN) and Instrument Line Shape (ILS) and scan model observing data was used mainly to develop calibration method and pre-processing prototype software. In this study 260 K scan model ECT observation data of only LWIR band interferogram was used during large noise in MWIR1 and MWIR2 bands. Effective improvements are expected in flight model manufacture and high precision all bands data can acquired.

C. *Quality control of data*

Original interferogram (IGM) may occurs some kinds of noise or errors that from fringe count error or noise spike. Data quality control scheme mainly includes three steps, first performs noise spike detection and correction by using a predefined impulse noise mask, as a correction, each detected spike will be substituted by mean of interferogram. Secondly aligning of IGMs of calibration source, ICT views IGMs and DS views IGMs are conducted through spectra correlation method. The last quality control technique is for aligning of earth view IGMs with that of ICT view and DS view.

3 Calibration Results

A total measured interferogram and the corresponding complex spectrum have two components; a normal phase component for radiance from the source and an anomalous phase component for radiance from the instrument itself. Of course, the radiance from the instrument can originate from many individual components with different phase characteristics, but they can be lumped together into one term with one phase. The complex calibration analysis is a generalization which properly accounts for a dual phase response. Long wave band phase spectra shown in Fig. 3. The phase spectra shows a parabolic curve that changed with spectra axis, uncalibrated spectra appears a strong absorbing peak at 800 cm^{-1} which may be related with some kind of absorbing gas and was still in the investigation.

Longwave band calibrated spectrum derived from above calibration method is shown in Fig. 4. The spectrum is presented as a brightness temperature to make any errors stand out as a deviation from the measured blackbody temperature of 260 K. For FOV1, 2 and 4 the bias are in range of less than 1 K for spectral location of 700–1100 cm^{-1} and are about 1–2 K for spectral range of less than 700 cm^{-1}, while for FOV 4 the bias sharply increased to more than 2 K. FOV3 presents bias of 2–4 K in whole LWIR band which may need more cases to verify it.

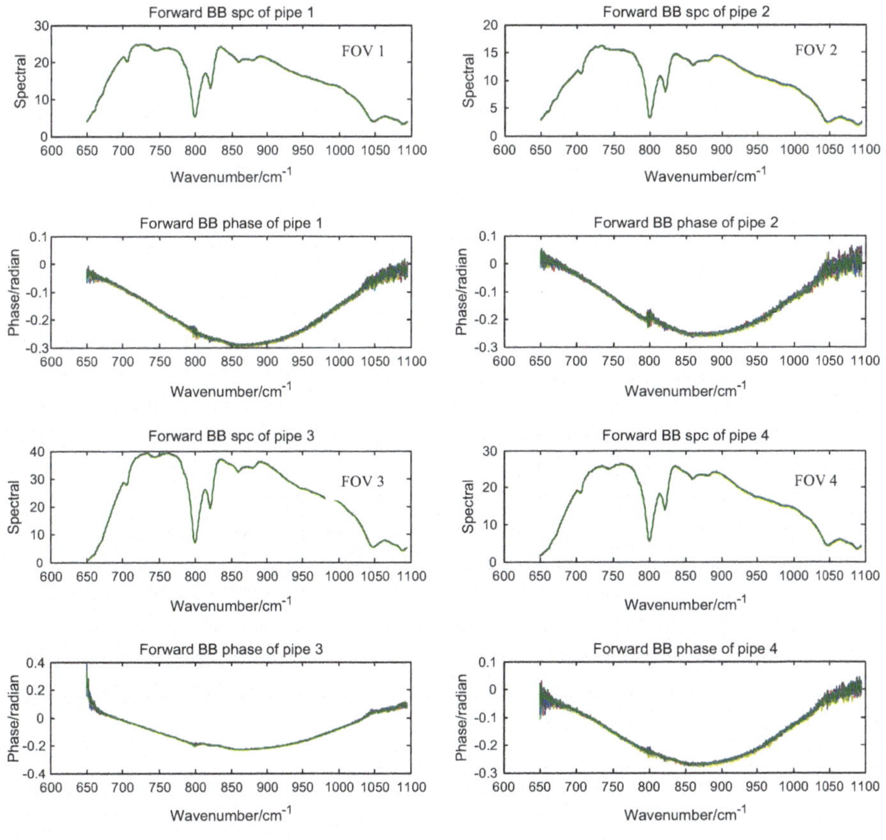

Fig. 3 Uncalibrated spectra of and phase spectra of blackbody sources for FOV1 ~4 in long wave band

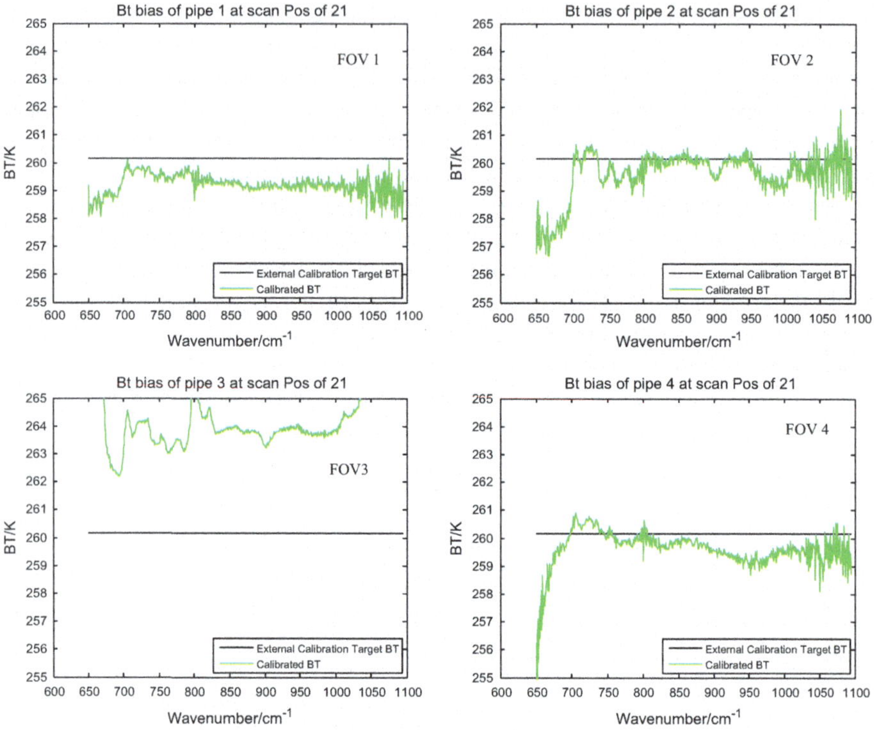

Fig. 4 Calibrated brightness temperature of FOV1∼4 when scene bt is at 260 K

Another two issues should be addressed here are non-linearity correction and instrument Instrument Line Shape (ILS) correction. Non-linearity levels in the LWIR and MWIR detectors are high enough to require application of a non-linearity correction to the uncalibrated spectra in order to reduce the radiometric error introduced by the detector non-linearity. When the radiation source in a Michelson interferometer is located in front of or behind the focal plane of the collimator, distortion arises in the spectral line shape. For interferometer, this well-known self-apodization induced by the instrument (ILS is distorted from a pure sync) shall be removed from the scene radiance in each channel before being handed off to the satellite production development algorithms. In status of Engineering model, these two correction has not been added to the data process yet and the biases are sure to decrease after these two correction be added in the next work of algorithm improvement.

4 Summary

HIRAS is designed and manufactured to measure the infrared emission spectrum of the earth and the primary use is atmospheric temperature and humidity sounding of high radiometric precision and accuracy. To characterize key performance parameters of instrument and develop and check out the test support equipment needed to test the HIRAS flight unit, HIRAS performs vacumm tests. Ground pre-processing scheme and method was tested based on the measurements of test data, radiometric calibration results revealed some interferogram quality and spectra absorbing peak appearance, validated the feasibility of complex calibration method and engineering model data. Future work will focus on non-linearity correction and ILS correction and also analysis of flight model data.

References

1. Dong C, Li J, Zhang P, et al, 2013: Methodology and application of satellite high spectral infrared atmospheric sounder. Publication of science, Beijing.
2. Han Y, Revercomb H, Cromp M, et al, 2013: "Suomi NPP CrIS measurements, sensor data record algorithm, calibration and validation activities, and record data quality, "[J]. *Journal of Geophysical Research: Atmospheres*, 118(22): 12734–12748.
3. Gu M, 2015: Engineering model design description of High Spectral Infrared Atmospheric Sounder on-board FY-3(04) satellite (Technical document).
4. Henry E. Revercomb, H. Buijs, Hugh B. Howell, et al. 1988: "Radiometric calibration of IR Fourier transform spectrometers:solution to a problem with the High-Resolution Interferometer Sounder," [J]. *Applied Optics*, 27(15): 3210–3218.
5. Bernard T, Denis B, "IASI Level 0 and 1 processing algorithms description".
6. Griffiths P.R. and De Haseth J.A. 1986, "Fourier transform infrared spectrometry", New York.

Design and Verification of Ratioing Radiometer Parameters

Li Mengfan, Zou Peng, Meng Binghuan, Shi Hailiang, Hong Jin and Qiao Yanli

Abstract Ratioing radiometer, which is a key part for improving the precision of on-board calibration facility, can be used to track and correct on-orbit degradation of solar diffuser. In this paper, an on-board calibration facility with solar diffuser is presented and the model of BRDF correction for solar diffuser is established based on the principle of ratioing radiometer. For the purpose of optimizing the performance of ratioing radiometer, the output of solar diffuser view and sun view should be matched. Considering that the solar illumination angle changes in different season, the ratio between signals from solar diffuser view to sun is designed to range from 0.899 to 1.692, and 1.296 at intermediate incident zenith angle. In order to verify the designed ratio, testing equipment is built on the basis of relative lamp-illuminated plaque system, where the lamp and plaque simulate the sun and solar diffuser, respectively. The result shows that the ratio is within 0.5% error of the designed value. Factors that could induce the testing error are discussed. The above results show that the design parameters of ratioing radiometer are reasonable which meet the requirements of on-orbit degradation monitoring for solar diffuser.

Keywords On-board calibration · Ratioing radiometer · Solar diffuser · BRDF

L. Mengfan (✉) · Z. Peng · M. Binghuan · S. Hailiang · H. Jin · Q. Yanli
Anhui Institute of Optics and Fine Mechanics, Chinese Academy of Science, Hefei, Anhui, China
e-mail: limengfan1989@163.com

L. Mengfan
University of Science and Technology of China, Hefei, Anhui, China

L. Mengfan · Z. Peng · M. Binghuan · S. Hailiang · H. Jin · Q. Yanli
Key Laboratory of Optical Calibration and Characterization, Chinese Academy of Science, Hefei, Anhui, China

H.P. Urbach and G. Zhang (eds.), *3rd International Symposium of Space Optical Instruments and Applications*, Springer Proceedings in Physics 192,
DOI 10.1007/978-3-319-49184-4_22

1 Introduction

Due to the vibration and space irradiation effect during and after launch, the quality of calibrated data for space camera (SC) gradually decreases. Over the past few decades, lots of efforts have been made for quantitative remote sensing by developing on-board calibration methods [1, 2]. By observing diffusely reflected sunlight from solar diffuser (SD) with the known bidirectional reflectance distribution function (BRDF), on-board calibration with entire aperture and optical path can be achieved. However, SD BRDF degrades especially at short wavelength as a result of solar exposure, which will lead to invalid calibration data if not corrected. Ratioing radiometer (RR) has been proposed [3, 4] to track SD degradation by taking alternate view of sun and SD. Because the solar radiance is known and stable, the ratio between the diffuser and sun views can be used to determine the SD BRDF degradation, which can significantly improve the precision of on-board calibration. The SD and RR system, which has been succeeded in maintaining remote sensing data quality at missions such as MODIS and its following instrument VIIRS [5], is becoming one of the main methods for on-board calibration.

Quantitative application of remote sensing data is developing rapidly in China, Anhui Institute of Optics and Fine Mechanics (AIOFM) has been studying SD and RR on-board calibration systems for several years. For MODIS and VIIRS RR system, there exists a rotating prism to accomplish sun view, SD view and dark view alternately. In this paper, a new kind of RR with no moving part is proposed and designed. Ports for sun view and SD view are also fixed, which enables the RR to perform more stable. RR response to sun and SD view are detected by the same electronics, in order to minimize the influence of the detector non-linearity within dynamic range, the energy received by the sun and SD view ports should be matched well, which means that how to keep the ratio between sun view response to SD view response about 1 is a critical problem. In the next chapters, a brief introduction of SD and RR system is presented, the model of SD BRDF correction is obtained, design of main optical and mechanical parameters of RR is illustrated, and corresponding test is carried out to verify the ratio between two view responses. The comparison between designed ratio and tested ratio and the analysis of error sources provide important guide to estimate the rationality of RR design.

2 Principle and Design

2.1 Principle of Calibration

Data product of the SC is top of atmosphere radiance at 760, 1575, 1650 and 2050 nm, containing the information of CO_2 content. As illustrated in Fig. 1, the on-board calibration for SC is achieved by taking a view of the standard diffusely reflected radiance from the SD with a known BRDF parameter by testing system in

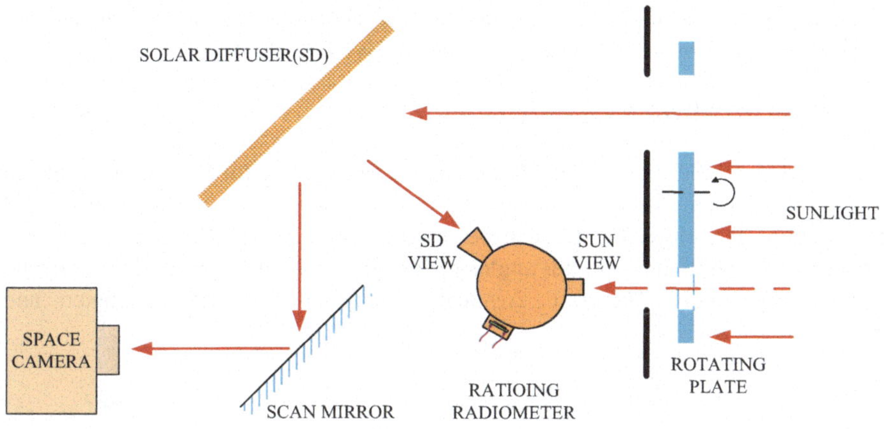

Fig. 1 Schematic of on-board calibration using solar diffuser (SD) and ratioing radiometer (RR) system

[6]. A rotating plate, which has an entrance for sunlight to illuminate the SD while on-board calibration, is placed at the front-end of the optical path. Through the preloaded command for the rotating plate, sunlight can be controlled to enter the calibrator or not. Numerous studies on SD applications suggested that the reflectance properties slowly degrade on-orbit as a result of solar exposure, which reduce the quality of on-board calibration. SD BRDF at shorter wavelength shows larger degradation, and the degradation is negligible at above 900 nm. RR is used in tandem with the SD to track and correct the degradation of SD. RR consists of a small-sized integrating sphere with a filtered detector at 760 nm to match the shortest band of SC and two fixed ports detecting sun view and SD view respectively. By making use of the rotating plate where another entrance for RR is designed, instead of designing a special moving part for RR, the RR is enabled to perform sun view, SD view and dark view successively for several times during each on-board calibration activity, which significantly improve the reliability of on-board calibration system. Except for the sun incidence angle that changes in different seasons, the position of SD and RR are fixed, and the same integrating sphere and electronics are used by RR to detect different views. Therefore, the on-orbit degradation of SD can be derived from the relative change in the ratio of SD view response to sun view response.

The radiance of SD that reaches the scan mirror can be expressed as Eq. (1)

$$L = E_{sun,t} \cos(\theta_{SD,\mathrm{t}}) \rho_{lab}(\theta_{SD,\mathrm{t_0}}, \phi_{SD,\mathrm{t_0}}; \theta_{SC,\mathrm{t_0}}, \phi_{SC,\mathrm{t_0}}) \Delta_{SD} \tag{1}$$

where L is SD radiance, $E_{sun,t}$ is the solar irradiance outside the atmosphere, ρ_{lab}, which is assumed to be constant till the first time of on-board calibration t_0, is the SD BRDF measured at prelaunch testing, θ_{SD} and ϕ_{SD} are the zenith and azimuth angles of the incident sunlight in SD coordinate system, θ_{SC} and ϕ_{SC} are the zenith and azimuth angles of scan mirror direction in SD coordinate system, Δ_{SD} is the correction factor for SD degradation which is given by Eq. (2).

$$\Delta_{SD} = \rho_{\mathrm{t}}(\theta_{SD,\mathrm{t_0}}, \phi_{SD,\mathrm{t_0}}; \theta_{SC,\mathrm{t_0}}, \phi_{SC,\mathrm{t_0}}) / \rho_{lab}(\theta_{SD,\mathrm{t_0}}, \phi_{SD,\mathrm{t_0}}; \theta_{SC,\mathrm{t_0}}, \phi_{SC,\mathrm{t_0}}) \quad (2)$$

According to the research of MODIS SD reflectance factor [7], the relationship of SD BRDF among different incidence and reflectance angles is stable during SD degradation. Therefore, Eq. (2) can be rewritten as Eq. (3). Where θ_{RR} and ϕ_{RR} represent the zenith and azimuth angles of RR SD view direction in SD coordinate system, respectively. Therefore, Δ_{SD} can be calculated by RR parameters and responses.

$$\begin{aligned}\Delta_{SD} &= \rho_t(\theta_{SD,\mathrm{t_0}}, \phi_{SD,\mathrm{t_0}}; \theta_{RR,\mathrm{t_0}}, \phi_{RR,\mathrm{t_0}}) / \rho_{lab}(\theta_{SD,\mathrm{t_0}}, \phi_{SD,\mathrm{t_0}}; \theta_{RR,\mathrm{t_0}}, \phi_{RR,\mathrm{t_0}}) \\ &= \rho_{\mathrm{t}}(\theta_{SD,\mathrm{t}}, \phi_{SD,\mathrm{t}}; \theta_{RR,\mathrm{t}}, \phi_{RR,\mathrm{t}}) / \rho_{lab}(\theta_{SD,\mathrm{t}}, \phi_{SD,\mathrm{t}}; \theta_{RR,\mathrm{t}}, \phi_{RR,\mathrm{t}})\end{aligned} \quad (3)$$

Figure 2 illustrates the schematic of RR. The direction of sun view port is in accordance with the intermediate sun incidence angle and a pinhole is placed at the bottom of the port to receive the irradiance of the sunlight. SD view port with aperture stop and field stop is toward the centre of the SD to get the radiance of diffused sunlight from the SD. The filtered detector, which is embedded in the integrating sphere and away from the direct light of two view ports, receives the diffusely reflected light of the sphere.

RR response to SD view and sun view at different calibration time can be expressed as Eqs. (4)–(7)

$$d_{SD,t}(\theta_{SD,\mathrm{t_0}}, \phi_{SD,\mathrm{t_0}}) = \cos(\theta_{SD,\mathrm{t_0}}) E_{sun,t} \rho_t(\theta_{SD,\mathrm{t_0}}, \phi_{SD,\mathrm{t_0}}; \theta_{RR,\mathrm{t_0}}, \phi_{RR,\mathrm{t_0}}) \Omega_{SD} S_{SD} \eta_{SD,t} R_t \quad (4)$$

$$dn_{SD,t_0}(\theta_{SD,t_0}, \phi_{SD,t_0}) = \cos(\theta_{SD,t_0}) E_{sun,t_0} \rho_{lab}(\theta_{SD,t_0}, \phi_{SD,t_0}; \theta_{RR,t_0}, \phi_{RR,t_0}) \Omega_{SD} S_{SD} \eta_{SD,t_0} R_{t_0} \quad (5)$$

$$dn_{sun,t}(\alpha_{\mathrm{t_0}}) = E_{sun,t} \cos \alpha_{\mathrm{t_0}} S_{sun} \eta_{sun,\mathrm{t}} R_t \quad (6)$$

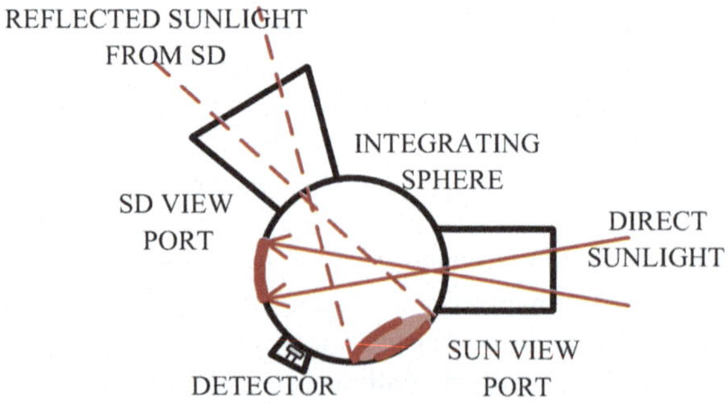

Fig. 2 Schematic of ratioing radiometer (RR)

$$dn_{sun,t_0}(\alpha_{t_0}) = E_{sun,t_0} \cos \alpha_{t_0} S_{sun} \eta_{sun,t_0} R_{t_0} \tag{7}$$

where dn_{SD} and dn_{sun} are RR SD and sun view responses, α is the angle between sunlight and direction of RR sun view port, S_{sun} and S_{SD} are the area of pinhole and aperture stop of sun view port, Ω_{SD} is the solid angle of SD view port, η_{SD} and η_{sun} are optical functions of the integrating sphere for SD and sun view which are assumed to be equal during the mission, and R is the response function of the detector.

Relative change in the ratio between Sun and SD view response for RR can be calculated using Eq. (8)

$$\frac{dn_{SD,t}(\theta_{SD,t}, \phi_{SD,t})/dn_{sun,t}(\alpha_t)}{dn_{SD,t_0}(\theta_{SD,t_0}, \phi_{SD,t_0})/dn_{sun,t_0}(\alpha_{t_0})} = \frac{\cos \theta_{SD,t} \rho_t(\theta_{SD,t}, \phi_{SD,t}; \theta_{RR,t}, \phi_{RR,t})}{\cos \theta_{SD,t_0} \rho_{lab}(\theta_{SD,t_0}, \phi_{SD,t_0}; \theta_{RR,t_0}, \phi_{RR,t_0})} * \frac{K(\alpha_{t_0})}{K(\alpha_t)} \tag{8}$$

where K is the sun view response at different α normalized to the sun view response for intermediate sun incidence angle, which can be expressed as $K(\alpha) = dn_{sun}(\alpha)/dn_{sun}(0°)$ and is measured prelaunch in the lab as in [8]. Then Δ_{SD} can be evaluated by Eq. (9).

$$\Delta_{SD} = \frac{dn_{SD,t}(\theta_{SD,t}, \phi_{SD,t})/dn_{sun,t}(\alpha_t)}{dn_{SD,t_0}(\theta_{SD,t_0}, \phi_{SD,t_0})/dn_{sun,t_0}(\alpha_{t_0})} * \frac{\cos \theta_{SD,t_0} \rho_{lab}(\theta_{SD,t_0}, \phi_{SD,t_0}; \theta_{RR,t_0}, \phi_{RR,t_0})}{\cos \theta_{SD,t} \rho_{lab}(\theta_{SD,t}, \phi_{SD,t}; \theta_{RR,t}, \phi_{RR,t})} * \frac{K(\alpha_t)}{K(\alpha_{t_0})} \tag{9}$$

2.2 Parameter Design of RR

For Eq. (8), long term degradation of electronics and optical factors of RR can be eliminated and have no impact on SD BRDF correction. To minimize non-linearity error caused by RR electronics, effort should be made to narrow down the dynamic range, which means that the energy through SD and sun view port should be matched for different sun incidence angle.

Ratio of SD view response to sun view response is proportional to the ratio between the energy received by two view ports. The solid angle of SD view port is evaluated by Eq. (10), where β is half field of view of SD view port. Then the ratio of SD view response to sun view response can be calculated by Eq. (11), where r_0 is the designed ratio.

Table 1 Main parameter of ratioing radiometer (RR)

Parameter	Value	Error
D_{sun}	0.6 mm	±0.005 mm
D_{SD}	4 mm	±0.05 mm
β	13.8°	±0.2°
D_{IS}	40 mm	±0.05 mm

$$\Omega_{SD} = \pi \sin^2(\beta) \tag{10}$$

$$r_0 = \frac{\cos(\theta_{SD})\rho_{lab}(\theta_{SD}, \phi_{SD}; \theta_{RR}, \phi_{RR})\pi \sin^2(\beta)S_{SD}}{\cos(\alpha)S_{sun}} \tag{11}$$

Sun incidence angle in SD coordinate system and the range of α for RR sun view port can be derived from the geometry of SD and RR on the satellite. θ_{SD} ranges from 50° to 70°, while α varies from −10° to 10°, and SD BRDF at intermediate sun angle of incidence is 0.327.

r_0 should be larger than 1 considering SD degradation. Meanwhile, size of some purchased product is restricted. To optimize the performance of RR, r_0 is designed to range from 0.90 to 1.692 and to be 1.296 at the intermediate sun incidence angle. The main optical and mechanical parameters are given by Table 1. Where D_{SD}, D_{sun} and D_{IS} are the diameter of aperture stop of SD view port, the pinhole of sun view port and the integrating sphere, respectively.

3 Test Principles

The radiance of sunlight on-orbit that reaches the SD and RR sun view port is considered to be equal. Both light source with extended, uniform and high energy in the laboratory and testing system outdoor that takes sunlight as light source are costly and difficult to achieve for RR and SD on-board calibration system to perform r_0 test.

A simplified testing system is proposed to simulate the sun and SD by a halogen lamp and a pressed plaque. The plaque is made by Teflon with 0°/45° BRDF derived from directional/hemisphere reflectance (DHR) traceable to National Institute of Metrology (NIM) of China. Figure 3 shows the testing principle of r_0. The test is carried out by separately making lamp and lamp illuminated plaque view using RR sun and SD view port. Both pinhole of RR sun view port for (a) and plaque for (b) are equidistant from the halogen lamp and collimate with it. RR SD view port makes plaque view with a zenith angle of 45°. The collimation is achieved by a laser and the mirror attached to the target face.

The ratio of RR sun view to SD view can be derived from tested ratio of plaque view response to halogen lamp view response by correcting the zenith angle of incidence and reflectance factor of the plaque using Eq. (12)

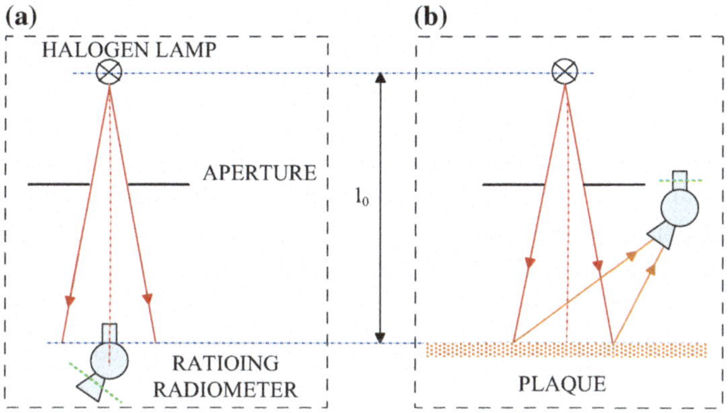

Fig. 3 Principle of r_0 test for **a** sun view port and **b** SD view port

Table 2 Main parameters of the test

Description	Value
Power of the lamp	1 kW
Size of the filament	20 mm * 7 mm
l_0	700 mm
ζ	0.994 @ 760 nm

$$r_0 = \frac{dn_p}{dn_h} * \frac{\rho_{lab}(\theta_{SD}, \phi_{SD}; \theta_{RR}, \phi_{RR})}{\zeta/\pi} * \cos(\theta_{SD}) \tag{12}$$

where dn_p is plaque view response, dn_h is halogen lamp view response, and ζ is the reflectance of the plaque. Table 2 shows the main parameters of this test.

4 Results and Discussion

The calculated ratio by Eq. (12) is 1.303 at intermediate incidence angle, and the ratio ranges from 0.905 to 1.701 in different seasons with repeatability of less than 0.4%. Results show that the tested ratio is accord with the designed value within 0.5% error. The error is mainly caused by two factors, the one is uncertainty of the test, and the other is the error of RR optic parameters listed in Table 1.

Test uncertainty can be induced by many factors during the test, such as distance error, collimation error, reflectance error, etc. The irradiance that actually reaches the pinhole of RR sun view port and plaque in Fig. 3a, b is different. It can be evaluated by Eqs. (13) and (14), respectively, where l_1 and l_2 are the distance from the halogen lamp to pinhole of RR sun view port in Fig. 3a and plaque in Fig. 3b, and E_0 is the corresponding irradiance for l_0.

$$E_1 = l_0^2/l_1^2 \cdot E_0 \tag{13}$$

$$E_2 = l_0^2/l_2^2 \cdot E_0 \tag{14}$$

According to Eqs. (11)–(14), the tested ratio, r_1, can be rewritten as Eq. (15), where $\theta_{\Delta p}$ and $\alpha_{\Delta h}$ are the error of halogen lamp incidence angle for the plaque in Fig. 3b and pinhole of RR sun view port in Fig. 3a, S'_{sun}, S'_{sun} and β' are actual values of S_{sun}, S_{SD} and β, which are constants after RR assembling.

$$r_1 = \frac{l_2^2 \zeta \sin^2(\beta') S'_{SD} \cos(\theta_{\Delta p})}{l_1^2 \cos(\alpha_{\Delta h}) S'_{sun}} * \frac{\rho_{lab}(\theta_{SD}, \phi_{SD}; \theta_{RR}, \phi_{RR})}{\zeta} * \cos(\theta_{SD}) \tag{15}$$

Equation (15) shows that the uncertainty is induced by the l_0 error, ρ_{lab} testing error, plaque 0°/45° BRDF error, collimation error, non-stability and non-uniformity of the halogen lamp irradiance, and non-stability of RR electronics. Table 3 shows the relative combined standard uncertainties for r_1 test. It can be inferred that the main source of uncertainty is reflectance factors of the solar diffuser and the plaque.

The maximum error of the ratio induced by RR parameter error is listed in Table 4. The diameter of pinhole of sun view port is small, so even an error of 0.004 mm can lead to 1.7% error of the ratio, but the error of RR optic parameters will have little influence on the performance of the RR.

Table 3 Combined testing uncertainty for r_1

Description	Uncertainty (%)
l_1	0.23
l_2	0.35
ρ_{lab}	0.70
Plaque 0°/45° BRDF	0.56
$\alpha_{\Delta h}$	0.05
$\theta_{\Delta p}$	0.05
Non-stability of halogen lamp	0.30
Non-uniformity of halogen lamp irradiance	0.38
Non-stability of RR electronics	0.13
Combined	1.11

Table 4 Error of ratio induced by RR parameters

Description	Error (%)
D_{sun}	1.7
D_{SD}	1
β	1.4

Result shows that the error of the ratio is much smaller than the uncertainty of the test and the maximum error induced by RR parameters, so the testing system introduced in this paper can effectively verify the parameters of RR.

5 Conclusion

On-board calibration system based on solar diffuser and ratioing radiometer is the guarantee of long term and high precision quantify application of space camera. A ratioing radiometer without any moving mechanism is proposed and the model for solar diffuser BRDF correction has been introduced. RR parameters should ensure the ratio between sun and SD view responses about 1 in order to optimize the performance of ratioing radiometer electronics. Testing device built on the base of relative lamp-illuminated plaque system can effectively verify the ratio between sun and SD view responses. The tested ratio between RR SD and sun view port responses is in agreement with the designed value with 0.5% error, and the uncertainty of the testing ratio is 1.11%. The above results indicate that RR design is reasonable, which is the foundation of RR application in future.

References

1. Zheng Xiaobing, "Some respects for development of radiometric calibration techniques for optical remote sensing satellites," *Journal of Atmospheric and Environmental Optics*, vol: 9, no. 1, pp. 2–8, 2014.
2. Li Xiaohui, Yan Changxiang, "Onboard calibration technologies for hyperspectral imager," *Chinese Journal of Optics and Applied Optics*, vol: 2, no. 4, pp. 309–315, 2009.
3. Slater P N and Palmer J M, "Solar-diffuser panel and ratioing radiometer approach to satellite sensor on-board calibration," *Proc. SPIE*, vol: 1, pp. 100–105, 1991.
4. Palmer J M and Slater P N, "Ratioing radiometer for use with a solar diffuser," *Orlando'91, Orlando, FL. International Society for Optics and Photonics*, pp. 106–117, 1991.
5. Xiaoxiong Xiong, Amit Angal, Junqiang Sun, Taeyoung Choi and Eric Johnson, "On-orbit performance of MODIS solar diffuser stability monitor," *Journal of Applied Remote Sensing*, vol. 8, no. 1, pp. 083514-083514, 2014.
6. Li Lingjun, Zhang Liming, Chen Hongyao, Si Xiaolong, Wang Jixiang, et al, " Development of BRDF absolute measurement device," *Acta Optical Sinica*, vol. 34, no. 5, pp. 0528002, 2014.
7. Xiaoxiong Xiong, Junqiang Sun, Joe Esposito, Xiaojin Liu, William L. Barnes, et al, "On-orbit characterization of a solar diffuser's bidirectional reflectance factor using spacecraft maneuvers,"*Proc. SPIE 5151, Earth Observing Systems VIII*, pp. 375–383, doi:10.1117/12.504802, November 13, 2003.
8. Li Mengfan, Xu Weiwei, Zou Peng, Zhang Liming, Meng Binghuan, et al, "Study of the response characteristic of ratioing radiometer," *Octa Optical Sinica*, vol. 36, no. 2, pp. 0212005,2016.

Polarization Measurement of the Grating Spectrograph Imager

Jingyi Wang, Yongxiang Guo, Junyu Ke and Yongqiang Li

Abstract With advantages of high-resolution, high-precise and wide spectral band, grating spectrograph imager was widely applied in aerospace remote field. In this paper, a kind of depolarizer is designed based on Muller matrix in front of the optical system to satisfy the requirement of grating spectrograph. The polarization sensitivity of depolarizer and grating spectrograph (with depolarizer) at 0 Fov, ±0.5 Fov in 4 bands between 0.4–0.95 μm are measured. The polarization measurement model is independent developed to achieve full-aperture testing. Through analyzing, test precision is better 0.5%. All polarization sensitivity results show two-cycle variation as expected. The polarization sensitivity of depolarizer is better than 10%, and that of the whole grating spectrograph is better than 2%. The experimental results are consistent with theoretical analyses. It is validated that the proposed depolarizer has obvious depolarization effect and restrained the polarization sensitivity of grating spectrograph effectively.

Keywords Grating spectrograph · Polarization measurement · Depolarizer · Muller matrix

1 Introduction

The imaging spectrograph supplies spectral images in one spectral dimension and two spatial dimensions simultaneously, which provides formation, evolution, internal structure of observed object and atmospheric composition. Based on the imaging spectrograph, precise physical, chemical and dynamical information can be obtained [1, 2].

Diffraction grating is the key optical element of grating spectrograph imager to split the multi-spectra because of its broad spectrum range and good linearity. But it has high polarization sensitivity and affects the precision of radiometric calibration.

J. Wang (✉) · Y. Guo · J. Ke · Y. Li
Beijing Institute of Space Mechanics and Electricity, Beijing, China
e-mail: applewinebar_1986@sina.com

H.P. Urbach and G. Zhang (eds.), *3rd International Symposium of Space Optical Instruments and Applications*, Springer Proceedings in Physics 192,
DOI 10.1007/978-3-319-49184-4_23

While the reflective light of ocean water or cloud contain evident polarization characteristics, it is necessary to eliminate the polarization of the light in front of the optical system and make sure the accurate radiance data [1–7].

The method of polarization suppression can be categorized into three types, such as, spectrum depolarization, time depolarization and space depolarization. Spectrum depolarization method utilize that the different wavelengths of incident light have different polarization states after passing the depolarizer. This method does not suit spectrograph. The principle of time depolarization is that the polarization states are different at different time by rotating the depolarizer. It would increase structural complexity and decrease stability of grating spectrograph. Space depolarization utilize that incident light at different apertures have different polarization states after passing the depolarizer. Lyot depolarizer is a kind of space depolarizer, and it has been widely used in the grating spectrometer based on the character of simple structure and good stability.

Based on Muller matrix, the crucial parameters of the Lyot depolarizer can be gotten in order to assure the scientific objectives of the imaging spectrograph. The theoretical analyses can be validated by the polarization measurement of the depolarizer and imaging spectrograph.

2 The Analysis and Design of Polarization Suppression

Figure 1 shows the configuration of Lyot depolarizer, the angle between the optical axes of two crystals is 45°, the ratio of the thickness is 2:1, and the materials are the same. After incident light go through the depolarizer, the polarization state of light at different aperture are different. The depolarizer reduces the polarization sensitivity of incident light. Use Muller matrix, the process of polarization suppression can be analyzed.

The incident and exit light of depolarizer can be denoted as *Vi* and *Ve*. Their stokes matrix are $V_i = [i_i, m_i, c_i, s_i]^T$ and $V_e = [i_e, m_e, c_e, s_e]^T$, where i is the power

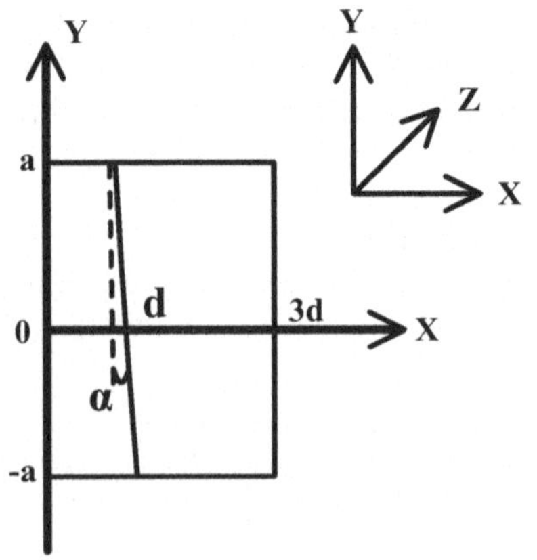

Fig. 1 The structure schematic of Lyot depolarizer

of polarized light, m is the power of horizontal component, c is the power of +45° direction component and s is the power of right-hand direction component. The phase retards of the two crystals are δ_1 and δ_2,

$$\delta_1 = \frac{2\pi \Delta \mathrm{n}}{\lambda}(\mathrm{d} - \mathrm{a}\tan\alpha), \tag{1a}$$

$$\delta_2 = \frac{2\pi \Delta n}{\lambda}(2d + a\tan\alpha), \tag{1b}$$

where $\Delta n = n_e - n_o$, n_e and n_o respectively expressed the refractive index of crystal extraordinary light and ordinary light, λ is the wavelength, d is the center thickness of the first crystal, α is the wedge angle, a is half of diameter of the depolarizer [8].

Suppose the incident light are linear polarized with the angle of φ to Y axis, the Stokes matrix is $V_i = [1, \cos 2\varphi, \sin 2\varphi, 0]^T$. Based on the matrix optics, the Muller matrix of Lyot depolarizer is M, the exit light in the area S can be expressed as follow,

$$V_e = \int_{\lambda}^{\lambda+\Delta\lambda} \int_{s} M V_i \, ds\, d\lambda$$

$$\begin{vmatrix} i_e \\ m_e \\ c_e \\ s_e \end{vmatrix} = \int_{\lambda}^{\lambda+\Delta\lambda} \int_{s} \begin{vmatrix} 1 & 0 & 0 & 0 \\ 0 & \cos\delta_2 & \sin\delta_1 \sin\delta_2 & -\cos\delta_1 \sin\delta_2 \\ 0 & 0 & \cos\delta_1 & \sin\delta_1 \\ 0 & \sin\delta_2 & -\sin\delta_1 \cos\delta_2 & \cos\delta_1 \cos\delta_2 \end{vmatrix} \cdot \begin{vmatrix} 1 \\ \cos 2\varphi \\ \sin 2\varphi \\ 0 \end{vmatrix} ds\, d\lambda \tag{2}$$

where

$$\begin{aligned} i_e &= \int_{\lambda}^{\lambda+\Delta\lambda} \int_{s} 1\, ds\, d\lambda \\ m_e &= \int_{\lambda}^{\lambda+\Delta\lambda} \int_{s} (\cos 2\varphi \cos\delta_2 + \sin 2\varphi \sin\delta_1 \sin\delta_2)\, ds\, d\lambda \\ c_e &= \int_{\lambda}^{\lambda+\Delta\lambda} \int_{s} (\sin 2\varphi \cos\delta_1)\, ds\, d\lambda \\ s_e &= \int_{\lambda}^{\lambda+\Delta\lambda} \int_{s} (\cos 2\varphi \sin\delta_2 - \sin 2\varphi \sin\delta_1 \cos\delta_2)\, ds\, d\lambda \end{aligned} \tag{3}$$

The polarization sensitivity of the exit light can be expressed as follow,

$$P = \frac{\sqrt{m_e^2 + c_e^2 + s_e^2}}{i_e}. \tag{4}$$

where $\int_s ds = \int_{-a}^{a} \int_0^A dzdy$. Because the integrand function does not contain variable z in Eq. (3), the integration of z can be ignored.

Substitute (1a) and (1b) into (3) and simplify [8],

$$\begin{aligned}
i_e &= 2a\Delta\lambda \\
m_e &= 2a\cos(2\varphi)\int_{\lambda}^{\lambda+\Delta\lambda} \cos(\frac{2\pi\Delta n}{\lambda}2d)\sin c(\frac{2\Delta n}{\lambda}a\tan\alpha)d\lambda - a\sin(2\varphi)\int_{\lambda}^{\lambda+\Delta\lambda} \cos(\frac{2\pi\Delta n}{\lambda}3d)2ad\lambda \\
&\quad + a\sin(2\varphi)\int_{\lambda}^{\lambda+\Delta\lambda} \cos(\frac{2\pi\Delta n}{\lambda}d)\sin c(\frac{2\Delta n}{\lambda}2a\tan\alpha)d\lambda \\
c_e &= 2a\sin(2\varphi)\int_{\lambda}^{\lambda+\Delta\lambda} \cos(\frac{2\pi\Delta n}{\lambda}d)\sin c(\frac{2\Delta n}{\lambda}a\tan\alpha)d\lambda \\
s_e &= 2a\cos(2\varphi)\int_{\lambda}^{\lambda+\Delta\lambda} \sin(\frac{2\pi\Delta n}{\lambda}2d)\sin c(\frac{2\Delta n}{\lambda}a\tan\alpha)d\lambda - a\sin(2\varphi)\int_{\lambda}^{\lambda+\Delta\lambda} \sin(\frac{2\pi\Delta n}{\lambda}3d)d\lambda \\
&\quad + a\sin(2\varphi)\int_{\lambda}^{\lambda+\Delta\lambda} \sin(\frac{2\pi\Delta n}{\lambda}d)\sin c(\frac{2\Delta n}{\lambda}2a\tan\alpha)d\lambda
\end{aligned} \tag{5}$$

The grating spectrograph need to be optimized in this paper, whose entrance pupil is about Φ26 mm. So the diameter of Lyot depolarizer is Φ32 mm to ensure that it can cover the entrance pupil. Choose the material as quartz.

The wedge angle affects the polarization sensitivity of the depolarizer and the separate angle between ordinary light and extraordinary light in the crystal. The separate angle could debase the resolution of the optical system, so the wedge angle should satisfy the condition: $\alpha < \varepsilon/|n_e - n_o|f$, where ε is the pixel dimension and f is the focal length [8]. ε is 13 μm and *f* is 104 mm, so set the wedge angle as 0.5°.

The Muller matrix of the grating spectrograph can be seen as the product of the Muller matrices of the depolarizer and optical system. So the polarization sensitivity of the grating spectrograph can be estimated as the product of the polarization sensitivity of the depolarizer and optical system. Considering the design of the grating spectrograph and optical system, set the polarization sensitivity of depolarizer as 10%. To satisfy the requirement of the polarization sensitivity between 0.4 and 0.95 μm, the thickness need be greater than 2.6 mm through Eq. (5). So set thickness as 3.3 mm.

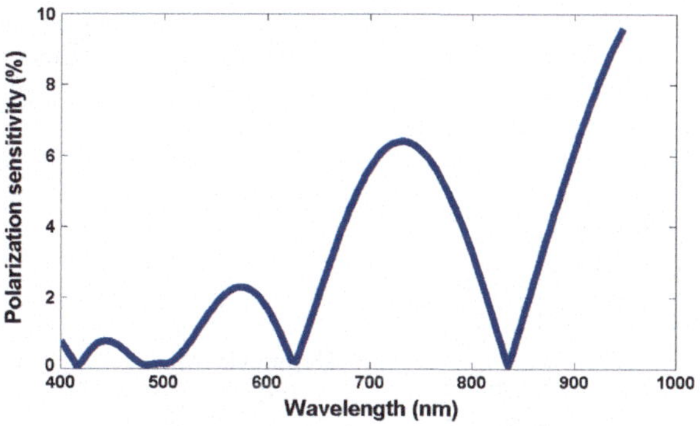

Fig. 2 Polarization sensitivity theoretical variation with wavelengths (at $\varphi = 0°$)

After the material, wedge angle, diameter and thickness of depolarizer being determined, it can get the theoretic relationship of polarization sensitivity and wavelengths of the exit light. When $\varphi = 0°$, the polarization sensitivity theoretic value vary with different wavelengths between 0.4 and 0.95 μm as shown in Fig. 2. It can be seen that the theoretic value is lower than 10% which is satisfy the design requirements.

3 The Polarization Measurement of Depolarizer and Grating Spectrograph

The polarization measurement model is independent developed to achieve full-aperture testing. The stability of Si detector and circuit is 0.01%. The stability of halogen lamp source is better than 0.02%. Through analyzing, test precision is better 0.5% in spectral range 0.4–0.95 μm. The test model is similar as ground support equipment (GSE) Polarized Source Assembly (PSA) validated by the Moderate Resolution Imaging Spectroradiometer (MODIS) [9]. The polarization measurement model is illustrated in Fig. 3. The polarization state can be rotated automatically to any direction from 0° to 360° by rotating the polarizer prism. The spectrum band can be chosen through inserting the corresponding filter.

The polarization sensitivity of depolarizer was tested first. As shown in Fig. 2, insert testing stop and focusing system and adjust the pitch angle and position of the all components in light path. Insert the filters one by one, whose center wavelength are 442, 570, 660, 670, 690, 860 nm respectively and spectral bandwidth are 20 nm. Polarization sensitivity can be expressed by the contractual definition of instrument [10].

$$P = \frac{I_{max} - I_{min}}{I_{max} + I_{min}}, \tag{6}$$

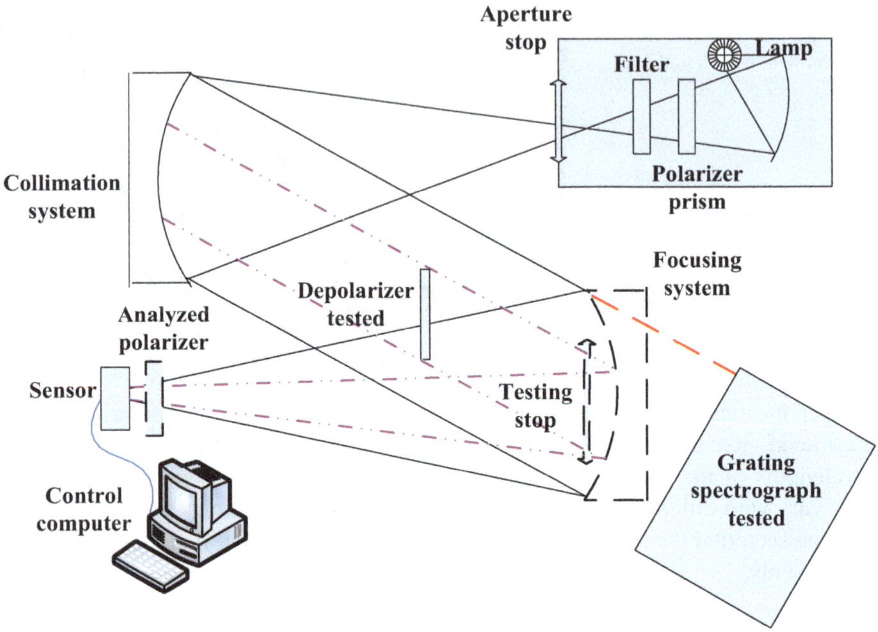

Fig. 3 The schematic of polarization measurement model

where *I* is the response signal of the sensor. When the polarizer prism rotates φ, the incident light rotates and the response signal of the sensor is $I(\theta)$ which changes with exit direction (θ) of light to Y axis. Rotating the analyzed polarizer from 0° to 180°, there can get I_{max} and I_{min}. So the results of polarization sensitivity are different at different φ. The whole field of view of grating spectrograph is 7.1°, so 0 Fov, ±0.5 Fov and ±1 Fov are be chosen to test. The theoretical and test result of depolarizer at different center wavelengths under 0° field of view condition is shown in Table 1. The polarization sensitivity theoretical variation with different deviation angle at 0 Fov under different center wavelength and the test result of polarization sensitivity variation with different deviation angle at different field of view under center wavelength 670 nm are given in Figs. 4 and 5, respectively.

From the Table 1 and Figs. 4 and 5, it can be seen that the test values are similar as expected. The results in Fig. 4 are concluded from Eq. (5). The curve of polarization sensitivity variation at 0 Fov in Fig. 5 shows two-cycle ($\cos 2\varphi$ function) as the theoretical response in Fig. 4. This phenomenon has been validated

Table 1 Polarization sensitivity of depolarizer

Center wavelength		442 nm	570 nm	660 nm	670 nm	690 nm	860 nm
P(0 Fov, $\varphi = 0°$)	Theoretical value	1.64%	4.71%	3.74%	0.84%	3.79%	3.12%
	Measured value	2.33%	1.76%	2.89%	1.33%	3.12%	4.64%

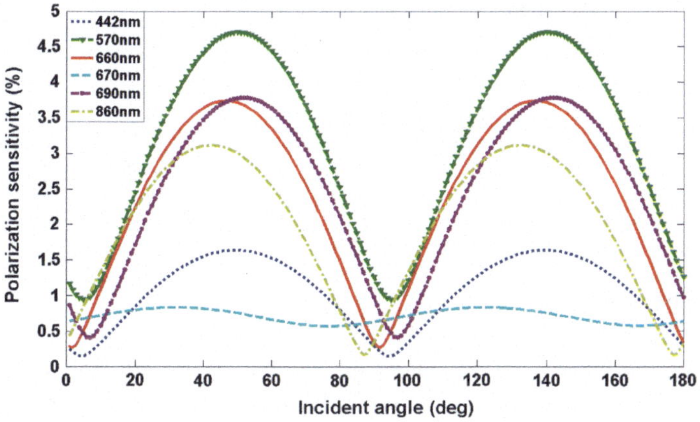

Fig. 4 Polarization sensitivity theoretical variation with different deviation angle at 0 Fov under different center wavelength

Fig. 5 Polarization sensitivity variation with different deviation angle at different field of view under center wavelength 670 nm

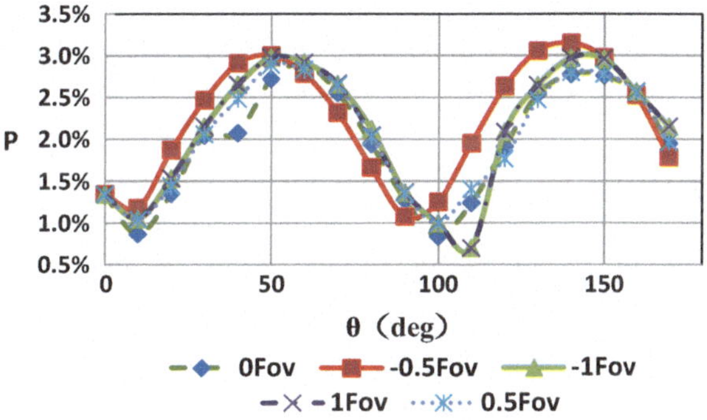

by the polarization performance of MODIS [9]. It illuminates that the experimental results are valid and not affected by external stray radiation. The polarization sensitivity variations at different field of view and the max polarization sensitivity are similar in Fig. 5. It shows that the depolarizer tested can suppress the polarization sensitivity in the whole field of view effectively.

The polarization measurement of grating spectrograph is similar as that of depolarizer. As shown in Fig. 3, move optical devices 6–10 out of light path, put the grating spectrograph in the light path and adjust it. Substitute the $DN(\theta)$ of the grating spectrograph for $I(\theta)$ to record and calculate the polarization sensitivity. The test result of grating spectrograph at different center wavelengths under different field of view is shown in Table 2.

From the Table 2, it can be seen that the polarization sensitivity of grating spectrograph all below 2%, satisfy the design requirement. The depolarizer plays a significant role in polarization suppression.

Table 2 Polarization sensitivity of grating spectrograph

Center wavelength	−1 Fov (%)	-0.5 Fov (%)	0 Fov (%)	0.5 Fov (%)	1 Fov (%)
442 nm	1.71	1.45	1.24	1.30	1.46
570 nm	1.90	1.70	1.66	1.55	1.61
680 nm	1.81	1.33	1.42	1.66	1.90
860 nm	1.47	1.76	1.80	1.88	1.63

4 Conclusion

A fair amount of work went into understanding the results and the uncertainties of the grating spectrograph polarization measurements. The results of this work indicate that spatial transmission variations in the depolarizers can be described by Muller matrix. The depolarizer and the grating spectrograph show good polarization characteristics over the needed wavelength range, and test result within the required measurement uncertainty limits. Preliminary tests hopefully show that these theoretical analyses of polarization suppression and test method can be used in other remote sensors with optical element which have obvious polarization characteristics.

References

1. Rolf Mager. Wolfgang Fricke, John P. Burrows, Johannes Frerick and Heinrich Bovensmann. A new-generation of hyperspectral remote sensing instrument[J]. Proc. of SPIE, vol. 3106, pp. 84–94, 1997.
2. Johan de Vries, Gijsbertus H.J. van den Oord, Ernest Hilsenrath, Maurice te Plate, Pieternel Levelt and Ruud Dirksen. Ozone Monitoring Instrument(OMI) [J]. Proc. of SPIE, vol. 4480, pp. 315–325, 2002.
3. M. Vedel, S. Breugnot and N. Lechocinski. Spatial calibration of full stokes polarization imaging camera [J]. Proc. of SPIE, vol. 9099, pp. 90990I-1–13, 2014.
4. Yunfeng Bai, Linjun Li, Zhelong He, Yanwei Liu, Cheng Ma, Guang Shi and Lu liu. Polarization measurement through combination polarizers [J]. Proc. of SPIE, vol. 9142, pp. 91421L-1–6, 2014.
5. Yalong Gu, Alexander Gilerson, Carlos Carrizo, Amir Ibrahim and Samir Ahmed. Polarization analysis of target imaging in underwater environment[J]. Proc. of SPIE, vol. 9099, pp. 90990X-1–12, 2014.
6. Weiwei Feng, Yanjun Ji and Ligang Chen. The impact of signal–noise ratio on degree of linear polarization measurement[J]. Optik, vol.124, pp. 192–194. 2013.
7. Carol J. Bruegge, Mary L. White, Nadine C.L. Chrien, Enrique B. Villegas and Virginia G. Ford. Multi-angle Imaging SpectroRadiometer (MISR) design issues influenced by performance requirements [J]. Proc. of SPIE, vol. 1939, pp. 104–113,1993.
8. Chen Fang, Li Ming, Xu Pengmei .The numerical analysis method of depolarizer for broadband imaging spectrometer [J]. Proc. of SPIE, vol. 8532, pp. 1–7, 2012.
9. James Young, Ed Knight and Cindy Merrow. MODIS polarization performance and anomalous four-cycle polarization phenomenon [J]. Proc. of SPIE, vol. 3439, pp. 247–256, July 1998.
10. Eugene Waluschka. Sensitivity of VIIRS polarization measurements [J]. Goddard space flight center, August 2010.

Possibility of Applying SLAM-Aided LiDAR in Deep Space Exploration

Yuwei Chen, Jian Tang, Ziyi Feng, Teemu Hakala, Juha Hyyppä, Chuncheng Zhou, Lingli Tang and Chuanrong Li

Abstract This paper discusses the possibility of applying LiDAR based SLAM technology into deep space exploring research. Nowadays, a new round of lunar exploration upsurge comes, USA and Russia resumed their previously lunar projects, while China, ESA (European Space Agency), Japan and India all have own/joint future plans of exploring moon for distinct missions. Recently, ESA proposed to construct a lunar-base by the 2030s through international collaborations. It's predictable that more state-of-art technologies will step out from the laboratory, support more scientific deep space research and play considerable roles in space exploration in future. LiDAR is a technology which is used widely for distance measuring and unknown environment surveying. LiDAR has the features including long distance detection, high range measurement resolution and robustness in complex ambient lighting situation. Currently, LiDAR is intensively utilized in various astronautics researches such as rover landing, docking in planetary orbit and atmospheric composition detection. This paper introduces two 2D and one 3D mapping systems using SLAM-aided LiDAR, presents all the three systems currently performance in indoor and forest cases, and finally discusses the possibility of applying the technology into various applications such as unknown environment mapping, navigation, BIM (Building Information Modelling) and damage detection in deep space exploration.

Keywords Laser scanning · SLAM · Deep space exploration · NAVIS

Y. Chen · C. Zhou · L. Tang · C. Li
Key Laboratory of Quantitative Remote Sensing Information Technology, Academy of Opto-Electronics, Chinese Academy of Science, Dengzhuangnan Road 9, Beijing 100190, China

Y. Chen · J. Tang · Z. Feng (✉) · T. Hakala · J. Hyyppä
Department of Remote Sensing and Photogrammetry, Finnish Geospatial Research Institute, Geodeetinrinne 2, 02431 Kirkkonummi, Finland
e-mail: ziyi.feng@maanmittauslaitos.fi

J. Tang
GNSS Research Center, Wuhan University, Luoyu Road 129, Wuhan 430079, China

H.P. Urbach and G. Zhang (eds.), *3rd International Symposium of Space Optical Instruments and Applications*, Springer Proceedings in Physics 192,
DOI 10.1007/978-3-319-49184-4_24

1 Introduction

The first exploration beyond earth orbit is the mission "Pioneer Zero" launched in 1958 by US [1]. In the following 20 years, the lunar exploration had a rapid development as results of space race between US and Soviet Union. Various Lunar missions including impacting, lunar surface photography, landing, sample obtaining have been completed Nevertheless, scientists believe that there is still plenty of scientific information on the moon deserving to be explored, and the nature resource in the moon can be utilized [2]. Therefore, the lunar missions have attracted more attention worldwide after 1990s, and more countries involve in the exploration activities [3, 4]. Lunar surface detection and the environment mapping are significant both for the moon exploration and rover movement. When constructing the lunar-base which is proposed by ESA (European Space Agency), accurate topographic map, instrument localization and movement control are obviously critical. Through previous studies, LiDAR can contribute in solving these issues.

LiDAR is an import scientific tool for planetary exploration, which is widely used for autonomous precision landing and hazard avoidance of spacecraft, terrain reconstruction and autonomously navigating of mobile robotics. The Japanese Hayabusa spacecraft landed at the Muses Sea on Itokawa using optical images and LiDAR topography [5]. In 2013, Chinese Chang'E-3 firstly carried out the autonomous hazard avoidance and precision landing on celestial body outside Earth using the LiDAR topography [6]. In addition, NASA developed a 3D space-borne scanning laser radar system for shuttle surface inspection which is critical to prevent the launching accidents caused by cracks or missing pieces [7]. Therefore, the laser technology and corresponding instruments were already employed in various space missions.

As said before, accurate terrain map and real time localization are significant for further scientific research and nature resource exploration on the moon. There is a possibility to acquire the information by combining the SLAM (Simultaneous Localization and Mapping) technology and LiDAR. SLAM is a technology which enables a rover to generate a consist map of a scanned unknown environment, and localize itself into the map simultaneously [8, 9]. A SLAM scan-matching algorithm called IMLE (Improved Maximum Likelihood Estimation) is proposed and developed in FGI (Finnish Geospatial Research Institute) which improves the mapping accuracy [10]. Two 2D SLAM-aided LiDAR mapping systems were also developed by FGI. With the proposed algorithm, the system positioning resolution reaches 0.03 m in indoor case and 0.32 m in dense forest whose environment is much more complex. In addition, another 3D mapping system using same matching algorithm was developed, which can reach centimeter measurement accuracy in indoor situation.

In the following chapters, the three SLAM-aided mapping systems and the system performances in both indoor and outdoor cases are presented in more detail in Chap. 2. Furthermore, the possibilities of applying SLAM technology combining with LiDAR in the astronautics research are discussed in Chap. 3. Some conclusions are drawn in the end.

2 System Introduction

The two 2D mapping systems consist high-end version system ROAMER 2 and low-end version system NAVIS. Comparing with these two systems NAVIS-3D mapping system equips with two laser scanners. All the three systems can utilize IMLE (Improved Maximum Likelihood Estimation) for location and mapping. However, the distinct hardware affects measurement results accuracy. In this chapter, basic information and performance of the three systems are introduced.

2.1 System Overview

Figure 1 demonstrates the overview of the two mapping systems FGI ROAMER R2 and NAVIS-2D. It can be seen that the ROAMER R2 was mounted on an all-terrain-vehicle (ATV) when working in the forest, while the NAVIS-2D was fixed on a cart which can be moved manually.

The ROAMER R2 system contains a NovAtelSPAN navigation system and a high-speed phase-shift LiDAR (FARO Focus3D 120S), in the navigation system

(a) FGI ROAMER R2 mobile mapping system

(b) NAVIS indoor mapping hardware platform

Fig. 1 **a** ROAMER 2 in ATV with mounted devices including GNSS antennas [10] and receivers, IMU and LiDAR and **b** NAVIS-2D on cart with IMU and laser scanner [11] there are a GNSS receiver (NovAtel Flexpak6) and an IMU (NovAtel UIMU-LCI) [10]. With UIMU-LCI, the Gyro and Accelerometer Bias Instability are less than 1.0°/h and 1.0 mGal respectively theoretically. The mapping system has the capability of 3D forest mapping which is based on the sensors installed on the platform. The LiDAR sensors utilized in ROAMER R2 is small footprint laser scanners—FARO Focus3D X330 which are maintained horizontally in the backward position. As the LiDAR sensor beam divergence is 0.011° (0.19 mrad), and the beam diameter at exit is 2.25 mm. When conducting forest mapping, the maximum measured range was 25 m, therefore the maximum footprint size is around 7 mm.

NAVIS-2D is a similar IMU-aided slam mapping system with a lower cost IMU compared with ROAMER R2 [11]. The employed IMU is Xsens MTi-G 710 which is MEMS (Micro Electro Mechanical systems) with six DOF (Degree of Freedom). Compare with the IMU adopted in ROAMER R2, in Xsense the Gyro and Accelerometer Bias Instability are much larger which are 200°/h and 2000 mGal respectively. The laser scanner utilized is Hokuyo UTM-30LX-EW which is multiple-echoes detection enabled laser scanner. And the multiple-echoes detection functionality can help to collect more environment information comparing with single echo model. As shown in Fig. 1, the laser scanner and IMU measurement results are recorded by the laptop. It's possible for NAVIS to detect and filter out the dynamic objects such as moving car, which brings more possibilities of applications based on NAVIS.

When the system works, the laser scans and attitude from LiDAR and IMU respectively are synchronized. With attitude data, the scans are projected from 3D spatial space to 2D plane. The projection process details can be referred in [10]. After the projection, SLAM positioning and mapping should be performed. Considerable studies about SLAM matching algorithm have been done to improve the final mapping result accuracy. Generally, in scanning matching two or more successive scan points frames are utilized to determine the platform movement via the matching algorithms. Some popular algorithms are classical Iterative Closest Point (ICP), Iterative Closest Line (ICL), and Maximum Likelihood Estimation (MLE) [12]. However, the IMLE algorithm has been developed by FGI and applied in all three mapping systems. More details and analysis about the IMLE algorithm can be referred in [10].

In NAVIS-2D a so-called greater value rule is adopted to address the overwrite conflict, in which whenever the new scan projected on an occupied grid in the map grid cell, the maximized likelihood value of this scan is given to be 0.9. In addition, the other matching algorithms could be employed together to further improve the matching accuracy. For more information around the algorithm applied in indoor mapping system can be referred in [12].

Both ROAMER R2 and NAVIS-2D are 2D mapping systems. NAVIS-3D mapping system is developed based on the NAVIS-2D system. In the original 2D mapping system, it generated 2D horizontal map with one horizontal-scan LiDAR. In NAVIS-3D mapping system another LiDAR was equipped vertically. 3D modelling of the target area can be generated based on the information collected

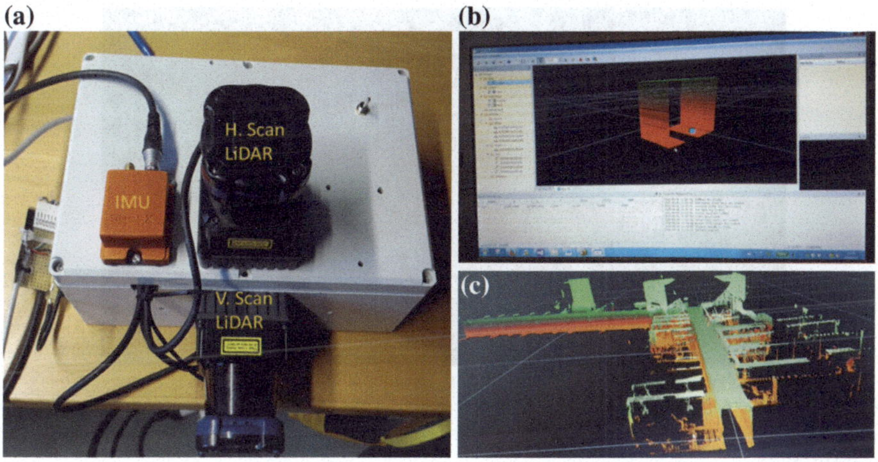

Fig. 2 **a** NAVIS-3D mapping system **b** NAVIS program for data collecting and **c** the final constructed 3D modelling

simultaneously from both horizontal and vertical scanning. However, the same matching algorithm which has been described before was adopted in the system. The system overview and experiment results can be referred in Fig. 2. Via diagram (a) in Fig. 2, it can be seen that the two LiDARs are fixed at mutually right-angled position.

2.2 System Performance

The ROAMER R2 system had been tested in EVO Southern Finland (61°19′ N, 25° 11′ E). It has various features to be a forest study site, e.g. it provides cross-section from natural to managed southern boreal forests, and the dominant tree species are Scots pine and Norway spruce. There are 224 trees along the test route have been measure by a total station and RTK GPS in advance to obtain the ground truth data. Part of scanning results could be observed in Fig. 3. Positioning errors are presented by the RMSE (Root Mean Square Error). The general results can be found in Table 1, and it can be figured out that the mapping accuracy can reach 0.32 m in the best case. More details around the test in forest and results analysis can be referred in [10].

The NAVIS-2D system was tested in the FGI second floor to generate the map of the whole floor. The mapping result can be referred in Fig. 4. The system was supposed to be used for UGVs (Unmanned Ground Vehicles) [11]. Through the test results in [11] the conclusion can be drown that the indoor positioning accuracy can reach 0.037 m with the NAVIS-2D system. Consequently, it can be reliable method to achieve quick mapping goal.

Fig. 3 The SLAM result in a dense boreal forest with ROAMER 2, the *red point* is the reference stem center measured by a total station with known coordinate, and the *white point* cloud is the SLAM result

Table 1 The positioning accuracy in different forest

Forest condition	Easting (m)	Northing (m)	2D map results (m)
Open forest	1.73	2.33	2.90
Dense forest	0.16	0.27	0.32

The NAVIS-3D mapping system was tested as well in the same corridor with NAVIS. The modelling result can be viewed in (c) of Fig. 2. It can be seen that the whole model of the corridor is clearly. Comparing diagram (c) in Figs. 2 and 4 which was obtained through NAVIS system, it is liable to figure out that the 3D modelling can be matched with the 2D positioning mapping. Through measuring test, the modelling results can reach centimetre accuracy.

3 Possibility Discussion

It's self-evident that the topography and hazard maps are critical for spacecraft landing site selection. LiDAR sensors were utilized to build the surface maps. The Lunar Orbiter Laser Altimeter (LOLA) instrument used on NASA's Lunar Reconnaissance Orbiter (LRO) mission assists in selecting the landing sites for future exploration activities, and detecting water ice on lunar surface [13]. Similar technology was employed on Chang'E-1 mission launched by China National Space Administration (CNSA) to assist in generating lunar surface 3-D image [14]. However, this technology could be improved via combining the SLAM technology. With the SLAM-aided LiDAR, more accurate surface map could be generated for landing site selection. Moreover, 3D model of unknown environment can be built when the rover works on the planetary through 3D LiDAR sensors. Through the

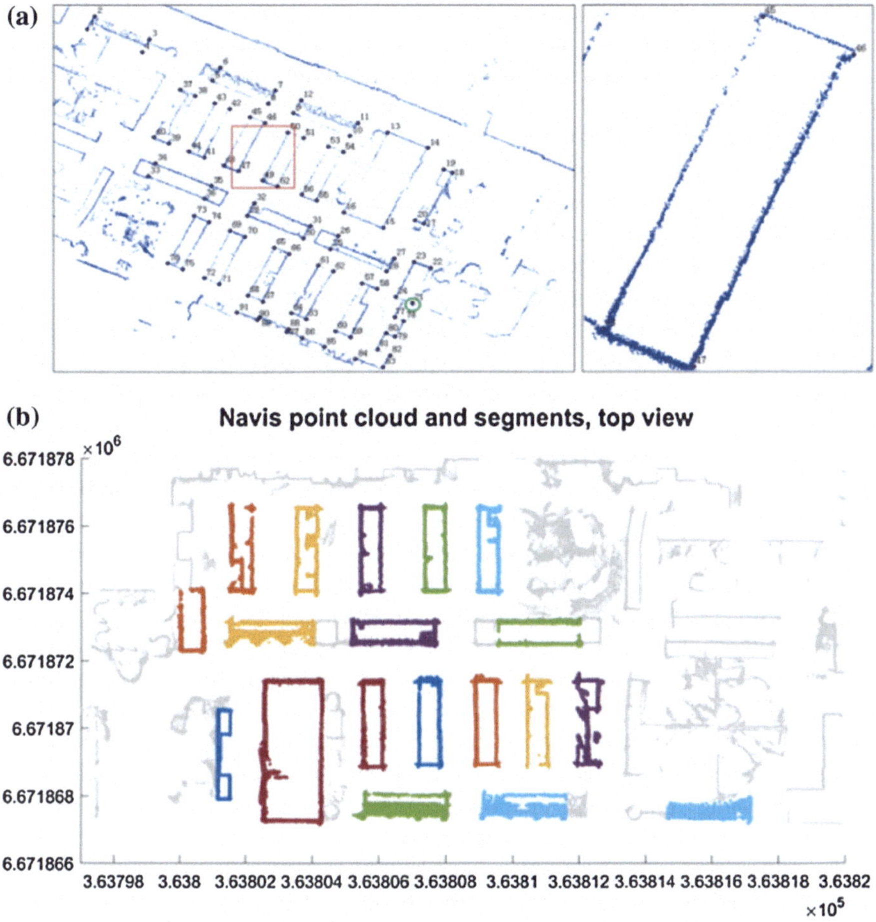

Fig. 4 **a** The SLAM result of FGI library generated by NAVIS-2D **b** generated indoors map with NAVIS-2D

NAVIS-3D mapping system introduction it is clearly that the 3D model of unknown environment can be constructed during the scanning. Therefore, 3D map can be generated as the rover moves on the planet during extra-terrestrial exploration. The 3D map will present a more intuitive view of the explored environment for landing site selection and further scientific study. Another significant issue in extra-terrestrial exploration is navigation on the planetary. LiDAR sensors were already applied for rover navigation. NASA has developed an autonomous navigation system based on LiDAR sensing, which is deployed on NASA's autonomous rover and Lunar Electric Vehicle to increase the terrain hazard situational awareness [15]. Canadian Space Agency has developed an autonomous planetary mobile robotics based on LiDAR sensing, while uses two different LiDAR sensors, a

ILRIS-3D sensor from Optech for outdoor terrain reconstruction and a SICK sensor for obtaining detailed terrain models for localization and path planning [16, 17]. As the mapping is conducted simultaneously of positioning in an unknown environment in the SLAM-aided LiDAR system, this LiDAR will assist the rover in navigation with less computation which leads to more efficiency position data transmission back to earth for navigation decision. Moreover, the 2D/3D models of the unknown environment will be constructed while exploration. Therefore, the introduced LiDAR based SLAM technology can be studied for rover autonomous navigation system development and terrain model construction.

As said before, constructing a moon base is imperative. It will be significant to acquire exhaustive and accurate information of the buildings under construction. It can be imaged that various kinds of robots will be involved into constructing. Mobile robotics is the foundation for information collection, and LiDAR sensor is an ideal eye of the mobile robotics due to its accuracy, long range and robustness in various lighting conditions. Using the SLAM-aided 3D LiDAR modelling system can be an efficient approach to acquire the space village visible BIM. The BIM will assist researchers in making decisions about robots operations, and furthermore it will be indispensable when developing autonomous decision-making robots.

The technology can be applied in damage estimation of spacecraft as well. It's described in [7] that how the object surface damage can be detected based on the scanned distance deviation with laser scanner. However, in the previous detection method the principle is laser triangulation. Now, the target position and distance can be measured by SLAM based LiDAR directly. Through comparing the new constructed target 3D model and the pre-saved original 3D model, the damage situation can be calculated and obtained. The unique benefit of SLAM based LiDAR system is its positioning function in unknown environment which enables the damage detection being ubiquitously conducted. It will be possible to design regular damage detection routing to enable researchers have more reaction time, before the worst situation happens when any damage happens during space missions.

4 Conclusions

In this paper, Three SLAM-aided LiDARs which are used for unknown environment mapping are introduced. One system was tested in forest area, while the other two systems which are 2D mapping system and 3D system respectively were tested in indoor environment. The result proves that the system positioning accuracy reaches 0.03 m in indoor situation and 0.32 m in forest for 2D system, and in 3D system the accuracy reaches centimetre level as well. The introduced SLAM-aided LiDARs demonstrate how the technology is applied on ground mapping system and its positioning process, as well as accuracy. Based on the achievable features, the possibilities of applying the SLAM-aided LiDAR mapping system in lunar exploration was discussed. Four possibilities including 3D map generation, navigation, BIM gathering and damage detection and estimation are presented. Among those possibilities, various technologies based on LiDAR have already been applied in most aspects. Through previous analysis, theoretically, combining the SLAM technology can make the existed LiDAR systems perform better and more functions. In addition, the features of SLAM enable itself possible to play an important role in future when constructing lunar village.

What's more, it is possible to reduce the explorer mass by integrating the mentioned multiple functions into limited devices which is significant in deep space exploration. Consequently, applying SLAM-aided LiDAR might be a direction to further improve the deep space exploration systems.

Acknowledgements This study was financially supported by the Chinese Academy of Science (181811KYSB20130003), by Chinese Ministry of Science and Technology (2015DFA70930) and the National Nature Science Foundation of China (41304004), by the Academy of Finland projects "Towards Precision Forestry", "Centre of Excellence in Laser Scanning Research (CoE-LaSR) (272195)", "Interaction of Lidar/Radar Beams with Forests Using Mini-UAV and Mobile Forest Tomography".

References

1. http://nssdc.gsfc.nasa.gov/nmc/spacecraftDisplay.do?id=ABLE1,visited on 13th, June, 2016.
2. http://phys.org/news/2009–07-moon-potential-goldmine-natural-resources.html,visited on 13th, June, 2016.
3. China Nation Space Administration," Chinese Lunar Exploration Program", June, 2014.
4. Y. Takizawa, S. Sasaki, M. Kato, "KAGUYA (SELENE) mission overview," Proceedings of the 26th ISTS (International Symposium on Space Technology and Science), Hamamatsu City, Japan, June 1–8, 2008.
5. H. Yano, T. Kubota, H. Miyamoto, T. Okada, D. Scheeres, Y. Takagi, K. Yoshida, M. Abe1, S. Abe, O. Barnouin-Jha, A. Fujiwara, S. Hasegawa, T. Hashimoto, M. Ishiguro, M. Kato, J. Kawaguchi, T. Mukai, J. Saito1, S. Sasaki, M. Yoshikawa, "Touchdown of the Hayabusa Spacecraft at the Muses Sea on Itokawa", *Science*, vol. 312, pp. 1350–1353, 2006.
6. Zhang H H, Liang J, Huang X Y, Zhao Y, Wang L, Guan Y F, Cheng M, Li J, Wang P J, Yu J, Yuan L, "Autonomous hazard avoidance control for Chang'E-3 soft landing (In Chinese)". *Sci Sin Tech*, vol, 44, pp. 559–568, 2014.
7. Lamoreux, James C., Siekierski, James D., Carter, J. P. N., "Space Shuttle thermal protection system inspection by 3D imaging laser radar", Proc. SPIE, Vol, 5412, PP, 273–281, 2014.
8. H. Durrant-Whyte and T. Bailey, "Simultaneous localization and mapping: part I," IEEE Robotics & Automation Magazine, vol. 13, pp. 99–110, 2006.
9. T. Bailey and H. Durrant-Whyte, "Simultaneous localization and mapping (SLAM): part II," Robotics & Automation Magazine, IEEE, vol. 13, no. 3, pp. 108–117, 2006.
10. J Tang, Y Chen, A Kukko, H Kaartinen, A Jaakkola, E Khoramshahi,T Hakala, J Hyyppä, M Holopainen, H Hyyppä, "SLAM-Aided Stem Mapping for Forest Inventory with Small-Footprint Mobile LiDAR", *Forests*, vol. 6, pp. 4588–4606, 2015.
11. J Tang, Y Chen, M Lehtomäki, H Kaartinen, L Zhu, K Risto, Y Wang, J Hyyppä, H Hyyppä, "The Accuracy Comparison of Three SLAM based Indoor Mapping Technologies", unpublished, 2016.
12. J Tang, Y Chen, A Jaakkola, J Liu, J Hyyppä and H Hyyppä, "NAVIS-An UGV Indoor Positioning System Using Laser Scan Matching for Large-Area Real-Time Applications", *Sensors*, vol. 14, pp. 11805–11824, 2014.

13. H. Riris, J. Cavanaugh. X. Sun, P. Liiva, M. Rodriguez, G. Neuman, NASA GSFC, USA, Sigma Space, "The lunar orbiter laser altimeter (LOLA) on NASA's lunar reconnaissance orbiter (lro) mission", International Conference on Space Optics, 2010.
14. Wang Jianyu, Shu Rong, Chen Weibiao, Jia Jianjun, Huang Genghua, Wang Binyong, Hou Xia, "CE-1 Satellite based Laser Altimeter", *Science China press*, Vol 40, issue 8, pp. 1063–1070, 2010.
15. L Pedersen, M Allan, H Utz, M Deans, X Bouyssounouse, Y Choi, L Flückiger, S Y. Lee, V To, J Loh, W Bluethmann, R R. Burridge, J Graf, K Hambüchen "Tele-operated Lunar Rover Navigation Using LiDAR". International Symposium on Artificial Intelligence, Robotics and Automation in Space (I-SAIRAS), Turin, Italy, 2012.
16. I Rekleitis, J L Bedwani, E Dupuis. "Autonomous Planetary Exploration using LIDAR data", IEEE International Conference on Robotics and Automation Kobe International Conference Center Kobe, Japan, 2009.
17. M W Hussein, J W Tripp. "3D Imaging Lidar for Lunar Robotic Exploration", Proc. SPIE 7331, Space Exploration Technologies II. Orlando, 2009.

In-Orbit Calibration Method for Sun Sensor Based on Sun Ephemeris and Star Sensor

Qiao-yun Fan and Jia-wen Peng

Abstract An in-orbit calibration model has been adapted to improve the sun sensor's in-orbit accuracy. This model consists of 2 sub-models. One is the measurement model of the sun sensor, the other is pseudo measurement model. In the measurement model, variety of in-orbit factors have been taken into consideration, and the measurement model has been modified to compensate the errors caused by the in-orbit factors. While in the pseudo measurement model, the accurate sun ephemeris and the high-precision output of star sensor are combined together to calculate the true sun vector in spacecraft-body coordinate system. The accurate sun ephemeris is used to calculate the true sun vectors in GCRS (Geocentric Celestial Reference System), the attitude data from star sensor is used in coordinate system transformation of the sun vector. The pseudo measurement model has been analyzed considering varies errors, and the result shows that the errors would not have effect on the calibration model. The least squares regression algorithm is applied in this calibration model to optimize the internal and external parameters. The calibration model has been verified on a sun sensor which was installed on a satellite and launched in September 2015. The result shows, without in-orbit calibration, the accuracy of the sun sensor in-orbit decreased to 0.22° in x direction and 0.27° in y direction, while the accuracy was 0.11° and 0.14° before launching. With the in-orbit calibration method applied, the accuracy of the sun sensor would rise to 0.15° and 0.18°.

Keywords Sun sensor · In-orbit calibration · Sun ephemeris

The original version of this chapter was revised: Spell errors in the first author's name has been corrected. The erratum to this chapter is available at doi: 10.1007/978-3-319-49184-4_50

Q. Fan · J. Peng (✉)
Key Laboratory of Precision Opto-Mechatronics Technology,
Ministry of Education, School of Instrument Science and Opto-Electronics
Engineering, Beihang University, Beijing 100191, People's Republic of China
e-mail: penguinleo@163.com

H.P. Urbach and G. Zhang (eds.), *3rd International Symposium of Space Optical Instruments and Applications*, Springer Proceedings in Physics 192,
DOI 10.1007/978-3-319-49184-4_25

1 Introduction

Sun sensor is an important attitude sensor. And it has been widely used in the space mission. Figure 1 shows a digital sun sensor based on the linear CCD developed by our laboratory. This type sun sensor has been applied to 3 small satellites [1].

The accuracy of the sun sensor could achieve the expected accuracy after the ground calibration. But the installation error, the vibration during launching, the temperature in space all could decrease the accuracy of the sun sensor. To reduce the effects of these factors, the in-orbit calibration method is introduced. The sun ephemeris and the accurate data of star sensor are combined in this method. Not only on the sun sensor in Fig. 1, but also on other types sun sensors, could this in-orbit calibration method be applied.

2 In-Orbit Calibration Model

2.1 Measurement Model of Sun Sensor

After ground calibration, sun sensor could measure sun angle accurately. But the sun sensor has to experience dramatic impact during launching, intense temperature changing in orbit. These could reduce the measurement accuracy. What's more, the installation error would have great impact on the accuracy. So it is necessary to calibrate in orbit for sun sensor [2–4].

In-orbit error model is founded on the error model used in ground calibration. Considering various in-orbit factors, there are 5 inner parameters in the in-orbit measurement model. These parameters are sensitive to the factors mentioned above and could reduce the accuracy of the sun sensor, they are listed as following in (3).

The in-orbit factors have impact on installation accuracy too. Suppose that the sun sensor is installed on the surface of satellite, and the ideal installation matrix is marked as $\boldsymbol{R}_{inst0}$. Euler angle a_x, b_x, c_x is introduced to describe the installation error

Fig. 1 Sun sensor

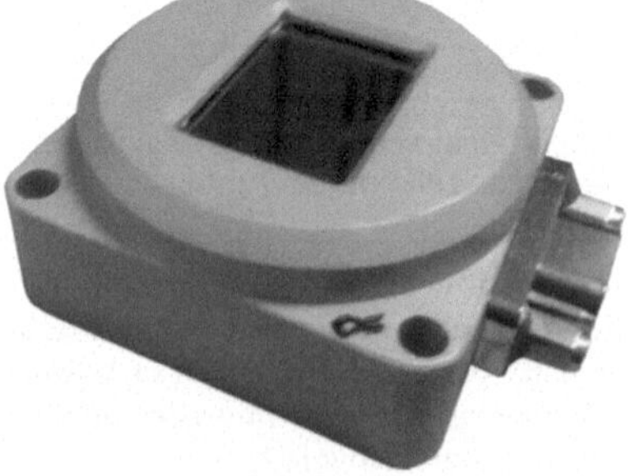

matrix. And corresponding basic rotate matric were denoted as $\boldsymbol{R}_a$, $\boldsymbol{R}_b$, $\boldsymbol{R}_c$. Then the installation matrix R_{inst} could be written as (1).

$$\boldsymbol{R}_{\mathrm{inst}} = \boldsymbol{R}_{\mathrm{inst0}}\boldsymbol{R}_{\mathrm{c}}\boldsymbol{R}_{\mathrm{b}}\boldsymbol{R}_{\mathrm{a}} \tag{1}$$

Define the output of the sun sensor as $\boldsymbol{S}_{\mathrm{mes}} = [\tan\alpha \quad \tan\beta \quad 1]^{\mathrm{T}}$, the sun light vector in the satellite body coordinate could be expressed by an equation containing $\boldsymbol{S}_{\mathrm{mes}}$.

$$\boldsymbol{S}_{\mathrm{b}} = \boldsymbol{R}_{\mathrm{inst}} \cdot \frac{\boldsymbol{S}_{\mathrm{mes}}}{|\boldsymbol{S}_{\mathrm{mes}}|} \tag{2}$$

Equation (3) is the error model of the sun sensor. As it shows in (3), the error model contains 8 parameters. There are 5 inner parameters and 3 external parameters.

$$\begin{cases} \alpha = F_1'(X_1' \quad X_2' \quad b \quad c \quad T_1 \quad T_2 \quad T_3 \quad a_x \quad b_x \quad c_x) \\ \beta = F_2'(X_1' \quad X_2' \quad b \quad c \quad T_1 \quad T_2 \quad T_3 \quad a_x \quad b_x \quad c_x) \end{cases} \tag{3}$$

The α and β is the sun angle in the satellite body coordinate. The X_1' and X_2' are the facula positions measured by the image sensor. The rotate angle b and c, the dislocation vector $[T_1 \quad T_2 \quad T_3]^{\mathrm{T}}$ are used to describe the position of the image sensor in the mask coordinate. The position relationship between the mask and the image sensor is the foundation of the measurement principle. And the detail error analysis is showed in the Ref. [1].

2.2 *Pseudo Measurement Model*

In the in-orbit calibration model, the accurate output of the star sensor is used as the true attitude value. But the true value of the sun angle could not be calculated with the accurate attitude data from star sensor alone. So the exact sun ephemeris is taken into the model. The combination of attitude data from star sensor and sun ephemeris could easily calculate the true value incident angle [5].

i. *Coordinate definition*

Geocentric Celestial Reference System (GCRS) is marked as $O_G - X_G Y_G Z_G$.

Geocentric inertial coordinate system is marked as $O - XYZ$, O is located at the center of earth. The X axis pointed to the mean equinox of J2000.0. the Z axis was perpendicular to the mean equatorial plane of J2000.0 and pointed to North pole of celestial. The Y axis was defined by right hand rule.

Center of mass inertial coordinate system of the satellite was marked as $O_i - X_i Y_i Z_i$, its origin is located in the centroid of satellite. And the axis of this frame are parallel to the axis of geocentric inertial coordinate system respectively.

The satellite coordinate is denoted $O_b - X_bY_bZ_b$, the origin is put at the center of the satellite, the axis of the coordinate are fixed on the body of satellite.

ii. *Model of the Pseudo measurement*

The pseudo measurement is used to calculate the true value of the sun angle in the calibration model. The inputs of the pseudo measurement are the accurate output of the star sensor and the measurement time in ephemeris time scale. This section describes the pseudo measurement model.

The model is shown as Fig. 2. And the calculation steps of the model are listed below.

- Time scale conversion. The time scale of the satellite is Coordinated Universal Time(UTC), while the time scale of the DE405 ephemeris is Ephemeris Time.
- Sun vector calculation. After the time scale conversion, the time point at measurement could be substituted into the DE405 Ephemeris. Then the sun vector in GCRS would be returned.
- Coordinate transformation. The sun vector in GCRS should be transformed to the sun vector in Geocentric inertial coordinate.
- Attitude calculation. Fitting and interpolation methods are used in attitude data processing to calculate the best fitted attitude at the measuring point.
- Coordinate transformation. The attitude at measuring point could be used to transform the sun vector from Geocentric inertial coordinate to satellite coordinate.

The sun vector in satellite coordinate could be transformed to the measurement coordinate of the sun sensor.

Time scale conversion is an important part in the pseudo measurement. UTC is the time scale using in satellite, while Barycentric Dynamical Time (TDB) and Ephemeris Time is adopted in DE405 Ephemeris. DE405 is developed by JPL. In DE405, TDB and Ephemeris Time have the same meaning. In this paper, time zone conversion and leap second compensation are introduced to transform UTC to International Atomic Time (TAI). Then according to the rule set by International Astronomical Union (IAU), Terrestrial Time (TT) could be get from TAI by adding a constant 32.184 s. At last, Fairhead-Bretagnon sequence was introduced to calculate the bias between TT and Ephemeris Time [6]. Fairhead-Bretagnon sequence is shown in reference [6].

$$\boldsymbol{S} = \boldsymbol{B} \cdot \boldsymbol{S}_G \tag{4}$$

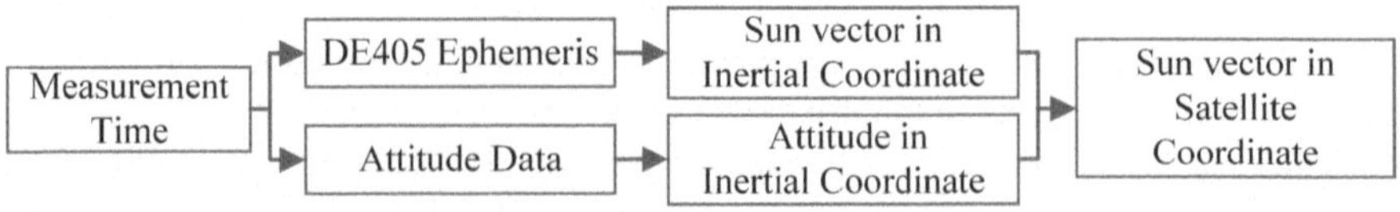

Fig. 2 The model of pseudo measurement

The most important process is the coordinate transformation. Figure 3 describes sun vector in different coordinate, and the relationships among them. In the left part of Fig. 3, $\boldsymbol{S}_{\mathrm{G}}$ was the sun vector calculated by DE405 in GCRS. According to the IAU's definition of GCRS and J2000.0 inertial coordinate, there is a conversion matrix $\boldsymbol{B}$ between these coordinates. The rotation matrix $\boldsymbol{B}$ was a constant matrix which is independent of time and is named framework deviation matrix. So it is easy to transform the $\boldsymbol{S}_{\mathrm{G}}$ to $\boldsymbol{S}$ in the J2000.0 inertial coordinate system through the rotation matrix $\boldsymbol{B}$.

Compared with the distance between the earth and the sun, the position of the satellite in the J2000.0 inertial coordinate is too small, there is no need to consider its affection. Then according to the coordinate definition, the sun vector in Center of mass inertial coordinate is equal to the $\boldsymbol{S}$.

$$\boldsymbol{S}_{\mathrm{i}} = \boldsymbol{S} \tag{5}$$

Take advantage of the accuracy data from star sensor, the attitude of the satellite could be calculated out and the rotation matrix $\boldsymbol{R}_{\mathrm{ib}}$ could help to transform the sun vector into the satellite coordinate.

$$\boldsymbol{S}_{\mathrm{b}} = \boldsymbol{R}_{\mathrm{ib}}^{\mathrm{T}} \cdot \boldsymbol{S}_{\mathrm{i}} \tag{6}$$

In summary, the sun vector could be calculated out by measurement time and the accurate attitude data. As long as the attitude and the time data were accurate enough, the output of the pseudo measurement could be a true value in calibration.

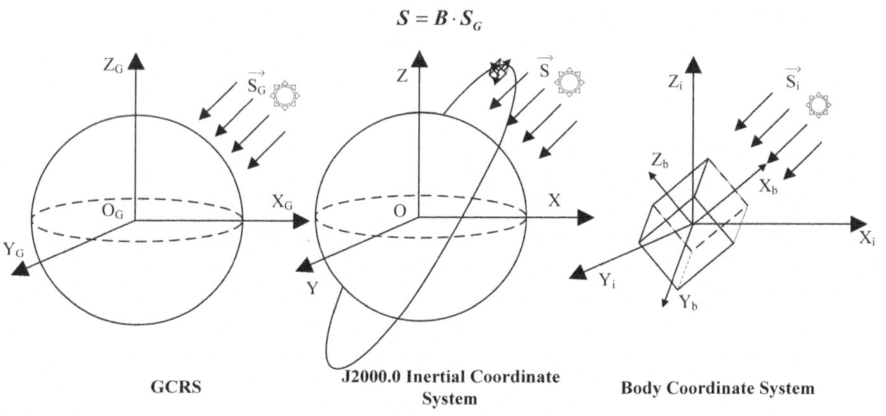

Fig. 3 Transformation between coordinate

But actually, the asynchronous measurement between the sun sensor and the star sensor would introduce some errors in the calibration model. The pseudo measurement could contain some errors.

2.3 Error Analysis of the Calibration Model

The outputs of sun sensor and star sensor contain time delay more or less. In other words, if the computer on the satellite require attitude data, these sensors would return the results in last measurement period. Then the delay is less than one measurement period. If the measurement time delay of the sun sensor was substituted into the DE405 Ephemeris in calculation, the sun vector in pseudo measurement model would contain error. So the sun vector error sensitivity analyzation is used to evaluate the error caused by the sun sensor measurement delay. And the measurement delay time of the star sensor could result in the attitude error, then the attitude error sensitivity analyzation is used to evaluate the influence of the star sensor measurement delay.

i. *Sun vector error*

DE405 Ephemeris is a precise ephemeris which is used to describe the movement of the planets in solar system. The errors in ephemeris are ignored. But because of the measurement delay, the sun vector calculated from DE405 could contain errors. In ground test, the minimum update rate of the sun sensor is about 10 Hz. So the maximum measurement delay is 100 ms. Due to the accuracy limitation of the sun sensor, and the huge distance between sun and earth, it is obvious that the changing of sun vector in measurement delay could be ignored, in other words, the changing of the sun vector is less than the resolution of the sun sensor.

ii. *Satellite attitude error*

In this paper, the rotation order was 'YZX'. Star sensor is an accurate attitude sensor, so it was possible to calibrate the sun sensor by the output data of the star sensor. But as mentioned above, sun sensor data and star sensor data contained measurement delay more or less. Fitting and interpolation methods are the best way to get the best fitted data.

Figure 4 is the attitude data. X axis was time in second. The dotted line was the fitting result, those dots on the line were the attitude data from star sensor. These attitude data are intercepted from the whole data. It is obvious that the satellite was rotating around the Y axis. The RMSE in Y was 1.21e−4, the RMSE in X was 3.17e−05, and the RMES in Z was 7.75e−05. The fitting results shows that the movement of the satellite is very stable, and the star sensor is working normally with high accuracy.

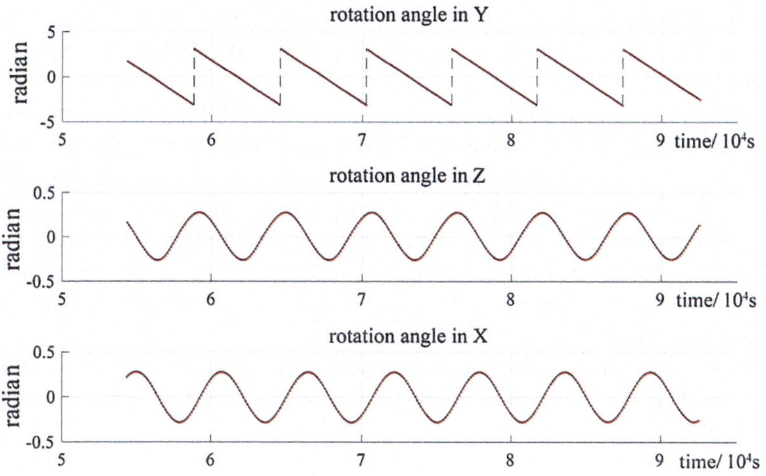

Fig. 4 Attitude data

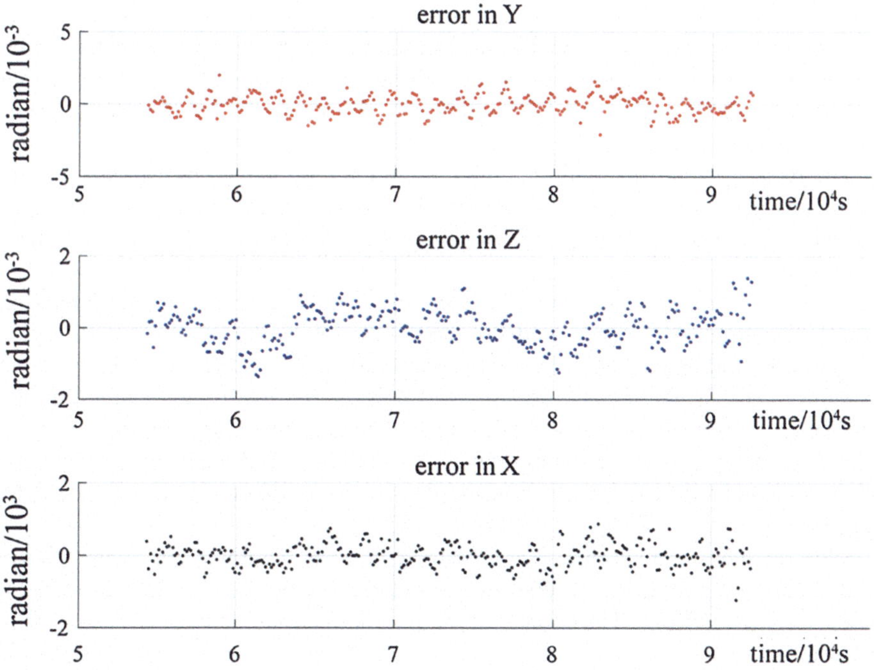

Fig. 5 Fit error

Figure 5 is the attitude fit error in three direction. The maximum error was less than 5e−4 rad. The attitude error was far less than the measurement accuracy of the sun sensor.

According to the attitude data, the rotation of the satellite was very slow, the angular velocity in Y direction was −0.0011 rad/s. Considering the measurement delay of the sun sensor and the star sensor, the attitude data and the measurement time could be misalignment. The minimum update rate is 10 Hz, so the maximum delay is 200 ms. In other words, the maximum angle changing caused by delay is 0.00022 rad. It means that the during the attitude delay, the change of attitude was very small, smaller than the resolution of the sun sensor. And the change could be ignored in this paper.

All above, the pseudo measurement model is accurate enough to calibrate the sun sensor. And the calibration model could reduce the affection caused by the in-orbit factors.

2.4 Parameter Optimization Algorithm

In this paper, there are 8 parameters needs to calibrate. 5 internal parameters and 3 external parameters. In order to achieve high-precision error compensation, internal and external parameters will be calibrated simultaneously by the least-squares regression algorithm.

The measurement time and the output of sun sensor at each measurement point are recorded by the computer in satellite. The outputs of sun sensor are $\tan(\alpha_{\mathrm{mes}})$ and $\tan(\beta_{\mathrm{mes}})$. The α_{mes} and β_{mes} are components of the sun angle in measurement coordinate.

From (3), it is clear that the output of the sun could not be used in the calibration model. The inputs of the error model of the sun sensor are the locations of the facula measured by the image sensor. So before calibration, it's necessary to calculate the location of the facula at each measurement point. After the ground calibration the parameters of the measurement model are known accurately, and the measurement model has downloaded in the sun sensor. With the output of the sun sensor and the measurement model in the sun sensor, the locations of facula X_1', X_2' could be approached.

With the measurement time record by the computer and the attitude data, the sun angle of pseudo measurement could be calculated out by the pseudo measurement model $\tan(\alpha_0)$, $tan(\beta_0)$. The outputs of the pseudo measurement are the standard in calibration.

Substitute X_1', X_2' into (3). The output of the sun sensor in-orbit error model and the output of the pseudo measurement model differ from each other. The output of the pseudo measurement is standard; the measurement model of the sun sensor is under test. Take advantage of the least-squares regression algorithm in calibration, the parameters would approach to a best fitting value.

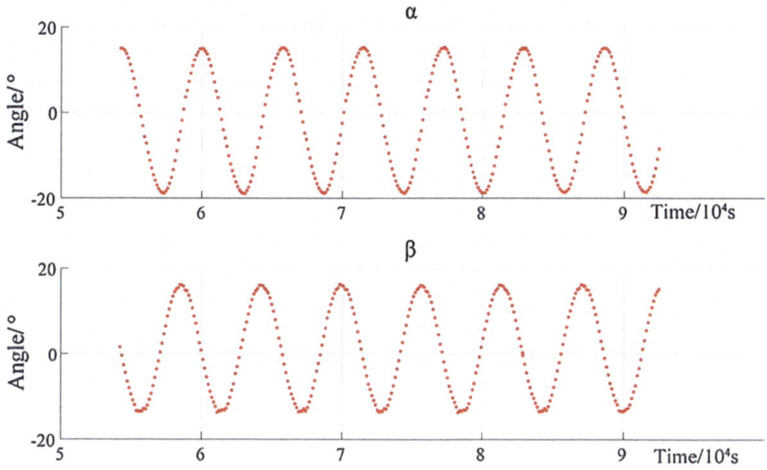

Fig. 6 Output of sun sensor

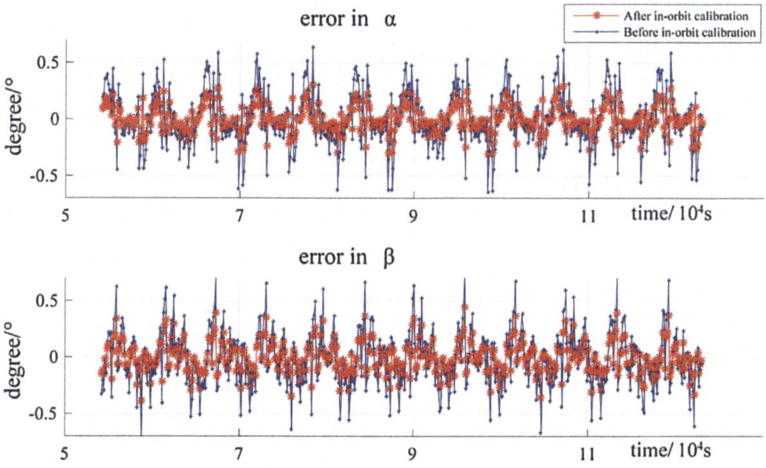

Fig. 7 Error of sun sensor

3 Result

Figure 6 shows the output data of the sun sensor applied in a satellite which was launched in September 2015. The upper line titled α is the component of the incident angle in x direction, while the line titled β is the component in y direction.

Figure 7 is the error of the sun sensor. The error is defined by the difference between the output of the sun sensor and the output of the pseudo measurement model. The upper subfigure in Fig. 7 shows the error in α which is the component of the sun angle in x direction, while the lower subfigure shows the error of β which is the component of the sun angle in y direction. The lines with little dots and greater amplitude in Fig. 7 is the error of the original output of the sun sensor before in-orbit calibration. The line with bigger dots and smaller amplitude is the output of the sun sensor after in-orbit calibration. The RMSE in x direction and y direction of the sun sensor was 0.22° and 0.27° before in-orbit calibration. The RMSE of the sun sensor after in-orbit calibration is 0.15° in x direction and 0.18° in y direction. It is obvious that the calibration method could decrease the error efficiently.

This calibration model could be extended on other sun sensors with different measurement models. It could reduce the impact of the negative factors in orbit, and help the sun sensor compensate the errors caused by these negative factors.

References

1. Qiao-yun Fan, Xiao-feng Tan, "An error compensation method for a linear array sun sensor with a V-shaped slit", Measurement Science and Technology, 26(2015), 115009.
2. J. Enright, Godard, "Advanced sun-sensor processing and design for super-resolution performance", Aerospace Conference, 2006 IEEE.
3. J. Enright, Godard "Design optimization of a digital sun sensor for use with parametric processing", IEEE transactions on instrumentation and measurement, VOL. 57, NO. 10, October 2008, 2188–2195.
4. J. Enright, Doug Sinclair, Chris Li "Embedded algorithms for the SS-411 digital sun sensor", Acta Astronautica 64 (2009), 906–924.
5. J. R. Wertz "Spacecraft Attitude Determination and Control", Dordrecht: Springer Netherlands, 1978.
6. Michael Soffel, Ralf Langhans "Space-Time Reference Systems", Springer Science & Business Media, 2012.

Study of Gain Test Method for Charge Coupled Device

Shanshan Cui, Binghuan Meng, Zhenwei Qiu, Pingping Yao, Donggen Luo and Jin Hong

Abstract The photoelectric performance of CCD is an important project, the test of it is the key technology to realize precision remote sensing. The CCD gain is a required parameter, which can be used to calculate the quantum efficiency, account the dynamic range and so on. The traditional gain test methods at present are the photon transfer curve method and the Fe55 method, the former is more popular because of its simple test equipment. This paper derives the gain test principle based on the photon transfer curve method, analyses the mathematical relationship between the shot noise and the gain, suggests the limitation of the traditional single pixel statistical method, and proposes a new method to calculate the temporal noise based on uniformity correction. In addition, this paper also builds the gain testing system for CCD, and obtains two sets of experimental data using the two methods, the test data indicates that there is a good agreement with the manufacturer's value, the traditional method has 0.94% difference, while the new method has 1.96% difference, which guarantees the accuracy and availability of the new method.

Keywords CCD · The photoelectric performance · Gain test · Noise · Uniformity correction

1 Introduction

The charge coupled device (CCD) is one of the main components in Directional Polarization Camera (DPC), the target gets across the optical system and images to the CCD incept surface, the output image is used as a reference for retrieving the object information. CCD gain test is one of the important parts in DPC development process, which can establish the quantitative relationship between the photo-induced

S. Cui (✉) · B. Meng · Z. Qiu · P. Yao · D. Luo · J. Hong
Key Laboratory of Optical Calibration and Characterization, Anhui Institute of Optics and Fine Mechanics, Chinese Academy of Sciences, Hefei, China
e-mail: cuishanshan13731@163.com

H.P. Urbach and G. Zhang (eds.), *3rd International Symposium of Space Optical Instruments and Applications*, Springer Proceedings in Physics 192,
DOI 10.1007/978-3-319-49184-4_26

gray value and photo-electron number, it is mainly used to account the quantum efficiency or the dynamic range of the optical remote sensor.

The major difference for gain test is the processing method of photo-electron number. Essentially there are two kinds of idea at present, one is called the Fe^{55} method [1–3], whose idea is that it can generate constant photo-electron number when the high-energy photons irradiate the silicon substrate. With this method, the gain is calculated. The method has an advantage of high accuracy and stability, but with more complex test equipment and higher cost. Another method is called photon transfer method [4], the main idea is to use the linear relationship between the photo-induced gray value and the shot noise, and the gain can be fitted by the least square method. Compared with the first method, this method has simple test equipment and becomes more popular, but it need to create a large statistical sample, and require long test time and large amount of calculation, besides, the external noise can be easily introduced during the testing time due to the fluctuation of the test light source, which is difficult to distinguish with the noise to be measured.

This paper addresses the problem that the photon transfer method is time-wasting and may be dominated by gradual variations to illumination source, then proposes the array of pixels statistical method based on uniformity correction. A contrast gain test experiment has been done based on the NIMO CCD55-20 used in DPC, and the result shows the agreement with the manufacturer's value.

2 Test Principle for Photon Transfer Method

The CCD gain K is defined as the ratio of photo-induced gray value to the number of accumulated electrons, the process is assumed to be linear and can be described by Eqs. (1):

$$\mu_y = K(\mu_e + \mu_d) = \mu_{y\cdot dark} + K \cdot \mu_e \tag{1}$$

Photons hit CCD photo-sensitive area and generate electrons, the process is usually random, independent and continuous, the number of electrons fluctuates statistically, this tiny fluctuation is called shot noise, and the probability is Poisson distributed. Therefore the variance of the fluctuations is equal to the mean number of accumulated electrons:

$$\sigma_e^2 = \mu_e \tag{2}$$

With Eqs. (1) and (2), the gain K can be written as Eq. (3), it is related to the shot noise and the photo-induced gray value.

$$K = \frac{u_y - u_{dark}}{\sigma_e^2} \tag{3}$$

As illustrated in Fig. 1 [5], the total noise includes shot noise, dark noise, amplifier noise, readout noise, quantization noise and so on. Likewise, the shot noise is the parameter to be solved, the dark noise is thermally induced electrons which is influenced by CCD temperature and exposure time, the partial circuit noise can be improved by the design of related circuit and device optimization, the readout noise and the quantization noise are hard to eliminate, and can be used as the background noise [6, 7]. Based on this, the total noise can be expressed by

$$\sigma^2 \approx \sigma_e^2 + \sigma_d^2 + \sigma_r^2 + \sigma_q^2 \tag{4}$$

In this equation, all the noise is electronic noise and cannot be measured, and it is necessary to convert to the parameters that can be measured, it can be written as

$$\sigma_y^2 = K^2(\sigma_e^2 + \sigma_d^2 + \sigma_r^2) + \sigma_q^2 = K^2\sigma_e^2 + K^2\sigma_d^2 + K^2\sigma_r^2 + \sigma_q^2 \tag{5}$$

With Eqs. (3) and (5), the gain can be written as

$$\sigma_y^2 = \underbrace{K}_{slope}(\mu_y - \mu_{dark}) + \underbrace{K^2\sigma_d^2 + K^2\sigma_r^2 + \sigma_q^2}_{offset} \tag{6}$$

In these equations, μ_y is the gray value of the bright image, $\mu_{y\cdot dark}$ is the gray value of the dark image, μ_d is the number of dark noise electrons, μ_e is the number of photon generated electrons, σ_e^2 is the shot noise, σ_d^2 is the dark noise, σ_r^2 is the readout noise, σ_q^2 is the quantization noise, and σ_y^2 is the total gray value noise.

Equations (6) is central to the characterization of CCD. From the linear relationship between the noise σ_y^2 and the mean photo-induced gray value $\mu_y - \mu_{dark}$, it is possible to determine the system gain K from the slope. This method is known as the photon transfer method [8]. Besides, Except for the small effect caused by $\mu_y - \mu_{dark}$, the accuracy of the gain is mainly related to the precision of the noise σ_y^2.

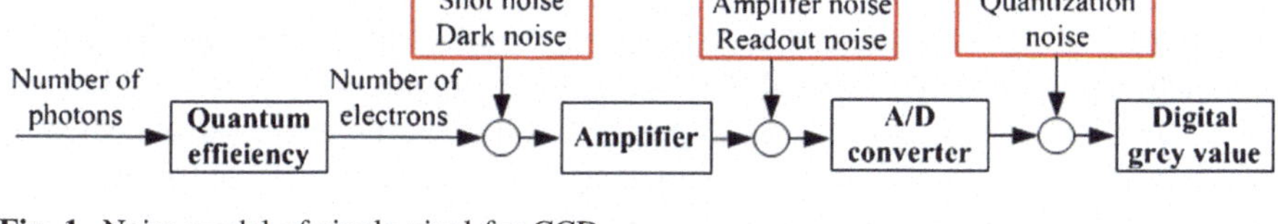

Fig. 1 Noise model of single pixel for CCD

3 Noise Test Method

3.1 Single Pixel Statistical Method

The model discussed so far considered only a single pixel, and noise measurement need to provide sufficient samples (such as 500 times) [9]. Assume the number of samples is m, and the noise for a single pixel can be written as Eqs. (7)–(9):

$$\sigma_y^2(i,j) = \frac{1}{m-1}\sum_{k=1}^{m}\left(\mu_y(i,j,k) - \overline{\mu}_y(i,j)\right)^2 \tag{7}$$

$$\mu_y(i,j,k) = \mu_{y\cdot bright}(i,j,k) - \frac{1}{m}\sum_{k=1}^{m}\mu_{y\cdot dark}(i,j,k) \tag{8}$$

$$\overline{\mu}_y(i,j) = \frac{1}{m}\sum_{k=1}^{m}\mu(i,j,k) \tag{9}$$

In these equations, i and j are the row and the column indices of the CCD array, $\mu_y(i,j,k)$ is the photo-induced gray value of the k^{rd} sample, $\overline{\mu}_y(i,j)$ is the mean value of $\mu_y(i,j,k)$. And the mean noise of the whole CCD array is given by:

$$\sigma_{y-all}^2 = \frac{1}{M \times N}\sum_{i=1}^{M}\sum_{j=1}^{N}\sigma_y^2(i,j) \tag{10}$$

With this method, a large number of statistical samples are measured, and it leads to a longer test time and huger data processing, this is the disadvantage that cannot be ignored. In addition, the fluctuation of the light source in the test time may also introduce into the uncertainty, which is not easy to eliminate.

3.2 Array of Pixels Statistical Method based on Uniformity Correction

Considering of the limitation of the single pixel statistical method, an alternative approach is proposed. The central idea is that the temporal noise is reflected in the spatial area, i.e., the array of the CCD pixels. In other words, the noise collected in the same pixel potential well is converted to the noise that collected in different potential well. However, because of the spatial nonuniformity between the pixels, this method can also introduce external noise, fortunately this error is fixed, and it can be eliminated through uniformity correction.

Essentially, there are two basic nonuniformities, one is called dark signal nonuniformity (DSNU), the other one is called photon response nonuniformity (PRNU). DSNU has little influence, which can be eliminated by subtracting the background gray value. This can be expressed by

$$\mu_y(i,j) = \mu_{y.bright}(i,j) - \frac{1}{n}\sum_{k=1}^{n} \mu_{y.dark}(i,j,k) \tag{11}$$

The symbol n is the total number of samples, i and j are the row and the column indices of the array, $\mu_{y.bright}$ is the gray value of the bright image and $\mu_{y.dark}$ is the gray value of the dark image with the same exposure time. The mean of the dark images can remove the temporal noise and then only the effect of DSNU is remained.

PRNU is the main error source in this method. A test result of PRNU for NIMO CCD55-20 at 865 nm is given in Fig. 2 at 50% saturation, and the value is about 1.03%. With different gray value, the result will be different, the maximum is about 1.65%. If the value is not corrected, the accuracy of the method will be submerged.

There are many uniformity correction methods at present [10, 11], "two-point correction method" is used widely because it is simple and easy to implement. With this correction method, at least two test points are needed, a low illumination point and a high illumination point, the output image is used to calculate the correction factor of each pixel. This can be expressed by Eqs. (12)–(14). The noise is then obtained by Eqs. (15) and (16).

$$Y_{correction}(i,j) = G(i,j) \times \mu_y(i,j) + O(i,j) \tag{12}$$

$$G(i,j) = \frac{\overline{\mu}_H(i,j) - \overline{\mu}_L(i,j)}{\mu_H(i,j) - \mu_H(i,j)} \tag{13}$$

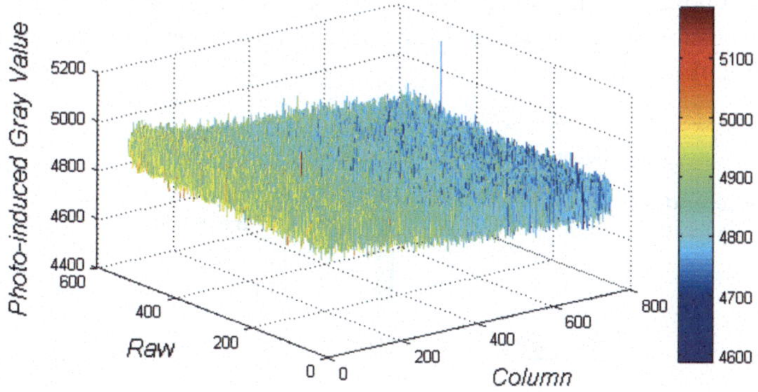

Fig. 2 The result of PRNU for CCD at 865 nm

$$O(i,j) = \frac{\mu_H(i,j)\overline{\mu}_L(i,j) - \mu_L(i,j)\overline{\mu}_H(i,j)}{\mu_H(i,j) - \mu_L(i,j)} \tag{14}$$

$$\sigma_y^2 = \frac{1}{M \times N - 1}\sum_{i=1}^{M}\sum_{j=1}^{N}\left(Y_{correction}(i,j) - \overline{Y}_{correction}\right)^2 \tag{15}$$

$$\overline{Y}_{correction} = \frac{1}{M \times N}\sum_{i=1}^{M}\sum_{j=1}^{N} Y_{correction}(i,j) \tag{16}$$

In these equations, $\mu_y(i,j)$ is the gray value obtained by Eqs. (11), the symbol G and O are the PRNU correction factor, $Y_{correction}(i,j)$ is the gray value after correction, $\mu_H(i,j)$ and $\mu_L(i,j)$ are the photo-induced gray value of the high illumination and low illumination respectively, $\overline{\mu}_H(i,j)$ and $\overline{\mu}_H(i,j)$ are the mean value of $\mu_H(i,j)$ and $\mu_L(i,j)$, $\overline{Y}_{correction}$ is the mean value of $Y_{correction}(i,j)$.

4 The Gain Test Experiment

4.1 The Gain Test

Figure 3 shows the schematic diagram of the gain test, it includes the band type uniform illumination light source, CCD and the computer. The former provides an uniform input signal to CCD, likewise all the pixels can receive approximately the same irradiance in the exposure time, in this way, the residual difference of the gray value is caused by PRNU. During the testing time, the temperature stability of CCD can be guaranteed by blowing nitrogen.

A contrast experiment has been done based on the NIMO CCD55-20 at 865 nm, two sets of data has been collected in the experiment process. One is used in single pixel statistical method, in order to provide sufficient samples, 500 images has been collected at different exposure time. The other one is used in the new method that the paper proposed, it only need collect one bright image and 30 dark images at different exposure time, the value 30 mainly eliminates the influence of temporal noise from dark signal.

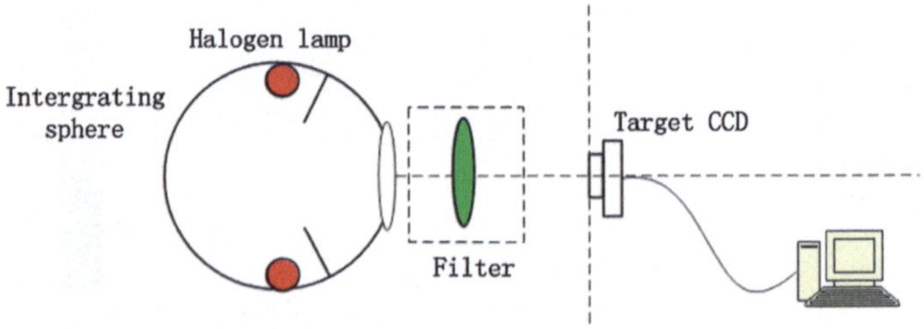

Fig. 3 The schematic diagram of the gain test

Using Eqs. (7)–(10) and Eqs. (11)–(16) respectively to calculate the noise and the results are given in Fig. 4. The gain measured by the single statistical method is 0.02337, while the gain measured by the array of pixels statistical method is 0.02313. The residual error is relatively higher in Fig. 4b, especially the second and the eighth point, that is because they are the low illumination and high illumination point respectively, and the PRNU after correction is relatively lower, which lead to low noise, if the two value are removed, the fitted value is about 0.02359.

The CCD manufacturer E2 V has tested the gain with its testing method, and the value is about 0.02315.

4.2 *The Analysis of Results*

With our test method, the fluctuation of the test light source is the main uncertainty for single pixel statistical method. Figure 5 shows the actual test curve for the light source in 30 min, the peak-to-peak value is about 0.053%.

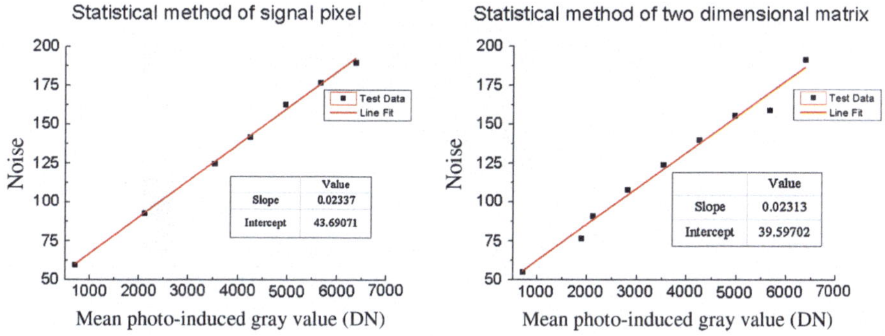

Fig. 4 The contrast of the gain test results

Fig. 5 The stability of the test light source

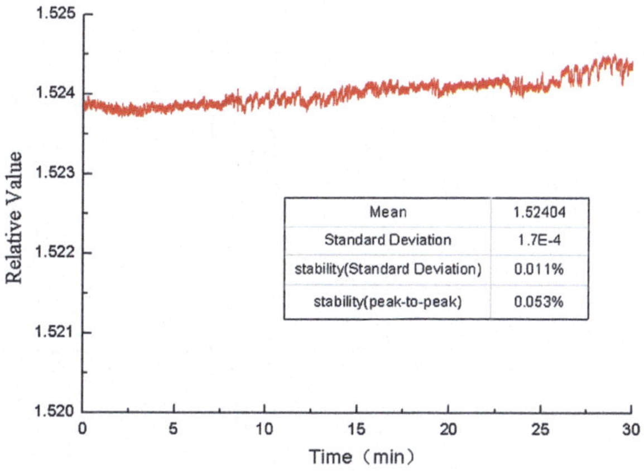

The PRNU correction accuracy is the primary factor for array of pixels statistical method, the selection of the correction method is very important. This paper choose “two-point correction method” as mentioned in section III(B), and Fig. 6 shows the result after correction at 50% saturation, the value is about 0.26%. With different gray value, the result will be different, the maximum is about 0.46%.

All of the two methods are based on the assumption that the CCD response property is linear, and in fact, the linearity of our CCD is really well (Fig. 7), the value of nonlinearity is about 0.06%.

In summary, compared with the data the CCD manufacturer gives, the error for single pixel statistical method is approximately 0.94%, and the error for array of pixels statistical method is about 1.86%. Although the uncertainty is reduced, the test time and the amount of calculation are greatly decreased in the range of allowable error.

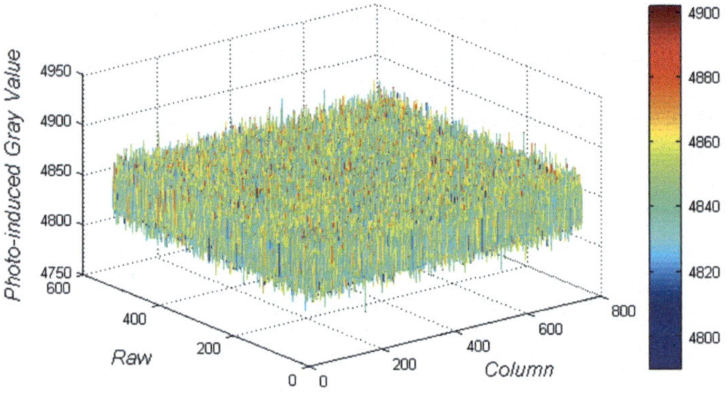

Fig. 6 The result of PRNU for CCD after correction

Fig. 7 The result of nonlinearity for CCD

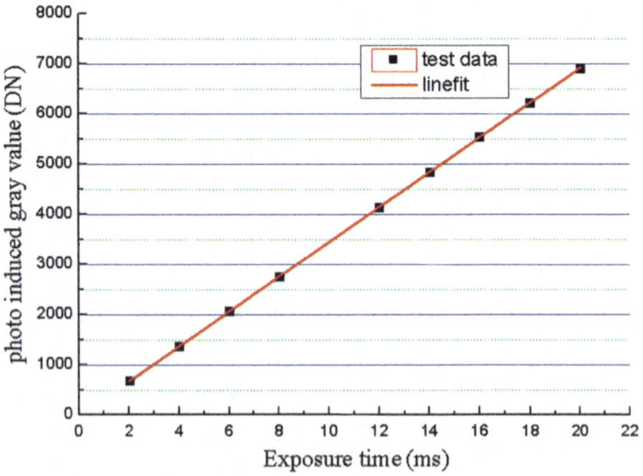

5 Conclusion

The photon transfer curve method is widely used at present. It is based on the rigorous signal transmission theory, and the temporal noise is the main factor which affects the accuracy of CCD gain test. The single pixel method has higher accuracy if the illumination source is relatively stable, but it is time-wasting and complicated. According to the investigation, this paper put forward a new idea that converts the temporal noise to spatial region, the method does not need to measure a large number of samples, and the test process is relatively simple. Both the methods are adopted for comparative analysis through gain test experiment, the result indicates that the new method is about 1.86% difference with the manufacturer's method, which verifies that this method is effective.

References

1. Steckert J, "Design, implementation and setup of an automated quantum efficiency measurement system for charge-coupled devices," *LBNL diplom thesis*, 2005.
2. Groom D. E, Bebek C. J, Maximilian Fabricius, Armin Karcher, William F. Kolbe, Natalie A. Roe, et al, "Quantum efficiency characterization of LBNL CCD's," *SPIE*, 6068:101–111, 2006.
3. Lesser M. P, McCarthy B. L. "Quantum efficiency characterization of scientific CCDs," *SPIE*, 2654B:278–286.
4. Shang Yuanyuan, Guan yong, Zhang Weigong, Zhao Xiaoxu. "Research on evaluation method of CCD image sensor," *Acta optica sinica*, 28(12):317–322, 2008.
5. Sun Hui, Xu Shuyan, Sun Shouhong, Li Junlin. "Research on evaluation method of optical imaging sensors' photon response non-uniformity noise," *Laser & Optoelectronics Progress*, 52:042302-1–042302-6, 2015.
6. Xu Xiuzhen, Li Zitian, XueLijun. "Analysis and processing of CCD noise," *Infrared and Laser Engineering*, 33(4):343–357, 2004.
7. Li Yunfei, Li Minjie, Si Guoliang, Guo Yongfei. "Noise analyzing and processing of TDI-CCD image sensor," *Optics and Precision Engineering*, 15(8):1196–1202, 2007.
8. EMVA Standard 1288: Standard for characterization of image sensors and cameras [S], 2005.
9. Li Zaifeng, Li Junlin, Lan taiji. "Research of the CCD camera's electronics gain test method," *Electronic measurement technology*, 37(12):49–52, 2014.
10. Hu Guihong, Chen Qian, Shen Xiaoyan. "Research on the Nonlinearity of Infrared Focal Plane Arrays," *Journal of Optoelectronics Laser*, 14(5):489–492, 2003.
11. Feng Lin, Liu Shuang, Zhao Kaisheng, Guan Anquan. "Method of nonuniformity correction for IRFPA with nonlinear response," *J. Infrared Millim Waves*, 25(3):221–224, 2006.

Measurement, Correction and Validation of Out-of-Band Response for Multi-spectral Remote Sensing Instruments

Yinlin Yuan, Xiaobing Zheng, Haoyu Wu, Wenchao Zhai, Honghu Qian, Donggen Luo, Weifeng Yang and Jin Hong

Abstract Out-of-band response (out-of-band light leaks) is a significant uncertainty factor in multi-spectral remote sensing instruments radiometric calibration. In this paper, a method using a spectrum-tunable reference light source (STIS) which possess new functions of continuous and discrete spectral coverage to measure, correct and validate the 443 nm band of a multi-spectral camera has been mentioned. Adjusting STIS output two discrete radiance spectrum with its spectral width of about 50 nm in the out-of-band spectral region to directly measure out-of-band radiance responsivity. Correcting out-of-band response via spectral radiance measured by reference spectral radiometer and out-of-band radiance responsivity with the help of out-of-band response correction model. Validating out-of-band response correction accuracy via contrasting the relative departure between band-weighted radiance measurement of 443 nm band with out-of-band response correction and reference spectral radiometer measurement, under the conditions of measuring the test continuous spectrum produced by STIS. The result shows that the relative departure between the band-weighted radiance and the reference is decreased from 8.8 to 3.1% before and after the out-of-band response correction. The results demonstrate that STIS is suitable for measuring, correcting and validating the out-of-band response of multi-spectral remote instruments.

Keywords Out-of-band response · Out-of-band · Correction · Multi-spectral · Spectrum-tunable source · Radiometric calibration

Y. Yuan (✉) · X. Zheng · H. Wu · W. Zhai · H. Qian · D. Luo · W. Yang · J. Hong
Key Laboratory of Optical Calibration and Characterization,
Chinese Academy of Sciences, Hefei 230031, Anhui, China
e-mail: ylyuan@aiofm.ac.cn

H.P. Urbach and G. Zhang (eds.), *3rd International Symposium of Space Optical Instruments and Applications*, Springer Proceedings in Physics 192,
DOI 10.1007/978-3-319-49184-4_27

1 Introduction

Multi-spectral remote sensing instruments (MSRS) commonly use narrow band filter for spectral selection, of which the ratio between bandwidth $\Delta\lambda$ and center wavelength λ is about $0.1(\Delta\lambda/\lambda \approx 0.1)$ [1]. For an idea MSRS, the spectral response is completely limited in the selected wavelength region. Due to the manufacturing process limits, there are out-of-band light leak from the cutoff spectral region which are scattered out of the narrow band filter constructed under less than ideal deposition conditions. The out-of-band light leak will produce out-of-band response. Previous researches and applications show that out-of-band response is about 2–7% of the integrated standardized in-band response [2], which is a significant uncertainty factor in MSRS radiometric calibration [3–6]. In the high-precision observation applications, such as in the condition of the absolute uncertainty should be limited within 5%. Therefore, measurement and correction of out-of-band response for MSRS is desirable.

Recent years, the tunable-laser-based sources from National Institute of Standards and Technology (NIST) [7] are used to measure out-of-band response for MSRS via fine spectral scanning technique from instrument-level test. However, out-of-band response measurement for MSRS will take a long time by tunable-laser-based calibration sources which isn't simple to operate and maintain [8].

At present, the out-of-band response for MSRS is still obtained by spectral radiance of target and absolute spectral responsivity from component-level test in China. The absolute spectral responsivity is retrieved from the transmission of narrow band filter [9, 10] and the detector responsivity of MSRS at 1 nm intervals from 350–1100 nm. The uncertainty of the detector responsivity measurement is less than 1.2% [11] which can meet the requirements of absolute spectral responsivity, but the transmission measurements of narrow band filter is unsatisfied. Previous researches showed that some narrow-band filter were susceptible to excessive high angle scatter, which were not captured by spectrophotometer or monochromator designed to collect the specular component of transmitted light. There are some difference between the specular and total transmission measurement, the absolute spectral responsivity is usually lower than the actual situation [12] as well as out-of-band response.

In this paper, the method of out-of-band response measurement, correction and validation is proposed using the STIS [13]. The out-of-band radiance response of the 443 nm band of a multi-spectral camera [14] was corrected by directly measuring against the STIS. The detailed measurement, correction and validation process can be found in the sections below.

2 The Method of Out-of-Band Response Measurement and Correction

For a MSRS, it may be possible to correct, in part, for the effect of a band's out-of-band spectral response, since the radiometer provides radiances in the out-of-band region [15, 16]. As such, We divide the out-of-band spectral region of channel i into several spectral regions (j is from 1 to N) associated with MSRS channel j (see Fig. 1).

The region between λ_l and λ_u, for which the system transmittance is consistently greater than 1%, is defined to be the in-band region. R_i is the in-band absolute responsivity of the given channel i of MSRS, as shown in (1).

$$R_i = \int_{\lambda_l}^{\lambda_u} R(\lambda_i)d\lambda \tag{1}$$

where $R(\lambda_i)$ is the absolute spectral responsivity of the given channel i in MSRS. The wavelength region outside this range is referred to as the out-of-band region accordingly. The out-of-band radiance responsivity δ_j could be given similarly, as shown in (2).

$$\delta_j = \int_{\lambda_{l,j}}^{\lambda_{u,j}} R(\lambda_i)d\lambda \tag{2}$$

where $\lambda_{l,j}$ and $\lambda_{u,j}$ are the low and upper limited wavelength point the spectral region j, respectively.

Correspondingly, the response of spectral region j could be shown in (3).

$$s_j = \int_{\lambda_{l,j}}^{\lambda_{u,j}} L_s(\lambda_i)R(\lambda_i)d\lambda \tag{3}$$

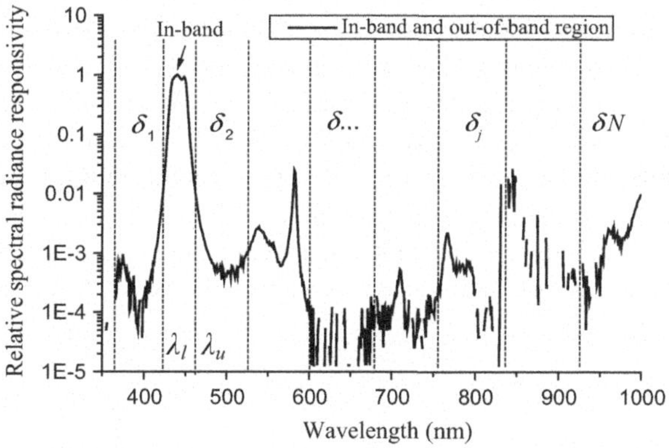

Fig. 1 Dividing method of out-of-band spectral radiance responsivity of band i in MSRS

The total out-of-band response is shown as below.

$$S_i^{oob} = \int_{\lambda\min}^{\lambda_l} L_s(\lambda_i)R(\lambda_i)d\lambda_i + \int_{\lambda u}^{\lambda_{\max}} L_s(\lambda_i)R(\lambda_i)d\lambda_i = \sum_{j=1\&j\neq i}^{N} s_j \tag{4}$$

where $\lambda_{\min}$ and $\lambda_{\max}$ are the low and upper practical limited wavelength point of channel i, respectively.

For channel i, the band-average radiance $L_{BSR}(\lambda_i)$ is commonly used to represent target radiance of selected spectral region, as shown in (4).

$$L_{BSR}(\lambda_i) = \int_{\lambda_l}^{\lambda_u} L_s(\lambda_i)R(\lambda_i)d\lambda_i / R_i \tag{5}$$

where $L_s(\lambda_i)$ is spectral radiance of target.

The total response $DN_{\lambda i}$ of channel i could be expressed by the absolute responsivity R_i, the out-of-band response S_i^{oob} and the band-weighted radiance $L_{BSR}(\lambda_i)$, as shown in (5).

$$DN_{\lambda i} = R_i \times L_{BSR}(\lambda_i) + DC_{\lambda i} + S_i^{oob} \tag{6}$$

The total out-of-band response S_i^{oob} could be corrected by the out-of-band radiance responsivity δ_i and by band-average radiance $L_{BSR}(\lambda_j)$ associated with channel j [15, 16], as shown in (7).

$$S_i^{oob} = \sum_{j=1\&j\neq i}^{N} s_j \approx \sum_{j=1\&j\neq i}^{N} L_{BSR}(\lambda_j)\delta_j \tag{7}$$

For most MSRS, there are no band associated with some spectral regions spanning 910–1150 nm. We can assume that the radiance for region from 910 to 1150 nm is half that of the original radiance for adjacent channel N, according to target radiance of previous work [15, 16].

The correction accuracy of out-of-band response S_i^{oob} depends on the measurement of the out-of-band radiance responsivity δ_j measurement in the divided spectral region. The measurement method of out-of-band radiance responsivity δ_j measurement is described as follows. Adjusting STIS output discrete spectrum in the divided spectral region j ($j = 1 \sim N, j \neq i$) to measure out-of-band response S_j in turn, meanwhile, using a reference spectral radiometer to measure spectral radiance $L_c(\lambda_j)$, as shown in the left side of Fig. 2.

According to previous researches [15, 16], the band-weighted radiance $L_{BSR}(\lambda_i)$ in the divided spectral region j could be replaced by the band-average radiance

$L_{BAR}(\lambda_i)$, so the out-of-band radiance responsivity δ_j could be calculated by the ratio between out-of-band response S_i^{oob} and the band-average radiance, as shown in (7).

$$\delta_j = S_j / L_{BAR}(\lambda_j) \tag{8}$$

where $L_{BAR}(\lambda_i)$ could be expressed by spectral radiance $L_c(\lambda_i)$ (see Eq. 8).

$$L_{BAR}(\lambda_j) = \int_{\lambda_{l,j}}^{\lambda_{j,u}} L_c(\lambda_i) d\lambda_i / (\lambda_{j,u} - \lambda_{j,l}) \tag{9}$$

Carefully adjust the output spectral radiance $L_c(\lambda_i)$ of STIS in the divided spectral region j is very crucial for the out-of-band radiance responsivity δ_j measurement with a high degree of accuracy. The output integrating radiance of STIS should be enough for normalization of the out-of-band response. For some MSRS, one of the specifications suggests being that there shall be no response greater than 0.001 once 50 nm in the out-of-band spectral region [17], compared with the in-band response. Take the SeaWiFS with 12 bits as an example, the Saturation radiance is range from 30 to 100 μW cm^{-2} sr^{-1} nm^{-1}. When the ratio between out-of-band radiance response and in-band response is 0.001, to measure a response, the minimum integrating radiance in the out-of-band spectral region requirement in the out-of-band spectral region is 9.8–24.4 μW cm^{-2} sr^{-1} nm^{-1}.

3 Experiment and Data Analysis of Out-of-Band Response Measurement, Correction and Validation

To carry out experiment of out-of-band response measurement, correction and validation, a multi-spectral camera [18], STIS and a reference spectral radiometer are used, as shown in the right side of Fig. 2. The main experimental step are described as follows:

(1) Adjusting STIS to output two discrete spectra with spectral width of about 50 nm in the out-of-band spectral region. Using the sample multi-spectral camera to directly measure against the STIS to get the out-of-band response of 443 nm band, meanwhile, using a reference spectral radiometer to measure band-average radiance;
(2) Adjusting STIS to output a continuous spectrum spanning 350–1100 nm, and using the multi-spectral camera and the reference spectral radiometer to directly measure against the STIS.
(3) Calculating the relative difference between band-weighted radiance measurement by the 443 nm band of the MSC with out-of-band response correction and reference spectral radiometer measurement.

The reference spectral radiometer was used to measure spectral radiance of STIS output spectra as the reference. In the spectral region from 410 to 950 nm, the

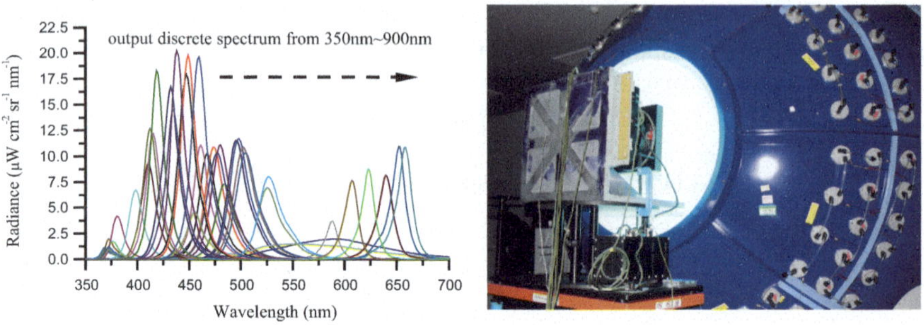

Fig. 2 The method of measuring out-of-band radiance (*left*) and objective picture of STIS (*right*)

measurement signal-to-noise ratio (SNR) is about 2000, and the out-of-band response is no more than 0.5%, both of which are better than the multi-spectral camera [19].

The out-of-band spectral region of band 443 nm in the multi-spectral camera was divided into 5 spectral regions, as shown in column 1 of Table 1. The two discrete spectrums are shown in the left side of Fig. 3, which were used to directly measure the out-of-band response and responsivity in spectral region 3 and 4, respectively.

We also adjust STIS to output a discrete spectrum which fully covers the in-band spectral region, to measure the in-band responsivity of 443 nm band. The in-band response is 1408.39 (raw digit number), and the band-weighted radiance measured by reference spectral radiometer is 20.0621 μW cm^{-2} sr^{-1} nm^{-1}. The band-weighted radiance is calculated both by spectral radiance of the in-band discrete spectrum, and by in-band relative spectral responsivity of 443 nm band which is from component -level test (see Fig. 3. right). For there is no out-of-band response, so we can get the in-band responsivity 70.201DN/(μW cm^{-2} sr^{-1} nm^{-1}).

The out-of-band responses measured directly in spectral region 3 and 4, are 3.5 and 30.1 respectively (see Fig. 4). The integrating radiance are 925.65 and 693.25 μW cm^{-2} sr^{-1} nm^{-1} respectively. According to Eq. 7, the out-of-band responsivity and measured directly in spectral region 3 and 4, are 0.4158 and 4.2966 respectively.

The total-band response measured directly by the continuous sample spectrum is 1862.148, (see Fig. 5). If we don't correct the out-of-band response, we can get the uncorrected band-weighted radiance 26.52 μW cm^{-2} sr^{-1} nm^{-1}. The reference band-weighted radiance measured by spectral radiometer is 24.365. The relative departure between uncorrected band-weighted radiance measurement of 443 nm and the reference band-weighted radiance is about 8.8%.

For the test continuous spectrum, we use out-of-band responsivity of region 1–5 which are from component-level test with the related band-average radiance, to calculate total out-of-band response 39.17. We also use out-of-band responsivity measured directly of region 3 and 4, and use out-of-band responsivity of region 1, 2 and 5 which is from component -level test, with the related band-average radiance,

Table 1 Out-of-band response of 443 nm channel in multi-spectrum camera

Wavelength Range (nm)	Out-of-band responsivity from component -level test	Direct out-of-band responsivity measurement	Reference (μW cm^{-2} sr^{-1} nm^{-1})	Out-of-band response from component-level test	Direct out-of-band response measurement
350–422	0.0582	–	12.929	0.75	0.75
463–599	0.7722	–	17.416	13.45	13.45
600–710	0.0323	0.4158	14.872	0.48	6.18
711–810	0.0154	4.2966	12.545	0.19	53.9
811–1050	2.667	–	9.136	24.3	24.3
Total out-of-band response by calculation				39.17	98.58
Raw (DN)				1862.148	
443 nm Absoluted in-band response R_i (DN/μW cm^{-2} sr^{-1} nm^{-1})				70.201	
Uncorrected band-weighted radiance				26.52	
Corrected band-weighted radiance (μW cm^{-2} sr^{-1} nm^{-1})				25.96	25.12
Reference band-weighted radiance (μW cm^{-2} sr^{-1} nm^{-1})				24.365	
Relative departure				6.5%	3.1%

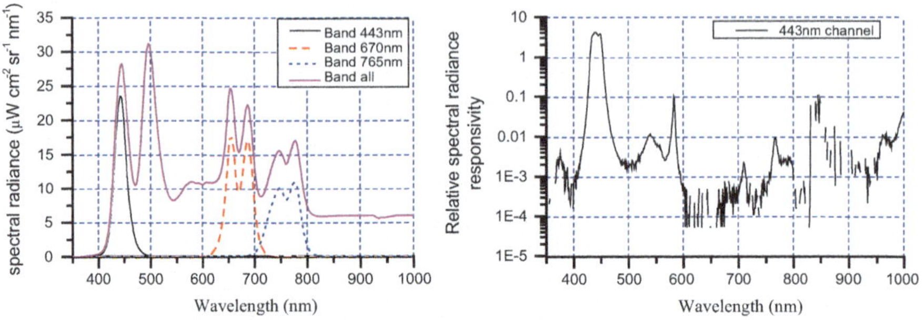

Fig. 3 The discrete (*blue* and *red lines*), continuous spectra (*magenta line*) produced by STIS and the relative spectral responsivity of band 443 nm

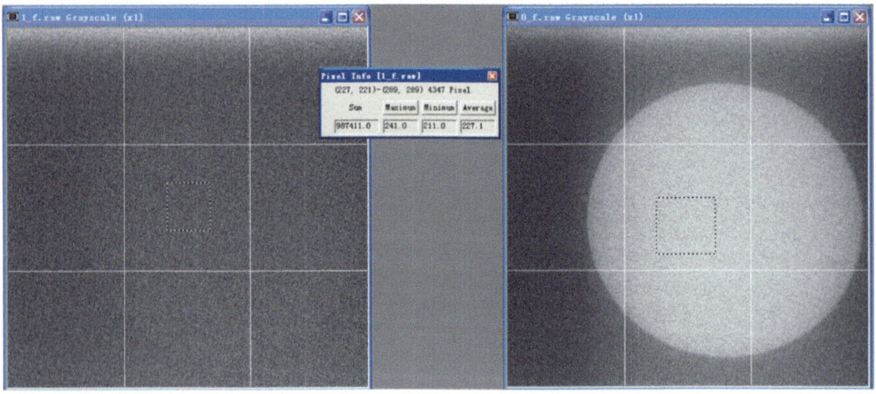

Fig. 4 The out-of-band responses measured directly in spectral region 3 and 4

to calculate total out-of-band response 98.58. According to Eq. 6, the corrected band-weighted radiance using out-of-band responsivity from component-level test and from both directly measured and component-level test are 25.96 and 25.12 respectively, compared with the reference band-weighted radiance, the departure of which are 6.5 and 3.1% respectively, as shown in the last column of Table 1.

There are two main reasons resulting in that the corrected band-weighted radiance is larger than the reference band-weighted radiance.

(1) It is not conducive to directly measure the out-of-band responsivity of region 1 and 2 by STIS, which use dense LED arrays illuminating modules to adjust both spectral distribution and radiation dynamic range. For the pick wavelength of LED arrays illuminating modules in spectral region 1 and 2, there are some weak radiant in the in-band spectral region, which is harmful to the out-of-band responsivity measurement. There are not available LED arrays illuminating modules in spectral region 5, where the out-of-band responsivity couldn't be measured.

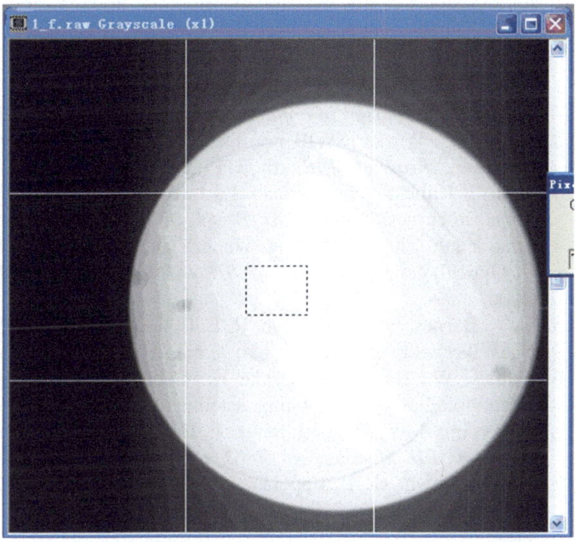

Fig. 5 Measurement result of the test continuous spectrum

(2) The out-of-band responsivity from component-level test of region 1, 2 and 5, are lower than the actual situation. From the data analysis of out-of-band responsivity, we can find that there are some spectral radiance responsivity that are less than zero, that is to say, the signal-to-noise ratio of spectral radiance responsivity measurement are very low, when the spectral transmission of narrow-band filter in the multi-spectral camera band 443 nm are less than 0.0001.

Although the band-weighted radiance could not be calibrated in all-band, the effect of out-of-band response measurement and correction is obvious.

4 Summary and Conclusions

The out-of-band response is a significant uncertainty factor in MSRS radiometric calibration, and its measurement and correction is necessary. In this paper, the method of out-of-band response measurement, correction and validation is proposed by using the STIS, and related experiment was carried out for a multi-spectral camera.

The experiment results showed that the band-weighted radiance measurement before and after the out-of-band response correction were decreased from 8.8 to 3.1%, which demonstrated that the STIS was applicable for out-of-band response measurement and correction in multi-spectral remote sensing instruments.

References

1. Zheng Xiaobing, Yuan Yinlin, Xu Qiuyun, et al. "New reference sources for radiometric calibration," *Journal of applied optics*, vol. 33(1), pp. 101–107, 2012.
2. Edward C. Wack. "SWIR non-linear radiometric response due to errors in relative spectral response," *Proc. of SPIE*, vol. 3117, pp. 217–224, 1997.
3. John E. Hubbsa, john Garciab, Eustace L. Dereniakb. "Method to validate relative spectral response curves," *Proc. of SPIE*, vol. 3379, pp. 510–519, 1998.
4. Qiao Yanli, Zheng XiaoBing, Wang Xianhua et al. "Whole-Process Radiometric Calibration of Optical Remote Sensors," *JOURNAL OF REMOTE SENSING*, vol. 10(5), pp. 616–623, 2006.
5. R. A. Barnes, Eueng-nan Yeh, R. E. Eplee, et al. SeaWiFS technical report series volume 39 SeaWiFS calibration topics, part 1, *NASA Tech. Memo*, vol. 39, pp. 104566, 1996.
6. Barnes, W., Brown, S., Lykke K., Guenther, B., Xiong, X., et al. "Comparison of two methodologies for calibrating satellite instruments in the visible and near infrared Earth Observing Missions and Sensors: Development, Implementation, and Characterization," *Proc. of SPIE Meeting*, vol. 7862, pp. 78620C-1–78620C-19, 2010.
7. S. W. Brown, G. P. Eppeldauer, and K. R. Lykke, "Facility for spectral irradiance and radiance responsivity calibrations using uniform sources," *Appl. Opt.*, vol. 45, pp. 8218–8237, 2006.
8. Steven W. Brown, Robert D. Saunders, Zhigang Li., et al. "An absolute detector-based spectral radiance source," *Proc. of SPIE*, vol. 7807, pp. 78070A-1–78070A-9, 2010.
9. Zhou Lei, Wu Haoyu, Zheng Xiaobing. "Calibration for Relative Spectral Responsivity of Multichannel Radiometer," *Acta Metrological Sinica*, vol. 28(1), pp. 37–41, 2007.
10. Qiuyun Xu, Xiaobing Zheng, Zhengqiang Li, et al. "Absolute spectral radiance responsivity calibration of Sun photometers," *Review of Scientific Instrument*, vol. 81, pp. 033103-1–033103-7, 2010.
11. Zhang Jianmin, Lin Yandong, Shao Jing et al. "A standard facility for spectral response measurement of silicon photodiodes," *Acta Metrological Sinica*, vol. 19(3), pp. 194–198, 1998.
12. P.D. Fuqua, C. J. Panetta, J.D. Barrie, et al. "Out of Band Scatter Measurements from OLI Optical Bandpass Filters," *Proc. of SPIE*, vol. 8510, pp. 851009-1–851009-14, 2012.
13. Yuan Yin lin, Xu Jun, Zhai Wen Chao, et al. "Design and Test of a Spectrum-tunable Integrating Spheres Reference Light Source with Large Exit Aperture," *Acta Optica Sinca*, vol. 33(7), pp. 0712004-1–0712004-8, 2013.
14. Tang Weiping, Hong Jin, Wang Yuanjun, et al. "Airborne Directional Polarization Camera and Its Optical System Design," *JOURNAL OF ATMOSPHERIC AND ENVIRONMENTAL OPTICS*, vol.3(3), pp. 212–216, 2008.
15. Robert A. Barnes, James J. Butler. "Modeling spectral effects in Earth-observing satellite instruments," *Proc. of SPIE*, vol. 6744, pp. 67441 K-1–67441 K-21, 2007.
16. R. A. Barnes, A. W. Holmes, W. L. Barnes, et al. "SeaWiFS technical report series volume 23 SeaWiFS prelaunch radiometric calibration and spectral characterization," *NASA Tech. Memo*, vol. 23, pp. 104566, 1994.
17. Julia A. Barsi, Brian L. Markham, Jeffrey A. Pedelty. "The Operational Land Imager: Spectral Response and Spectral Uniformity," *Proc. of SPIE*, vol. 8153, pp. 81530G-1–81530G-11, 2011.
18. Tang Weiping, Hong Jin, Wang Yuanjun, et al.. "Airborne directional polarization camera and its optical system design," *Journal of Atmospheric and Environmental Optics*, vol. 3(3), pp. 213–216, 2008.
19. David C. Hatchell. "ASD Technical Guide 3rd Ed. Section 0-1," *Analytical Spectral Devices, Inc.*, www.asdi.com.

Development of Self-Calibration Spectral Radiometer of Correlated Photons on Orbit

Jianjun Li, Yan Liu, Dongyang Gao, Youbo Hu, Yuanyuan Guo, Wenchao Zhai, Fangang Meng, Jing Yan and Xiaobing Zheng

Abstract Optical sensor on orbit is usually calibrated by on-board calibrator or some astronomical objects such as moon, sun or other stars in the visible and near infrared band, primary standard and standard transfer train are the most important sources of uncertainty during the process of calibration. Currently many satellite's calibrators are needed to be installed on orbit. Calibrator can be used to calibrate satellite sensor periodicity, it is helpful to improve the quality of science data products derived from sensor observations, but it's difficult to evaluate itself change of calibrator, calibration accuracy on orbit is difficult to less than 3%.A new self-calibration spectral radiometer of correlated photons is developed based on spontaneous parametric down-conversion (SPDC) effect, including radiometric observation and self-calibration mode, radiometric observation can be used to measure radiance of optical sensor and transfer to other satellite sensor using cross calibration method. Self calibration mode can calibrate response change of optics and electronic on orbit and modify calibration coefficient of optical sensor periodicity, compared with other calibration method on orbit, correlated photons is absolute and not dependent on primary standard and standard transfer train, only by measuring the ratio of the number of coincidence events to the number of trigger detection, trigger channel optical transmittance, the quantum efficiency of trigger detector is not need to be konwn.Principle of the new spectral radiometer is introduced, optics system of eight channel radiometer is designed, electronic and signal processing is analysed. ICMOS low light sensor is used to observe of space and spectral distribution of correlated photons, some measurement results are introduced in detail, its uncertainty is assessed, combine uncertainty is less than 0.35%. It is hopeful that the self-calibration spectral radiometer will be used in the fields of on-orbit calibration, cross calibration of other satellite sensor, field observation instrument calibration and climate change observation in future, etc.

J. Li (✉) · Y. Liu · D. Gao · Y. Hu · Y. Guo · W. Zhai · F. Meng · J. Yan · X. Zheng
Key Laboratory of Optical Calibration and Characterization,
Anhui Institute of Optics and Fine Mechanics,
Chinese Academy of Sciences, Hefei, China
e-mail: jjli@aiofm.ac.cn

H.P. Urbach and G. Zhang (eds.), *3rd International Symposium of Space Optical Instruments and Applications*, Springer Proceedings in Physics 192,
DOI 10.1007/978-3-319-49184-4_28

Keywords Self-calibration · Correlated photons · Spectral radiometer · Uncertainty

1 Introduction

Long-term climate data records often consist of observations made by multiple sensors, It is, therefore, extremely important to have instrument overlap, to be able to track instrument stability, to quantify measurement uncertainties, and to establish an absolute measurement scale traceable to the International System of Units (SI) [1].

On-orbit calibration of sensor play an important role because calibration efficient of lab will cause change when sensor is launched in space, many calibration methods are developed including on-orbit calibration and site calibration, on-orbit calibration utilize solar diffusing panels and/or lamps in the solar reflectance wavelength range, ~300–2500 nm, site calibration is required to select large area, uniform ground reference site, when satellite sensor overpass reference site and simultaneously measure atmosphere and field parameter, thus based on the model of atmosphere transfer, calibration coefficient of satellite sensor can be obtained in the same conditions of time, spectral and observing geometry, the calibration coefficient can be transfer to other satellite sensor by cross calibration.

On-orbit calibration and site calibration can play an important role in the life cycle of satellite sensor, which can revise and update on-orbit calibration coefficient and improve accuracy of data products. Unfortunately, two calibration methods are both need a primary standard and standard transfer train, the increase of transfer train will cause the decrease of accuracy of calibration. For example, on-orbit calibration accuracy of MODIS sensor is less than 5%, which employ method of solar diffusing panels. Calibration accuracy of FY-3 satellite sensor is also less than 5% using vicarious method [2].

A new spontaneous parametric down-conversion (SPDC) calibration method is initially developed in the 80s of the last century, many theory and experiment results are reported [3–7], the principle effect and measurement accuracy is validated, a UV pump laser is interaction with nonlinear crystal, when meeting the condition of phase-matching, correlated photons can be produced, which have character of time, space, polarization, etc. Therefore, detection of one photon of a correlated pair, provides both spatial and temporal location information of the other photon of the paironce a photon is detected, which can calibrate the quantum efficiency of photodetector and spectral radiance of infrared source, a relative standard uncertainty for the correlated-photon method of 0.18% ($k = 1$) is obtained by NIST, compared with traditional substitution method, two independent calibration techniques differ is 0.14% [8]. SPDC calibration apparatus is also set up in our lab (AIOFM) in China, many characters including time and space correlation, photon rate distribution, etc., are researched [9, 10]. Based on SPDC principle,

SPDC calibration procedure is independent of primary standard and standard transfer train, the method is intrinsically absolute.

A self-calibration spectral radiometer of correlated photons is developed with the requirements of sensor on-orbit calibration, the measurement principle and optics design is introduced, electronic and signal processing is analyzed, spectral distribution photo and time correlation is observed in lab, and the DUT channel detection efficiency is measured and its uncertainty is evaluated, it is possible that self-calibration spectral radiometer of correlated photons have an opportunity to become on-orbit calibrator in future.

2 Principle of Self-Calibration Spectral Radiometer

To make a detection efficiency measurement, a trigger single-photon detector (SPD) is set to intercept some of the downconverted light. The single-photon detector under test (DUT) is positioned to collect all the photons correlated to those seen by the trigger detector. The DUT channel detection efficiency is the ratio of the number of coincidence events to the number of trigger detection events in a given time interval. A coincidence is defined as when both the trigger and the DUT detectors fire within a given time window due to detection of both photons of a downconverted pair.

Principle of correlated photons calibration has been introduced in many papers [5, 6], where we describe it in brief, the general quantum efficiency of signal channel and idle channel can be expressed using the following equation.

$$\eta_{G1} = M_c/M_2 \tag{1}$$

$$\eta_{G2} = M_c/M_1 \tag{2}$$

where M_1, M_2 and M_c is photon counts of signal channel, idle channel, and coincidence of two channel respectively.

When self-calibration spectral radiometer of correlated photons is launched in space, the transmittance, center wavelength and bandwidth of optics component will be changed, spectral responsivity of photodetector will also be occured degradation in the space environment., which is shown in Fig. 1, Eqs. 1 and 2 can be revised using the following Eqs. 3 and 4 respectively.

$$\eta_{G1} = \frac{M_c}{M_2} = \tau_1\delta_1\eta_1 \tag{3}$$

$$\eta_{G2} = \frac{M_c}{M_1} = \tau_2\delta_2\eta_2 \tag{4}$$

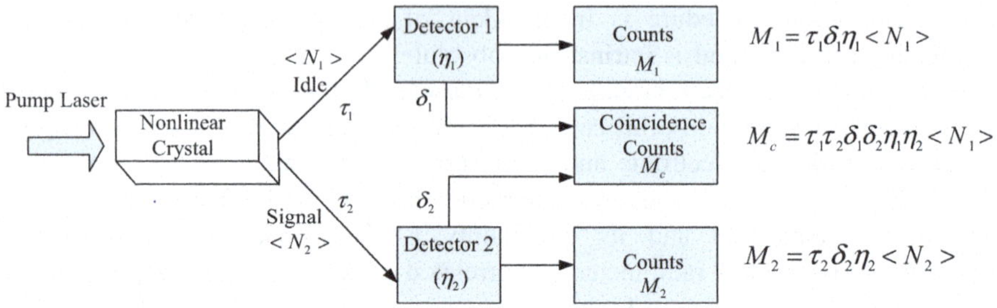

Fig. 1 Total detection efficiency calibration based on SPDC method

where τ_1, τ_2 is transmittance of signal channel and idle channel, δ_1, δ_2 is electronic response change of signal channel and idle channel. η_1, η_2 is quantum efficiency of photon detector of signal channel and idle channel, respectively.

Spectral radiometer of correlated photons receive photon flux can be expressed.

$$\Phi = L \cdot \Omega \cdot A \cdot \tau \tag{5}$$

where, L is radiance of entrance pupil, Ω is solid angle, A is effective aperture area, τ is total transmittance of optics system.

Photon counter receive photon number can be expressed.

$$N = \delta \cdot \frac{\Phi \cdot \lambda}{hc} \cdot \eta \tag{6}$$

where, N is output pulse counts of photon counter, λ is input wavelength, η is absolute quantum efficiency of photon detectors, h is Plank constant, c is speed of light in vacuum, δ is the circuit loss.

Combined with Eqs. 5 and 6, the following expression can be obtained.

$$N = \eta \cdot \delta \cdot \tau \cdot \frac{\Omega \cdot A \cdot \lambda}{hc} \cdot L \tag{7}$$

where h, c is physical constant, Ω, A is geometry parameter of optics system, which can be measured with high accuracy in lab, $\eta \cdot \delta \cdot \tau$ can be obtained using self-calibration mode, which can be evaluated by Eq. 6.

Equation (7) is the principle expression of self-calibration mode of spectral radiometer of correlated photons, by which the relation between photon counts and radiance of entrance pupil can be obtained. Instrument response change $\eta \cdot \delta \cdot \tau$ can be calibrated using self-calibration mode of correlated photons.

3 Instrument Design

3.1 Optics System Design

Figure 2 shows principle schematic of optics system of self-calibration spectral radiometer, which can be divided two modes, including radiometric observation and self-calibration, two modes can be switched by scanning mirror, when instrument work in the mode of radiometric observation, object source enters telescope system and focus on the location of field stop, and then collimated, splitting, and focused on the photodetector. When instrument work in the mode of self-calibration, a 355 nm pump laser (Coherent, CX355) enters half-wave plate and pumps nonlinear I type BBO crystal, half-wave plate can be used to adjust the polarization of pump laser, crystal size is 8 mm (Height) × 8 mm (Broad) × 10 mm (Length) and cutting angle of crystal is 37°, correlated photons is emitted from the geometry center of crystal, collimated and refected by scanning mirror, and then enters the same optical path as same as the mode of radiometric observation. Eight wavelengths (B1–B8) of correlated photons are detected (4 pairs of correlated photons), which covers the wavelength range of visible to near infrared.

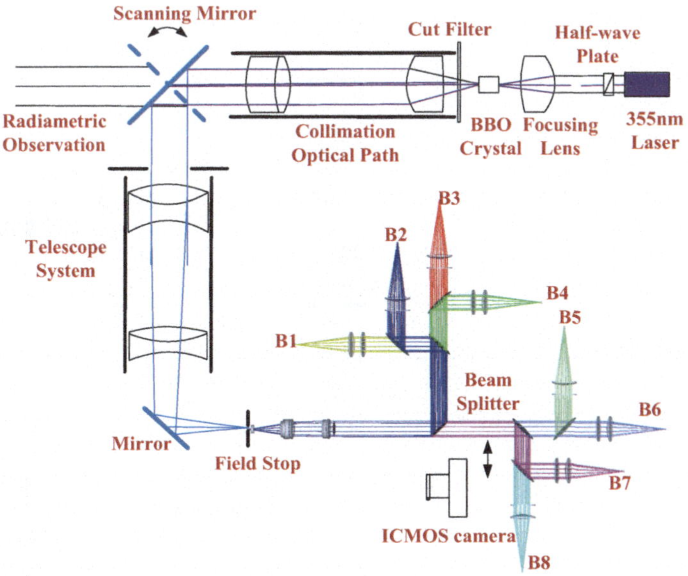

Fig. 2 Principle schematic of optics system of self-calibration spectral radiometer

4 Electronic and Signal Processing

Figure 3 shows the schematic of electronic and signal processing, when correlated photons entering single photon counting detector, a photon is converted a pulse signal by some probability, trigger channel is directly entering the Start input of time-amplitude convertor (ORTEC TAC), the DUT channel passing the delay and entering the Stop input of TAC, TAC can record time interval of photons of two channels and convert different amplitudes, TAC output is connected to photon counter (ORTEC 974) and multichannel analyzer (ORTEC MCA). A typical histogram of intervals between counts of the trigger and the DUT can be shown in MCA software. Trigger channel and the DUT channel are also needed to connect to counter. In order to evaluate the DUT channel detection efficiency, measurement data can be processed by PC software.

In order to revise the dead time of SPD detectors, effective trigger output of TAC is connected to counter, when half-wave plate is rotated by 45°, SPDC process is turned off, background trigger counts and accidental coincidence counts can be recorded by counter, which can be significant for the future high accuracy uncertainty evaluation of detection efficiency.

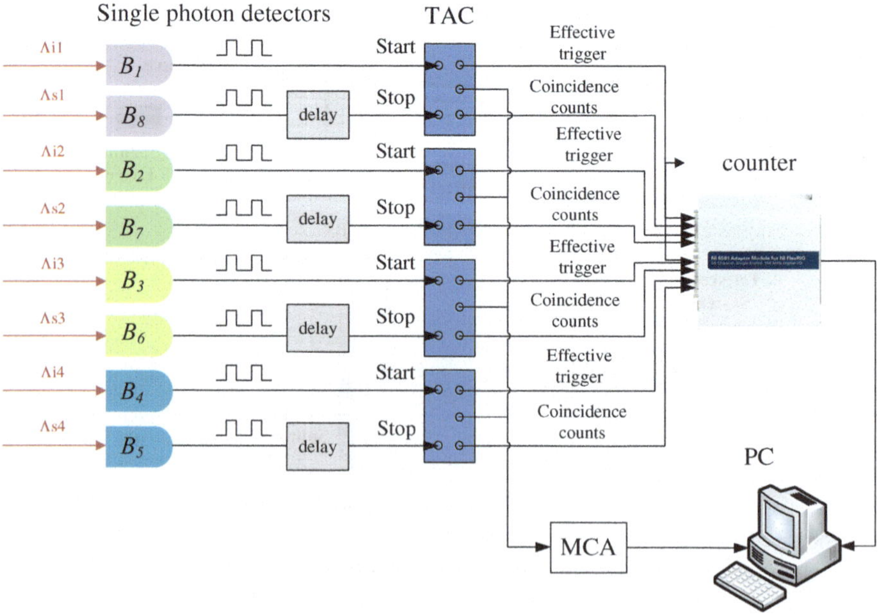

Fig. 3 The schematic of electronic and signal processing

5 Measurement Results and Uncertainty Analysis

5.1 Observation of Correlated Photons Rings

In order to observe broad spectral character of SPDC process, we put ICMOS camera in front of the first beam splitter (see Fig. 2), cut filter is put between nonlinear crystal and collimation optical path in order to decrease the rest pump laser, narrow filter is located in front of ICMOS camera, which can select different spectral rings of correlated photons. Figure 4 shows spectral photo of correlated photons rings, different narrow filter is inserted before ICMOS camera, such as Fig. 4b 430 nm/10 nm, Fig. 4c 550 nm/10 nm, Fig. 4d 685 nm/10 nm, Fig. 4e 725 nm/40 nm and Fig. 4f 860 nm/10 nm expect Fig. 4a is full spectrum when no filter is inserted, different spectral radiance of correlated photons ring is observed, which prove the broad spectral range of SPDC process. We also insert a narrow filter of 1000 nm before ICMOS camera, no ring will be observed, the reason that ICMOS camera is no response to this wavelength. The scatter light ring around correlated photons ring is mainly caused by pump laser interaction with nonlinear crystal and other optical components. It is proven when we rotate half-wave plate, correlated photons ring will be disappeared gradually, however, pump scatter light are also appeared in the photo of Fig. 4.

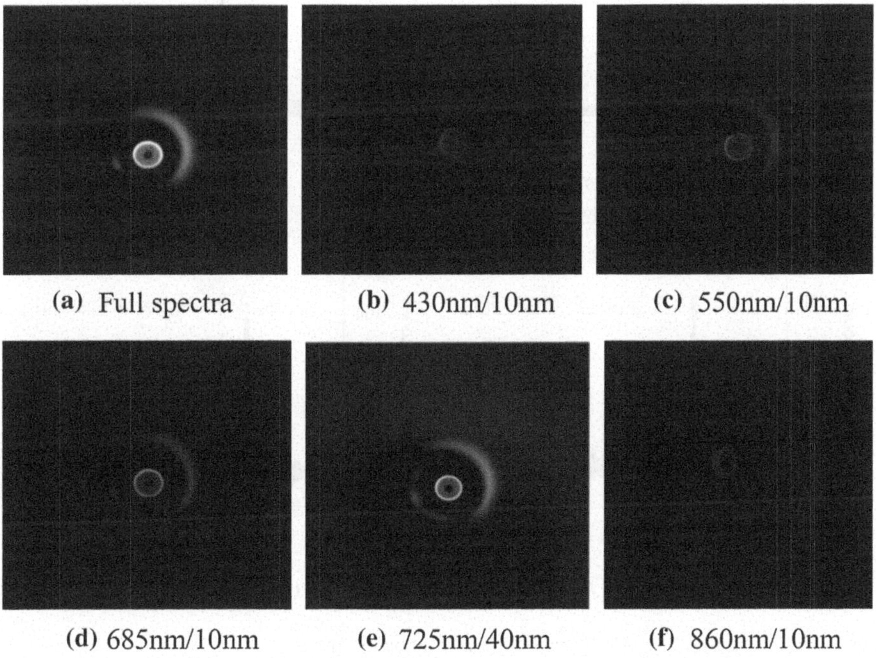

Fig. 4 Spectral photo of correlated photons rings

6 Time-Correlation of Correlated Photons Ring

For a self-calibration spectral radiometer, it's important to produce many pairs of correlated photon, a TAC coincidence measurement instrument can be used to validate time correlation of correlated photons ring (see Fig. 3), when 355 nm pump laser is interaction with BBO crystal, under the constraints of energy and momentum conservation, many correlated photons rings are produced simultaneously, time-correlation of correlation photons pair is observed by TAC and MCA, TAC can convert time interval of different correlated photons to different pulse amplitude, MAC can collect different pulse amplitude and store in different channel, a real correlated photon events will form coincidence peak in MCA channel. Different coincidence peaks of correlated photon rings are observed and shown in Fig. 5, when pump laser power is 1 mW, coincidence peak of 685 nm and 736.9 nm is shown in Fig. 5a, when pump laser power is 20 mW. coincidence peak of 633 and 808 nm, 645 and 790 nm, 700 and 720 nm is observed in Fig. 5b–d respectively both the accident coincidence events and no correlation events can also

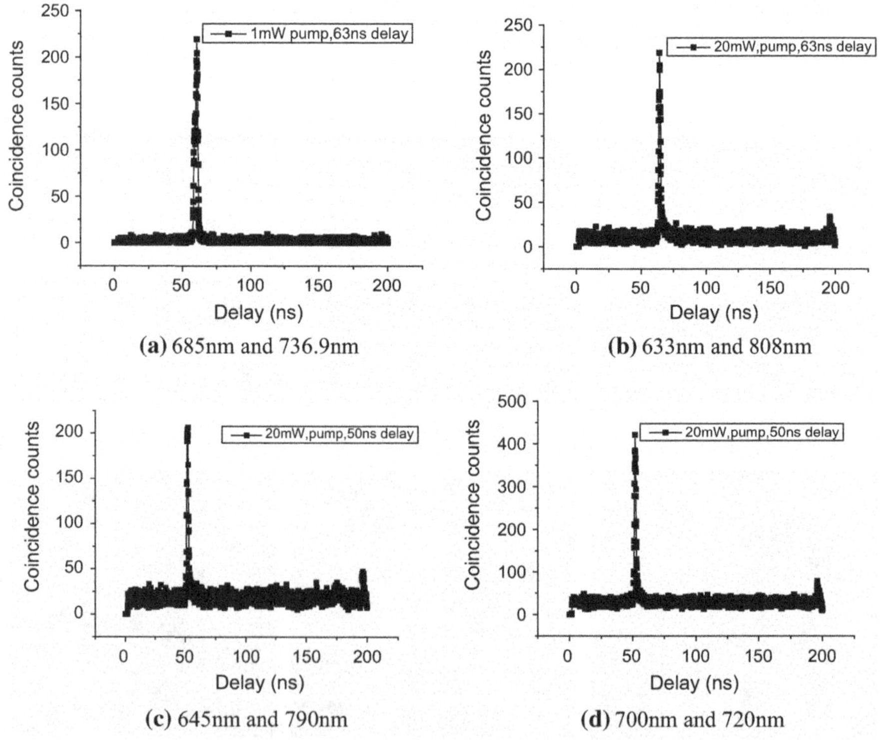

(a) 685nm and 736.9nm (b) 633nm and 808nm (c) 645nm and 790nm (d) 700nm and 720nm

Fig. 5 Coincidence curve of different correlated photons rings be observed in Fig. 5, uncorrelated firings of the two channels result in a flat histogram. Which prove the above four pair of photons have character of time-correlation.

7 The DUT Channel Detection Efficiency and Uncertainty Assessment

From Eq. (1) we know, The DUT channel detection efficiency can be defined the ratio of the number of coincidence events to the number of trigger detection events in a given time interval. In order to receive the total optical flux of correlated photons ring, PMT (Hamamatsu R2949) photon detector is used to receive the correlated photons ring, the size of focus of correlated photon ring is about 2 mm, pump laser power is set by 0.1 mW, coincidence counts, accident coincidence counts, valid trigger background counts, valid trigger counts, is recorded by photon counter respectively, The DUT channel detection efficiency is calculated, which is shown in Table 1. Based on measurement uncertainty transfer law [11], measurement uncertainty of the DUT channel detection efficiency can be expressed using Eqs. (8). The uncertainty of 10 times measurement was shown in Table 1. We obtained a relative standard uncertainty of 0.35% of the DUT channel detection efficiency for the correlated-photon method.

$$\begin{aligned} u^2(\eta_{DUT}) = {} & c_{M_c}^2 \cdot u^2(M_c) + c_{M_{et}}^2 \cdot u^2(M_{vs}) + c_{M_N}^2 \cdot u^2(M_N) + c_{M_{ac}}^2 \cdot u^2(M_{ac}) \\ & + 2 \cdot r(M_{et}, M_c) \cdot c_{M_{et}} \cdot c_{M_c} \cdot \sqrt{u^2(M_c) \cdot u^2(M_{et})} \end{aligned} \tag{8}$$

where M_c is coincidence counts within a given time window, M_{ac} is accidental coincidence counts, M_{et} is effective trigger counts, M_N is background counts of

Table 1 The DUT channel detection efficiency and uncertainty assessment

Description	Average value	Sensitivity coefficient c_i	Standard uncertainty $u(x_i)$	Uncertainty contribution $c_i \times u(x_i)$
Coincidence counts	523.45 s^{-1}	2.06×10^{-6}	7.13 s^{-1}	1.47×10^{-5}
Valid trigger counts	48892.77 s^{-1}	-1.12×10^{-8}	159.88 s^{-1}	1.79×10^{-6}
Accidental coincidence counts	284.06 s^{-1}	-2.06×10^{-6}	5.34 s^{-1}	1.09×10^{-5}
Valid trigger background counts	2697.82 s^{-1}	1.12×10^{-8}	18.67 s^{-1}	2.09×10^{-7}
Correlation coefficient between coincidence and trigger counts	0.19	–	–	3.16×10^{-6}
Combined standard uncertainty	1.81×10^{-5}			
Detection efficiency	0.52%			
Relatively standard uncertainty	0.35%			

trigger channel. c_{M_c}, $c_{M_{et}}$, c_{M_N}, $c_{M_{ac}}$ is sensitivity coefficient of the above measurement parameter, $u(M_c)$, $u(M_{et})$, $u(M_c)$, $u(M_c)$ is standard uncertainty, $r(M_{et}, M_c)$ is correlation coefficient between coincidence and trigger counts.

8 Conclusion

A new self-calibration spectral radiometer of correlated photons is developed, its principle, optical and electronic design is introduced, electronic and signal processing is analyzed. some important experiment including broad spectral coverage, time-correlation of correlated photons ring, and the DUT channel detection efficiency are done and give us some confidence to explore feasibility, limited by current measurement installation, we observe broad spectral ring image only in the range of 430 nm and 860 nm, the low light camera cannot detect at other spectral wavelength, time-correlation of correlated photons ring in other wavelength range will be researched later, the DUT channel detection efficiency and measurement uncertainty is evaluated. We are building toward a protocol that will allow the method to be moved out of the metrology lab and into the user community.

Acknowledgments The work was supported by National 863 project funding under Grant Nos.2015AA123702.the National Natural Science Foundations of China under Grant Nos. 61275173.

References

1. Xiao Xiaoxiong, Jim Butler, MODIS solar reflective calibration traceability, Proc. SPIE, 7452, Earth observing system XIV, 74520 K.
2. Hu Xiuqing, Sun Ling, Liu Jingjng, et al. IEEE Transaction on Geoscience and Remote Sensing, 2012, 50(12):4915.
3. D.N. Klyshko. Sov. J. Quantum Electron, 1980, **10**(9):1112.
4. C.K. Hong and L. Mandel. Physical Review A, 1985, **31**(4):2409.
5. A. Migdall, R. Datla, A. Sergienko et al. Metrologia, 1996, 32:479.
6. G. Brida, S. Castelletto, C. Novero, et al. Metrologia, 1998, **35**:397.
7. So-Young Baek and Yoon-Ho Kim. Physical Review A, 2008, **77**:043807.
8. Sergey V. Polyakov and Alan L. Migdall. Optics Express, 2007, **15**(4):1390–1407.
9. Li Jian-jun, Zheng Xiao-bing, Lu Yun-jun, et al. Chinese Optics Letters, 2008, **6**(7):472.
10. LI Ping-fu, LI Jian-jun, XIA Mao-peng, et al. Journal of Atmosheric and Environmental Optics, 2013, **8**(3):1.
11. B.N. Tayor and C.E. Kuyat, NIST Technical Note 1297(1994 edn.).

A Novel Study on the Technique for Deriving O/N_2 from Thermospheric Far Ultraviolet Dayglow Emissions

Yongchao Zhang, Jun Zhu, Huan Yin and Xiaoxin Zhang

Abstract The far ultraviolet dayglow emissions of the thermosphere, produced by the photoelectrons impact on the atomic oxygen or the nitrogen molecules, are the most important emissions used as remote sensing signals in the far ultraviolet emission range. The column density of the atomic oxygen relative to the nitrogen molecules, denoted by the quantity O/N_2 which is used as the monitor for the space weather of the upper atmosphere, can be derived from simultaneous measurements of the disk-viewing OI 135.6 nm and the LBH dayglow emission. In this paper, the solar activity and geomagnetic activity are both considered and researched for the technique of making use of these two far ultraviolet dayglow emissions to obtain O/N_2. An important finding is that the intensity ratio 135.6/LBH is seen to vary linearly with the intensity of the solar activity and the geomagnetic activity. Then, the reason why the solar activity and the geomagnetic activity have different effects on the O/N_2 retrieval curves was briefly analyzed and discussed. Finally, the effects of the solar activity and the geomagnetic activity to the O/N_2 retrieval technique are verified. And the result shows that the solar activity can result in dispersal for the O/N_2 retrieval curve, and the geomagnetic activity has no effect on the O/N_2 retrieval curve. This work provides a solid basis for the development of the O/N_2 retrieval technique, and this technique will play an important role in monitoring the space weather of the thermosphere.

Keywords Remote sensing · Far ultraviolet · Dayglow emissions · 135.6/LBH · O/N_2

Y. Zhang (✉) · J. Zhu · H. Yin
DFH Satellite Corporation Limited, Beijing 100094, China
e-mail: zyc09optics@sina.com

X. Zhang
China Meteorological Administration, National Center for Space Weather, Beijing 100081, China

H.P. Urbach and G. Zhang (eds.), *3rd International Symposium of Space Optical Instruments and Applications*, Springer Proceedings in Physics 192,
DOI 10.1007/978-3-319-49184-4_29

1 Introduction

The Far Ultraviolet (FUV) dayglow emissions are produced by photoelectrons impact on the upper atmosphere, and they provide important optical signals for remote sensing of the major constituents (N_2, O, O_2) in thermosphere [1, 2]. Imaging the FUV dayglow emissions from space, which can avoid the severe absorption of the lower atmosphere, can provide diagnostic information on thermospheric response to geomagnetic storms, and may be a powerful method for monitoring the space weather of the upper atmosphere [3–5]. A number of imaging instruments have been launched to observe the thermospheric FUV emissions, such as DE/SAI, Polar BEAR/AIRS, IMAGE/WIC, DMSP/SSUSI and TIMED/GUVI [6–10]. OI 135.6 nm and LBH are both used for these FUV instruments, because OI 135.6 nm is considered the best airglow emissions for remote sensing of the atomic oxygen density in the thermosphere, and LBH is the most prominent molecular emission among all the thermospheric FUV emissions [1, 11]. In China, two FUV instruments aimed at observing OI 135.6 nm and LBH band in the range of 140–180 nm will be carried by the satellite FY-3D, and these two FUV emissions will be used for monitoring the dayglow activity and the space weather of the thermosphere.

Simultaneous measurements of the disk-viewing OI 135.6 nm and LBH dayglow can be used to derived the column abundance of thermospheric O relative to N_2 (O/N_2), and there has been much interest and research in this problem for the past decades [12–16]. The FUV dayglow data of DE/SAI has been used to estimate the O/N_2 [13], and the data of DE/SAI which has a wide transmission band include OI 130.4 nm, OI 135.6 nm and LBH band from 140 to 170 nm. Y. Zhang et al. have made use of the FUV dayglow data of TIMED/GUVI to derive the O/N_2 [14], and the data of TIMED/GUVI which are recorded by separate channels include HI 121.6 nm, OI 130.4 nm, OI 135.6 nm, LBHs (140–150 nm), LBHl (165–180 nm). They both use the derived O/N_2 to study the thermospheric response to the geomagnetic storms. The data of the Polar BEAR/AIRS (OI 135.6 nm, LBH band from 155 to 170 nm) which are also recorded by separate channels have been used to derive the thermospheric O/N_2 [15, 16], and they illustrate the potential of monitoring thermospheric dynamics through latitudinal and temporal variations in O/N_2.

There are mainly two methods which have been used to derive the thermospheric O/N_2, and they can be distinguished by the recorded type of the FUV dayglow data. One method is used for wide-band type data, for example the data of DE/SAI [6]. The method relies on a first principles calculation of the intensity for the given band-pass, and then builds a functional relationship between the intensity of the wide band and O/N_2. But this method has a large error owing to errors in cross sections, the assumed solar EUV flux, and calibration [13]. The other method is used for separate-channels type data, such as the data of Polar BEAR/AIRS and TIMED/GUVI. This method uses the intensity ratio of OI 135.6 nm and LBH band (i.e. 135.6/LBH) to obtain the parameter O/N_2, and it can effectively reduce the errors in cross sections, the assumed solar EUV flux, and calibration. So in this paper we make use of this method to derive the thermospheric O/N_2, and we will talk about it in detail in Sect. 3.

The method which uses the intensity ratio 135.6/LBH to derive the thermospheric O/N_2 has been introduced for the first time by Strickland et al., and the FUV dayglow emissions which they use are OI 135.6 nm and LBH from 155 to 170 nm. They studied the dependencies of the nadir column emission rates of these two FUV emissions on the solar activity, and found that the intensity ratio of OI 135.6 nm to LBH (155–170 nm) keeps constancy with the solar activity. And they also explained the constancy of 135.6/LBH with the solar activity by the fact that the altitude distribution of solar EUV energy deposition does not change significantly with the solar activity. However, as a matter of fact, the altitude distribution of solar EUV energy deposition depends on the altitude distribution of the major constituents (N_2, O, O_2) in the thermosphere. So, the factors which affect the altitude distribution of the thermospheric species, such as the solar activity and geomagnetic activity, may significantly change the altitude distribution of solar EUV energy deposition.

In fact, different FUV emissions show different radiative properties which also rely on the solar activity and geomagnetic activity [1]. The absorption effect of the molecular oxygen is different for different LBH bands. As for the LBH band from 140 to 160 nm, the absorption effect of the molecular oxygen is much stronger than the effect for the LBH band from 160 to 180 nm. And the altitude distribution of the major constituents (N_2, O, O_2) in the thermosphere changes with the variation of the solar activity and geomagnetic activity [1, 17, 18], especially in the created area of the FUV emissions (90–500 km), so the altitude distribution of the FUV dayglow emissions may change significantly. So, for different FUV dayglow emissions, we should reconsider the relation between 135.6/LBH and the factors which may significantly change the altitude distribution of solar EUV energy deposition. So far, the OI 135.6 nm and the LBH band from 140 to 180 nm have not been studied for the technique of deriving the thermospheric O/N_2.

The purpose of this paper is to study the key geophysical factors that may affect the technique for making use of the intensity ratio of the OI 135.6 nm and the LBH band from 140 to 180 nm dayglow emissions to obtain thermospheric O/N_2. We reconsider and study the relationship between 135.6/LBH and the solar activity or the geomagnetic activity, and then discuss the reason why these two factors affect O/N_2 retrieval technique in Sect. 2. Section 3 verifies the effects of the solar activity and geomagnetic activity to the O/N_2 retrieval technique. Section 4 completes this paper with a brief summary.

2 Fuv Dayglow Emissions Simulation

It is well known that OI 135.6 nm and the LBH band from 140 to 180 nm dayglow emissions are produced by photoelectrons impact on the major species (O, N_2) in the thermosphere, and the equations calculating column emission rates of these two FUV dayglow emissions for disk-viewing the Earth from space are [19]

$$I_{OI135.6} = 10^{-6} \int_{Z_l}^{Z_u} j_{OI135.6}(z) T(z, \mu) e^{-t(z)/\mu} dz/\mu \tag{1}$$

$$I_{LBH} = 10^{-6} \int_{Z_l}^{Z_u} j_{LBH}(z) f_\lambda e^{-t_\lambda(z)/\mu} dz/\mu \tag{2}$$

where Z_u is the upper boundary of calculated emission region (500 km); Z_l is the lower boundary of calculated emission region (90 km); $j_{OI135.6}$ is the volume emission rate of OI 135.6 nm including the multiple scattering effects; j_{LBH} is the total volume emission rate of LBH; $T(z)$ is the transmission function for self-absorption by O; $t_\lambda(z)$ is the optical depth for absorption by O_2 at λ; f_λ is the fraction of j_{LBH} at λ; μ is the cosine of the viewing angle (0° for nadir); 10^{-6} is to give units of Rayleigh.

These two equations are appropriate to simulate the column emission rates of OI 135.6 nm and LBH dayglow emissions disk-viewed by instruments on satellite FY-3, the altitude of which is 830 km.

2.1 FUV Dayglow Emissions and Solar Activity

In order to study the dependencies of nadir column emission rates of OI 135.6 nm and LBH from 140 to 180 nm on the solar activity, (1) and (2) are used to calculate these two FUV dayglow emissions under a series of solar activities using the MSISE-00 model atmosphere. The solar activity levels from low to high are characterized using the Hinteregger model. Variation of solar activity is described by the solar 10.7 cm flux ($F_{10.7}$). Nadir viewing OI 135.6 nm and LBH from 140 to 180 nm column emission rates versus the solar activity is presented in Fig. 1.

Essentially, there seems to be a linear relationship between the column emission rates of these two FUV emissions and the solar activity in Fig. 1. When the solar activities vary from low to high, the photoelectrons are produced more. And the OI 135.6 nm and LBH (140–180 nm) are produced by photoelectrons impact on major constituents, so it is reasonable that these two FUV dayglow emissions vary linearly with the solar activity. Using the intensity data of these two FUV dayglow emissions, it is forward to get the intensity ratio 135.6/LBH vary with the solar activity. And the intensity ratio 135.6/LBH versus solar activity is presented in Fig. 2. The ratio 135.6/LBH is also seen to vary linearly with the solar activity. It is an important finding in this work, because changes in 135.6/LBH can not be unambiguously interpreted in terms of changes in O/N_2. So, the solar activity is an important factor which should be considered in the process of O/N_2 retrieval technique.

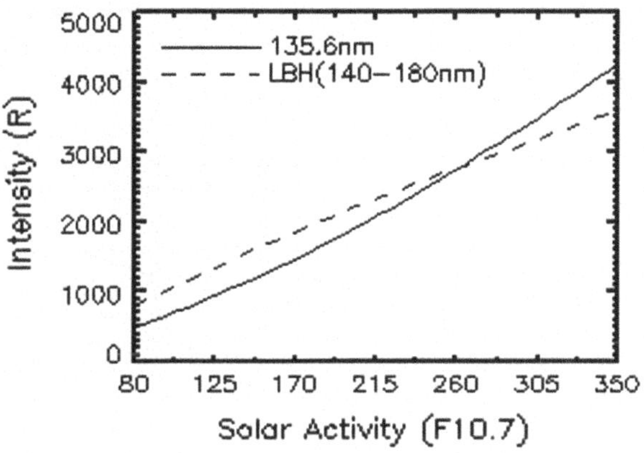

Fig. 1 Column emission rates versus solar activity

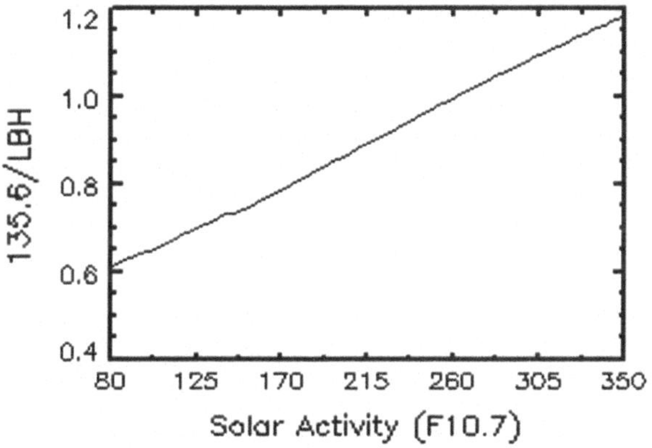

Fig. 2 135.6/LBH versus solar activity

2.2 FUV Dayglow Emissions and Geomagnetic Activity

We also consider the relationship between FUV dayglow emissions and the geomagnetic activity. And the geomagnetic activity is described by the Ap index, which is used in the MSISE-00 model atmosphere. It is well known that the major species in thermosphere vary with the geomagnetic activity, so the column emission rates may change with this geophysical factor. And we study the dependencies of the column emission rates of these FUV emissions on the geomagnetic activity. The column emission rates of nadir viewing OI 135.6 nm and LBH from 140 to 180 nm versus the geomagnetic activity is presented in Fig. 3.

Clearly, the column emission rates of these FUV emissions also seem to vary with the geomagnetic activity in Fig. 3. This can be explained by the fact that the major species in the thermosphere vary with the geomagnetic activity, so the column emission rates of OI 135.6 nm and LBH (140–180 nm) change with the geomagnetic activity. And the intensity ratio 135.6/LBH versus the geomagnetic

Fig. 3 Column emission rates versus geomagnetic activity

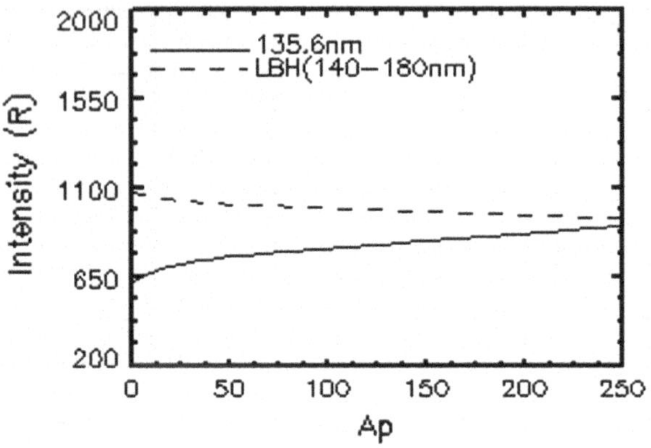

Fig. 4 135.6/LBH versus geomagnetic activity

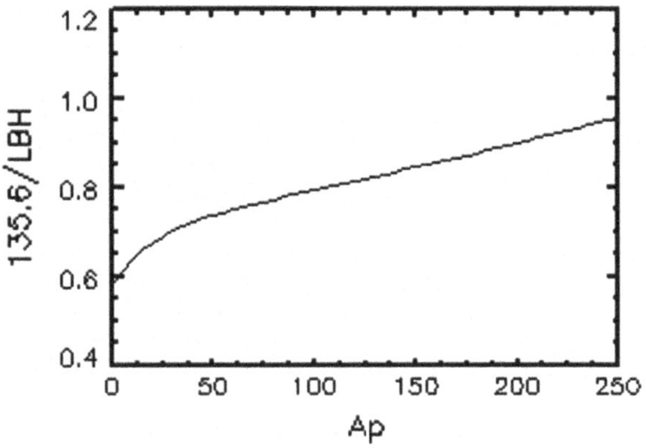

activity is presented in Fig. 4. The intensity ratio 135.6/LBH is also seen to vary with the geomagnetic activity. It is also an important finding in this work, and we will reconsider and study the effect of geomagnetic activity to the O/N_2 retrieval technique.

2.3 *Discussion*

In this section, we will discuss the effects of solar activity and geomagnetic activity to the O/N_2 retrieval technique. To interpret variations in the intensity ratio 135.6/LBH in terms of column density ratio O/N_2, one must treat the problem in column densities rather than in density profiles versus altitude, because there is no altitude information available from disk-viewing FUV emissions. The intensity ratio 135.6/LBH can be represented by

$$\frac{135.6}{LBH} = \frac{\int_{N_{T_0}}^{0} f_O(N_T) g_{OI135.6}(N_T) dN_T}{\int_{N_{T_0}}^{0} f_{N_2}(N_T) g_{LBH}(N_T) dN_T} \tag{3}$$

where N_{T_0} must be large enough to effectively account for the emission as seen from above the emitting layer, N_T is the total column density of major species (N_2, O, O_2), and f_l is the mixing ratio of the l species. The column density ratio O/N_2 is simply

$$\frac{O}{N_2} = \frac{\int_{N_T(N_{N_2}^{ref})}^{0} f_O(N_T) dN_T}{\int_{N_T(N_{N_2}^{ref})}^{0} f_{N_2}(N_T) dN_T} \tag{4}$$

where $N_{N_2}^{ref}$ is the N_2 reference depth (10^{17} cm^{-2} in this work).

It is clear that changes in 135.6/LBH result from changes in composition and g-factors. Before relating the intensity ratio 135.6/LBH to the column density ratio O/N_2, we note the property of the g-factors ($g_{OI135.6}$ and g_{LBH}) presented in (3). Their shapes are essentially identical, and (3) can be treated as $f_O(N_T)$ and $f_{N_2}(N_T)$ weighted by g-factors, so we present (3) in polynomials to first order.

$$\frac{135.6}{LBH} = \frac{\int_{N_{T_0}}^{0} f_O(N_T) g_{OI135.6}(N_T) dN_T}{\int_{N_{T_0}}^{0} f_{N_2}(N_T) g_{LBH}(N_T) dN_T} \approx k \frac{\int_{N_T}^{0} f_O(N_T) dN_T}{\int_{N_T}^{0} f_{N_2}(N_T) dN_T} = k \frac{O}{N_2} \tag{5}$$

Essentially, 135.6/LBH is seen to vary linearly with O/N2, and the coefficient k is correlated with g-factors. Then, we can explain the effects of solar activity and geomagnetic activity to the O/N2 retrieval technique according to (5). In this work, we use MSISE-00 model atmospheric model to calculate the density of the major species in thermosphere. The geomagnetic index Ap just correlates with the column density ratio O/N2, so the geomagnetic activity has little effect on the retrieval curves. The solar 10.7 cm flux ($F_{10.7}$), which we use to describe the variation of solar activity, has effects on the coefficient k. And for different solar activities, the coefficient k is different, so the solar activity does lead to the dispersal of the retrieval curves.

3 O/N_2 Retrieval Technique

The process to get the O/N_2 retrieval curve from OI 135.6 nm and LBH (140–180 nm) dayglow emissions will be described briefly in the following section, as presented in the Fig. 5.

We choose the MSISE-00 model atmosphere, and the input parameters of MSISE-00 include Date (Year and Day), Time (Universal Time), Geomagnetic activity index (Ap), Solar 10.7 cm flux ($F_{10.7}$) and its 81-day average value ($\langle F_{10.7} \rangle$), etc. Giving values to these input parameters and integrating the O and N_2 densities down to an N_2 depth of 10^{17} cm^{-2} [15], we can get the value of O/N_2. Then setting up the density scale factor of O, we can obtain a series of O/N_2 values from 0.4 to 1.2.

Now, we will check the effects of solar activity and geomagnetic activity to the O/N_2 retrieval technique. In the process of obtaining the O/N_2 retrieval curve, we use the MSISE-00 atmospheric model to calculate the density of the major species in the thermosphere. Firstly, we choose the date and time. For the purpose of comparison, two cases are built and presented in Table 1.

Secondly, we set up the values of geomagnetic index (Ap) and Solar 10.7 cm flux ($F_{10.7}$). In order to study the effect of solar activity to the O/N_2 retrieval technique separately, we fix the geomagnetic index at a quiet level (e.g. Ap = 5), and set up Solar 10.7 cm flux ($F_{10.7}$) at six different levels, such as $F_{10.7}$ = 80, 120, 160, 200, 250, 300. Then, we can calculate the retrieval curves for the six different levels of solar activity using the five steps described in the above chapters. Here, we just calculate these two disk-viewing FUV dayglow emissions in nadir and the solar zenith angle is 0°. The results of the six O/N_2 retrieval curves are shown in Fig. 6.

Clearly, Fig. 6 shows that the six O/N_2 retrieval curves in each of the two cases are dispersed for the six different $F_{10.7}$ values, and it is the proof that the O/N_2 retrieval technique is affected by the solar activity. If we ignore the effects of the solar activity, large errors should be obtained in the grieved O/N_2 values. The physical explanation will be discussed in Sect. 4.

In order to check the effects of the geomagnetic activity to the O/N_2 retrieval technique, we fix the Solar 10.7 cm flux at a quiet level (e.g. $F_{10.7}$ = 70), and six Ap

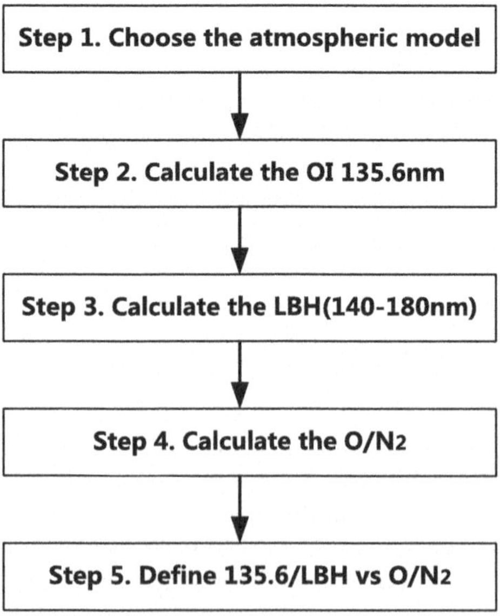

Fig. 5 Flowchart for calculating the O/N_2 retrieval curve

Table 1 Parameters for model atmosphere

Parameters	**Case 1**	**Case 2**
Year	1980	1997
Day	81	265
UT	43,200	43,200

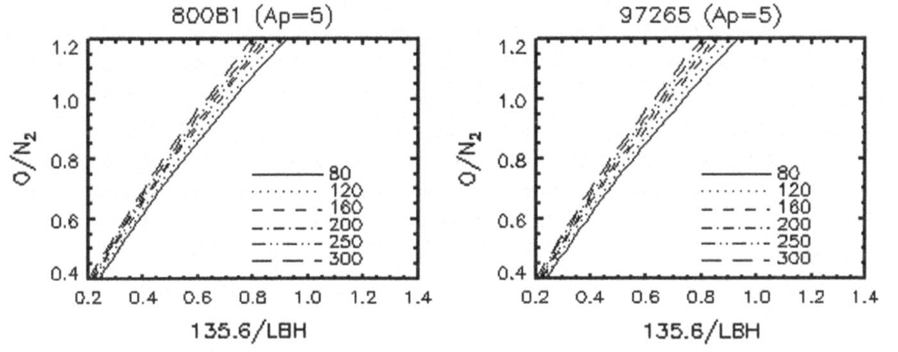

Fig. 6 O/N_2 retrieval curves for different solar activities

values (10, 50, 100, 150, 200, and 250) are used to drive the MSISE-00 atmospheric model. Then the retrieval curves for different levels of geomagnetic activity are also used the five steps described in Sect. 3. Here, nadir viewing geometry and zero degree of solar zenith angle are also assumed to calculate these two FUV dayglow emissions. The O/N_2 retrieval curves for different geomagnetic activities are presented in Fig. 7.

Figure 7 shows that the six O/N_2 retrieval curves in each of the two cases are seen to superposed for different geomagnetic activities, and it is the evidence that the geomagnetic activity has little effect to the O/N_2 retrieval technique. So in the technique of deriving O/N_2 from the intensity ratio 135.6/LBH, the impact of geomagnetic activity on the O/N_2 values could be ignored.

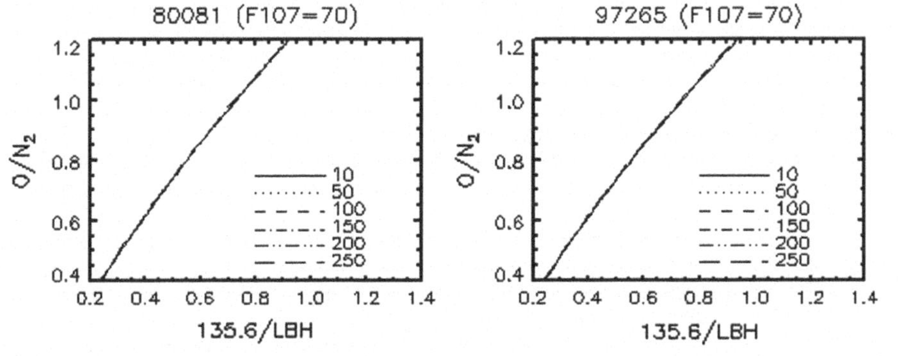

Fig. 7 O/N_2 retrieval curves for different geomagnetic activities

4 Summary and Future Work

Two key geophysical factors are studied for the technique of deriving O/N_2 from OI 135.6 nm and LBH (140–180 nm). We find that these two FUV dayglow emissions are both seen to vary with the solar activity and geomagnetic activity. Then, we study and check the effects of these two factors to the O/N_2 retrieval technique, and the result shows that the solar activity can lead to the dispersal of the retrieval curves, and the geomagnetic activity has little effect to the O/N_2 retrieval technique. So, we should pay attention to the solar activity and ignore the geomagnetic activity in the process of deriving O/N_2 from these two FUV dayglow emissions.

In this work, only nadir disk-viewing FUV dayglow emissions are studied for the O/N_2 retrieval technique, and non nadir disk observations where spherical effects are important have not been studied. We will continue this work, and the goal of us is to develop the O/N_2 image retrieval technique, which will be useful in now casting and forecasting of the space weather in the thermosphere.

References

1. Meier, R. R., "Ultraviolet spectroscopy and remote sensing of the upper atmosphere," *Space Science Reviews*, vol.58, pp. 1–185, 1991.
2. Robert E. Huffman, *Atmospheric Ultraviolet Remote Sensing*, Academic Press, 1992.
3. Meier, R. R., Picone, J. M., "Retrieval of absolute thermospheric concentrations from the far UV dayglow: An application of discrete inverse theory," *Journal of Geophysical Research*, vol.99, pp. 6307–6320, 1994.
4. L. J. Paxton, D. Morrison, D. J. Strickland, M.G. Mcharg, Y.l. Zhang, B. Wolven, et al., "The use of Far Ultraviolet Remote Sensing to Monitor Space Weather," *Advances in Space Research*, vol.31, pp. 813–818, 2003.
5. G. A. Germany, M. R. Torr, D. G. Torr, and P. G. Richards, "Use of FUV auroral emissions as diagnostic indicators, *Journal of Geophysical Research*," vol.99, pp. 383–388, 1994.
6. Frank, L.A., J.D. Craven, K.L. Ackerson, M.R. English, R.H. Eather, and R.L. Carovillano, "Global auroral imaging instrumentation for the Dynamics Explorer mission," *Space Science Instrumentation*, vol.5, pp. 369–393, May, 1981.
7. S. B. Mende, H. Heetderks, H. U. Frey, M. Lampton, S.P. Geller, R. Abiad, et al., "Far Ultraviolet Imaging from the IMAGE Spacecraft. 2. Wideband FUV Imaging," *Space Science Reviews*, vol.91, pp. 271–285, June, 2000.
8. Israel Oznovich, A. Ravitz, Moshe Tur, R.E. Huffman, R.E. Eastes, and A. F. Quesada, "Far Ultraviolet Remote Sensing of Ionospheric Emissions by Polar BEAR," *IEEE Transactions on Geoscience and Remote Sensing*, vol.31, pp. 931–945, September, 1993.
9. L. J. Paxton, C. I. Meng, G. H. Fountain, B.S. Ogorzalek, E.H. Darlington, S.A. Gary, et al., "Special Sensor Ultraviolet Spectrographic Imager (SSUSI): An Instrument Description," *SPIE*, vol.1745, pp. 2–15, 1992.
10. L. J. Paxton, A. B. Christensen, D. C. Humm, B.S. Ogorzalek, C.T. Pardoe, D. Morrison, et al., "Global ultraviolet imager (GUVI): measuring composition and energy inputs for the NASA Thermosphere Ionosphere Energetics and Dynamics (TIMED) mission," *SPIE*, vol.3756, pp. 265–276, 1999.
11. R. W. Eastes, "Emissions from the N_2 Lyman-Birge-Hopfield Bands in the Earth's Atmosphere," *Physics and Chemistry of the Earth*, vol.25, pp. 523–527, 2000.

12. Kil, H., W. K. Lee, J. Shim, L.J. Paxton, "The effect of the 135.6 nm emission originated from the ionosphere on the TIMED/GUVI O/N_2 ratio", *Journal of Geophysical Research*, vol.118, pp. 1–7, 2013.
13. D. J. Strickland, R. J. Cox, R. R. Meier and D. P. Drob, "Global O/N_2 derived from DE 1 FUV dayglow data: Technique and examples from two storm periods," *Journal of Geophysical Research*, vol.104, pp. 4251–4266, 1998.
14. Zhang, Y., L. J. Paxton, D. Morrison, B. Wolven, C.I. Meng, S.B. Mende, et al., "O/N_2 changes during 1–4 October 2002 storms: IMAGE SI-13 and TIMED/GUVI observations," *Journal of Geophysical Research*, vol.109, pp. 1–16, August, 2004.
15. D. J. Strickland, J. S. Evans, L. J. Paxton, "Satellite remote sensing of thermospheric O/N_2 and solar EUV, 1. Theory," *Journal of Geophysical Research*, vol.100, pp. 12,217–12,226, 1995.
16. J. S. Evans, D. J. Strickland, L. J. Paxton, "Satellite remote sensing of thermospheric O/N_2 and solar EUV, 2. Data analysis, *Journal of Geophysical Research*, vol.100, pp. 12,227–12,233, 1995.
17. A. E. Hedin, "MSIS-86 Thermospheric Model," *Journal of Geophysical Research*, vol.92, pp. 4649–4662, 1987.
18. A. E. Hedin, "Extension of the MSIS thermosphere model into the middle and lower atmosphere," *Journal of Geophysical Research*, vol.96, pp. 1159–1172, 1991.
19. D. J. Strickland, J. Bishop, J. S. Evans, T. Majeed, P.M. Shen, R.J. Cox, et al., "Atmospheric Ultraviolet Radiance Integrated Code (AURIC): Theory, Software architecture, inputs, and selected results," *Journal of Quantitative Spectroscopy & Radiative Transfer*, vol.62, pp. 689–742, 1999.

Egress Mechanism Color Image Segmentation Based on Region and Feature Fusion in Mars Exploration

Ying Li, Wei Rao, Jing Peng, Ying Du, Linzhi Meng and Zheng Gu

Abstract During the egress phase of Mars exploration, the egress mechanism image segmentation is the basis of the egress mechanism vision measurement analysis. In this paper, an egress mechanism image segmentation method based on region and feature fusion is proposed. Firstly, preliminary image segmentation based on color feature is carried out in the scene image. Then, according to the difference between the reference image and the real time scene image in image scale, rotation and 3D view angle, the rotation, scale and affine invariant feature points are extracted and used as the seed pixels to complete egress mechanism image segmentation using region growing. Lastly, by constructing environment for experiments, the method of egress mechanism image segmentation has been verified. The method is robust to the change of the view angle, the change of the egress mechanism slope and the change of the illumination. The experimental result shows that it can be used in the egress mechanism image segmentation in the Mars exploration.

Keywords Mars exploration · Image segmentation · Region growing · MSER

1 Introduction

During the egress phase of Mars exploration, the slope of the egress mechanism is one of the decisive elements. To realize vision measurement of the egress mechanism, the primary step is to segment the egress mechanism image from the whole scene image.

The difficulties on egress mechanism image segmentation lie in two aspects. The first, according to the change of the imaging view angle and the slope of egress mechanism, the geometric deformation will exist in the real time scene image from the reference image. In addition, the illumination condition is also different. How to

Y. Li (✉) · W. Rao · J. Peng · Y. Du · L. Meng · Z. Gu
Institute of Spacecraft System Engineering CAST, Beijing, China
e-mail: angelina-cast@hotmail.com

H.P. Urbach and G. Zhang (eds.), *3rd International Symposium of Space Optical Instruments and Applications*, Springer Proceedings in Physics 192,
DOI 10.1007/978-3-319-49184-4_30

recognize the egress mechanism to be measured in the real time scene image and carry out image segment is the problem to be solved in this paper.

On the other hand, in terms of image segmentation, currently, there are four main methods [1]. The method based on features is simple and efficiency, but the region consistency and spatial continuity is not taken into account. The method based on region can keep the spatial relations between pixels and meanwhile compute the features similarity, but the result is always split or under segmentation. The method of edge is fast, but the process is easily affected by noise leading to inaccurate segmentation result. The method based on model optimization can provide satisfactory results, but the search progress of global optimal solution is inefficiency.

In this paper, a novel method based on fusion of the region and features in color images segmentation of egress mechanism image is presented in Mars exploration. Firstly, the color feature is used for preliminary image segmentation. Then, the fusion method of region and features is used to segment the egress mechanism. The method is robust to the change of imaging view angle, the slope of egress mechanism and the illumination condition.

2 Preliminary Segmentation Based on the Color Feature

The color image of the scene can be obtained from the mast cameras of the rover. Compared to the grayscale image, more abundant characteristic information can be extracted from the color image. In this chapter, color feature is used for the preliminary segmentation of the real time scene image.

Using three combined variables can make up a kind of color space. There are some different color spaces. Description of the same object in different color space is measured from different perspectives. In current, the most commonly used color space is RGB model. From the RGB model, other color space can be deduced from linear transformation and non-linear transformation. For example, YIQ and YUV color space can be obtained from linear transformation and the standard RGB, CIE and HSV color space can be got from non-linear transformation [2].

The three components of the RGB model are highly relevant. As long as the brightness changes, the corresponding change will be caused in the three components. As a result, it is suitable for the display system but not suitable for the image segmentation.

HSV color space is a kind of user-oriented symbol color model [3]. It is much closer to the human color perception system. H is the abbreviation of Hue, S is Saturation and V is Value. The three components are independent of each other. And it is invariance to the illumination condition to some extent. The hue is the basic feature of the color nature and the hue and the saturation are corresponding to human perception, so the model is very useful in image processing algorithms.

H is measured by angle with the range of 0°–360°, calculated from the red counter clockwise direction. Red is 0°, while green is 120° and blue is 240°. The range of S is 0–1, the greater value represents the more saturated of the color. V is the level of the brightness with the range of 0–1. The hue and the saturation are only related to absorption and rejection character of the object, and have little relationship with the environment brightness. As a result, the certain color information can be segment by adjusting the value of H and S in the HSV color space. As the color feature of egress mechanism is different from the other part of the scene image, the RGB color space is firstly changed to the HSV color space and the preliminary image segmentation is carried out by adjusting the value of H and S.

3 Image Segmentation of the Fusion of Region and MSER Features

Using color feature, the preliminary image segmentation can be obtained, but there is not only the egress mechanism information needed but also other parts of environment noise are included. This chapter is to recognize and segment the egress mechanism measured in the real time scene image based on the preliminary segmentation result. On account of short-distance imaging, change of the view angle of the egress mechanism and relative distance of the imaging system will cause difference in scale, rotation and 3D view angle between the reference image taken beforehand and the real time scene image, as shown in Figs. 1 and 2. Moreover, the illumination environment is also different on Mars and cause the change of the pixels grayscale. As a result, the scale, rotation and affine invariant feature points needed to be found, and these feature points should be robust to the change of illumination.

Among the local feature extraction algorithm, Lowe's scale invariant feature transform [4] (widely known as "SIFT") uses the same basic principle as for the Harris and Hessian-based operators, but different in using Difference of Gaussians (DoG) instead of the Laplacian of Gaussians (LoG). It's restricted to a scale invariant version and also robust to the rotation and illumination, but has low computing efficiency. The main detectors in the past 40 years is inducted and compared [5], and it indicates in these corner operator, blob operator and region operator, the maximally stable extremely regions [6] (known as "MSER") is scale,

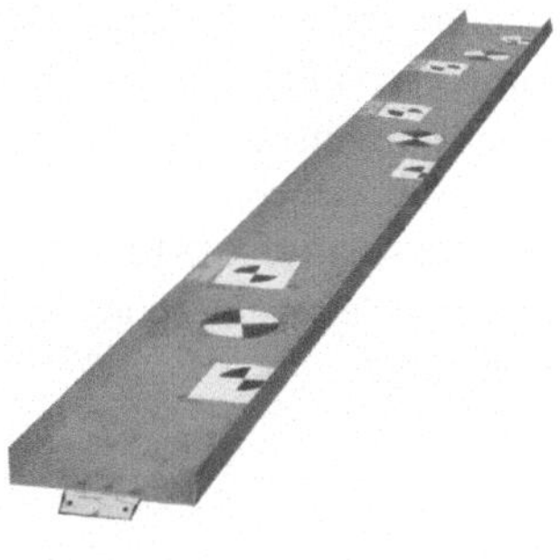

Fig. 1 The reference image of egress mechanism to be measured

Fig. 2 A scene image

rotation and affine invariant, and has a good performance in repeatability, accuracy, robust and efficiency, and shows a strong adaptability to change of illumination, the imaging view angle and the scale.

In this chapter, MSER operator is used in the reference image and the real time scene image, respectively. Carry out feature matching to get the feature points of the egress mechanism to be measured in the real time image. Then, use these feature points as seed pixels, and complete egress mechanism image segmentation through region growing.

3.1 MSER Detection

MSER detection was proposed by Matas et al. It detects a set of connected regions from an image, where each region is designed by an extremal property of the intensity function within the region to the values on its outer boundary. MSERs are detected by analyzing a unique grayscale image representation denoted as component tress. The method starts by taking pixels of zero intensity and progressively adding pixels with higher intensity levels, at each stage largest connected regions or "connected components" will represent extremal regions. As more and more gray levels are added, the connected component regions will grow and some initially separate ones will merge. Maximally stable extremal regions are those connected components that are close to stable over a range of intensities. Relative area change is an affine-invariant property, so finding MSER regions guarantee both scale and affine invariance.

Intrinsically, MSERs have arbitrary shapes. For matching purposes, SURF descriptor is used to convert them into ellipses of appropriate areas, orientations and moments.

3.2 SURF Descriptor

The speeded-up robust features (SURF) method of Bay et al. [7] required a smaller descriptor than the large one employed by SIFT. In order to increase speed, the integral image approach was used to perform rapid computation of the Hessian and was also used during scale-space analysis. Sums of Haar wavelets were used in place of gradient histograms, resulting in a descriptor dimensionality of 64-half that of SIFT. By maintaining a rigorous, robust design, the operator was described as outperforming SIFT, and also proved capable of estimating 3D object orientation within fractions of a degree and certainly more accurately than SIFT, Harris-Laplace, and Hessian-Laplace.

3.3 Feature Matching

Construct SURF descriptors of the reference image and the real time scene image, respectively, and then establish the matching relationships between these features, using Euclidean distance as the similarity judgment measures. Set a threshold for the nearest neighbor and the secondary near neighbor of the feature points to reduce the false match. Compute the fundamental matrix of the matching points using MSAC (M-estimator Sample Consensus) and reject the exterior points. MSAC is an improvement method of RANSAC (Random Sample Consensus), and Torr and Zisserman [8] indicate it is improved on efficiency and precision compared with RANSAC.

3.4 Region Growing

Region growing considers the relations of the pixel and its neighborhoods. Pixels of like intensity or other suitable property [9] are successively grouped together to form larger and larger regions until the whole image have been segmented.

The region growing algorithm can be described as follows:

1. Find seed pixels as the starting points of the growing in the image areas to be segmented.
2. Merge the neighborhood pixels of the seed pixels if they have the same or the similar characteristics in the area.
3. Consider the new pixels merged as the seed pixels and repeat the growing process, until all the pixels satisfying the constraint are included.

After features matching between the reference image and the real time scene image, matching points of the egress mechanism can be extracted and we can use these matching points of the egress mechanism as the seed pixels.

The region growing scheme is as follows.

$$|G_p - G_s| < T \tag{1}$$

In the formula, the value of is the gray value of the pixel to be measured. The value of G_p is the gray value of the seed pixel. And the value of G_s is the threshold.

4 Experimental Results and Analysis

Set up the experimental environment to give experimental results and analysis. Use a pair of cannon 5D cameras, with the CCD scale is 36 mm × 24 mm. The image resolution is 2784 × 1856. The cameras are set up on a mast about 1.5 m. The egress mechanism is a pair of rigid steels and each size is about 2.5 m × 0.25 m.

Firstly, use the color feature for preliminary segmentation, shown as Fig. 3. From the picture, we can see that most of the terrain interferon can be segment out.

Then, find the feature points in the reference image and in the real time scene image respectively, using MSER operator, describing them in SURF descriptor and establishing the matching relationships between these features, shown as Figs. 4, 5 and 6.

Fig. 3 Preliminary segmentation result

Fig. 4 Extract feature points in the reference image

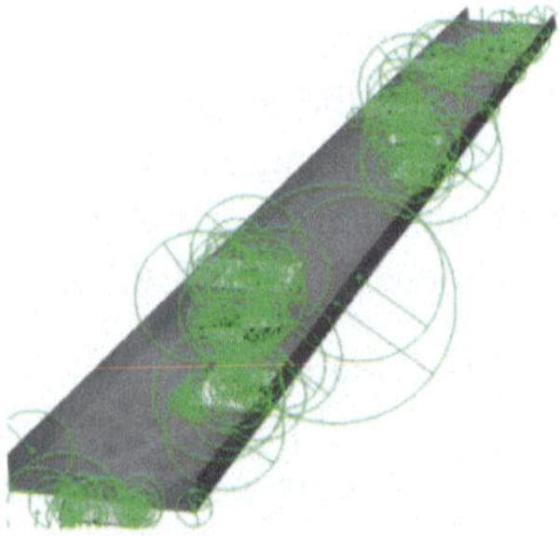

Fig. 5 Extract feature points in the real time scene image

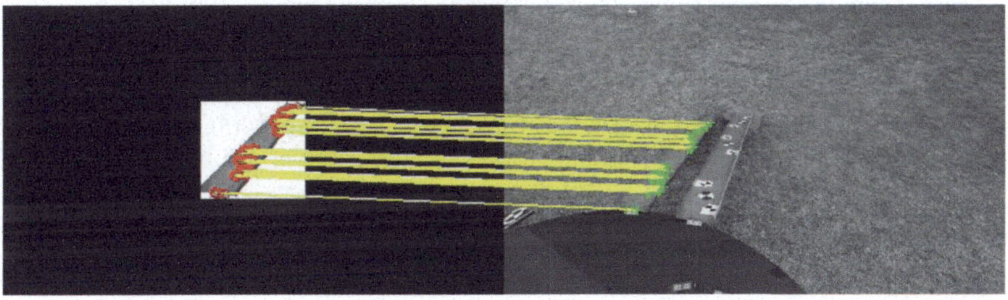

Fig. 6 Establish the matching relationships between the features

Fig. 7 The left egress mechanism image segmentation result

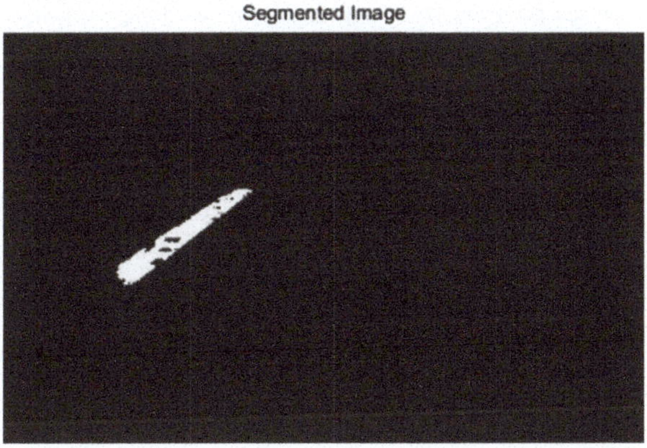

Consider the matching points in the real time image as the seed pixels and use region growing to segment the egress mechanism we want. The segmentation result is shown as Fig. 7.

5 Conclusion

A new method of the egress mechanism image segmentation in the Mars exploration egress phase is proposed in this paper. Using color feature, fusion of the region and MSER feature, egress mechanism image can be segment from the real time scene image. The method is robust to the change of the view angle, the change of the egress mechanism slope and the change of the illumination. The experimental results and analysis show that the method can be used in the egress mechanism image segmentation in the Mars exploration. In the future, the research of adaptability, the robust and the accuracy of the algorithm will be continued.

References

1. Zhang Yujin, image engineering, Beijing: Tsinghua university press, 2012:120–121.
2. Wang Jianan, A color image segmentation method based on automatic seed region growing. Changchun: the northeast normal university, 2007.
3. Yu wangsheng, Hou zhiqiang, Song jianjun. Color image segmentation based on marked-watershed and region-merger. Acta electronic sinica, 2011, 39(5): 1007–1012.
4. Lowe D G. Distinctive image features from scale-invariant keypoints [J]. International Journal of Computer Vision, 2004, 60 (2): 91–110.
5. Tuytelaars T, Mikolajczyk K. Local invariant feature detectors: a survey. Foundations and trends in Computer Graphics and Vision, 2008, 3(3):177–280.
6. Matas J, Chum O, Urban M, et al. Robust wide-baseline stereo from maximally stable extremal regions [J]. Image and Vision Computing, 2004, 22(10): 761–767.
7. Bay H, Ess A, Tuytelaars T, et al. Speeded-up robust features (SURF) [J]. Computer Vision and Image Understanding, 2008, 110(3): 346–359.
8. Torr P H S, Zisserman A. MLESAC: A new robust estimator with application to estimating image geometry [J]. Computer Vision and Image Understanding, 2000, 78(1): 138–156.
9. Duan shan, Wang yingfei, Color images feature extraction based on region growing. Journal of south-central university for nationalities (Nat. Sci. Edition), 2012, 31(2):104–108.

An Automatic Precise Registration Method Based on the Relative Geometric Calibration Between Bands for Satellite Multi-spectral Image

He Wei, Long Xiaoxiang, Yu Jing and Zhang Chi

Abstract A registration method for satellite multi-spectral image is proposed, which is based on the relative geometric calibration between bands. This method precisely calibrate the relative geometric distortion of reference and non- reference band, and ensure that the exact geometric imaging models of different band images have great relative geometric accuracy. Based on the exact geometric imaging model and the same-name pixels positioning consistency, the exact same-name pixels conversion relationship of reference band and non-reference band can be obtained, which can guarantee that the accurate automatic multi-spectral image registration can be realized without image matching. This method is a real geometrically registration, and the correction model between bands is exact geometric imaging model. It is theoretically rigorous, and the registration result couldn't be affected by the image quality. In order to verify the correctness and feasibility of this method, the experiment has been carried out on ZY-3 satellite multi-spectral image. By now, this method has been successfully applied in the satellite images pre-treatment process.

Keywords Satellite multi-spectral image · Registration · Relative geometric calibration · Object positioning consistency · Exact geometric imaging model

1 Introduction

The calibration of multi-spectral pictures between different band image is the key process during satellite images pre-treatment, and its precision directly affect the following processing and application quality [1, 2]. Due to the fluctuation, the

H. Wei (✉) · Y. Jing · Z. Chi
Institute of Spacecraft System Engineering, CAST, Beijing, China
e-mail: hewei501_cast@163.com

L. Xiaoxiang
China Center for Resource Satellite Data and Application, Beijing, China
e-mail: 825725689@qq.com

H.P. Urbach and G. Zhang (eds.), *3rd International Symposium of Space Optical Instruments and Applications*, Springer Proceedings in Physics 192,
DOI 10.1007/978-3-319-49184-4_31

distortion of camera, and other factors, non-linear geometric deformation often exist in the original satellite multi-spectral images between different bands [3]. The bands registration of multi-spectral image only achieved by simple bands panning couldn't meet the high accuracy requirements of product quality. Therefore, the high precision registration method of satellite multi-spectral image has been a hot topic to research in the field of remote sensing.

The purpose of image registration is to eliminate the relative geometry deformation, and determine the corresponding relation of the same-name pixel [4, 5]. The current image registration methods are mostly based on image matching. Through image matching, the same-name images are obtained, and then corresponding relations of the same-name images are established by the geometric correction model (such as affine transformation model) [6]. In the early 90s of last century, Keller Y. proposed an image registration method based on the maximum gradient magnitude [7]; Based on SIFT calculation, Harris calculation, wavelet transform, SUSAN calculation, surf calculation and a variety of remote sensing image registration method has also been extensively and in-depth researched [8–11]. But this method cannot represent the complex geometric deformation due to the ignorance of the analysis on the geometric deformation between images. Meanwhile, for those texture features which are not obvious, and the regions of different radiation, this method is no longer applicable. Because each registration should be done before image matching, which would reduce the efficiency, so in the satellite image pre-treatment, it is necessary to study a new method not depending on the image matching registration to avoid such problems. Yanming etc., fit a fixed band registration model to realize the automatic registration of multi-spectral image, through the statistics of each band of the same name pixel image deviation, and obtained the sub-pixel registration accuracy in BJ-1 and DMC uk2 small satellite multi-spectral image [12]. Because this method is only the simple statistics of each band image corresponding pixel deviation, but not from the exact geometric imaging model, and not the analysis of the relative geometry deformation between various band image, so it is not strict theoretically.

In this paper, it puts forward an automatically accurate registration method of satellite multi-spectral images based on relative geometric calibration between bands, according to the characteristics of satellite multi-spectral camera imaging. By this method, accurate automatic multi-spectral image registration can be realized without image matching, can improve the processing efficiency greatly. This method is a real geometrically registration, and the correction model between bands is exact geometric imaging model. It is theoretically rigorous, and the registration result couldn't be affected by the image quality.

In order to verify the theoretical correctness and practice feasibility of the proposed method, and the experiments on ZY-3 satellite multi-spectral image were carried out. ZY-3 satellite multi-spectral camera includes four bands as B1, B2, B3, B4, [13] each band image resolution and width were the same. In the focal plane, the four band line array along the flight direction, as shown in Fig. 1.

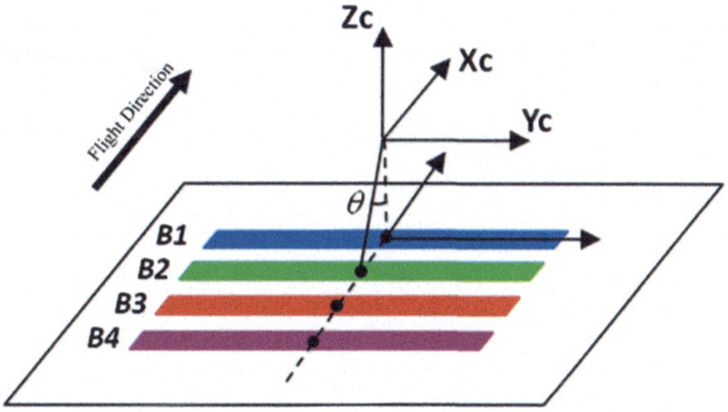

Fig. 1 Focal plane of ZY-3 satellite multi-spectral camera

2 Registration Principle

2.1 Geometry Model Between Bands

The registration method is based on the construction of the imaging geometry model. Take B1, B2 band in Fig. 1 as an example. It is shown in Fig. 2, supposing P is a certain object point in the area overlapped by the B1 and B2 image, B1, B2 has imaged on it and the image point was P_1 and P_2, S_1 and S_2 was the center of projection respectively. According to the constraint relationship of the same-name point intersection, light S1P1, S_2P_2 inevitably intersect to the object point P. According to geometric imaging model of the satellite multi-spectral camera [14], there are:

$$\begin{cases} \begin{bmatrix} Xp_1 \\ Yp_1 \\ Zp_1 \end{bmatrix}_{WGS84} = \begin{bmatrix} Xs_1 \\ Ys_1 \\ Zs_1 \end{bmatrix}_{WGS84} + m_1 R_{T1} R_{BJ1} R_{BS} \begin{pmatrix} x_{c1} \\ y_{c1} \\ -f \end{pmatrix} \\ \begin{bmatrix} Xp_2 \\ Yp_2 \\ Zp_2 \end{bmatrix}_{WGS84} = \begin{bmatrix} Xs_2 \\ Ys_2 \\ Zs_2 \end{bmatrix}_{WGS84} + m_2 R_{T2} R_{BJ2} R_{BS} \begin{pmatrix} x_{c2} \\ y_{c2} \\ -f \end{pmatrix} \end{cases} \quad (1)$$

As shown in formula (1), $(Xp_1, Yp_1, Zp_1)^T$ and $(Xp_2, Yp_2, Zp_2)^T$ were the coordinate of object point P in WGS84 coordinate system, which is obtained based on exact geometric imaging model solution, by using the imaging geometry parameters and object height information of band B1 and B2. $(x_{C1}, y_{C1}, -f)$ and $(x_{C2}, y_{C2}, -f)$ are coordinates of the same-name image points in the camera coordinate system; m_1 and m_2 are photographic scale factor; (X_{S1}, Y_{S1}, Z_{S1}) and (X_{S2}, Y_{S2}, Z_{S2}) are coordinates of projection center S_1 and S_2 in WGS84 coordinate system; R_{BS} is the installation matrix for the camera in the satellite body coordinate system; R_{BJ1} is the transformation matrix between satellite body coordinate system and J2000 system when P_1 was imaging, and R_{T1} is the transformation matrix

Fig. 2 Band-to-band geometric imaging relationship

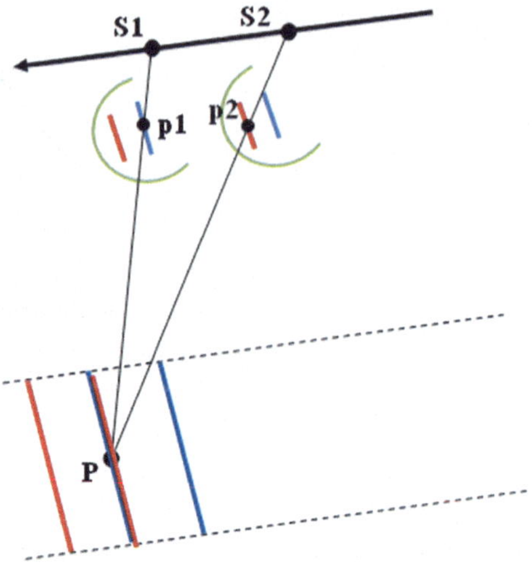

between J2000 coordinate system and WGS84 system. R_{BJ2} and R_{T2} are the conversion matrix when P_2 is imaging.

The positive and negative conversion between image coordinates and the ground point coordinates of the earth can be achieved by using exact geometric imaging model and object height information. Specific conversion formula was not listed to simplify the method description, mapping relation (2)–(3) were used to represent the coordinate conversion involved in this paper. f_1 represents under the exact geometric imaging model, using object height information to project image point (x, y) positively to the object, and obtained the object point coordinates (X, Y, Z); f_2 represents to project object point coordinates (X, Y, Z) negatively to the image, and obtained the image point (x, y).

$$(x, y) \xrightarrow{f_1} (X, Y, Z) \tag{2}$$

$$(X, Y, Z) \xrightarrow{f_2} (x, y) \tag{3}$$

2.2 Registration Method

Take band B2 band as a reference, using geometry imaging parameters of non-reference band B1 (similar to the B3 and B4) and object height information, the coordinates of the object P, $(Xp_1, Yp_1, Zp_1)^T$, was obtained, by executing the formula f_1 on the same-name image points P_1. Using the geometry imaging parameters of band B2, after the substitution to put the coordinate P $(Xp_1, Yp_1, Zp_1)^T$ into the

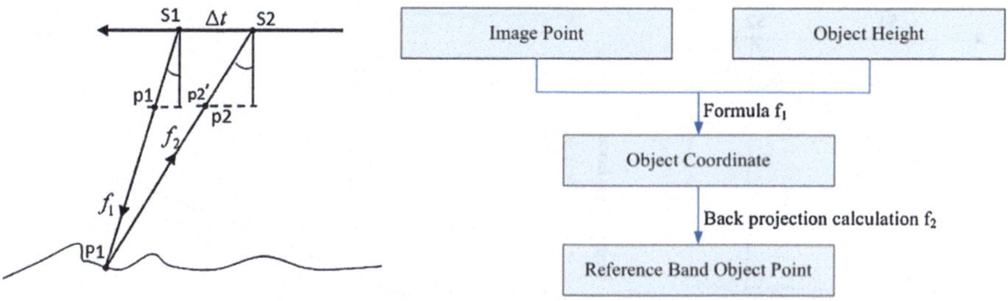

(a) Registration Principle **(b)** Coordinate Conversion of the Same-name Pixel

Fig. 3 The principle and coordinate conversion of the same-name pixel

formula (3) and implementation of the back projection calculation f_2, the coordinates of the same-name image point P2 $\left(x'_{C2}, y'_{C2}\right)$ was obtained. Ideally, when the band geometric imaging parameters and object height information were accurate, the calculated result for above of the coordinate P_2 $\left(x'_{C2}, y'_{C2}\right)$, should equals its real coordinate (x_{c2}, y_{c2}), namely $\left(x'_{C2}, y'_{C2}\right) = (x_{c2}, y_{c2})$. Therefore, the relationship of the same-name image point between B1 and B2 band image can be established, and the multi-spectral image registration can be achieved, this is the multi-spectral image registration method based on object positioning consistency , as shown in Fig. 3. Based on this method, for the image point P_1 which is on the B1 band image, the coordinates of the same-name image point P_2 which is on the B2 can be determined by the implementation as shown in the graphic, to achieve automatic registration of multi-spectral images (Fig. 3).

2.3 *Errors Analysis*

Actually, due to the geometric imaging parameters and object height information errors, $\left(x'_{C2}, y'_{C2}\right)$ cannot be equal to (x_{c2}, y_{c2}). So analysis must be done on the factors which influence the same-name pixel registration error, and negative influence should be eliminated to ensure the accurate registration of the same-name pixel between bands.

1. Height Errors

Take ZY-3 satellite as an example, B2 band as a reference band, analysis was made about the effects brought by height errors of other band to band B2, as shown in Fig. 4. Seen from the graph, the linear relationship was found between the height error and registration error. Therefore, the height error influence on the accuracy of registration cannot be ignored.

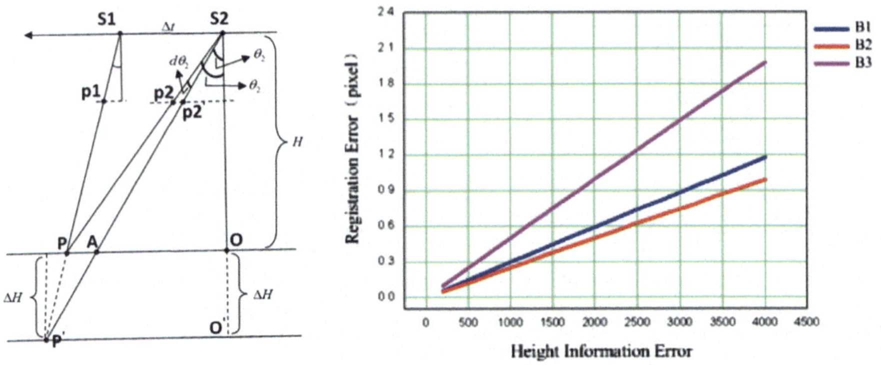

Fig. 4 Relationship between height error and registration error

Using the global ASTER data G-DEM as object height information can solve this problem [15, 16]. Found in Aster G-DEM, as the disclosed Global DEM data in 2009, the precision of planar sampling is up to 30 m, of height sampling is up to 10 m, which can fulfill the requirements of satellite multi-spectral image registration accuracy.

2. Relative Geometric Distortion

If geometric distortion (dx_1, dy_1), (dx_2, dy_2) exist in the image unit between B1 and B2 bands at the points of the same-name pixel P_1&P_2, the calibration error (v_x, v_y) could be obtained as formula (4).

$$\begin{cases} v_x = r_x = dx_1 - dx_2 \\ v_y = r_y = dy_1 - dy_2 \end{cases} \tag{4}$$

(r_x, r_y) is the relative geometric distortion between bands. As shown in formula (4), the relative geometric distortion between bands can leads to registration errors of the same size. Therefore, the compensation of relative geometric distortion among image units and the recovery of the precise relative position relationship is key point and necessary.

3 Geometric Calibration of Bands

3.1 Geometric Calibration Model of Bands

In order to compensate the relative geometric distortion between the B1, B2 image units, additional parameters Δx and Δy were introduced into the parameter model of the non-reference band B1 image unit, and the self-calibration adjustment model was established based on the extended co-linearity equation, which is the common method of linear array camera geometry calibration [17]. At this time, the exact geometric imaging model of band B1 was transformed into formula (5)

$$\begin{bmatrix} X_{S1} \\ Y_{S1} \\ Z_{S1} \end{bmatrix}_{WGS84} + m_1 R_{T1} R_{BJ1} R_{BS} \begin{pmatrix} x_{c1} + \Delta x_1 \\ y_{c1} + \Delta y_1 \\ -f \end{pmatrix} = \begin{bmatrix} Xp \\ Yp \\ Zp \end{bmatrix}_{WGS84} \tag{5}$$

3.2 Calculation of Geometric Calibration Model

Followed by the establishment of relative geometric calibration model on non-reference band, calculation of the model parameters is implemented. To select a multi-spectral images with a better image quality, first of all, a certain number of points (p_1^i, p_2^i) $(i = 1, \ldots, n)$ with uniform distribution on the B1 and B2 two band images (p_1^i, p_2^i indicates a pair of same-name image points in the B1 and B2 bands respectively); For each of point (p_1^i, p_2^i) $(i = 1, \ldots, n)$, by using B2 imaging geometry parameters and object height information, implement the formula to p_2^i, and $(X_{p^i}, Y_{p^i}, Z_{p^i})_{WGS84}$, the coordinate of the object P^i is obtained. At last, using the coordinate of point P^i obtained by step 2 as control point, based on formula (5), the relative geometric calibration model parameter of B1 band image unit is calculated, and the relative geometric calibration of B1, B2 band image unit can be realized. The specific formulation would not be described in this paper.

After getting the relative geometric calibration parameters of each non-reference band, it is recorded and used for the subsequent image registration, according to the above method.

4 Experiments and Analysis

4.1 Experiments Swatch

In order to fully verify the theoretical correctness and practice feasibility, 3 multi-spectral images experiments are carried out on ZY-3 satellite at different time and orbit, which are recorded as ZY3MUX1, ZY3MUX2, ZY3MUX3, and the geometry imaging parameter files are included. In addition, according to previous analysis, this experiment uses ASTER G-DEM data as object height model.

Table 1 Registration accuracy of same-name pixel

Image	Landform	Bands	Registration accuracy/pixel		
			x	y	Total
ZY3MUX2	Mountainous region	B1–B2	0.212	0.152	0.260
		B3–B2	0.209	0.163	0.265
		B4–B2	0.215	0.188	0.285
ZY3MUX3	Flatlands	B1–B2	0.188	0.134	0.230
		B3–B2	0.174	0.185	0.253
		B4–B2	0.192	0.183	0.264

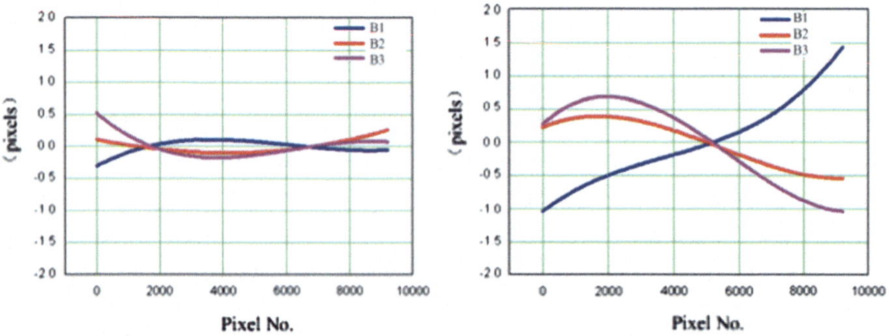

(a) Flight Direction (b) CCD Direction

Fig. 5 Relative geometric distortion curve between non-reference and reference band

4.2 Experiments Method and Result

Based on the described method, image ZY3MUX1 is selected to set relative geometric calibration, in which B2 is as a reference band and B1, B3, B4 as non-reference band. Then the object registration experiments are carried out on ZY3MUX2, ZY3MUX3 by using results of the calibration above. Table 1 shows the registration results of ZY3MUX2, ZY3MUX3.

In order to test the registration accuracy, a number of image points are extracted from the multi-spectral bands to obtain the registration errors. Figure 5 shows the geometric distortion curve between non-reference and reference band.

4.3 Experiments Result Analysis

The following conclusions can be drawn from the above experiment results:

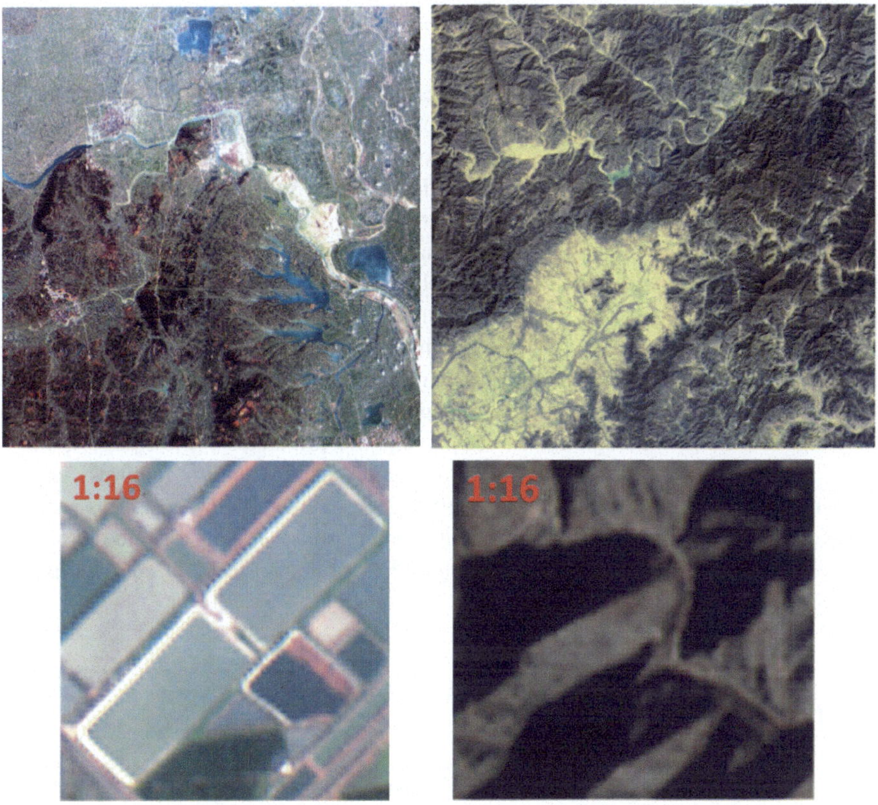

(a) ZY3MUX2 **(a)** ZY3MUX3

Fig. 6 Overview of images after registration

(a) For ZY-3 satellite multi-spectral camera, it is shown in Fig. 5, among B1, B3, B4 and B2 image units, the relative geometric distortion along the track direction is small, basically within 0.2 pixels; the distortion along the CCD direction is larger which can reach 1.5 pixels. Therefore, the relative geometric calibration among bands must be carried out in order to achieve the sub-pixel level registration accuracy.
(b) Amplifying 4 times (1:16) of the multi-spectral images after registration, no obvious band dislocation is found by visual judgment (Fig. 6).
(c) It's seen in Table 1, among all the same-name image points, the object registration deviation between bands is better than 0.5 pixels, which shows that the registration accuracy of this method can completely fulfill the requirements of sub-pixel accuracy.

5 Conclusion

In this paper, an automatic registration method for satellite multispectral image is proposed, which is based on the relative geometric calibration between bands, regarding the imaging characteristics of satellite multi-spectral camera. This method precisely calibrate the relative geometric distortion of reference and non- reference band, and ensure that the strict geometric imaging models of different band images have great relative geometric accuracy. Based on this approach, band registration experiments are carried out on ZY-3 satellite multi-spectral image. Experimental results show that it can achieve better than 0.5 pixel accuracy of band automatic registration on ZY-3 satellite without matching, and greatly enhance the processing efficiency. In addition, this method is a real geometrically registration, and the correction model between bands is exact geometric imaging model. It is theoretically rigorous, and the registration result can't be affected by the image quality, even in the waters, deserts and mountains area where texture features are not abundant and matching quality is difficult to be guaranteed. By now, this method has been successfully applied to the satellite image preprocessing system developed by the State Key Laboratory of Surveying and mapping information engineering in Wuhan University.

References

1. WANG Honghai, XU Jianyan, LONG Xiaoxiang. An Automated Evaluation Method of Sub-Pixel Band-To-Band Registration Accuracy for Satellite Multi-spectral Image [J]. Spacecraft Recovery & Remote Sensing, 2010,31(5):32–37.
2. XU Liyan, WANG Jing, et al. Multi-Spectral Remote Sensing Image Registration Based on Feature Point [J]. Computer Science, 1998, 23(4):320-323.
3. Chander G, Saunier S, et al. SSTL UK-DMC SLIM-6 Data Quality Assessment [J]. IEEE Transactions On Geoscience and Remote Sensing, 2009, 47(7) : 2380–2391.
4. LI Jun, ZHOU Yueqing, LI Deren. A New Fusion Approach Based On Local Histogram Matching Filtering Techniques [J]. Acta Geodaetica et Cartographica Sinica, 1999, 28(3):226–231.
5. Alexander W, David A. Clausi. ARRSI: Automatic Registration of Remote-Sensing Images [J].IEEE Transactions on Geoscience and Remote Sensing. 2007, 45(5):1483–1493.
6. YANG Huachao, ZHANG Lei, et al. An Automated Image Registration Method with High Accuracy Based on Local Homography Constraint [J]. Acta Geodaetica et Cartographica Sinica, 2012, 41(3):401–408.
7. Keller Y, Averbuch A. Multisensor Image Registration via Implicit Similarity. [J]. IEEE Transactions on Pattern Analysis and Machine Intelligence, 2006, 28(5): 794–801.
8. YANG Changqing, WANG Xiaotong, et al. Automatic Registration of Aerophotos Based on Feature Space [J]. Acta Geodaetica et Cartographica Sinica, 2009, 38(4):302–310.
9. LI Qiaoliang, WANG Guoyou, et al. Robust Scale-invariant Feature Matching for Remote Sensing Image Registration [J]. IEEE Geoscience and Remote Sensing Letters, 2009, 6(2): 287–291.
10. HONG G, ZHANG Y. WAVELET-BASED Image Registration Technique for High-Resolution Remote Sensing Images [J]. Computer & Geosciences, 2008, 34:1708–1720.

11. ZHANG Qian, LIU Zhengkai, et al. Automated Registration of Aerophotos Based on SUSAN Operator [J]. Acta Geodaetica et Cartographica Sinica, 2003, 32(3):245–250.
12. YAN Ming, WANG Zhiyong, et al. Multi-Spectral Image Registration Method Based on CCD Geometric Bias Model [J]. Journal of Remote Sensing, 2012, 16(6):1145–1156.
13. TANG Xinming, ZHANG Guo, et al. Triple Linear-array Imaging Geometry Model of Ziyuan-3 Surveying Satellite and Its Validation [J]. Acta Geodaetica et Cartographica Sinica, 2012, 41(2):191–198.
14. YUAN Xiuxiao, YU Junpeng. Calibration of Constant Angular Error for High Resolution Remotely Sensed Imagery [J]. Acta Geodaetica et Cartographica Sinica, 2008, 37(1):36–41.
15. Hirt C, Filmer M S, et al. Comparison and Validation of the Recent Freely Available ASTER GDEM Ver1,SRTM Ver4.1 and GEODATA DEM-9S Ver3 Digital Elevation Models over Australia [J]. Australian Journal of Earth Sciences, 2010, 50:337–347.
16. Nikolakopoulos K G, et al. SRTM vs ASTER Elevation products. Comparison for two regions in Crete, Greece [J]. International Journal of Remote Sensing, 2006, 27(21): 4819–4838.
17. HU Fen. Research on Inner FOV Stitching Theories and Algorithms for Sub-images of Three Non-collinear TDICCD Chips [D]. Wuhan: Wuhan University, 2010.

Shadow Extraction from High-Resolution Remote Sensing Images Based on Gram-Schmidt Orthogonalization in Lab Space

Jianhua Guo, Fan Yang, Hai Tan and Bing Lei

Abstract This chapter proposes a new method of high resolution optical remote sensing image shadow extraction algorithm based on Gram-Schmidt orthogonalization addressing the issues of current shadow extraction algorithm's complex modelling and low efficiency in computation, the basic idea of which is that the inner product of different orthogonal vectors equals to zero. The basic steps are as follow: First, the transform of the image from RGB space to Lab space is conducted based on the principle of human visual system, then the values of L/a/b shall all be confined in the range of [0, 255], which shall be stored in the three channels of RGB respectively for the following process. Colour inversion of the shadow is necessary due to its cool-colour darkness on the image and the inverted warm colour is convenient for the follow process. Take samples from the colour inverted image and the three corresponding channel values of image pixels will be used as feature vectors. Then obtain the mean values of samples of the same class for each channel, which will be processed as the feature vector for that class. Finally, the Gram-Schmidt orthogonalization computation shall be conducted for the eigen-vector of each class to obtain the corresponding orthogonalization vector, and then use the corresponding orthogonalization vector of shadow characteristic vector as weight to enhance the colour feature vectors of the shadow meanwhile reducing the colour feature vectors of non-shadowed in order to extract the shadow from the image. Comparing with the current shadow extraction algorithm, the effectiveness of the method proposed in the chapter is verified.

Keywords Height-resolution remote sensing · Shadow extraction · Gram-Schmidt orthogonalization · RGB · Lab space

J. Guo (✉) · F. Yang · H. Tan · B. Lei
School of Geomatics, Liaoning Technical University, Fuxin 123000, Liaoning, China
e-mail: nkszjx@163.com

J. Guo · H. Tan · B. Lei
SASMAC, National Administration of Surveying, Mapping and Geoinformation, Beijing 100048, China

H.P. Urbach and G. Zhang (eds.), *3rd International Symposium of Space Optical Instruments and Applications*, Springer Proceedings in Physics 192,
DOI 10.1007/978-3-319-49184-4_32

1 Introduction

The shadow of a remote-sensing image means the ground object (huge buildings, woods and mountain ranges etc.) shading the sunlight forming a dark imaging zone in the back of the ground object. The shadows of ground object generated brings positive and negative impacts on the image interpretations: on one hand, the shadows could be reflections of the height of buildings, used for estimations of building volumetric fraction and extractions of 3D informations of the buildings, identifications of targets (such as cloud detection, building positioning), computations of hypsography and gradient etc., on the other hand, the existing shadows often produce lots of difficulties in image processing, which includes object shape distortions in shaded zone, false colours, feature information loss, unapparent ground object by the vagueness of shadows noises, thus impacting the matching precisions of digital images and the information extractions of speed and ground object. Due to these negative impacts, it is often necessary to effectively eliminate the shadow on the image before any imaging analysis process conducted. Therefore, the priority is to effectively detect the shadow, identify its location, compute its range whether it's gonna be used or eliminated.

The current shadow detection algorithms could be divided into two main categories: one is model-based, which is to detect the shaded zone based on geometric shapes of occluded objects, DSM data and sunshine direction, sensors and other parameters; the other is feature-based, extracting the shaded zone by the commonalities and differentiations between shaded zones and non-shaded zones. Prior knowledges regarding scenes, targets and lighting conditions are necessary for the model-based algorithm, which usually are used for processing specific scenes. Therefore, it is subject to certain major limitations. The Attribute-based algorithm makes use of spectrum and geometric characteristics of shaded zone to detect the shadow. Lots of papers focusing on shadow-feature-based shadow detection by researchers, which have achieved satisfying results, take advantage of the feature that the luminance value of shaded zone is lower than that of non-shaded zone on the image in early stage and use the threshold method for shadow detection. Etenmadmia [1] proposed the homomorphic filter-based detection algorithm. Salvador [2] proposed a detection algorithm based on colour invariability of C1C2C3 according to a colour transfer algorithm of invariant colour characteristic, which can be applied for the shadow detections of Static object (buildings) and dynamic objects (vehicles). Polodrio [3] proposed a threshold segmentation detection algorithm based on HSI model, in which component I minus component S in the HSI model with a given threshold value. Any result less than the given threshold value will be segmented as the shaded zone. Ma [4] proposed NDSVI based on HSV model. Yang et al. [5] proposed a detection algorithm based on $(S - I)/(S + I)$ model according to the HSI model and morphology theory. Zhou [6] and others proposed NDUI (Normalized Difference Umbra Index) concept for extracting shadow from shaded zone in the research of floral carbon-capturing models integrating the HSI model and NDVI shadow index for shadow extraction in the shaded zone. The above shadow detection algorithms have achieved good results under certain circumstances in certain area. However, low detection precision still exists, particularly in the complex ground object areas. Therefore, this chapter proposed a innovative shadow extraction algorithm with better detective precision, which is the shadow extraction arithmetic based on Gram-Schmidt orthogonalization [7]. It has been proved by the experiment results that the precision of the proposed algorithm by us is better than the traditional methods.

2 Study Data

In order to verify the validity of this algorithm, we choose the experimental image on Fig. 1 for shadow detection experiment. This image is of Quick Bird multispectral image, the resolution of which is 1.0 m and the size of the image is 656 × 832 pixels. This image is from a urban residential area full of complex and interfering surface features for shadow extraction, such as intercity high-ways in grey, flower nursery and other ground objects.

3 Methodolgy

A. The principle of Gram-Schmidt Orthogonalization
Assume the feature vector composed by the RGB three channels of image pixels in Euclid space V is $\alpha_1, \alpha_2, \ldots \alpha_s$ (where, s is the number of pixels, $a_i = (x_R^i, x_G^i, x_B^i)$, x_R^i, x_G^i, x_B^i is the corresponding value of the RGB three channels of the ith pixel). They are linearly independent and the definition of Gram-Schmidt orthogonal transformation is as follows:

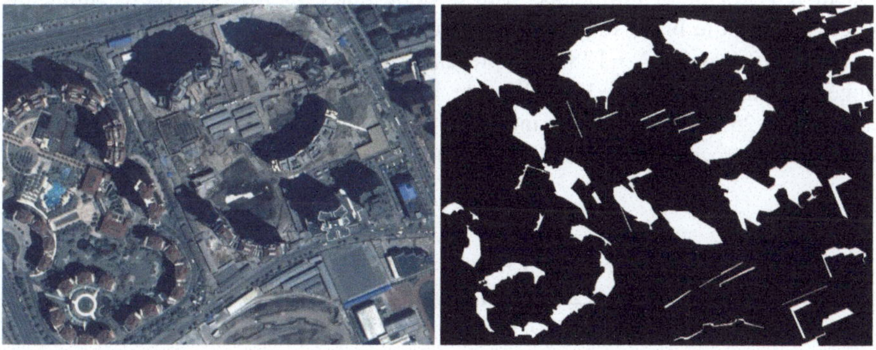

Fig. 1 The Quick Bird image with 1.0 m resolution. Where, the *left* is experiment image and the *right* is its reference image

$$
\begin{aligned}
\beta_1 &= \alpha_1; \\
\beta_2 &= \alpha_2 - \frac{\langle \alpha_2, \beta_1 \rangle}{\langle \beta_1, \beta_1 \rangle}\beta_1; \\
\beta_3 &= \alpha_3 - \frac{\langle \alpha_3, \beta_1 \rangle}{\langle \beta_1, \beta_1 \rangle}\beta_1 - \frac{\langle \alpha_3, \beta_2 \rangle}{\langle \beta_2, \beta_2 \rangle}\beta_2; \\
&\cdots\cdots \\
\beta_s &= \alpha_s - \frac{\langle \alpha_s, \beta_1 \rangle}{\langle \beta_1, \beta_1 \rangle}\beta_1 - \frac{\langle \alpha_s, \beta_2 \rangle}{\langle \beta_2, \beta_2 \rangle}\beta_2 - \cdots - \frac{\langle \alpha_s, \beta_{s-1} \rangle}{\langle \beta_{s-1}, \beta_{s-1} \rangle}\beta_{s-1}.
\end{aligned} \tag{1}
$$

where, $\beta_1, \beta_2, \ldots, \beta_s$ are non-zero vector and pairwise orthogonal, i.e.,

$$
\begin{cases} \langle \beta_i, \beta_j \rangle = 0, i \neq j \\ \langle \beta_i, \beta_j \rangle = \|\beta_i\|, i = j \end{cases} \tag{2}
$$

The normalized vectors form an orthonormal set which can be described as:

$$
\gamma_i = \frac{1}{\|\beta_i\|}\beta_i, \quad i = 1, 2\ldots, s \tag{3}
$$

Therefore, the result of inner product between the vectors $\alpha 1, \alpha 2, \ldots, \alpha s$ and γi is:

$$
\begin{cases} \langle \alpha_j, \gamma_i \rangle = 0 & j \in 1, 2 \cdots s,\ i \in 1, 2 \cdots s\, and\, j \neq i \\ \langle \alpha_j, \gamma_i \rangle = K & K \neq 0 \end{cases} \tag{4}
$$

B. Lab space

Lab colour space was established by CIE in 1931 based on international standards of colour measurement, which is a device-independence colour system and a physiological characteristics based colour system. In the Lab colour space, component L is for pixel luminance with a value range of [0, 100] representing the process from pure black to pure white; a is for the range of red to green, the value range of which is [127, −128]; b represents the range of yellow to blue, the value range of which is [127, −128]. We are normalizing the values of L/a/b into the range of [0, 255], which are put into the three channels o RGB for the convenience of following process. The specific definition is as follows:

$$
\begin{cases} L = 0.2126 \times R + 0.7152 \times G + 0.0722 \times B \\ a = 1.4749 \times (0.2213 \times R - 0.3390 \times G + 0.1177 \times B) + 128 \\ b = 0.6245 \times (0.1949 \times R + 0.6057 \times G - 0.8006 \times B) + 128 \end{cases} \tag{5}
$$

where, the value range of R/G/B and L/a/b are all within [0, 255].

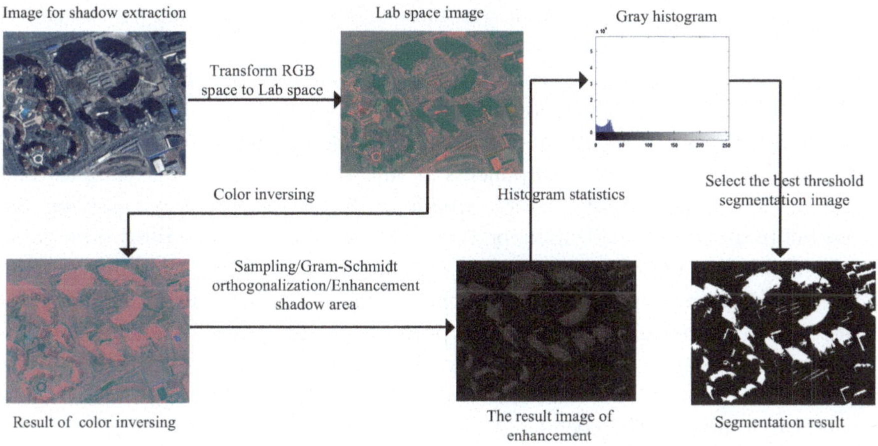

Fig. 2 The extraction flowchart of high resolution optical remote sensing image *shadow* based on Gram-Schmidt orthogonal transformation. First, transform the image from RGB space to Lab space based on the human visual system. Second, Colour inversion, transform the cold tone of the *shadow* shown on the image to warm tone. Then, take samples from the colour-inverted image and obtain the mean values of samples in same category in each channel. Finally, get the corresponding orthogonal vectors through the process of Gram-Schmidt orthogonal transformation, then use the corresponding orthogonalization vector of *shadow* feature vector as weight to enhance the colour feature vector of the shadow meanwhile reducing the colour feature vectors of non-shadowed. Last, extract the *shadow* area from the image through threshold segmentation

C. Image colour inversing

The image after the Lab conversion process can not be used directly due to the dark-blue cold tone of the shadow shown on the image. According to the formula (5), if we take the samples from the Lab space image and obtain the orthogonal vectors of the sample through the Schmidt orthogonal transformation, the corresponding orthogonal vectors of the shadow is used as the weight to enhance the shadow, which imposes no evident effect on the reduction of non-shaded zone and enhancement. So, what we need to do in this step is to transform the ground object with cold tone to that with warm tone and get the highest values among the corresponding values of R/G/B and the lowest value of that of non-shaded surface features. We use the concept of colour conversion in this chapter and the colour inversion expression is as follow:

$$I(i,j,:) = 255 - I(i,j,:) \tag{6}$$

where, $I(i,j,:)$ represents the corresponding channel values of R/G/B of (i,j)th pixel in image I.

D. Shadow extraction from high-resolution remote sensing images based on Gram-Schmidt Orthogonalization

The key step of shadow extraction from high-resolution remote sensing images based on Gram-Schmidt Orthogonalization is the calculation of the corresponding orthogonal vectors of the sample eigenvectors. A 3-D feature vector is defined for each pixel using the RGB components of the images. Select the samples and calculate their 3-D feature vectors, which are orthogonalized by applying the Gram-Schmidt Orthogonalization process. We set the corresponding orthogonal vector of the feature vectors of shadow samples as a weigh w. The inner product between weighting vector and the pixel vectors defined from the original high-resolution remote sensing image produces a composite image which has maximum intensity in green colored regions and minimum intensity in other regions. The composite image has only one channel, meaning that it is a panchromatic image. The specific steps are shown below chart (Fig. 2).

4 Results Analysis and Discussion

In order to verify the validity of this algorithm, we use the improved Tsai model proposed by Chung [8] replacing the global threshold with sequence threshold to get result of shadow detection with better precision; and the shadow detection model of Multi Features Integration (MFI) by Jiahang [9] and his shadow detection model of Self-adoptive feature selection (SAFA); and NDUI by Zhou [6]; and remote sensing image detection algorithm based on colour model by Zhao [10]. The experiment results of the comparison verifications between these five algorithms are as follows:

In order to have a more objective result of the comparison experiment, we carried on the statistics of the coefficient Kappa of results from each algorithm and the overall accuracy (OA), computation time respectively to compare the advantages and disadvantages of each algorithm. The result of experiment statistics is shown as Table 1:

The statistics in Table 1 and experiment results in Fig. 3 show that the Kappa of these 6 algorithms are: Ours > SAFS > NDUI > MFI > Zhao > Cheng, among

Table 1 Performance comparison of different algorithms

Algorithm	Omitted error	Committed error	Total error	Kappa (%)	OA (%)	Time/(s)
Chung	2450	45,730	48,180	74.56	91.17	4.61
MFI	16,881	7311	24,192	84.46	95.57	14.43
SAFS	10,510	10,906	21,416	86.78	96.08	22.12
Zhao	10,005	23,677	33,682	80.24	93.83	2.26
NDUI	12,247	11,141	23,388	85.48	95.71	2.01
Ours	10,572	9797	20,369	87.37	96.27	5.18

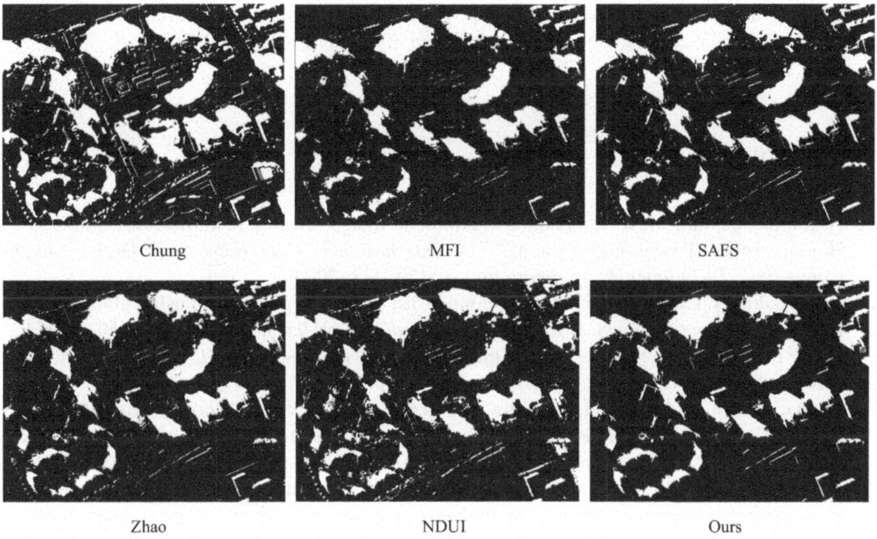

Fig. 3 The experiment results of different algorithms

which the algorithm of Cheng has the worst precision result among the six with its Kappa only of 87.37%; when it come to OA, the result of Ours > SAFS > NDUI > MFI > Zhao > Cheng demonstrates that the overall accuracy of our algorithm is also better than that of the remaining five algorithms; when it comes to the computational time, the result of SAFS > MFI > Ours > Cheng > Zhao > NDUI shows that NDUI algorithm has the shortest computational time of 2.01 s while SAFS algorithm has the longest computational time of 22.12 s. Although the computational time of the algorithm by us is 5.18 s, that is two times of that of NAUI algorithm; it is four times more efficient than that of SAFS. The above analysis shows that the proposed algorithm has better advantages in all aspects, especially the precision is the best among the six algorithms.

5 Conclusions

This chapter presents a shadow extraction algorithm of high resolution optical remote sensing image based on Gram-Schmidt orthogonalization, which highlights the shadow zone and suppresses the non-shadow zone from the image through the concept of the inner product of different orthogonal vectors equaling to zero. The image processing techniques of the transform from RGM space to Lab space and colour inversion are adopted for the convenience of the following enhancement of the shadow zone with Gram-Schmidt orthogonalization. The experiment result shows that the proposed algorithm has high extraction accuracy and less computation time.

Acknowledgments This work was supported by outstanding postgraduate development schemes of School of Geomatics, Liaoning Technical University under grant YS201503.

References

1. H. Etemadnia, and M. R. Etemadnia, "Automatic Image Shadow Identification using LPF in Homomorphic Processing System." in *International Conference on Digital Image Computing: Techniques & Applications*, pp. 429–438, 2003.
2. E. Salvador, A. Cavallaro, and T. Ebrahimi, "Cast shadow segmentation using invariant color features," *Computer Vision & Image Understanding,* vol. 95, no. 2, pp. 238–259, 2004.
3. A. M. Polidorio, F. C. Flores, N. N. Imai, A. M. G. Tommaselli, and C. Franco, "Automatic Shadow Segmentation in Aerial Color Images," pp. 270–277, 2003.
4. H. Ma, Q. Qin, and X. Shen, "Shadow Segmentation and Compensation in High Resolution Satellite Images." *IEEE International Geoscience & Remote Sensing Symposium, IGARSS 2008, July 8–11, 2008, Boston, Massachusetts, USA, Proceedings*, pp. 1036-1039, 2008.
5. J. Yang, "A Shadow Removal Method for High Resolution Remote Sensing Image," *Geomatics & Information Science of Wuhan University,* vol. 33, no. 1, pp. 17–20, 2008.
6. J. H. Zhou, Y. F. Zhou, X. H. Guo, and Z. Ren, "Methods of extracting distribution information of plants at urban darken areas and repairing their brightness," *Journal of East China Normal University,* vol. 39, no. 6, pp. 1–9, 2011.
7. S. H. Rezatofighi, H. Soltanian-Zadeh, R. Sharifian, and R. A. Zoroofi, "A New Approach to White Blood Cell Nucleus Segmentation Based on Gram-Schmidt Orthogonalization." 2009 International Conference on Digital Image Processing, pp. 107–111, 2009.
8. K. L. Chung, Y. R. Lin, and Y. H. Huang, "Efficient Shadow Detection of Color Aerial Images Based on Successive Thresholding Scheme," IEEE Transactions on Geoscience & Remote Sensing, vol. 47, no. 2, pp. 671–682, 2009.
9. L. Jiahang, "Study on Automatic Recognition of High Resolution Optical Remotely Sensed Images Based on Visual Feature," Shanghai Jiao Tong University, 2010.
10. X. F. Zhao, and H. U. Xiao-Wen, "Remote Sensing Images Shadow Detection Based on Color Models," Science Technology & Engineering, vol. 5, pp. 20–22, 2013.

Optimal Sensitivity Design of Multispectral Camera Via Broadband Absorption Filters Based on Compressed Sensing

Suixian Li and Liyan Zhang

Abstract Spectrum acquisition of imaging scenes with super spectral resolution can be realized by multichannel spectral camera with broadband absorption filters under the condition that the multichannel spectral camera sensitivity is optimized. Algorithms to broadband absorption filters selection to optimize the camera sensitivity proposed in the past have no strict theoretical guarantees on reconstruction accuracy. Consequently, the insight had not been uncovered until the Compressive Sensing (CS) theory has ripped in the last recent years. Combined the proofed datasets of published literature and sensing matrix design theory of CS algorithm to optimal the sensitivity of multispectral camera by filter selection is proposed and verified. The more variation of filter vectors can be selected, the more accuracy of the spectral reconstruction results can be acquired with super resolution.

Keywords Multispectral camera · Sensitivity optimization · Compressive sensing (CS) · Absorption filter · Coded aperture snapshot spectral imaging system (CASSI) · Compressed broadband multispectral imaging system (CBMIS)

1 Introduction

In hyperspectral observation such as remote sensing and microscopic spectral imaging, narrowband spectrum acquisition is a tradition mode by using the principle of scattering, diffraction and interference of the incident spectrums of the scene [1–3]. The narrowband hyperspectral imager (or spectrometer) with light splitting element such as prisms or gratings has been a highly degree of ripeness, take Hyperion imager for example, its system repeatability can reach 2% and

S. Li (✉)
Flying College, BinZhou University, Binzhou 256600, Shandong, China
e-mail: leesx_72@163.com

L. Zhang
College of Resources Environment and Tourism,
Capital Normal University, Beijing 100048, China

H.P. Urbach and G. Zhang (eds.), *3rd International Symposium of Space Optical Instruments and Applications*, Springer Proceedings in Physics 192,
DOI 10.1007/978-3-319-49184-4_33

radiometric precision upwards 3–5% after calibration [3]. However, Due to weakly allocated photon energy through the narrowband spectrum acquisition channels, traditional means of hyperspectral image acquisition encounters an inevitable consequence of the contradiction between large data quantity and spectral resolution, at the same time, the signal to noise ratio of hyperspectral image are limited [4]. Based on Fresnel double mirror interference, the total reflection imaging spectrometer has been gotten over the problems of narrow spectrum bandwidth constrained by wavelength response of the refraction elements, low signal-to-noise ratio due to limited luminous flux stopped by the optical slit etc., however, the problem of large data quantity has not been solved in principle, and the data reconstruction algorithm introduced makes higher computational complexity [4–6]. How to decrease the amount of data acquisition of hyperspectral scene with acceptable reconstructed data quality has been the bottleneck to increase the efficiency of hyperspectral imaging.

In multispectral color imaging which mainly used in the field of spectral color reproduction, spectral images with resolution not lower than 5 nm in laboratory were verified by a simply constructed multispectral system with several broadband spectral channels [7–11], which accuracy is much higher than most of the hyperspectral imager with narrowband type of spectrum acquisition channel loaded on space borne or airborne platform. In principle, narrow band acquisition has simple insight as it measures intensity of each wavelength sample of the spectrum or its spectrogram directly; however broadband multispectral color imaging system is not the case at all. The broadband spectral channel is formed by a usual camera with broadband transmittance filters amounted on front of it, so the detector of broadband spectral response channel integrates all the photoelectrons of wavelengths span of the incident light depended upon the spectrum of the scene and the wavelength transmittance curve of the filter. Although much successful in color science, the broadband spectral imaging has not been fully understood by people outside the multispectral color reproduction field until the emergency of the *Compressive Sensing* information theory and its application on imaging science [12–14].

The basic idea of CS is to utilize the special structures of the signal itself (or in a transformed domain) so that a small number of measured projections can be used to reconstruct the signal [12, 13]. There have been many studies in applications of CS, one of the first applications being the well-known single pixel camera [14], which measures a small number of random projections of an image from which it successfully reconstructs the image. In fact, the concept of broadband multispectral color imaging is in accordant with compressive sensing in that much less number of sample channels or say compressed spectral acquisition channels, which satisfied the fewer sample data at stage of signal acquisition and similar concept of signal reconstruction method. There are several literatures research spectral imaging based on CS via compounding spectrometer and random coded aperture with corresponding reconstruction algorithm [15–22], but the result of experiments indicates it still remains a long way ahead before utilized in practice.

In the following sections, We will review the theory of optimal the broadband spectral sensitivity of multispectral camera by filter selection published in the literature, then takes the view of CS to it and make comparison of performance of the two existing spectrum sample mode related above, so that a more theoretic framework of optimal method to setup of multispectral camera can be concluded.

2 Multispectral Camera Model

2.1 Traditional Multispectral Camera Model

The terminology and definitions referring the concepts of multispectral imaging and multi-channel imaging are sometimes confusing, with different meanings by different authors. In color imaging science, terminology referring to *multispectral imaging* is sometimes *called multi-channel visible spectrum imaging* (MVSI), or simply *spectral imaging* with more than three spectrum channel as traditional 3RGB camera, which was used to high fidelity color reproduction based spectral match to avoiding traditional color metameric match [23]. However, in remote sensing, *multispectral imaging system* has its specific meaning: spaceborne or airborne *multispectral camera* with more than 4 or less than several tens of the number of spectrum channels such as HJ1A has 4 spectral channels and the MODIS has 36, which spectral band spans from visible to Infrared light.

A common multispectral camera design concept features multispectral acquisition channels consisting of several optical bandpass filters. By shifting the filters sequentially into the optical path, the electromagnetic spectrum is acquired through the channels, thus making an approximate reconstruction of the spectrum feasible. For the sake of focusing on spectral transformation analysis, we neglects the 2D distribution of imaging scene and the geometric distortions occur [24] in each spectral channel due to the optical filters exhibit different thicknesses, refraction indices and may not be aligned in a perfectly coplanar manner. So, multispectral camera model can be described as the following [9, 25]:

$$c_k = \int_{\lambda_{\min}}^{\lambda_{\max}} l_R(\lambda) r(\lambda) o(\lambda) \phi_k(\lambda) s(\lambda) d\lambda + \varepsilon_k = \int_{\lambda_{\min}}^{\lambda_{\max}} r(\lambda) \omega_k(\lambda) d\lambda + \varepsilon_k \tag{1}$$

where c_k is response of the kth spectral channel of multispectral camera, $l_R(\lambda)$ is the spectral distribution of light source, $\phi_k(\lambda)$ is the spectral transmittance function, $o(\lambda)$ is the optics spectral characterization function, $s(\lambda)$ is the spectral sensitivity of the camera sensor, $r(\lambda)$ is the reflectance of object, ε_k is noise item. $\omega_k(\lambda)$ in formula (1) contains all of the parameter except reflectance vector that called *spectral sensitivity optimal vector*. In practice, only $\phi_k(\lambda)$ or $l_R(\lambda)$ can be feasibly changed by select a series of filters or light source, so that the sensitivity of multispectral camera can be optimized [9].

2.2 Multispectral Camera Model Based on Compressive Sensing

If imaging scene reflectance vector **R** is a n dimensional column vector, where n is the total spectral sampling number of a reflectance curve; and denote m as the number of channels, obviously $k \leq m \ll n$. The reflectance vector can be denoted linearly as:

$$\mathbf{R} = \boldsymbol{\Psi}\mathbf{r} \tag{2}$$

where $\boldsymbol{\Psi}$ is a n × m matrix and its rows is the basis of reflectance R, and therefore r is column vector with m elements of coefficient; From the compressive sensing (CS) point of view, $\boldsymbol{\Psi}$ is a sparse orthogonal matrix [12, 13]. In practice of multispectral imaging, Ψ is derived from SVD or wavelet representation and any other sparse basis decomposition of the training reflectance set of imaging scene [11].

Regardless of the noise and rewriting formula (1) as matrix, the equivalent response model of multispectral camera can be represented as:

$$\mathbf{C} = \boldsymbol{\Phi}\mathbf{R} = \boldsymbol{\Phi}\boldsymbol{\Psi}\mathbf{r} = \boldsymbol{\Theta}\mathbf{r} \tag{3}$$

where **C** (m dimensional column vector) is camera response vector; $\boldsymbol{\Phi}$ is the transform matrix form **R** to which is called "*sensing matrix*" in CS; $\boldsymbol{\Theta}$ is the transform matrix (or sensing matrix) form **r** to **C**. In multispectral imaging practice, $\boldsymbol{\Phi}$ is decided by all the elements that impact the response of the sensor, such as filter, illumination, and any other spectral features of the optics of the system, including noise item.

The model described in (3) is based on the fact that reflectance of natural spectral imaging scene can be sparsely represented, or say, can be represented as a linear combination of little number of bases approximately. By this model, high spectral definition of spectrum can be reconstructed by just few spectral acquisition channels of multispectral imaging system [7–11].

Compressive sensing (CS) is an emerging field that exploits the underlying sparsity of a signal to perform sampling at rates below the Nyquist-criterion. The key idea in CS is the realization that most signals encountered in practice are sparse in some sense and the theory of CS exploits such sparsity to dictate that far few sampling resources than traditional approaches are needed [12, 13]. The practice in multispectral imaging in literature is much more similar as the well-developed and verified compressive sensing information theory. So, the objective function of optimal multispectral imaging can be deduced under the guidance of CS, and it will be deduced and verified in the following section.

3 Optimal Sensitivity of Multispectral Camera Via Absorption Transmittance Filter

3.1 Two Way of Optimization in Practice

Because the spectral sensitivity function of filters or spectrum distribution of illuminations can be selected in a large space more availably, optimization of the sensitivity of multispectral camera was performed by selecting filters among a larger number of filters set [8, 9, 26–29] or light source spectrum distribution by modulating a large numbers of LEDs [30, 31]. We focus filter selection in this paper because the tendency of utilizing multispectral camera techniques to remote sensing [10] where the imaging scene is illuminated by natural light that cannot be controlled.

3.2 Filter Selection Methods in Literature and Data Collecting

In this section, we will review the method of filter selection in setup of multispectral camera briefly, and focus on the optimal spectral sensitivity data collection, which will be carefully treated in the following section with an aim to find insight to the optimal filters selected, if there are, to explore if the verified data in the literature meet the optimal criterion in CS theory.

Algorithms to broadband absorption filters selection to optimize the camera sensitivity proposed in the past have no strict theoretical guarantees on reconstruction accuracy. Consequently, the insight had not been uncovered until the Compressive Sensing (CS) theory has ripped in the last recent years. There are two kinds of algorithms to filter selection. One is *filter analysis method*, which focusing on the analysis of filter set itself to explore if some mathematical features of transmittance curve of the corresponding subset perform well according to competent metrics in practice [8, 9]. In Ref. [9], Several types of filters usually in practice are considered to optimization operation by comparison of their performances, including discrete and overlapped Gauss curve shaped filters, filters selected by different selection methods of *Maximum Linearity Independence* and *Maximum Orthogonalization*, concluded the overlapped filter set and filters selected by *maximum linearity independence* performs better; With same number of filters, broadband transmittance absorption filter sets selected by *maximum linearity independence* performs best. The other is *systematic recursion method*, which searching for an optimal set of spectral sensitivity functions among the complete combinations of the given filter components, the optimal design of camera spectral sensitivity functions are correspondent to the highest competent metric [26–29]. In practice, the number of all the combination of filters is always very large, and it is impossible to complete recursion of all the filters in acceptable time cost. Some pre-analysis on the filter information has to be carried out in order to finish optimal filter search. Connah consider the condition number of filer set is a optimization criterion [26], Quan optimized spectral sensitivity [27] by selecting filters with two step to minimize the computational cost by selecting the set with highest μ-factor value [28] from the sub-optimal collection obtained with UMG [32], where the candidate matric of μ-factor is a metric based on geometrical difference, and Unified Measure of Goodness (UMG) simultaneously considers the imaging noise and its propagation.

Several authors published optimized filters data together with spectral sensitivity of imaging sensor data in their researches, these data sets paves the way to our research, we wonder if these data that was verified optimized consistent with the system optimization criterion of CS with a aim to generalize the CS theory to *Compressed Broadband Multispectral Imaging System* (*CBMIS*). The details of data background cannot be related due to the limited length and can be refer to the corresponding literature [10, 32]. Table 1 lists Data collection specifications from the literatures.

3.3 Optimal Multispectral Imaging in the View of CS

In the view of *compressive sensing* information theory, a criterion for determining the quality of sensing matrix, $\mathbf{\Theta} = \mathbf{\Phi\Psi}$ in formulate (3), is the *worst coherence*. Formally, denote the Gram matrix $\mathbf{G} = \mathbf{\Theta}^{T}\mathbf{\Theta}$ when the column of $\mathbf{\Theta}$ have unit norm and let

$$\mu(\mathbf{\Theta}) \equiv \max_{1 \leq i,j \leq N i \neq j} \left|\mathbf{G}_{i,j}\right| \tag{4}$$

Table 1 Data collection specifications for comparison from literatures

Origin of dataset	Filter transmittance curve	Spectral sensitivity of sensor	Illuminates spectral distribution	Description of pairs to comparison
Ref. [32]	No	Yes	A/D65/F11	Less optimal (fabricated set 1) versus optimal 3RGB (IBM Pro/3000) sensitivity by UMG, a metric of spectral color matching with little metameric. Sensitivity includes CCD and filter transmittances of total camera
Ref. [10]	Yes	Yes	No (natural scene)	RGB digital camera without versus with optimized colored filters (G, G + O + MG2) to recover spectral data from natural scenes

be the largest off-diagonal element of the Gram matrix. A good goal in designing a sensing matrix is to choose Φ and Ψ so that μ is as close as possible to $M^{-1/2}$ [33, 34]. The main practical advantages of coherence-based theory is that it is possible to compute whether the CS system is optimized. In the following, we will examine the optimized system in literature by this criterion.

3.4 Verification by Data Collected in the Literature

Each of the four sets of sensitivity pairs listed in in the Table 1 is compared in the corresponding literature, so we can use the data to compute the $\mu(\boldsymbol{\Theta})$ to check the criterion related to above. As described in formula (3), the sensing matrix $\boldsymbol{\Theta}$ can be presented several spectral response vectors: illuminate **I**, camera sensitivity **S**, filter transmittance, scene reflectance eigenvectors derived by SVD (singular value decomposition). In the following, we will take dataset Ref. [32] for example to computer $\boldsymbol{\Theta}$ under the condition that using optimized colored filters **T** (G + O + MG2) and D65 standard illuminate. The dataset are illustrated in Fig. 1. All the data have m = 61 wavelength dimensions, sampled from 400 to 700 nm at every 5 nm of the corresponding vectors.

In the computation of Ref. [32] dataset, we normalized the illuminates by the maximum value equals 1, and sensing matrix is represented as:

$$\boldsymbol{\Theta} = \mathbf{S}^t \cdot diag(\mathbf{I}) \cdot \boldsymbol{\psi}_0 \tag{5}$$

where

$$\mathbf{S} = [\mathbf{S}_0, diag(\mathbf{T}) \cdot \mathbf{S}_0] \tag{6}$$

and the superscript $[]^t$ denotes transpose of a matrix, $diag()$ denotes matrix with diagonal elements of a vector. Formula (6) denotes the overall camera sensitivity contain with and without filters, each has 3 column vectors, where **T** is the vector composed by the three filters transmittances with every corresponding element multiplied, for the filters were mounted on the front of the camera by overlapped together. We use 1 order norm when normalize each columns of Θ, which is more effective predictor [33]. The computation for Ref. [10] dataset is similar to that of Ref. [32] described above. All the results are listed in Table 2.

From Table 2, we can see that $\mu(\Theta)$ of the optimal ones more closer to $M^{-1/2}$ except Ref. [10] data sets under A standard illuminate, perhaps fabricated set 1 camera more adaptive photograph under A similar light source in studio; $\mu(\Theta)$ under D65 present more consistency, it can be seen clearly the optimal ones is more close to the $M^{-1/2}$ (see Fig. 2), so under daylight illuminate we can use the criterion of quality of sensing matrix in compressive sensing information theory.

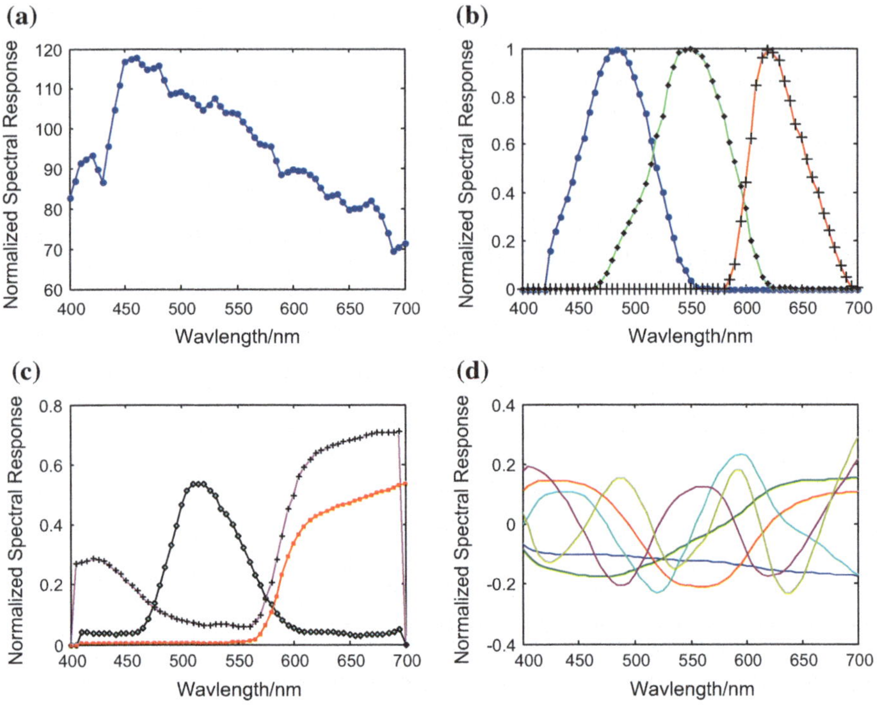

Fig. 1 The illustration of Ref. [32]. Dataset **a** D65 spectrum distribution **I**, **b** camera spectral sensitivity without filters amounted outside $\mathbf{S_0}$, **c** spectral transmittance **T** of filter G, O and MG2, and **d** the first six eigenvectors of reflectance of Macbeth Color Checker $\mathbf{\Psi_0}$

Table 2 Optimal sensing matrix index of broadband spectral imaging systems

Origin of dataset	Description	$M^{-1/2}$ (M)	$\mu(\Theta)$/A	$\mu(\Theta)$/D65
Ref. [32]	Less optimal (fabricated set 1)	0.5774(3)	0.2435	0.2255
	Optimal 3RGB (IBM Pro/3000)	0.5774(3)	0.2142	0.3052
Ref. [10]	RGB digital camera without filter	0.5774(3)	0.1227	0.1044
	Less optimized colored filters G	0.4082(6)	0.2060	0.1579
	Optimized colored filters (G + O + MG2)	0.4082(6)	0.2118	0.2681

4 Conclusion

In this paper, we have taken a comparative computation of the quality of spectral sensitivity of multispectral imaging camera with broadband filters under the view of *compressive sensing* for the first time. From the optimal multispectral camera in practice, we can see that Compressed *Broadband Multispectral Imaging System*

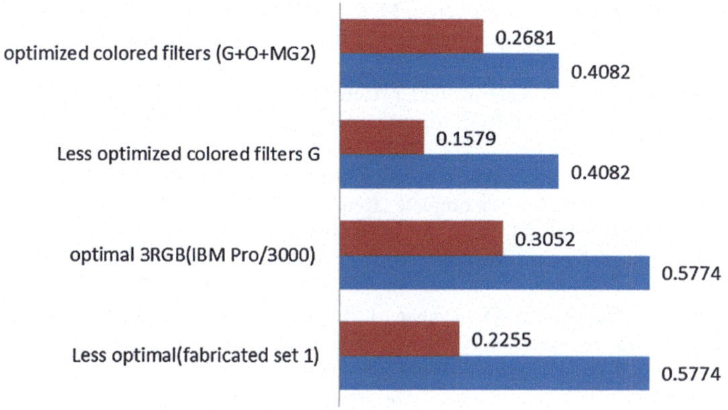

Fig. 2 Comparison of less optimal and the optimal sensitivity under D65 light source

(*CBMIS*) is probably to acquire natural hyperspectral imaging scene with small number of spectral channels. The *compressive sensing* theory is consistent with traditional filter selection optimal method, at the same time, it can be a guidance to the optimization of spectral sensitivity of *CBMIS*.

References

1. Yuval Garini, Ian T. Young, George McNamara. "Spectral Imaging: Principles and Applications", *International Society for Analytical Cytology: Cytometry*, Vol. 69(Part A), pp 735–747, 2006.
2. XIANG Li-bin, WANG Zhong-hou, LIU Xue-bin, YUAN Yan, JI Zhong-ying, LV Qun-bo, "Hyperspectral Imager of the Environment and Disaster Monitoring Small Satellite", R*emote Sensing Technology and Application*, Vol. 24, pp 257–262, 2009.
3. McCorkel, J.; Thome, K.; Ong, L. "Vicarious Calibration of EO-1 Hyperion", *IEEE Journal of Selected Topics in Applied Earth Observations and Remote Sensing*, Vol. 6(2), pp. 400–407, 2013.
4. WU Wen-min, LIAO Ning-fang, CHAI Bing-hua, "Study on all reflective imaging spectrometer based on Fresnel Double_mirror", *OPTICAL TECHNIQUE*, Vol. 32, Chinese, pp 431–433, 2006.
5. A. S. Filler, "Apodization and Interpolation in Fourier Transform Spectroscopy", *Journal of the Optical Society of America*, Vol. 54, pp 762–767, 1964.
6. D. A. Naylor and M. K. Tahic, "Apodizing functions for Fourier Transform Spectroscopy", *Journal of the Optical Society of America*, Vol. 24, pp 3644–3648, 2007.
7. Burns, Peter D., Berns, Roy S. "Quantization in Multispectral Color Image Acquisition", *Color and Imaging Conference*, Vol. 1999(4), pp 32–35, 1999.
8. Jon Y. Hardeberg. "Filter Selection for Multispectral Color Image Acquisition", *Journal of Imaging Science and Technology*. Vol. 48(2), pp 177–182, 2004.
9. LI Suixian, LIAO Ning, fang, SUN Yunan, "Optimal Sensitivity of Multispectral imaging system based on PCA", *Opto-Electronic Engineering*, Vol. 33(3), pp 127–132, 2006.

10. Eva M. Valero, Juan L. Nieves1, Sérgio M. C. Nascimento, Kinjiro Amano, David H. Foster, "Recovering spectral data from natural scenes with an RGB digital camera and colored filters" *Color Research & Application*, Vol. 32(5), pp 352–360, October 2007.
11. Juan L Nieves, Eva M Valero, Javier Hernández-Andrés, Javier Romero, "Recovering fluorescent spectra with an RGB digital camera and color filters using different matrix factorizations", *Applied Optics*, Vol. 46(19), pp 4144–4154, August 2007.
12. E. Candes, J. Romberg, and T. Tao, "Robust uncertainty principles: exact signal reconstruction from highly incomplete frequency information," *IEEE Transactions on Information Theory*, Vol. 52, no. 2, pp. 489–509, Feb 2006.
13. D. L. Donoho, "Compressed sensing," *IEEE Trans. Inform. Theory*, Vol. 52, no. 4, pp. 1289–1306, 2006.
14. M. Duarte, M. Davenport, D. Takhar, J. Laska, T. Sun, K. Kelly, and R. Baraniuk, "Single-pixel imaging via compressive sampling," *IEEE Transactions on Signal Processing Magazine*, Vol. 25, no. 2, pp. 83–91, March 2008.
15. D.J. Brady, M.E. Gehm, "Compressive imaging spectrometers using coded apertures". *Proceedings of the SPIE - The International Society for Optical Engineering*, Vol. 6246(1), pp 62460A–62460A-9, 2006.
16. M. E. Gehm, R. John, D. J. Brady, R. M. Willett; T. J. Schulz, "Single-shot Compressive Spectral Imaging with a Dual-disperser Architecture". Optics Express, Vol. 15(21), pp 14013–14027, 2007.
17. ZHOU Jiankang, CHEN Xinhua, JI Yiqun, SHEN Weimin, "Research on Principle and Experimentation of High-Resolution Optical Compressive Spectral Imaging", Acta Optica Sinica, Vol. 34(1), pp 107–112, 2014.
18. Laura Galvis, Henry Arguello, Gonzalo R. Arce, "Coded aperture design in mismatched compressive spectral imaging", *Applied Optics*, Vol. 54(33), pp 9875–9882, 2015.
19. Lu-Lu, Qian; Qun-Bo, Lü,Min, Huang; Li-Bin, Xiang. "Piecewise spectrally band-pass for compressive coded aperture spectral imaging", *Chinese Physics B*, Vol. 24(8), pp 80703–80708, 2015.
20. N. Diaz; H. Rueda; H. Arguello. "High-dynamic range compressive spectral imaging by grayscale coded aperture adaptive filtering", *IngenIlería e InvestiIgacilón* Vol. 35(3):pp 53–60, 2015.
21. Arce, G.R.; Brady, D.J.; Carin, L.; Arguello, H.;Kittle, D.S., "Compressive Coded Aperture Spectral Imaging: An Introduction", *IEEE Journal of Selected Signal Processing Magazine*, Vol. 31(1), pp 105–115, 2014.
22. Yuan, X.; Tsai, T.; Zhu, R.; Llull, P.; Brady, D.; Carin, L. "Compressive Hyperspectral Imaging with Side Information" ,*IEEE Journal of Selected Topics in Signal Processing*, Vol. 9(6), pp 964–976, 2015.
23. Imai, Francisco H, " Comparative Study of Metrics for Spectral Match Quality", *Proc. CGIV: The First European Conference on Colour Graphics, Imaging and Vision*, pp. 492–496, 2002.
24. Johannes Brauers, Nils Schulte, and Til Aach, "Multispectral Filter-Wheel Cameras: Geometric Distortion Model and Compensation Algorithms", *IEEE transactions on image processing*, Vol. 17, no. 12, December 2008.
25. Jon Y. Hardeberg, "Acquisition and Reproduction of Color Image: Colorimetric and Multispectral Approaches", *ISBN: 1–58112-135-0,* http://www.dissertation.com, USA, 2001
26. P. L. Vora and H. J. Trussell, "Measure of goodness of a set of color-scanning filters", J. Opt. Soc. Am. A, Vol. 10, No. 7, pp. 1499–1503, 1993.
27. Du-Yong Ng, Jan P. Allebach, "A subspace matching color filter design methodology for a multispectral imaging system", *IEEE Transactions on Image Processing*, Vol. 15(9), pp. 2631–2643, October 2006.
28. David Connah, Stephen Westland, and Mitchell G.A. Thomson, "A Computational Model for the Design of a Multispectral Imaging System", *IS&T/SID Ninth Color Imaging Conference: Color Science & Engineering: systems*, Vol. 2001, pp. 130–134, 2001.

29. S. Quan, N. Ohta and N. Katoh, "Optimization of camera spectral sensitivities", *Proc. of the IS&T and SID 8th Color Imaging Conference*, *IS&T*, Springfield, VA, pp. 273–277, 2000.
30. Muhammad Safdar, Ming Ronnier Luo, Yuzhao Wang, Xiaoyu Liu, "Multispectral Imaging System based on Tuneable LEDs", *Conference: Multispectral Color Science (MCS) Symposium, At Tokyo, Japan*, Vol. 2015, May 19–22, 2015.
31. Raju Shrestha, Jon Yngve Hardeberg, "Multispectral Imaging System based on Tuneable LEDs", *23rd Color and Imaging Conference Final Program and Proceedings, Society for Imaging Science and Technology*, pp. 36–40, 2015.
32. SX Quan, N Ohta, RS Berns, XY Jiang, N Katoh, "Unified measure of goodness and optimal design of spectral sensitivity functions", *Journal of Imaging Science*, Vol. 46(6), pp. 485–497, 2002.
33. E Candes, Y Plan, "Near-ideal model selection by ℓ_1 minimization", *Annals of Statistics*, Vol. 37(5A), pp. 2145–2177, 2008.
34. R. M. Willett, R.F. Marcia, J.M. Nichols, "Compressed sensing for Practical optical imaging systems: a tutorial", *Optical Engineering*, Vol. 50(7), pp. 586–598, 2012.

Research and Design of the Architecture of On-Orbit Remote Sensing Information Processing System

Lanzhi Gao, Chao Tan, Panfeng Wu and Qixing Zhu

Abstract In this paper, we designed an on-orbit remote sensing information processing system based on parallel multiprocessor, according the demands of fast processing of remote sensing information and quick target monitoring. This system has extensibility and flexible reconstruction ability, and can be able to complete the on-orbit remote sensing information processing and provide various intelligence data to various combat units. This paper provides the hardware architecture of on-orbit remote sensing information processing and the architecture of Software platform, and also provides the processing methods of information processing and the application strategy. This paper gives us a standardized, modular and extensible system design. The system supports different types of payload including CCD camera, CMOS plane array camera, synthetic aperture radar for remote sensing information input, and supports two kinds of processing flow: parallel processing and assembly line processing. This system is able to support on-orbit real-time remote sensing information processing for different satellites and satisfy all kinds of demands of target monitoring application.

Keywords Parallel multiprocessor · Remote sensing · Information processing

1 Introduction

Along with the needs of the target recognition and emergency response of real-time on-orbit space remote sensing information processing in the field of our country spaceflight earth observation, How to realize the standardization and modularization and the extendibility of the software platform for the orbit information processing system has become a important demand. According to the rapid response space ORS plan proposed by the United States as the reference, we should solve the

L. Gao (✉) · C. Tan · P. Wu · Q. Zhu
No. 5 Academy No. 513 Institute, China Aerospace Science and Technology Corporation, Yantai 264003, China
e-mail: glzdance@163.com

H.P. Urbach and G. Zhang (eds.), *3rd International Symposium of Space Optical Instruments and Applications*, Springer Proceedings in Physics 192,
DOI 10.1007/978-3-319-49184-4_34

quick responsiveness of space systems, development of low cost, flexibility, rapid response ability, and build a low cost and rapid response ability of space platform. The application of the satellite by the tactical application extension to disaster monitoring and emergency incident response, and we are able to provide near real-time on-orbit target detection and the transmission of typical elements to the ground.

In this article, according to the demand of the disaster monitoring and fast response for earth observation, we make research on the key techniques of low costs of the satellite, we summarized the new future requirements of the development of our country's earth observation and the application of remote sensing satellite, and as follows.

(1) The modularization and standardization of satellite payload platform
In the design of the satellite payload platform, the platform of the satellites payload are capable of integration using unified standard, and can realize of rapid assembly and testing. The trend of module standardization is formed with the development trend of the computer and telecommunication system. The basic components are designed by using the standard interface instead of each item. The cost and complexity of each project are reduced.
(2) The high-performance processing ability on orbit information
Satellite autonomous flight software can realize a variety of satellite payload to perform autonomous tasks assigned by the function. One side, the satellites need to have the ability to verify and tactical operations, real time signal positioning function, and realize the high performance parallel computing and image fusion based on image processing technology, and can identify potential targets. In the design of software platform, the software can be reprogrammed on orbit so as to capture a new target. This on orbit information processing ability of the satellite can realize the emergency response of the earth remote sensing and disaster monitoring.
(3) On orbit re-configurable software platform architecture
Software and hardware platform has re-configurable ability. The software platform can adapt to various mission requirements, support on orbit re programming tasks according to the need to capture different goals or access to different information.

2 Design Ideas

In order to meet the requirements of the future Earth observation satellite on orbit remote sensing information processing and fast target detection, and improve the system on-board processing capability and extensible reconfiguration capability, the remote sensing information processing system design follow the follows the design principles.

(1) Generic, scalable hardware architecture for remote sensing information processing

In order to adapt to different payload on orbit remote sensing information processing and target monitoring requirements on orbit remote sensing information processing system with strong data processing ability, can support the CCD push broom camera type load, CMOS array camera loads and SAR load of remote sensing data processing, also on the bus and interface design, using the standard bus and bus standard. The on orbit remote sensing information processing involves the overall design of the system according to the function of the modular design and standard interface design, inside of which the bus module hardware standardization and interchangeable design based on industry standard VPX bus, also external output interfaces support multiple satellite bus, Space Wire bus, can bus and 1553B bus, to meet the interface requirements of the different satellite platforms.

(2) Standardization and reconstruction of remote sensing information processing software platform

In order to meet the requirements of hardware platform, the software platform of on orbit remote sensing information processing system can be extended based on operating system and middle-ware technology to support extracting different levels of intelligence data by the image processing algorithms, including image enhancement, image frame, target identification and validation and slices of the target image processing algorithm. So as to establish the corresponding image processing algorithm knowledge base and the target feature library by the user, the user can complete the target identification and capture in the application software layer.

In the software architecture, the software platform support various image processing algorithms based on middle-ware technology and, and the payload bus interface driver and the communication protocol. Software platform design supports all hardware abstraction module for unified module types, thus in the design of device driver and ensure each hardware module have uniform interface and call methods, set to ensure that there is sufficient compatibility and portability.

3 System Architecture Design

3.1 Hardware Architecture Design

According to the requirement and design idea of on orbit remote sensing information processing, a parallel remote sensing information processing system architecture is put forward, which is based on the multi-level, pipeline, parallel processing and unified scheduling. It is composed of a number of data processing layers and an integrated management layer. The system architecture comprises an input interface module, an output buffer module, an information processing module, an integrated management unit and an information storage module.

Remote sensing information processing system is able to support large capacity on orbit remote sensing data processing, to complete the target detection and recognition, and generate data products for different users, in order to realize the remote sensing information processing of hierarchical, pipe-lining and parallelization process, can configure multiple data processing layer, each layer of data processing layer may include one or more of the image processing module, each data processing layer support level data parallel processing, data processing layer between the support pipe-lining, parallel processing, this can greatly improve the processing ability of the system, and ensure real-time information generated.

Each data processing layer is composed of one or more image processing modules according to the requirements of processing information and processing speed. Each module adopts the same design standard and can be extended to the internal structure and external interface. According to the requirement of remote sensing information processing method, each layer of data products requirements and process, all the data processing layer and data processing layer support parallel processing or pipe-lining mode, each layer processing flow and algorithm, by comprehensive management module in the comprehensive management unit to control and configuration. And support according to the requirements on orbit information processing tasks, the hardware system can achieve system reconfiguration and reactive power can, through the comprehensive management unit on each data processing layer image processing module arrangement and division, re allocation of each layer of the software process and image processing algorithm, realization of hardware and software on orbit reconfiguration and on orbit re-configurable. This will greatly enhance the versatility and flexibility of the device, in response to the complexity of the environment and the task of the environment, the realization of the hardware platform can be reconstructed in order to meet the needs of different information processing.

As shown in Fig. 1, the system components include: image data interface module, output buffer module, image processing module, integrated management module and information storage module. The functions of each module are described as follows.

- Image data interface module

In order to provide different types of loads and the remote sensing data input, the system provides data interface module, to achieve different types of remote sensing data types (such as optical remote sensing and SAR image data) and different types of input interface (Camera link, LVDS) to remote sensing information processing system input matching. According to different types of remote sensing data and the input interface, design the corresponding input interface module, realize matching of different input interface and communication protocol, remote sensing data provide information processing unit, so as to ensure the system's other main modules to meet the standardization and generalization.

- The image processing module

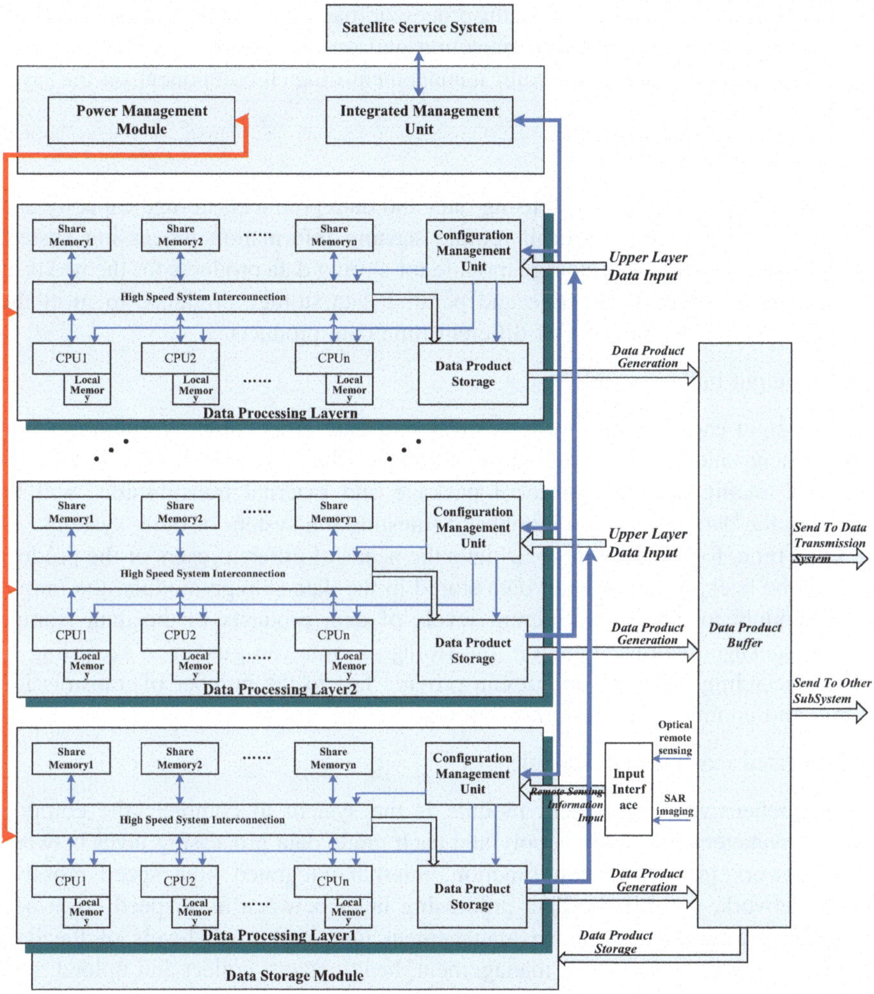

Fig. 1 On orbit remote sensing information processing system architecture based on multiprocessor

In the system architecture shown in Fig. 1, the image processing module is the most basic processing unit to complete the remote sensing image processing. Each layer of data processing layer the standardization of the architecture, by a plurality of image processing modules to form a layer of a layer of data processing is completed, each image processing module with multi-processor parallel system and configuration management unit composition. The multi-processor parallel system to achieve the on orbit of large capacity, high rate of remote sensing information processing and related information products generated, and provide on orbit re-configurable flexible ability; configuration management unit mainly on each level data processing layer of multi-processor parallel system task assignment, parameter configuration, software injection and on orbit re-configurable management, and fault detection and health management of each component on the layer data processing layer.

- Data storage module

In order to provide remote sensing data and data products, storage capacity and information trace-ability on orbit remote sensing information processing system also provides separate data storage module for storing data products for the needs of different users, support real-time and parallel data storage products, to meet the requirement of different users of different time data products.

- The output buffer module

The output cache module provides different data processing layer and external output cache and the uniform output interface, the implementation of different levels of intelligence data protocol package and external transmission, will be related to the data output to a number of transmission system or other systems. At the same time, for the satellite to achieve the needs of different users of the product data can be back, to the product data stored in the data storage module, the output cache module to complete different levels of data products to the data storage module packaged storage function, the storage format can guarantee according to the fast searching and fast output can provide fast to the number of transmission system and common data link transmission.

- Integrated management module

Comprehensive management module is the system to complete the comprehensive management, power supply and each grade data processing layer between cross network interconnection function, internal integrated high speed crossbar switch network, provide the data processing layer between high speed cross network. Integrated comprehensive management unit, and at all levels of the data processing layer configuration management, health status, collect and upload, also through the star provided system of CAN bus and telemetry channel, ground test and control system of satellite on orbit configuration and flight software on orbit injection. Thus, realize the mutual information on the ground and space, achieve different for different target monitoring and different operational requirements of data products generated.

3.2 Software Architecture Design

According to the future according to the operational requirements of satellite mission needs on orbit remote sensing information processing software architecture of the system adopts the hierarchical, extensible, configuration, user software and

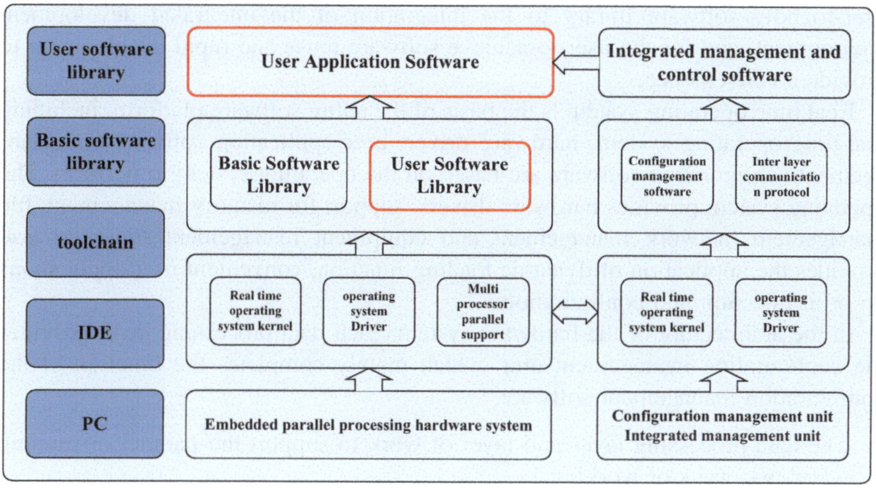

Fig. 2 Software platform architecture of remote sensing information processing system

the bottom software interface from architecture. Software architecture as shown in Fig. 2, construction of software platform based on real-time operating system, according to the user equipment and equipment development task division of roles and different, we will divide into platform software development environment, the basic software, software, configuration management software.

Development environment is used by the basic software and application software development of human-computer interaction integrated development environment, provides the user software development, compiling and debugging of the integrated development environment and provides software package to build tools, is convenient for users to build and integration can be reused software library.

The basic software is provided by the hardware provider, including basic embedded operating system, hardware drivers and basic software library, at the same time, including configuration management software and the integrated control software, configuration and function of integrated management system.

Application software is user development for on orbit target recognition and information extraction of on orbit software. Through calls to the basic software the software interface to achieve the operation of hardware, software library call user developed software library or integrated development environment to provide the complete image recognition algorithm and information extraction algorithm.

Configuration management software, including integrated management and control software and data processing layer configuration management software, the completion of the remote sensing information processing system management configuration, health management and reconfiguration management functions.

In the software architecture, integrated development environment provides the basic software and user software development of integrated development environment, support the user is the secondary development and integrated ability, user-friendly software library to the integration of the integrated development environment, and for the user to achieve software reuse and rapid development to provide a convenience.

Real time operating system is the basis of the entire software platform, including real-time operating system, hardware driver, user application software and management configuration software are based on the operating system to perform. The operating system provides hardware drivers, support for memory management, file management, network management and equipment management function, also provides the application of dynamic loading function, convenient user applications on orbit injection and configuration.

In the architecture of the hardware system, each data processing layer provides the configuration management unit, which mainly completes the function of the configuration management software.

- The data processing layer and layer of work to support the parallel or running water two ways of working.
- The task of the integrated management unit of the dispatch and distribution, the implementation of the image processing module of the software to load and control the running of the software.
- Monitoring the health status of each equipment module, according to the communication protocol and the integrated management unit for communication, and generate the project telemetry to the integrated management unit.

In the hierarchical structure of the system, the top layer of the integrated management module mainly completes the integrated management and work flow configuration of multiple data processing layer, and supports the process of parallel and pipeline processing. Comprehensive management system, through the CAN bus receive commands and data from the satellite platform, the ground uploaded the system configuration and the order parameters transfer to each level of data processing layer, to complete the configuration of the work mode of all levels of intelligence data processing layer and configuration parameters, collected from each series according to processing layer of the working conditions and health status, and the feedback through the telemetry channel the ground. Top integrated management and control software is mainly completed as follows:

- Interact with the satellite system, the system configuration and operation mode configuration command, and the running state of the system.
- Interact with the data transmission system, real-time or delay of the transmission of information at all levels of the ground.
- Management of each layer of internal data processing layer, including software filling, task allocation and work mode configuration.
- To achieve system fault detection and fault tolerance reconfiguration management, when a module failure, the autonomy of the task planning and re distribution, to ensure that the system can not be interrupted.

3.3 Design and Implementation of Multi-core Data Processing Strategy

According to the classification of parallel computer, this system may be regarded as a multiprocessor system based on message passing, the view of software as shown in Fig. 3, each PE processing unit consists of a processor and the landlords and the memory, it has its own address space, can only access the landlords and the memory of the unit. In message passing way to communicate between PE, it is through the communication of the nuclear Shared memory SHM. Each processing run its embedded operating system RTEMS, RTEMS nuclear stored in each PE unit in the PROM.

Analysis by above knowable, on-orbit remote sensing information processing system is a system with symmetrical structure of the whole system hardware, the symmetric structure is conducive to satellite affair management and other tasks of fault tolerance and parallel processing. In order to further utility of the advantage, we design the software system of on-orbit remote sensing information processing system using SPMD thought. Specific design is as follows:

Each node of the same physical entity, logically divided into a Master node (Master) and three from the section point (Slave). The master node is responsible for the star service management, the attitude control and key tasks such as data 0 knife from node is responsible for parallel computing tasks. Each node on the curing of exactly the same code, all nodes when it starts to get their own logical number, which is divided into master-slave node and run different initialization

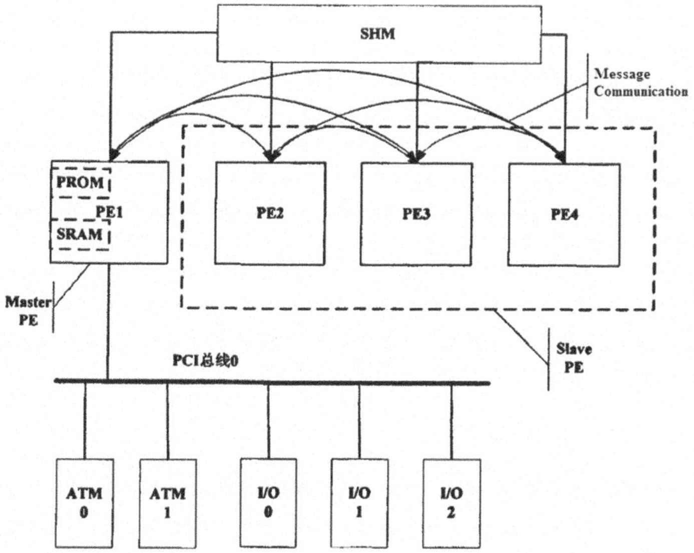

Fig. 3 Software architecture view

process and task, so each node has the ability to master node, this also is helpful for fault tolerance.

To manage the external devices, and the system initialization time, we specify one processor for Master PE, and the master node specified by the hardware, the principle is to elect connection most available peripherals, physics code which is the smallest in the normal state of nodes to the Master. However, in the process of system with electric start, some functions such as peripheral management, only the master node is initialized, although also have the ability, from node but because peripherals is managed by the master node, therefore, from the node is not the function modules of initialization.

Due to the operating system of PE on ROM nuclear are exactly the same, need to use logical number to specify the operating system for different PE nuclear different execution path. So you need to first hardware to physical number of PE, hardware According to the CPU and peripheral global state information and physical number of PE, in accordance with the contract agreement to sign Master PE Master node election. Operating system during initialization, reads the Master mark registers and local PE item number to determine the logical number. Logical number principle is: the Master node number is 1, the rest of the normal nodes according to the physical incremental numbering sequence are numbered consecutively starting from number 2.

4 Conclusion

At present in the field of China's space, the vast majority of satellite remote sensing data processing mainly depends on the under ground, processed by the ground processing system, which requires not only satellite data transmission system with transmission bandwidth and transmission rate, higher and in high resolution synthetic aperture radar remote sensing camera or as a pay load, the transmission of remote sensing data or near real-time processing becomes a choke point, and traditional data processing systems are unable to meet the rapid response for the future combat requirements of the on orbit satellite, so we carry out the imminent research on the data processing technology of remote sensing information. According to a large amount of remote sensing information processing and real-time data are generated on demand, we design a multi-processor data processing system with parallel processing architecture. We provide the on orbit system structure reconstruction, scalable system, to meet the different orbit remote sensing channel, at the same time, the demand of information processing satellite has the ability of two development and rapid integration of the system. In this paper, the research for the improvement of satellite on orbit remote sensing data processing ability and for different task demands of system reconfiguration management, and be used for our country high resolution remote sensing satellite on orbit data real-time processing technology, and we have not only provided a development direction, but also provided a new way for remote sensing data real-time on-board processing capability.

An Improved Side-Slither Method for On-Orbit Relative Radiometric Calibration

Chen ChaoChao, Wang Mi and Pan Jun

Abstract Pushbroom-style imaging systems are widely used in the high-resolution optical remote sensing. This style will lead to streaking and banding, because of inter-detector sensitivity differences, factors related to image formation. There are many traditional methods that are used to eliminate pixel-to-pixel uniformities, such as on-orbit calibration device, "Flat field" calibration and imaging statistical method with lots of images. This paper proposes an improved side-slither maneuver based on moment matching for on-orbit calibration of high-resolution satellite imagery. The disadvantage of the side-slither method is that the images captured with the sensor in a 90° orbit configuration must include uniformity sites such as desert and snow/ice regions. The improved method does not have the requirement that the content of the images for calibration must be uniformity sites. With a linear normalization model, a relative gain can be calculated using the moment matching method to adjust the side-slither maneuver. The approach is characterized by its reliability which is based on the purely statistical analysis of images without any sites constraint. And some indicators are used to assess the normalization residuals after the relative radiometric correction. It is noted that the accurate geometric registration of the successive line of sight projection is necessary and the total amount of pixels should be high enough to achieve a good precision.

Keywords Pushbroom · 90° · Yaw · Side-slither · Linear normalization model · Moment matching

C. ChaoChao (✉) · W. Mi · P. Jun
State Key Laboratory of Information Engineering in Surveying,
Mapping and Remote Sensing, Wuhan University, Wuhan 430072, China
e-mail: wangmi@whu.edu.cn

H.P. Urbach and G. Zhang (eds.), *3rd International Symposium of Space Optical Instruments and Applications*, Springer Proceedings in Physics 192,
DOI 10.1007/978-3-319-49184-4_35

1 Introduction

Currently, the focal plane design of the push-broom imaging mode has used widely because it has advantageous as it eliminates the need for cross-track motion when collecting data. At the same time, the push-broom style has higher signal-to-noise ratios (SNR) due to longer dwell times. Ideally every detector of the linear camera is electronically unique and would produce the same output signal when illuminated by the same input light source. However, this style will lead to same problems such as non-uniformity noise and dark current noise, because of inter-detector sensitivity differences, factors related to image formation. Therefore, an accurate radiometric normalization, as an important processing procedure, is essential. To enable this aim, relative radiometric calibration has been developed in the last two decades and applied with various methods [1].

Much work has gone into the development of various methods for obtaining relative radiometric coefficient for the remote sensing images. Using the laboratory integrating sphere, on-orbit calibration device, uniform calibration field and statistical method with lots of images [2], which is the main four kinds of categories. (1) The Characteristics of satellite would be measure prior to launch by using laboratory integrating sphere [3]. However, the response of sensors will be change in-orbit on account of environment changed, particulate contamination, etc. (2) The on-orbit calibration device can provide regular and high-accuracy radiometric coefficients. But the structure of device is complex and the precision of the coefficients would be descend with the aging of the device. (3) It need a large area of "Flat Fields" such as seas, deserts, glacier etc. by using the uniform calibration field method [4]. But it is difficult to find such a large a uniform ground scenery in full field of view. (4) The statistical method would acquire the high-accuracy radiometric coefficients if with lots of images and the stability of the typically detector responses [5]. It will be restrict with these conditions.

The side-slither maneuver, proposed by Bradley G. Henderson for QuickBird [6], is an efficient way to obtain the parameters of relative gains and offset from linear normalization model by using the imagery of yawing the satellite 90°. A side-slither maneuver rotates the array 90° on its yaw axis, meaning that every detector is aligned to image the same target as it moves along the velocity vector. In this chapter, we proposes an improved side-slither maneuver based on moment matching for on-orbit calibration of high-resolution satellite imagery. The method does not have the requirement that the content of the images for calibration must be uniformity sites. With a linear normalization model, the high precision relative radiometric calibration results can be obtained based on moment matching method. In this case, adjusting means (offsets) and standard deviations (gains) of each sensor's to a reference is sufficient. A relative gain can be calculated using the moment matching method to adjust the side-slither maneuver. The approach is characterized by its reliability which is based on the purely statistical analysis of images without any sites constraint. It does not need the calibration device, the "flat fields". More important, the method does not need lots of images, therefore each rail can achieve high precision relative calibration coefficients.

2 Model and Method

2.1 Linear Model

Assuming linear detector response, ideally when the camera observes the top-of-atmosphere radiance L, the initial output digital number (DN) Y(L) can be expressed as:

$$\mathrm{Y(L)} = A * L \tag{1}$$

where:

A is the absolute calibration coefficient that allows users to compute radiances.

But there are some problems such as the difference response and dark current noise in the CCD detector, so the output DN X(j, L) delivered by detector j is modelled by the equation:

$$\mathrm{X(j,L)} - C(j) = A * g(j) * L \tag{2}$$

where:

C(j) is the dark current of detector j (L = 0);
g(j) is the relative detector gain;
A is the absolute calibration coefficient that allows users to compute radiances [7].

With a linear model, normalization can be easily performed to get the normalized digital number Y as a function of X combing "Eqs. (1)" and "(2)":

$$\mathrm{Y(L)} = \frac{1}{g(j)} * X(j,L) - \frac{1}{g(j)} * C(j) \tag{3}$$

Renaming a few terms from Eq. (3), "$\frac{1}{g(j)}$" will be redefined as "a", $'' - \frac{1}{g(j)} * C(j)''$ will be redefined as "b":

$$\mathrm{Y(L)} = a * X(j,L) + b \tag{4}$$

As there is no particular issue to get these parameters, we will only focus on the computation of "a" and "b".

2.2 Pre-Launch Radiometric Calibration

A full radiometric calibration of the satellite was performed at laboratory during instrument level test. Each TDI level can acquires enough raw image at the low radiance, middle radiance and high radiance. Because of the availability of large aperture integrating spheres, we can acquire the high-accurate with the hypothesis of linear model. Each detector is assumed to have a linear response as a function of "Eq. (4)". The data from Table 1 are used for fitting the linear model by the method of least squares. Using the coefficient of laboratory, the result of radiometric correction are shown in Fig. 1.

2.3 Improved Side-Slither Method

The side-slither calibration is on-orbit maneuver that has been used to flat-field image data for the QuickBird, Pleiades-HR and RapiEye, Landsat-8 push-broom systems [8–10]. In side-slither imaging, the focal plane is rotated 90°, and the scanning is done parallel rather than perpendicular to the lineary array (Fig. 2 right). With such a mode, all the dectector can view the same lanscape and see the same amount of light [11]. If the geometric registration of the successive line of sight projection is high-accurate, all detectors will see the same strip of ground and therefore will have been exposed to the same amount of light. Rather than image the entire scene, the array sees just a thin slice of it.

Table 1 The mean digital number of raw image response for radiance L

Radiance [W/m2/sr]	TDI level–24 [DN]	TDI level–48 [DN]
5.16	40.9	81.2
10.50	82.5	164.3
31.43	247.5	490.3
58.67	458.4	896.3

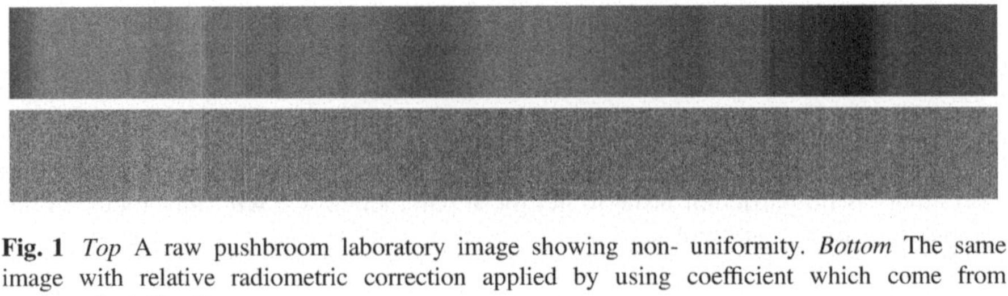

Fig. 1 *Top* A raw pushbroom laboratory image showing non- uniformity. *Bottom* The same image with relative radiometric correction applied by using coefficient which come from pre-launch clalibration

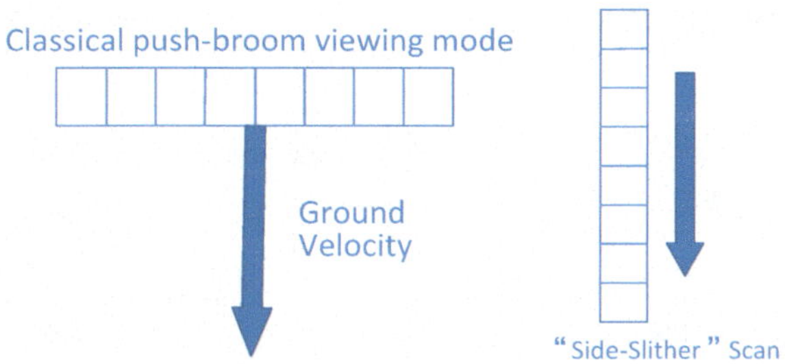

Fig. 2 *Left* Classical push-broom viewing mod; *Right* "Side-Slither" Scan

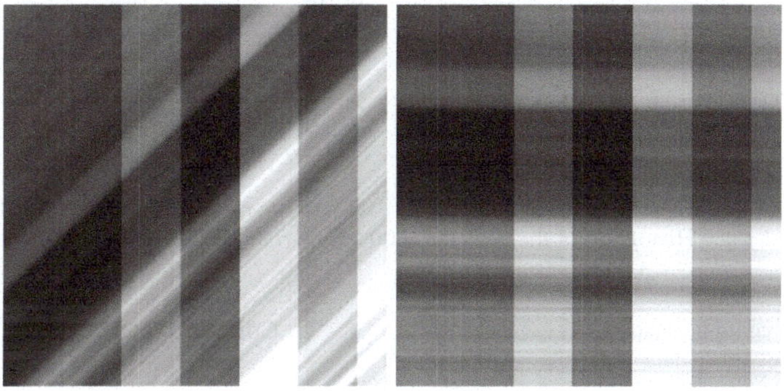

Fig. 3 *Left* a raw side-slither image; *Right* basic horizontal correction

The process of the side-slither maneuver contains three step. First, select the data to make a flat field. Second, put the same ground on a line by shift data. Last, for a given column, which corresponds to a single detector, the DN values are column averaged for every detector and all the column-averaged response are averaged for a given band, and then the "a" and "b" are computed by "Eq. (4)".

The disadvantage of the side-slither method is that the images captured with the sensor in a 90° orbit configuration must include uniformity sites such as desert and snow/ice regions. Figure 3 shows us the image may be not include uniformity, so we proposed an improved side-slither method by using moment matching to break through the limitations of sites. Moment matching is which matches the gain and offset of each sensor to typical values, and which is resistant to the effects of outliers. It can reduces the variance between sensors to a greater degree [12].

Figure 4 shows us the histograms of image captured by detector 1000, detector 2000, detector 3000 and the histogram of the all detectors. Figure 5 shows us the column mean of each detector. We can draw a conclusion the histogram is similar and the response of some detector group is different.

Now we are able to detail the method that can be used to compute normalization parameters [8]:

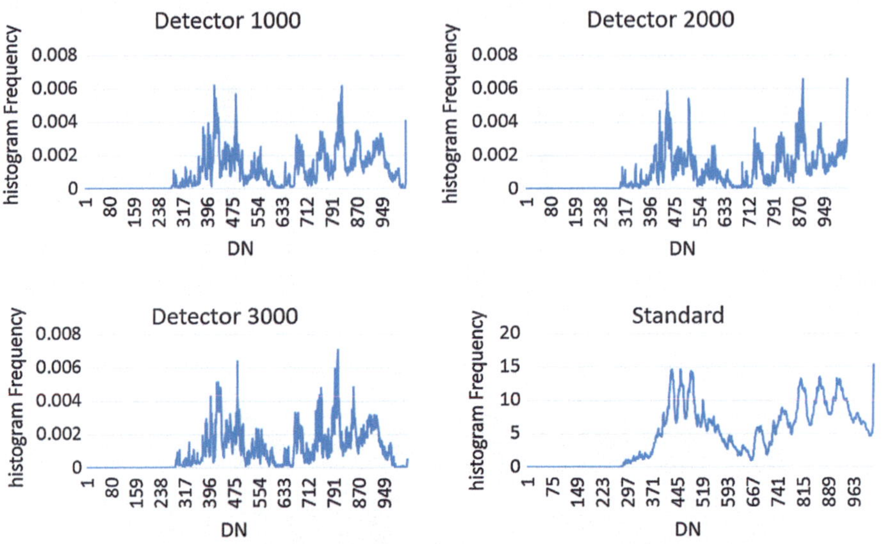

Fig. 4 *Upper-Left* the Histogram of the detector 1000; *upper-right* the Histogram of the detector 2000; *down-left* the Histogram of the detector 3000; *down-right* the Histogram of the All detector

Fig. 5 The column mean value of side-slither data

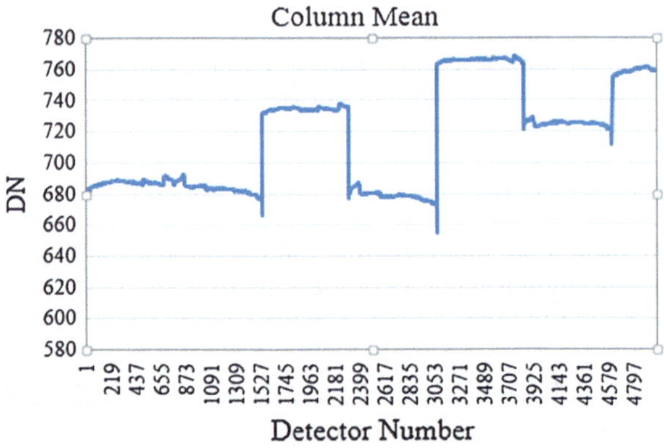

- Preprocessing of the raw image: shift of each column (Fig. 3);
- Remove leading and trailing zeroes and generate a histogram of pixel values for each of the detector;
- Truncate 1% of the histogram at both high and the low end, then calculate means and standard deviations of the truncate histograms for each sensor and the all detectors.
- Calculate the mean and standard deviation deviation of the whole image.
- Calculate the "a" and "b" as follows:

$$a = \frac{\sigma_{ref}}{\sigma_i} \quad (8)$$

$$b = \mu_{ref} - \frac{\sigma_{ref}}{\sigma_i} * \mu_i \quad (9)$$

One can easily see that we can run the previous method successfully without the limit of "Flat Field" sites.

3 Experimental Result

In Sect. 2, we purpose an improved side-slither method by using momnet matching to compute the relative radiometric parameters. For this work, we performed improved side-slither imaging using the Yaogan 26 satellite. Yaogan 26 was launched on 27 December 2014, from the Taiyuan Satellite Launch Center in

Fig. 6 *Left* original image of water; *middle* water image correction using lab coefficient; *right* water image coorection using the coefficient computed from the improved slide-slither method

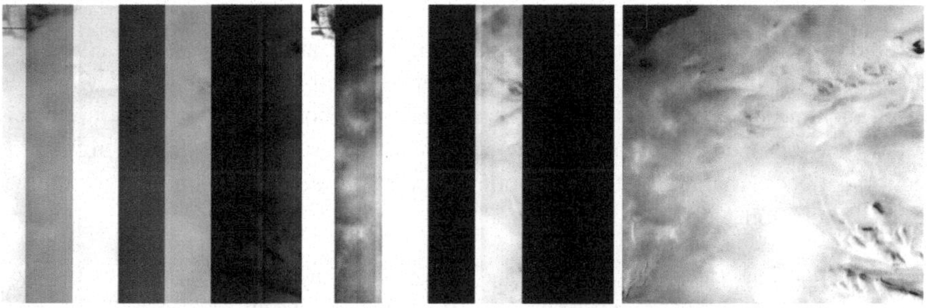

Fig. 7 *Left* original image of desert; *middle* desert image correction using lab coefficient; *right* desert image coorection using the coefficient computed from the improved slide-slither method

Fig. 8 *Left* original image of airport; *middle* airport image correction using lab coefficient; *right* airport image coorection using the coefficient computed from the improved slide-slither method

Table 2 Image quality

	Wa-0	Wa-Lab	Wa-SI	Des-0	Des-Lab	Des- SI	Ap-0	Ap-Lab	Ap-SI
Mean	93.970	94.437	88.226	457.245	452.685	603.971	284.889	282.931	268.652
Stdev	11.302	11.426	2.182	60.504	60.532	22.617	46.186	45.633	42.756
SNR	40.850	39.988	39.804	45.505	45.562	47.601	18.106	18.087	22.882
Entropy	3.185	3.299	1.642	4.930	4.968	4.138	5.181	5.168	4.104

Note "*Wa*" mean water image; "*Des*" mean desert image; "*Ap*" mean airport image
"0" mean original image; "*Lab*" mean using lab coefficient; "*SI*" mean improved slide-slither

China's northern Shanxi province. It is mainly used for scientific experiments, land survey, crop yield assessment and disaster monitoring.

Figures 6, 7, 8 show us the effect of the improved side-slither more effective than the lab on the water, desert, airport objects. Table 2 tells us the snr of water image is close, because the DN of water is low, so the noise is similar to true surface. Compare other quantitative indexs, the improved side-slither is better.

4 Conclusion

A relative gain can be calculated using the moment matching method to adjust the side-slither maneuver. The approach is characterized by its reliability which is based on the purely statistical analysis of images without any sites constraint. It is noted that the accurate geometric registration of the successive line of sight projection is necessary and the total amount of pixels should be high enough to achieve a good precision.

Acknowledgments This work was substantially supported by the National Basic Research Program of China 973 Program (grant numbers 2014CB744201) and the National Natural Science Foundation of China (grant numbers 91438203).

References

1. Y. Du, P. M. Teillet, and J. Cihlar, "Radiometric normalization of multitemporal high-resolution satellite images with quality control for land cover change detection," *Remote Sensing of Environment,* vol. 82, pp. 123–134, 2002.
2. Y. Duan, W. Chen, M. Wang, and L. Yan, "A relative radiometric correction method for airborne image using outdoor calibration and image statistics," *Geoscience and Remote Sensing, IEEE Transactions on,* vol. 52, pp. 5164–5174, 2014.
3. K. S. Krause, "WorldView-1 pre and post-launch radiometric calibration and early on-orbit characterization," in *Optical Engineering + Applications*, 2008, pp. 708116–708116-11.
4. B. K. Horn and R. J. Woodham, "Destriping Landsat MSS images by histogram modification," *Computer Graphics and Image Processing,* vol. 10, pp. 69–83, 1979.
5. F. L. Gadallah, F. Csillag, and E. J. M. Smith, "Destriping multisensor imagery with moment matching," *International Journal of Remote Sensing,* vol. 21, pp. 2505–2511, Aug 2000.
6. B. G. Henderson and K. S. Krause, "Relative radiometric correction of QuickBird imagery using the side-slither technique on orbit," in *Optical Science and Technology, the SPIE 49th Annual Meeting*, 2004, pp. 426–436.
7. P. Kubik and W. Pascal, "AMETHIST: A method for equalization thanks to HISTograms," in *Sensors, Systems, and Next-Generation Satellites Viii.* vol. 5570, R. Meynart, S. P. Neeck, and H. Shimoda, Eds., ed Bellingham: Spie-Int Soc Optical Engineering, 2004, pp. 256–267.
8. C. Anderson, D. Naughton, A. Brunn, and M. Thiele, "Radiometric correction of RapidEye imagery using the on-orbit side-slither method," in *SPIE Remote Sensing*, 2011, pp. 818008–818008-15.
9. F. Pesta, S. Bhatta, D. Helder, and N. Mishra, "Radiometric Non-Uniformity Characterization and Correction of Landsat 8 OLI Using Earth Imagery-Based Techniques," *Remote Sensing,* vol. 7, pp. 430–446, Jan 2015.
10. V. Martin, G. Blanchet, P. Kubik, S. Lacherade, C. Latry, L. Lebegue, et al. "PLEIADES-HR 1A&1B image quality commissioning: innovative radiometric calibration methods and results," in *SPIE Optical Engineering + Applications*, 2013, pp. 886610–886610-11.
11. H. Li and Y.-y. Man, "Relative radiometric calibration method based on linear CCD imaging the same region of non-uniform scene," 2014, pp. 929906–929906-9.
12. F. Ahern, R. Brown, J. Cihlar, R. Gauthier, J. Murphy, R. Neville, *et al.*, "Review article radiometric correction of visible and infrared remote sensing data at the Canada Centre for remote sensing," *International Journal of Remote Sensing,* vol. 8, pp. 1349–1376, 1987.

Remote Sensing Image Denoising with Iterative Adaptive Wiener Filter

Dan Wang, Xinfeng Zhang, Yong Liu, Zhiwei Zhao and Zhengji Song

Abstract Image denoising plays a significant role in the application of remote sensing images, since the noise not only deteriorates the visual quality, but also and more important, causes the performance drop of many computer vision algorithms, e.g., segmentation and object recognition. However, denoising is a quite challenging task, due to the complicated, nonlinear distribution of noises. In this paper, we propose an iterative adaptive Wiener filter for remote sensing image denoising, by exploring statistical characteristics of local similar patches. Given a noisy image, the proposed approach aims to pursuit a restored image, with sufficiently good quality. In the proposed method, firstly, a low-pass filter is applied to the observed noisy image. The resulted image is set as an initial version of the restored image, which is fed into the following iterative rounds and refined to progressively approximate to "noise-free" signal. In each round, we divide the image being processed into overlapping patches. Each one will be assigned into a group, by searching similar patches in its neighboring areas. Then the optimal Wiener filter model is estimated adaptively for each group, and performed on these involved patches. Since the sampled patches are overlapping, the resulted image is achieved by averaging on overlapped, filtered patches. After that, the resulted image will be processed in the same way in the next round. With the procedure repeated, the noises are gradually alleviated and the refined image is approached to "noise-free" one. Finally, the algorithm terminates when the image changes little of two neighboring iterations. The contribution of our paper lies in two aspects. First, we propose a novel Wiener filter strategy, which takes advantage of image self-similarity to estimate filter parameters adaptively. Second, iterative scheme can refine the results progressively, which significantly improve the image quality. Experimental results demonstrate that the proposed method outperforms state of the art methods and can significantly improve both the subjective and the objective quality of noisy remote sensing images.

D. Wang (✉) · Y. Liu · Z. Zhao · Z. Song
Beijing Institute of Spacecraft System Engineering, Beijing, China
e-mail: wangdan_ict_hit@163.com

X. Zhang
Institute of Computing Technology, Chinese Academy of Sciences, Beijing, China

H.P. Urbach and G. Zhang (eds.), *3rd International Symposium of Space Optical Instruments and Applications*, Springer Proceedings in Physics 192,
DOI 10.1007/978-3-319-49184-4_36

Keywords Remote sensing image · Wiener filter · Patch similarity · Denoisng

1 Introduction

Image denoising is an important technique for remote sensing images, which are easily contaminated by noise when acquired from satellites via remote sensor. The noise not only deteriorates the visual quality of remote sensing images, but also and more important, causes the performance drop of image processing algorithms, e.g., segmentation and object recognition, which limits the application of remote sensing images. In order to remove the noise, numerous image denoising algorithms [1–8] are proposed based on different image prior models and achieve significant improvement on image quality in the last thirty years. The Wiener filter is the optimal filter in the sense of the mean square errors (MSE), which is derived by minimizing the MSE between the estimated image and desired noise-free image. However, the Wiener filter assumes that the signal is stationary and known, which cannot be satisfied for noisy images in practice.

In recent years, many modifications on Wiener filters are proposed to improve denoising performance and make it applicable to practical noisy images. The adaptive Wiener filter is proposed in [3], which derives the filter parameters for every pixel by calculating the corresponding signal variance in a sliding window centered at the target pixels. In [4], Jin et al. further improve it by combining two adaptive Wiener filters in both spatial domain and wavelet domain. Considering the different characteristics of images, Zhou and Cheng [5] improve the wavelet domain Wiener filter by determining an optimal threshold and neighboring window size for every wavelet subband. Considering the image nonlocal similarity, the famous nonlocal means filter [6] estimated the latent noise-free pixel by a weighted average of its surrounding pixels, where the weights are based on the distances of the patch intensities between the target patch and the reference patches. In [7], Lin et al. applied the nonlocal means filter to the noisy images first to generate an initial estimation of the original image, and then derived the wavelet domain Winer filter with the initial estimation to remove noise by shrinking wavelet coefficients. In [8], Mohamed and Hardie proposed a novel multi-patch adaptive Wiener filter algorithm for image restoration, which uses a single-pass spatial-domain weighted sum of all pixels within all of the similar patches to form the estimate each desired pixel. The multi-patch based methods are very useful in denoising problem and also utilized in [9, 10].

In this paper, we propose an iterative adaptive Wiener filter (IAWF) to reduce noise for remote sensing images. The proposed IAWF can also be regarded as a multi-patch collaborative filter, which derives a specific Wiener filter for each group of similar image patches. For a noisy image, an initial estimation is first generated by applying a low-pass filter. Then, the initial estimated image is divided into different patches, and similar patches in a neighborhood are clustered into one group. For each group, the optimal Winer filter is derived based on the statistical characteristic of the similar patches in the initial estimation and the noise variance. Since these image patches are overlapping, the filtered image is reconstructed by averaging all the samples from different patches corresponding to same position. This procedure is performed iteratively by taking the previous output image as the latent original signals to calculate the signal covariance matrix and updating the noise level based on the difference between the noisy image and the output image from previous iteration. Along with the noise reduction, the derived Wiener filters approach the optimal ones and achieve significant performance improvement compared with other denoising methods. Experimental results on remote sensing images show that the proposed methods can obviously improve their subjective and objective quality.

The remainder of this paper is organized as follows. In Sect. 2, we first formulate the denoising problem and the optimal Winer filter. Section 3 introduces the framework of the proposed iterative adaptive Wiener filter and the details of the proposed algorithm. Experimental results are reported in Sects. 4 and 5 concludes the paper.

2 Denoising with Winer Filter

For an image $\mathcal{I}_x$, the corresponding noisy image contaminated by additive noise with identical independent distribution can be formulated as

$$\boldsymbol{y} = \boldsymbol{x} + \boldsymbol{n}\,. \tag{1}$$

Here $\boldsymbol{x}$, $\boldsymbol{y}$ and $\boldsymbol{n}$ are original image, noisy image and noise, which are organized into vector form. The Wiener filter, $\boldsymbol{f}$, is the optimal linear estimator by minimizing the mean square errors between original image and filtered image,

$$\boldsymbol{f} = \min_{\boldsymbol{f}} \|\boldsymbol{x} - \boldsymbol{f}\boldsymbol{y}\|_2^2. \tag{2}$$

Then the optimization problem can be solved as

$$\boldsymbol{f} = \boldsymbol{R}_x\left(\boldsymbol{R}_x + \sigma_n^2 \boldsymbol{I}\right)^{-1}, \tag{3}$$

where σ_n^2 is the noise variance, $\boldsymbol{I}$ is an unit matrix and $\boldsymbol{R}_x$ is the covariance matrix of the original signals, which is a real symmetric matrix. The eigen-expansion of $\boldsymbol{R}_x$ is given by

$$\boldsymbol{R}_x = E\left[\boldsymbol{x}\boldsymbol{x}^T\right] = \boldsymbol{U}\boldsymbol{\Lambda}\boldsymbol{U}^T = \sum_{k=1}^{N} \lambda_k \boldsymbol{u}_k \boldsymbol{u}_k^T\,, \tag{4}$$

where λ_k is the eigenvalue of $\boldsymbol{R}_x$, $\boldsymbol{u}_k$ is the corresponding eigenvector, N is the number of eigenvalues of $\boldsymbol{R}_x$, and $\boldsymbol{U} = [\boldsymbol{u}_1, \boldsymbol{u}_2, \ldots, \boldsymbol{u}_N]$ and $\boldsymbol{\Lambda} = diag[\lambda_1, \lambda_2, \ldots, \lambda_N]$. Then, the Wiener filter in Eq. (3) can be rewritten as

$$f = U\tilde{\Lambda}U^T = \sum_{k=1}^{N} \frac{\lambda_k}{\lambda_k + \sigma_n^2} u_k u_k^T, \tag{5}$$

and

$$\tilde{\Lambda} = diag\left[\frac{\lambda_1}{\lambda_1 + \sigma_{\mathbf{n}}^2}, \frac{\lambda_2}{\lambda_2 + \sigma_{\mathbf{n}}^2}, \ldots, \frac{\lambda_N}{\lambda_N + \sigma_{\mathbf{n}}^2}\right]. \tag{6}$$

Since U is actually the Karhunen-Loeve Transform (KLT) matrix, the Wiener filter in Eq. (5) is equivalent to applying, applies the soft-thresholding operation to KLT coefficients. This kind of filters is widely used image sparsity based denoising methods [11–13], which first transform the image from spatial domain to sparse domains, e.g., wavelet domain, and threshold the sparse signals to reduce noise. The KLT is the optimal transform for signal decorrelation to get its sparse representation. However, since it is a signal-dependent transform constructed with the eigenvectors of the original signal covariance, it is difficult to be applied to image denoising problem without original signals.

3 The Proposed Iterative Adaptive Wiener Filter

In this paper, we propose a novel iterative adaptive Wiener filter for remote sensing image denoising problem. Figure 1 shows the framework of the proposed method. The noisy image is first filtered with a low-pass filter to get an initial estimation for the original image, and then the initial estimated image is further divided into many groups with similar patches. Specifically, for each patch, the K most similar patches are acquired by searching in a local area surrounding it according to Euclidean distance of the image patch vectors,

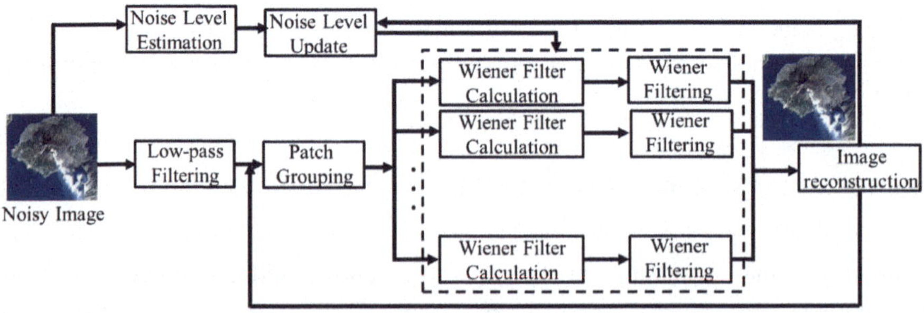

Fig. 1 The framework of the proposed iterative adaptive Wiener filter

$$\boldsymbol{d}_{ij} = \left\| \hat{\boldsymbol{x}}_i^{(0)} - \hat{\boldsymbol{x}}_j^{(0)} \right\|_2^2. \tag{7}$$

where $\hat{\boldsymbol{x}}_i^{(0)}$ is the ith image patch in the initial estimation image. For image patch group, $\boldsymbol{G}_{ij} = \left\{ \hat{\boldsymbol{x}}_{ij}^{(0)} | j = 1, 2, \ldots, K \right\}$, the covariance matrix $\boldsymbol{R}_x$ is estimated with Eq. (8) from these image patch vectors assuming they follows the same distribution.

$$\widehat{\boldsymbol{R}}_x = \frac{1}{K} \sum_{j=1}^{K} \hat{\boldsymbol{x}}_{ij}^{(0)} \left[\hat{\boldsymbol{x}}_{ij}^{(0)} \right]^T - \boldsymbol{\mu}\boldsymbol{\mu}^T, \tag{8}$$

and

$$\boldsymbol{\mu} = \frac{1}{K} \sum_{j=1}^{K} \hat{\boldsymbol{x}}_{ij}^{(0)}. \tag{9}$$

The noise variance of $\mathcal{I}_y$ is estimated using the robust median estimator in wavelet domain as proposed by Donoho in [11], which is calculated as

$$\hat{\sigma}_n = median(|\boldsymbol{c}|)/0.6745, \tag{10}$$

where $\boldsymbol{c}$ is the coefficient vector of the diagonal band of 2D wavelet in the first level. Based on the derived covariance matrix and noise level, the near-optimal Wiener filter for each similar patch group can be derived with Eqs. (4), (5) and (8). Then, the derived filters are applied to the corresponding image patches to reduce their noise. Finally, the image in the kth iteration is reconstructed by averaging the multiple estimates for each pixel,

$$\hat{\boldsymbol{x}}^{(k)}(i,j) = \frac{1}{|\Omega_{ij}|} \sum_{t \in \Omega_{ij}} \hat{\boldsymbol{x}}_t^{(k)}(i',j'), \tag{11}$$

where Ω_{ij} is the set composed of image patches that include the image pixel $\hat{\boldsymbol{x}}^{(k)}(i,j)$. Coordinates (i,j) and (i',j') indicate the same sample under the whole image and the image patch coordinate systems, respectively.

Since the initial estimation is not accurate, most of noise cannot be removed by the Wiener filters in the first iteration cannot reduce most of noise. Therefore, we perform the above procedure iteratively, and take the output of previous iteration as the estimation of original signal in current iteration. The noise level need be updated with the following equation,

$$\sigma_{\boldsymbol{n}}^{(k)} = s\sqrt{\left[\sigma_{\boldsymbol{n}}^{(0)}\right]^2 - var(\hat{\boldsymbol{x}}^{(k-1)} - \hat{\boldsymbol{x}}^{(0)})}, \tag{12}$$

where $var(\cdot)$ is the function to calculate the variance of vectors, and s is a scale factor.

4 Experimental Results

In this section, we carry out the proposed IAWF on noisy remote sensing images to verify its performance by comparing with other methods. The compared methods include adaptive Wiener filter (AWF) [14], OptNeighShrink method [5], BLS-GSM [15] and nonlocal means filter (NLM) [6]. We randomly download 15 remote sensing images captured by LandSat-7, from website of United States Geological Survey.[1] These images are captured by LAND SAT 7 ETM+ with spatial resolution 30 m. Gaussian noise is added to the images artificially, to verify the performance of the proposed method. Since the noise level is an important parameter for different denoising methods and is unknown in practice, we estimate it with the method in [11]. In the proposed method, we apply the AWF to the noisy images to generate the initial estimation, and set its parameter as $\sigma_n/20$ to avoid oversmoothing the image. The proposed algorithm is terminated when the mean absolution difference (MAD) of images between two iterations is smaller than 1.0, and the number of similar patches in one group is 50. The scale factor in Eq. (12) is 0.3.

The experimental results are measure by the Peak Signal-to-Noise ratio (PSNR) in decibels (dB), i.e.,

$$PSNR = 10 * \log_{10} \frac{255^2}{MSE}, \quad (13)$$

where MSE is the mean square error between the original signal and the distorted signal. The denoised image with high quality is closer to the original one when PSNR is higher. Table 1 shows the PSNR results on 15 remote sensing images contaminated by Gaussian noise with different noise levels. From the PSNR results, we can see that the proposed method can efficiently improve the quality of noisy remote sensing images. Our method achieves 5.87 and 9.41 dB improvement on average for these noisy images when noise standard deviations are 20 and 60. Compared with other denoising methods, the proposed method also achieves around 0.45–1.29 and 0.21–4.21 dB at σ_n = 20 and 60, respectively. Especially, our method takes the AWF as the initial estimation, and further improves its quality significantly. Since the PSNR quality metric is not well consistent with perceptual quality, we also take the widely used quality metric, Structural Similarity Index Metric (SSIM) [16], to further verify the performance of our proposed method. Based on the SSIM results in Table 2, our method also achieves the best quality compared with other methods. Figure 2 gives an example remote sensing image. It shows the subjective results for noisy images with noise standard deviation σ_n = 60 and the corresponding denoised images with different methods. We can see that the images restored by the proposed method show more visual pleasing results. Although the BLS-GSM and NLM methods can remove the noise obviously, they also smooth out many details in the original image. Our method achieves a good

[1]http://landsat.usgs.gov//gallery_view.php?category=nocategory&thesort=pictureId.

Table 1 PSNR results on images contaminated by Gaussian noise with different noise levels

Image number	$\sigma_n = 20$						$\sigma_n = 60$					
	Noisy images	AWF	OptNeigh-Shrink	BLS-GSM	NLM	Proposed	Noisy images	AWF	OptNeigh-Shrink	BLS-GSM	NLM	Proposed
1	22.78	26.74	27.19	27.28	27.18	27.81	13.84	19.25	20.55	21.88	18.73	22.54
2	22.62	25.35	25.42	25.08	24.88	26.01	14.08	19.01	19.87	20.16	18.34	20.41
3	22.31	27.17	28.01	28.15	27.82	28.31	13.56	20.35	22.71	22.82	18.70	22.95
4	22.13	26.22	26.60	26.71	26.62	27.17	13.18	20.10	22.15	22.18	18.22	22.32
5	22.27	26.05	26.45	26.11	26.21	27.02	13.74	19.77	21.40	21.68	18.62	21.92
6	22.15	26.24	26.97	27.16	26.90	27.55	13.28	20.14	22.34	22.38	18.41	22.51
7	22.50	27.09	27.44	27.76	27.10	28.00	13.98	19.89	21.70	22.21	19.14	22.41
8	22.23	26.67	27.11	27.46	26.98	27.71	13.54	20.14	22.22	22.43	18.68	22.54
9	22.59	25.19	25.21	25.00	24.77	25.72	13.79	18.78	19.72	19.99	17.99	20.14
10	22.14	28.13	29.30	29.56	29.16	29.83	13.45	20.92	25.63	25.92	19.36	26.06
11	22.21	28.13	29.41	29.94	29.42	30.45	13.92	20.91	24.37	24.88	19.79	25.13
12	22.29	26.06	25.85	26.13	25.93	26.77	13.76	19.73	21.10	21.61	18.68	21.80
13	22.24	27.55	28.35	28.51	28.27	28.87	13.36	20.56	24.32	24.58	19.02	24.72
14	22.14	28.54	30.44	31.16	29.97	31.52	13.32	20.88	25.19	25.85	19.22	26.00
15	22.33	28.48	29.64	30.20	29.37	30.31	14.16	20.88	23.97	24.34	19.98	24.54
Average	22.33	26.91	27.56	27.75	27.37	28.20	13.66	20.09	22.48	22.86	18.86	23.07

Table 2 SSIM results on Images with different noise levels

Image number	$\sigma_n = 20$						$\sigma_n = 60$					
	Noisy images	AWF	OptNeigh-Shrink	BLS-GSM	NLM	Proposed	Noisy images	AWF	OptNeigh-Shrink	BLS-GSM	NLM	Proposed
1	0.498	0.706	0.758	0.732	0.739	0.783	0.164	0.334	0.465	0.502	0.302	0.535
2	0.635	0.774	0.8	0.769	0.78	0.826	0.273	0.443	0.515	0.493	0.4	0.52
3	0.577	0.785	0.809	0.796	0.792	0.819	0.171	0.449	0.53	0.516	0.321	0.528
4	0.618	0.774	0.777	0.754	0.766	0.789	0.204	0.449	0.487	0.468	0.337	0.488
5	0.546	0.742	0.838	0.809	0.79	0.858	0.212	0.408	0.609	0.59	0.338	0.632
6	0.605	0.767	0.79	0.774	0.773	0.806	0.2	0.442	0.499	0.478	0.334	0.499
7	0.489	0.71	0.735	0.741	0.714	0.76	0.158	0.345	0.404	0.463	0.273	0.448
8	0.567	0.761	0.77	0.761	0.753	0.79	0.192	0.428	0.502	0.5	0.321	0.501
9	0.644	0.715	0.716	0.684	0.705	0.731	0.282	0.431	0.452	0.441	0.389	0.446
10	0.376	0.663	0.714	0.716	0.697	0.74	0.087	0.293	0.549	0.564	0.194	0.574
11	0.356	0.672	0.83	0.852	0.775	0.874	0.105	0.313	0.676	0.7	0.227	0.734
12	0.583	0.738	0.739	0.728	0.738	0.773	0.23	0.419	0.497	0.521	0.358	0.526
13	0.426	0.679	0.722	0.711	0.708	0.748	0.107	0.309	0.515	0.528	0.224	0.539
14	0.315	0.666	0.854	0.874	0.773	0.882	0.096	0.281	0.717	0.745	0.189	0.762
15	0.388	0.706	0.771	0.791	0.743	0.799	0.106	0.335	0.572	0.606	0.236	0.603
Average	0.508	0.724	0.775	0.766	0.750	0.799	0.172	0.379	0.533	0.541	0.296	0.556

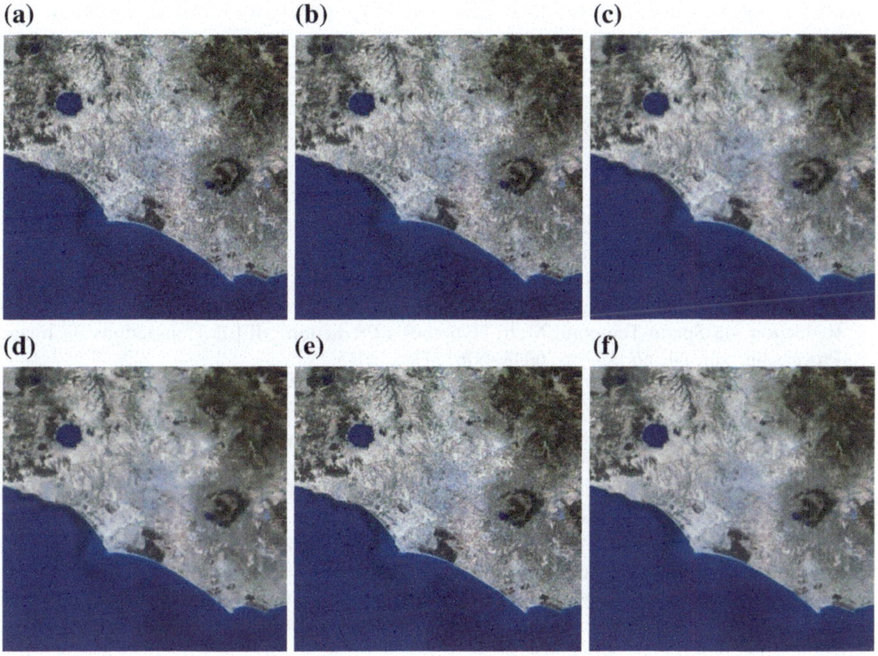

Fig. 2 Subjective image quality comparison for desnoised remote sensing images with different methods, **a** noisy image, **b** AWF results, **c** OptNeighShrink results, **d** BLS-GSM results, **e** NLM results, **f** the results of the proposed method

tradeoff between the noise reduction and image detailed structure preservation. Since our method estimate the threshold for singular values from similar patch group, the filtering strength can well adapt to image content. In addition, the overlapping of patches makes one pixel has multiple estimation independently, which is useful to noise reduction with independent distribution. Therefore, our method achieves a good tradeoff between the noise reduction and image detailed structure preservation.

5 Conclusion

In this paper, we proposed an iterative adaptive Wiener filter by utilizing the filtered similar image patches to predict the covariance matrix of original signals and derive the optimal Wiener filter for each group of image patches. The final reconstructed image is the average of all the filtered samples from different patches corresponding to the same position. The iterative strategy makes the derived Wiener filters approach the optimal ones with the noise reduction. Based on the experimental results, the proposed IAWF significantly improves the quality of noisy remote sensing images. Furthermore, it is also an very practical method since all the parameters are estimated from the noisy images.

References

1. X. Zhang, R. Xiong, X. Fan, S. Ma, and W. Gao, "Compression Artifact Reduction by Overlapped-Block Transform Coefficient Estimation With Block Similarity," *IEEE Transactions on Image Processing*, vol. 22, no. 12, pp. 4613–4626, Dec. 2013.
2. X. Zhang, R. Xiong, W. Lin, S. Ma, J. Liu, and W. Gao, "Video Compression Artifact Reduction via Spatio-Temporal Multi-Hypothesis Prediction," IEEE Transactions on Image Processing, vol. 24, no. 12, pp. 6048–6061, Dec. 2015.
3. D. T. Kuan, A. A. Sawchuk, T. C. Strand and P. Chavel, "Adaptive Noise Smoothing Filter for Images with Signal-Dependent Noise," in IEEE Transactions on Pattern Analysis and Machine Intelligence, vol. PAMI-7, no. 2, pp. 165–177, March 1985.
4. F. Jin, P. Fieguth, L. Winger and E. Jernigan, "Adaptive Wiener filtering of noisy images and image sequences," Image Processing, 2003. ICIP 2003. Proceedings. 2003 International Conference on, 2003, pp. III-349-52 vol. 2.
5. D. Zhou and W. Cheng, "Image denoising with an optimal threshold and neighbouring window," Pattern Recognition Letters, vol. 29, no. 11, pp. 1694–1697, Aug. 2008.
6. A. Buades, B. Coll, and J.-M. Morel, "A non-local algorithm for image denoising," in IEEE Computer Society Conference on Computer Vision and Pattern Recognition, 2005. CVPR 2005, 2005, vol. 2, pp. 60–65 vol. 2.
7. L. Lin and K. Lingfu, "Image Denoising Based on Non-local Means with Wiener Filtering in Wavelet Domain," Intelligent Information Hiding and Multimedia Signal Processing, 2009. IIH-MSP '09. Fifth International Conference on, Kyoto, 2009, pp. 471–474.
8. K. M. Mohamed and R. C. Hardie, "A collaborative adaptive Wiener filter for image restoration using a spatial-domain multi-patch correlation model," EURASIP J. Adv. Signal Process., vol. 2015, no. 1, pp. 1–23, Jan. 2015.
9. X. Zhang, R. Xiong, S. Ma, and W. Gao, "Reducing Blocking Artifacts in Compressed Images via Transform-Domain Non-local Coefficients Estimation," in 2012 IEEE International Conference on Multimedia and Expo (ICME), 2012, pp. 836–841.
10. X. Zhang, R. Xiong, S. Ma, and W. Gao, "Artifact reduction of compressed video via three-dimensional adaptive estimation of transform coefficients," in 2014 IEEE International Conference on Image Processing (ICIP), 2014, pp. 4567–4571.
11. D. L. Donoho, "De-noising by soft-thresholding," in IEEE Transactions on Information Theory, vol. 41, no. 3, pp. 613–627, May 1995.
12. S. G. Chang, B. Yu, and M. Vetterli, "Adaptive wavelet thresholding for image denoising and compression," IEEE Transactions on Image Processing, vol. 9, no. 9, pp. 1532–1546, Sep. 2000.
13. X. Zhang, W. Lin, J. Liu, and S. Ma, "Compression noise estimation and reduction via patch clustering," in 2015 Asia-Pacific Signal and Information Processing Association Annual Summit and Conference (APSIPA), 2015, pp. 715–718.
14. J. S. Lim, "Two-dimensional signal and image processing," Englewood Cliffs, NJ, Prentice Hall, 1990, 710 p.
15. J. Portilla, V. Strela, M. Wainwright, E. P. Simoncelli, "Image Denoising using Scale Mixtures of Gaussians in the Wavelet Domain," IEEE Transactions on Image Processing. vol 12, no. 11, pp. 1338–1351, November 2003.
16. Z. Wang, A. C. Bovik, H. R. Sheikh, and E. P. Simoncelli, "Image quality assessment: from error visibility to structural similarity," IEEE Transactions on Image Processing, vol. 13, no. 4, pp. 600–612, Apr. 2004.

Clutter and Noise Suppression Based on Match Filter

Dong-nan Chi and Li-na Xu

Abstract Background clutter and noise is the vital influence elements for objective detection from space remote sensing imaging data collected. Therefore, clutter and noise suppression is an issue technique to improve detection ratio and reduce false alarm ratio. A match filter design method based on logistical regression is present to solve target detection from rotating camera images with single sampling. According to the characters of targets in images, analyze the target character with single sampling adequately, and chose the learning sample covering all the characters. Considering that clutter and noise are confused into images, Single Cluster Noise Ratio (SCNR) is introduced to evaluate the effectiveness of match filter optimized. Optimization condition is the maximum of SCNR, and iteration optimization is utilized to obtain the module of the match filter. The simulation is completed to compare the performances of the match filter optimized using the method proposed in this paper. Experiment indicates that it is valid for the match filter optimized to process images.

Keywords Target detection in images · Match filter · Logistical regression

1 Introduction

The difficult to detect target against image background collected from infrared camera are resourced from the background noise and clutter. To increase detection rate and reduce false alarm rate, match filter is introduced to match image pixel with filter template successively and to realize target detection.

Match filter is to match target characteristics in space match filter with Fourier spectral of image data including target information input, and the corresponding data is output to accomplish target detection [1].

D. Chi (✉) · L. Xu
Beijing Institute of Space Mechanics and Electricity, Beijing, China
e-mail: chidongnan@163.com

H.P. Urbach and G. Zhang (eds.), *3rd International Symposium of Space Optical Instruments and Applications*, Springer Proceedings in Physics 192,
DOI 10.1007/978-3-319-49184-4_37

To obtain optimal match filter, this paper presents a method to iterate based on logistical regression through introducing the known target distribution in images. The overall and accuracy target characteristics applied in logistical regression algorithm will lead to optimize match filter template. At present, logistical regression is utilized in data excavating, computer vision, natural linguistics processing, biology character distinguish, searching engine, and robot application, and in this paper it is drawn into optimize match filter template.

In eighty, researchers in China had proposed to introduce match filter to detect linear feature of remote sensing image, and to quantify the parameters [2]. The literature [3] applies match filter to be digital filter into target detection, assumes electronic noise as white noise and defines background clutter as relevant Gaussian noise. Temporary characteristics are combined to derivate match filter with optimal signal noise ratio (SNR). To research target detection effectiveness from match filter in complex infrared background image, literature [4] describes target, background and noise consist of images in detail, and it presents the detection method of small maneuver target based on combined spatial filter with statistics character. It proposes a tri-relevant match filter based on dual relevant one to deal with the searching infrared target weak signal in literature [5]. The relevant result and SNR are improved obviously. Target distinguish ratio, target following accuracy and detecting distance are effective. Otherwise, focusing function of 2-dimension match filter built through adjusting reference position realizes multi-angle SAR imaging. Comparing with traditional SAR imaging algorithm, it breaks out the server requirement of sampling uniformity and continuity [6]. In this paper, logistical regression is included to optimize match filter, and experiment is implement to valid the effectiveness of template optimized.

2 Match Filter

The function for filter is to strengthen useful signal output, and restrain noise out of bandwidth [7]. Match filter is an appropriate linear filter for maximum SNR.

Amplitude-frequency and phase-frequency feature are two valid analysis means for digital signal and system, with amplitude characteristic describing frequency region and phase describes time region. For amplitude- frequency, match filter has uniform characteristic with input signal. Strong signal will lead to filter amplification and weak signal may be prevent obviously. And a variety of signals will pass whatever the noise is. The premise for match filter is white noise with flatness power spectral at each frequency point. For phase-frequency characteristic, it is opposite to phase character and input signal completely. After the signal passing match filter, its phase is zero, and it will iterate mutually. Noise phase is random, and the iteration is non-coherent. Therefore, it guarantees the maximum SNR at time region. The essence of match filter is to compensate input signal phase and it consists of match phase.

In fact, match filter can guarantee signal to pass as soon as possible and noise to restrained, whatever in time region and frequency region. And the output with maximum SNR is obtained.

In image processing, match filter is to compare image with match filter template to restrain clutter and noise [8]. Huge advantage is that the variance of target is small to lead small amplitude difference between pixel and next for a target. And this character will be advantage for tracking algorithm.

The objective of this paper is to realize target distinguish in image background, combing target distribution, match filter is iterated optimized. At first, the principle of match filter is clarified.

Assumption 1: S is image data, w is frequency variance, the upper label * is conjunction. The response function $H(\omega)$ and impulse response arc described as following:

$$H(\omega) = ke^{j\omega t_0} S^*(\omega)$$

$$h(t) = \frac{1}{2\pi} \int_{-\infty}^{\infty} H(\omega) e^{j\omega t} d\omega$$

with, k is constant, t_0 is the time of peak value, $t_0 = 0, j$ is the imaginary. The input of match filter is:

$$z(t) = s(t) + n(t)$$

with $s(t)$ and $n(t)$ are known signal and stable noise with zero mean value. Owing to linearity, match filter can be described through iterations between signal and noise [9].

(1) Effect from signal $s(t)$

Fourier transform for $s(t)$ is:

$$S(\omega) = \int s(t) e^{-j\omega t} dt$$

Output signal is present as,

$$s_o(t) = \frac{1}{2\pi} \int H(\omega) S(\omega) e^{j\omega t} d\omega$$

(2) Effect from noise $n(t)$

Assume the output noise is white noise n_0(t) and power spectral density $S_{n0}(\omega) = N_0/2$

$$S_{n_o}(\omega) = |H(\omega)|^2 S_n(\omega)$$

with, $S_n(\omega)$ is the power spectral of input noise. Therefore, the mean power of output noise is,

$$E\left[n_o^2(t)\right] = \frac{1}{2\pi} \int_{-\infty}^{\infty} S_{n_o}(\omega) d\omega$$
$$= \frac{1}{2\pi} \int_{-\infty}^{\infty} |H(\omega)|^2 S_n(\omega) d\omega$$

Filter output SNR is,

$$SNR_o = \frac{s_o^2(t)}{E\left[n_o^2(t)\right]} = \frac{\left(\frac{1}{2\pi}\int_{-\infty}^{\infty} H(\omega)S(\omega)e^{j\omega t}d\omega\right)^2}{\frac{1}{2\pi}\int_{-\infty}^{\infty} |H(\omega)|^2 S_n(\omega)d\omega}$$

To maximum the SNR, Schwarz inequality is utilized to deal with equation above. If the condition is satisfied,

$$H(\omega) = \frac{\alpha S^*(\omega)}{2\pi S_n(\omega)} e^{-j\omega t}$$

The inequality under is established,

$$SNR_o \le \frac{1}{2\pi} \int \frac{|S(\omega)|^2}{S_n(\omega)} d\omega$$

Let $\varepsilon = \frac{1}{2\pi}\int_{-\infty}^{\infty} |S(\omega)|^2 d\omega$ be signal power, and SNR with white noise can be described as,

$$SNR_o \le \frac{2\varepsilon}{N_0}$$

Therefore, the maximum SNR is $2\varepsilon/N_0$, and

$$H(\omega) = KS^*(\omega)e^{-j\omega t}$$

with, $K = \alpha/(\pi N_0)$ is arbitrarily constant.

3 Match Filter Template Optimism Based on Logistical Regression

3.1 Extracting Image Character

In the images obtained from remote sensing camera, each pixel is constitutive with targets, clutter and background. Therefore, to implement accuracy target detection, signal character must be ascertained firstly.

Owing to Point Spread Function (PSF) effect of optics system, the radiation energy of target is distributed in some pixels around projection center [10]. Gaussian PSF model can be described as

$$h(x,y) = \begin{cases} \frac{1}{2\pi\sigma^2} \int\limits_{x-\frac{1}{2}}^{x+\frac{1}{2}} \int\limits_{y-\frac{1}{2}}^{y+\frac{1}{2}} e^{-\frac{(x-x_0)^2+(y-y_0)^2}{2\sigma^2}} dxdy & (x,y) \in C \\ 0 & others \end{cases}$$

with, (x_0, y_0) is projection center of target, C is support region of PSF describing distribution range of target energy, σ is standard variance of Gaussian PSF distribution reflecting spreading range of target energy called PSF radius or spreading radius. Target energy will reduce with increase of PSF radius.

According to structure of single sampling, separate order with odd and even light-exploring unit in detection linear list is utilized to exposure. Based on the effect from camera optics PSF character, different projection position will induce great difference of energy distribution on imaging place. It is clarified in detail as following.

(a) Target projection center falls in main pixel center, target energy focus on a pixel, shown as Fig. 1a;
(b) Target projection center inclines 0.25×0.25 pixel to main pixel center, target energy distributes on a brighter pixel and three darker pixels, shown as Fig. 1b;

(a) In main pixel center **(b)** Incline 0.25×0.25 pixel **(c)** Incline 0.5×0.5 pixel

Fig. 1 Target projection positions

(c) Target projection center inclines 0.5 × 0.5 pixel to main pixel center, target energy distributes on four adjacent pixels, shown as Fig. 1c.

According to three characters, character vector models are built.

(a) With target falling in pixel center, gradient decrease method is applied to model character vector. $A(i,j)$ is defined as pixel center, and target energy will decrease with adjacent degree increase from $A(i, j)$, as shown in Fig. 2. Assume that energy at $A(i,j)$ and at adjacent eight pixel are separately E1 and E2. And energy at sixteen pixels next to adjacent eight pixels are $E3$, and so on. The weights of match filter template are represent with energy. Template is defined with 15 parameters symmetrically. Therefore, energies at $A(i - 7{:}\ i + 7, j - 7{:}\ j + 7)$ pixels are applied to logistical regression to optimize match filter template.
(b) Projection inclines 0.25 × 0.25 pixel to center. Assume that the size of each pixel is $2r$, and center coordinate is (0, 0). According to energy distribution, weights are shown with distance between each pixel and center (Fig. 3).
(c) With projection inclines 0.5 × 0.5 pixel to center, target energy distribute with four equal pixels as center and decrease with the adjacent degree increase. When minimum distance between pixel and center r_0 satisfy $2kr \le r_0 \le 2\sqrt{2}kr, r_0 = 2kr$ (k is a constant), r_0 presents match filter template weights (Fig. 4).

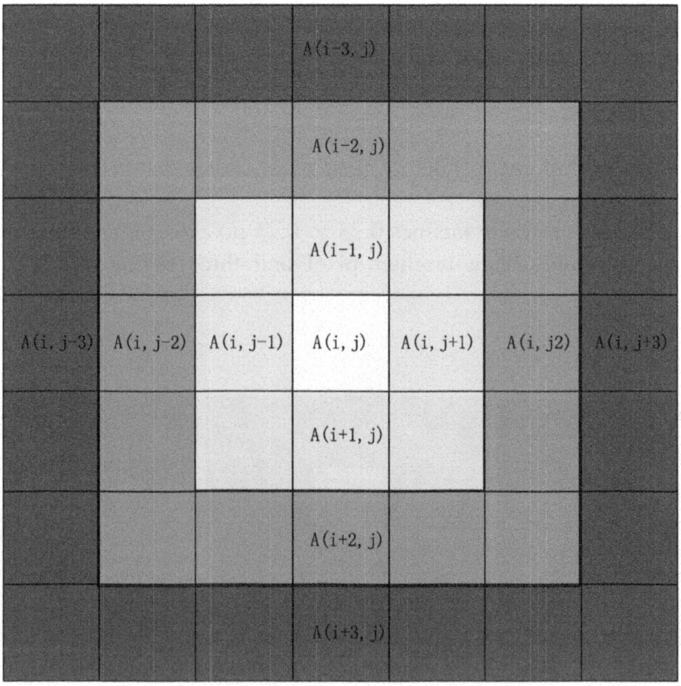

Fig. 2 Pixel distribution after sampling with projection at the center

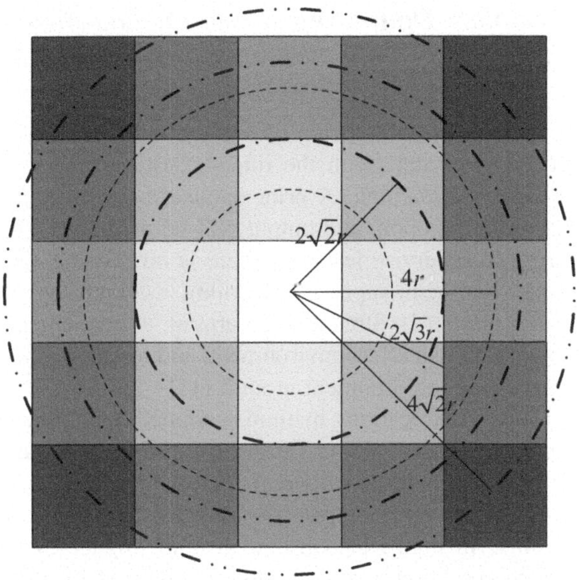

Fig. 3 Pixel distribution after sampling with projection inclining 0.25 × 0.25 pixel to the center

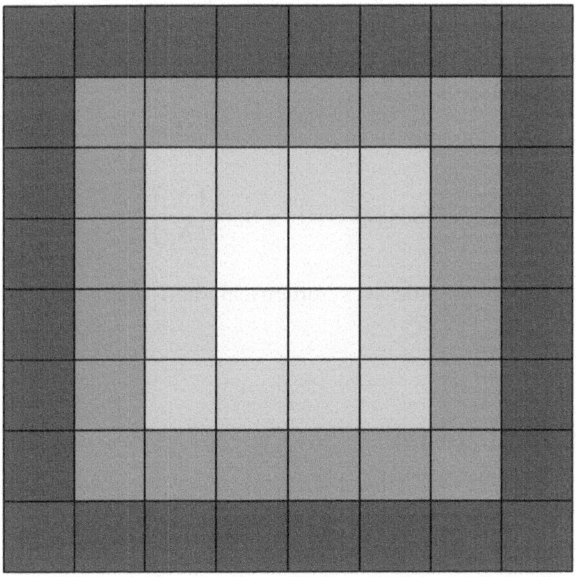

Fig. 4 Pixel distribution after sampling with projection inclining 0.5 × 0.5 pixel to the center

3.2 Match Template Optimizing Based on Logistical Regression

Logistical regression algorithm is a rule to analyze automatically from a class of data, and predict unknown data with the rule. Logistical regression algorithm is classified as: (a) supervise learning. A prior data set is given to build a function. Data can be input to the function, and output will be predictive. (b) non-supervise learning. Comparing to supervise learning, there is no training set, with a typical one-data cluttering. (c) semi-supervise learning is between supervised and non-supervised algorithm. (d) Strengthen learning is implement by observing operation. Each operation will affect environment, and learning object will judge on the basis of feedback from environment around [11].

In this paper, a supervised learning method is applied to optimize match module, in which, training set is set artificially. During target detection, power distribution characteristics in three situations described in Sect. 3.1 are input as training set, and determination results are training output. To consider the effect to target detection from clutter and noise in image background overall, Signal Clutter Noise Ratio (SCNR) is optimized condition, match module is obtained by regression analysis algorithm.

Definition 1

(a) Clutter and Noise (N_s): power standard deviation of image pixels 7×7 neighborhood around target;
(b) Target power S_t: difference from target power to mean of N_s;
(c) Signal clutter noise ratio: S_t/N_s.

Performance index function in logistical regression method is,

$$\max\{SCNR\} = \max\left\{\frac{S_t}{N_s}\right\}$$

Assume that match module is symmetrical, and distributes as normalization Gaussian pattern.

4 Verification of Result

In this paper, performance of 7×7 neighborhood is investigated by signal clutter noise ratio. Simulations are implemented as,

(a) Match filter is designed using image data, and match filter is obtained;
(b) Images are processed using match filter, comparing standard deviation before and after introducing match filter, the availability of background restrain is verified;

(c) Compare change with match filter and without match filter;
(d) Compare SCNR change with match filter and without match filter to detection ability.

Set parameters of PSF $\sigma = 0.3$, $C = 5$, describes deviate 0.3 and 5 pixels to target projection center (x_0, y_0) to check fluctuate standard deviation.

4.1 Match Module Optimization

Set target power distribution in Sect. 3.1 as character vectors, logistical regression method is introduced to optimize match filter, and a normalization symmetrical module with 15 parameters is obtained as [1; 0.16788; −0.15403; −0.09859; 0.09076; −0.039289; 0.008914; 0.007738]. And its distribution figure is shown as following (Fig. 5).

According to match filter principle, pattern with 15 parameters match filter leads to poor effect in beforehand and afterwards 7 series. Therefore, it is eliminated to investigate performance of match filter and following result is based on this condition.

4.2 The Performance of Match Filter

To compare match filter performances, target power is implemented firstly. 15 targets are added to an actual image collected from an infrared camera. To decrease complex background, target power is set to 50,000 (Table 1).

From the table above, target energy is strengthened with match filter. But it cannot ascertain whether it is beneficial to target detection just from energy strengthen (Table 2).

With match filter function, standard deviation of image background is amplified. Image collected from infrared camera is processed by signal sampling, and SNCRs

Fig. 5 Match filter module optimized

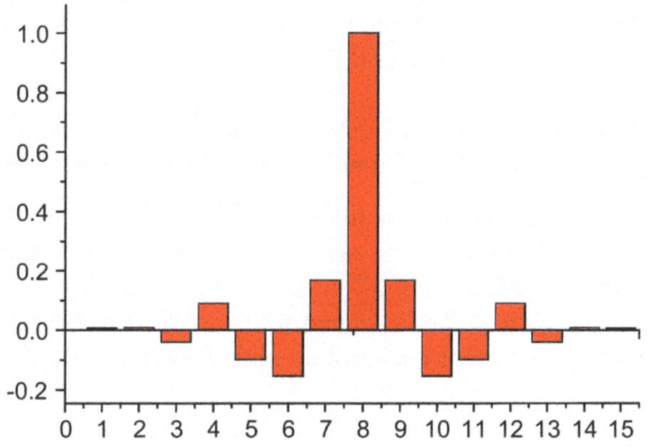

Table 1 Comparison of target energy

No.	Target energy	Target energy with match filter	Ratio
1	80420.5145	62616.2876	0.77861
2	95632.2629	61251.5135	0.64049
3	93200.6571	61427.6544	0.65909
4	60995.8396	57746.14767	0.94672
5	85477.7076	60291.0461	0.70534
6	74750.4076	58995.1768	0.78923
7	65306.7129	59744.0727	0.91482
8	76685.4735	61484.5821	0.80178
9	72310.7123	60296.2338	0.83385
10	77604.8481	61543.8711	0.79304
11	86586.5353	59216.57813	0.6839
12	68968.82613	58702.2359	0.85114
13	59894.2846	62079.0313	1.03648
14	48497.4691	58810.4652	1.21265
15	57738.6724	57885.5031	1.00254

Table 2 Comparison of standard deviation

No.	Description	Standard deviation of image
1	Without match filter	4197.7789
2	With match filter	3376.8327

with targets adding into image are obtained using and without match filter. When targets are strong correlation to background, that is SNCR is magnified, it is advantageous to target detection (Table 3).

SNCR is amplified with match filter introducing with analysis above. With data from experiment, conclusions are clarified as following:

(a) Meaning value ratio of target energy using signal sampling is 0.8433 with match filter introduced. It is illustrated that image power energy processed with signal sampling using match filter is weaken.
(b) Standard deviation ratio of image background is 0.8044 with match filter introduced. It is said that target energy is weaken with match filter processing the image.
(c) With match filter, SNCR of target energy in neighborhood is amplified, the ratio is 1.566632, and it is beneficial to target detection.

Six different match filter modules are shown in [3], and there are 15 weights which is symmetry in each module. The normalization weights are described as following (Table 4).

The images are separated to 16×16 sub-images. The maximum value is utilized to represent background clutter by standard deviation.

Table 3 Comparison of SNCR

No.	No match filter	Match filter
1	8.5220	13.2738
2	7.1353	13.5682
3	7.0891	13.9962
4	9.7861	14.4705
5	8.1218	13.9163
6	8.7593	13.6041
7	9.2395	13.0064
8	8.7868	12.4247
9	9.1017	12.7786
10	8.5063	13.2520
11	7.7434	13.0559
12	8.8482	13.1681
13	10.3307	14.2098
14	11.5170	15.2932
15	9.8899	12.3211

Table 4 Normalization weights of match filter module

No.	Match filter module
1	[1; 0.282; −0.368; −0.236; −0.112; −0.0429; −0.0144; −0.00429]
2	[1; 0.223; −0.432; −0.219; −0.0672; −0.00417; 0.00583; −0.00637]
3	[1; 0.149; −0.483; −0.155; −0; 0.0226; 0; −0.0335]
4	[1; 0.129; −0.468; −0.0978; 0.0185; 0.0046; −0.0245; −0.0618]
5	[1; 0.0339; −0.238; −0.135; −0.104; −0.107; −0.114; −0.140]
6	[1; 0.183; −0.337; −0.0408; −0.05; −0.073; −0.0673; −0.1157]

Table 5 Comparison of standard deviations

No.	Without filter		Without filter	
	Original image	16 × 16 sub-images (max)	Original image	16 × 16 sub-images (max)
1	161.7613	109.8737	22.9667	39.7139
2			20.8769	34.4525
3			19.2712	29.9653
4			19.2859	29.0613
5			99.0396	57.6547
6			23.0544	37.8969

From Table 5, standard deviations of images processing with match filter decrease obviously. It is shown that match filter is valid to restrain image background clutter and noise.

5 Conclusion

Match filter optimized based on logistical regression in this paper is valid to image background with standard deviation referred as fluctuation. With 7×7 neighborhood inspecting, SNCR is amplified with match filter and it is ability to target detection and target energy is decreased.

In conclusion, logical regression including in optimizing match filter module is feasible. Filter performance is improved obviously and is beneficial to target detection in remote sensing images.

References

1. Shen Xueju, Wang Yongzhong, Yao Guangtao. Design of matched filter based on analysis of background noise Fourier spectrum [J]. Journal of Applied Optics, 2008, 29(3): 343–346.
2. Yang Wunian. A Quantitative Equation for Selecting Optimum Parameters of Matching Filtering Template for Linear Feature Detection on Remote Sensing Images[J]. Journal of Chengdu College of Geology, 1985, l: 101–108.
3. Stephen A. Cota, Linda S. Kalman, Robert A. Keller. Advanced Sensor Simulation Capability [J]. Signal and Image Processing Systems Performance Evaluation, 1990, 1310: 134–149.
4. Sun Cuijuan, Yang Weiping, Shen Zhenkang. An Infrared Small Target System Based on Matched Filtering[J]. Laser and Infrared, 2006, 36(2): 147–150.
5. XU Fang, SHI Dinghe. Triple Correlation Matched Filter Used for Processing Faint Signals of Targets[J]. Fire Control and Command Control, 2001, 26(1): 65–68.
6. Zhou Hanfei, SU Yi, ZHU Yutao. Multi-Aspect SAR Imaging Using 2-D Matched Filtering [J]. ACTA Electronica SINICA, 2012, 40(12): 2426–2432.
7. Yuan Yun, Li Zhuang, Zhu Xianwei. An Improved SAR Image Speckle Noise Filter Based Oil Sigma Method[J]. Spacecraft Recovery & Remote Sensing, 2011, 32(4): 37–44.
8. David E. Norwood, Andrew E. Simpson, and Satish M. Mahajan. Impact Identification using Matched Filtering and Spectral Cross Correlation[J]. 2005, 11: 125–130.
9. Laura Burattini, Wojciech Zareba and Roberto Burattini. Adaptive Match Filter Based Method for Time vs. Amplitude Characterization of Microvolt ECG T-Wave Alternans[C]// Annals of Biomedical Engineering, Vol. 36, No. 9, September 2008: 1558–1564.
10. Louisa Varsano, Irena Yatskaer, Stanley R. Rotman. Temporal Target Tracking in Hyperspectral Images [J]. Optical Engineering, 2006, 45(12): 126–201.
11. O. Franzen, H. Blume, H. Schroder. FIR-Filter design with spatial and frequency design constraints using evolution strategies[J]. Signal Processing, 1998, 68: 295–306.

A Digital TDI Operation Method of Array CCD Based on Curve Fitting Algorithm

Lei Ning, Li Qiang, Hu Yuting, Bao Bin and Li Tao

Abstract With the rapid development of China's space and remote sensing industry, area array CCD has been widely used in space camera technology. In this chapter, on the basis that frame transfer area array CCD can work properly, according to the working principle of TDI CCD, a digital TDI operation mode of area array CCD based on curve fitting using FPGA is presented. Firstly, the displacement of two adjacent images is estimated; then corrects the displacement using Lagrange interpolation; and adds up the digital numbers of the same pixel in different exposure time then output. Experiment shows that this digital TDI operation mode can remarkably improve image SNR, obtain high sensitivity, high output rate and wide dynamic range under low illumination condition, greatly reduce the influence of harsh environment that lead to low SNR. In the meantime, digital TDI can improve the mismatching of pixel transfer rate and electric charge transfer rate and image quality reduction caused by earth rotation.

Keywords Array CCD · TDI CCD · Curve fitting · SNR

1 Introduction

CCD (Charge Coupling Device) is broadly used in photoelectric detection and imaging fields, since its low noise, large dynamic range, high quantum efficiency and high charge conversion efficiency [1]. Different types of CCD are used in remote sensing application, such as TDI (Time Delayed and Integration) CCD, area array CCD and etc. In this chapter, frame transfer area array CCD is used in remote sensing camera, and the traditional push broom imaging mode of TDI CCD system is applied. In each exposure time, the camera is pushed forward the distance of one pixel for exposure imaging. Therefore, the image data of two adjacent frames has a one pixel difference in the TDI push broom direction.

L. Ning (✉) · L. Qiang · H. Yuting · B. Bin · L. Tao
Beijing Institute of Space Mechanics and Electricity, Beijing, China
e-mail: nwpu_ln@sina.com

H.P. Urbach and G. Zhang (eds.), *3rd International Symposium of Space Optical Instruments and Applications*, Springer Proceedings in Physics 192,
DOI 10.1007/978-3-319-49184-4_38

To overcome the image quality degradation problem caused by the mismatch between pixel shift rate and transfer rate and earth rotation in traditional TDI technical, a digital TDI method based on curve fitting algorithm is proposed in this chapter. Firstly, displacement of adjacent two frame images output by area array CCD is estimated. Secondly, displacement is corrected used Lagrange interpolation polynomial. Finally, the corrected image data is output after digital time delay integration. In low luminance conditions, this method can gain high sensitivity, high output rate and large dynamic range. The negative factor of low SNR caused by the harsh environmental conditions is greatly improved.

2 The Working Principle of TDI CCD

TDI CCD is a special linear CCD. It adopts time delay integration (TDI) technique [2] and can obtain high sensitivity, high output rate and large dynamic range in low luminance condition. The working process of traditional analog TDI CCD is shown in Fig. 1. In the first integration period, the object is exposed and integrated in the first pixel of one column. The Photo-generated charge is not read out as ordinary CCD but transferred to the next pixel of this column. In the second integration period, object is moved, exposed and integrated in the second pixel of this column. The Photo-generated charge newly gained adding the charge transferred from previous pixel then transfer to the next pixel in this column, and so on. In the Nth integration period, object has moved to Nth pixel of this column to expose and integrate. After this integration time, photo-generated charge of the Nth pixel added the charge transferred from previous (N-1) pixels transfer to readout resister, and the whole charge is read out as the ordinary linear CCD.

In this imaging mode, pixel shift rate should match charge transfer rate and pixel shift direction must follow the TDI direction. Otherwise, the image degradation in TDI direction (column direction) and its vertical direction (row direction) is introduced. On the one hand, due to the multistage accumulation integration character of TDI CCD, the clear image can be obtained by ensuring the match of pixel shift rate

Fig. 1 Charge transfer schematic of TDI CCD

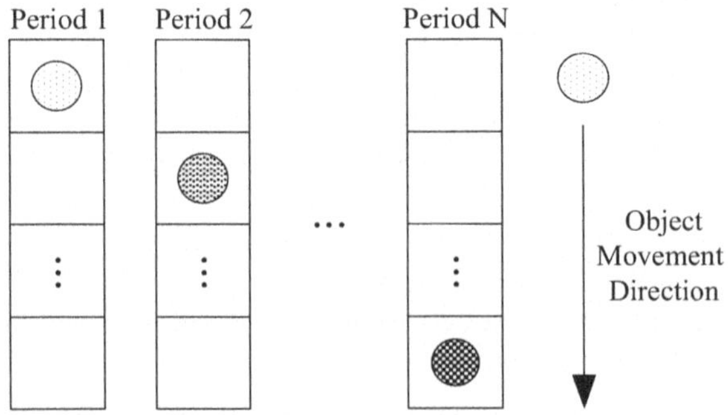

and charge transfer rate in a long time. On the other hand, because of earth rotation, the satellite moving direction is not same as the actual imaging direction which will cause image motion in one integration time and reduce image quality.

3 Digital TDI Working Mode

In this paper, frame transfer area array CCD is used in remote sensing camera, and the traditional push broom imaging mode used in TDI CCD system is applied. The imaging results when pixel shift rate and transfer rate is matching are given in Fig. 2, which assumes the size of area array CCD is 4×4. It can be seen from the figure that the image data of two adjacent frames has a one pixel difference in TDI direction. In this condition, the digital TDI working mode of area array CCD can be realized by adding the data corresponding to the same scene in different frame image.

As discussed in the previous section, the image degradation in push broom direction is introduced by the mismatch between pixel shift rate and transfer rate. When pixel shift rate is faster than transfer rate, the moving distance in one exposure time is larger than one pixel. This will cause an overlapping area between the adjacent two lines in each frame image. As shown in Fig. 3, the lower border of the first row of pixels and the upper border of the second row of pixels will sweep the gray area. Therefore, the second line in the first frame image and the first line in the second frame image are not corresponding to the same scene. In contrast, when pixel shift rate is slower than transfer rate, the moving distance in one exposure time is smaller than one pixel. This will cause an interspaced area between the adjacent tow lines in each frame image. As shown in Fig. 4, the lower border of the first row of pixels and the upper border of the second row of pixels both will not sweep the gray area.

And furthermore, due to the earth rotation, there is a displacement in the horizontal direction (which is vertical to the push broom direction) between adjacent frame images during remote sensing camera imaging.

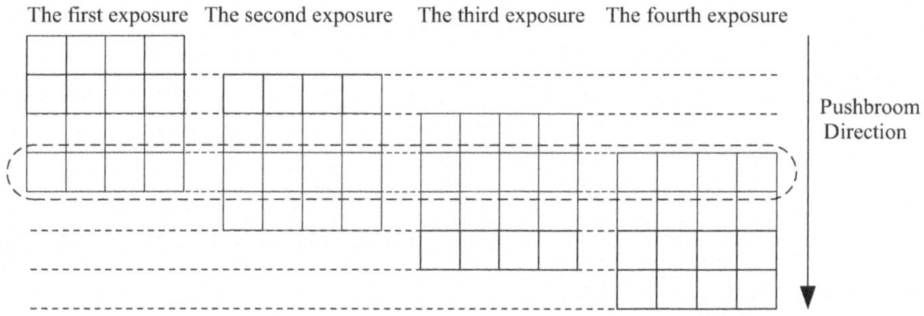

Fig. 2 The imaging results when pixel shift rate and transfer rate is matching

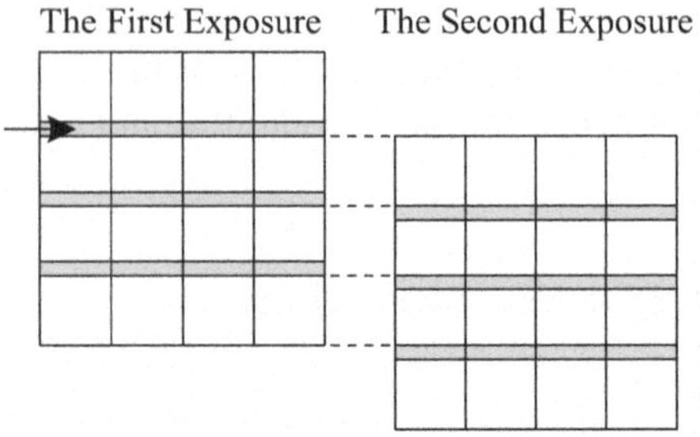

Fig. 3 The imaging results when pixel shift rate is faster than transfer rate

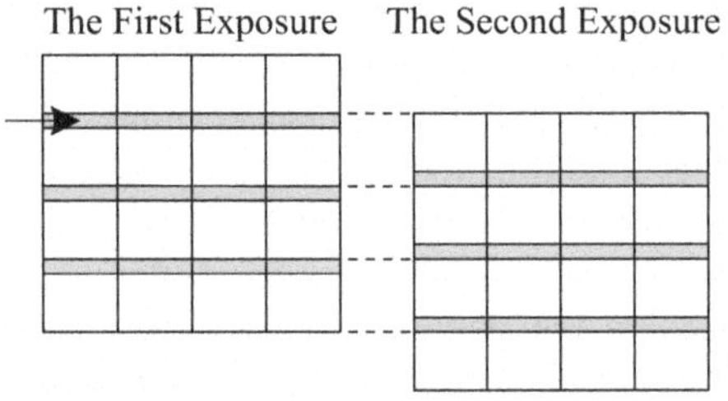

Fig. 4 The imaging results when pixel shift rate is slower than transfer rate

Figure 5 gives the flow chat of digital TDI working mode. To realize the digital TDI working mode of area array CCD, the displacement in push broom direction and its vertical direction between two adjacent frames of image is estimated firstly. Secondly, the image is interpolated in accordance with the calculated displacement. Finally, the data corresponding to the same scene in different frame image is added.

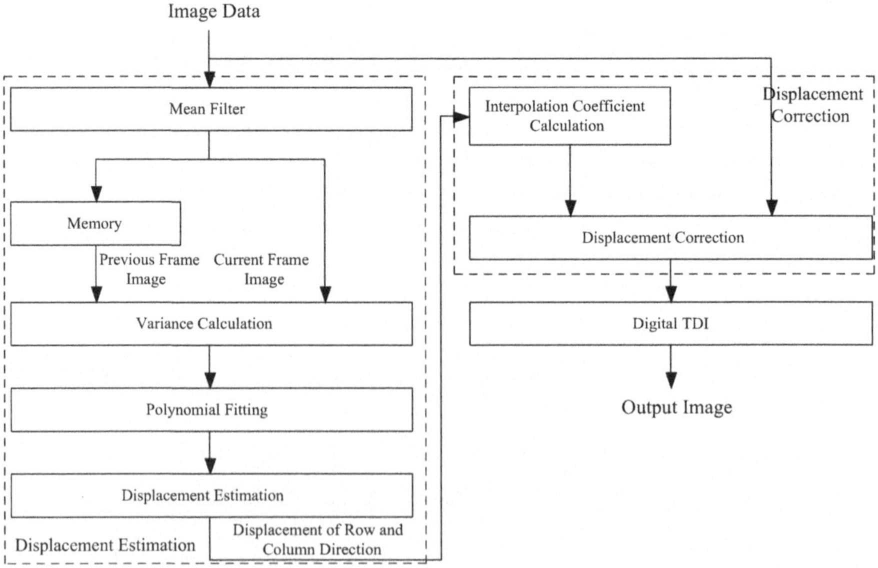

Fig. 5 The flow chart of digital TDI working mode

A. Displacement Estimation

Theoretically, the variance between two images corresponding to the same scene at different time reflects their relative movement. A displacement estimation algorithm is proposed based on this idea. The flow chat of displacement estimation algorithm is given in Fig. 6.

(1) Mean Filter

There is a variety of noises in the output signal of CCD [2], such as shot noise, dark current noise, fix pattern noise (FPN), thermal noise, read noise, and so on. The thermal noise and read noise are signal independent and like the white noise which mean value is 0. Therefore, in order to reduce the influence of random noise and improve the algorithm reliability, the mean filter is implemented to the image, the formula is as follows:

$$g(x,y) = \frac{1}{(2a+1)(2b+1)} \sum_{s=-a}^{a} \sum_{t=-b}^{b} f(x+s, y+t) \tag{1}$$

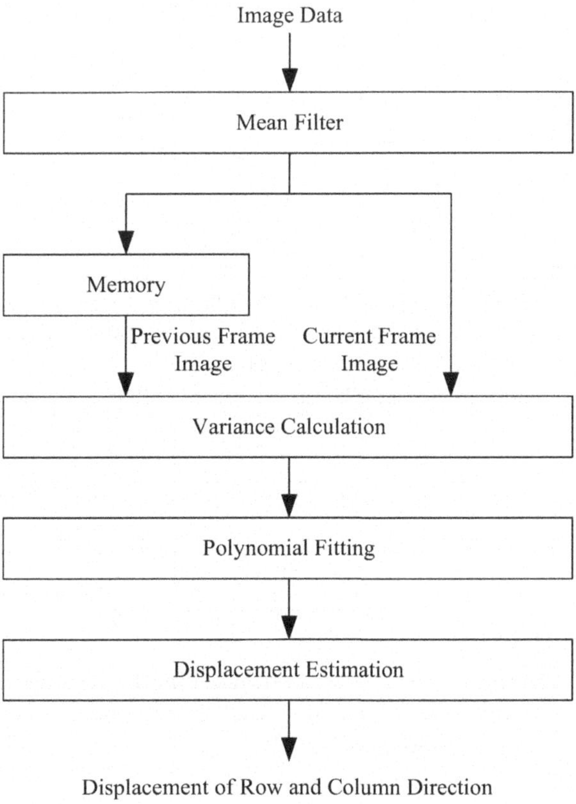

Fig. 6 The flow chart of displacement estimation algorithm

where, x is the TDI direction (row direction), y is the CCD direction (column direction), $f(x, y)$ is the input image, $g(x, y)$ is the image after mean filter, (2a + 1) and (2b + 1) is the size of mean filter window. In order to save hardware resource, the mean filter of row and column direction is separated and executed respectively.

(2) Variance Calculation

Theoretically, the variance between two images of different time corresponding to the same scene reflects their relative movement. When these two images are complete coincide, the variance is minimal. Therefore, the displacement of two image can be determined by calculated its variance in different phases.

The push broom method is used in this chapter camera. So the image data of adjacent frames have a one pixel difference in TDI push broom direction (shown in Fig. 2). The variance of adjacent frame images when they are completely overlap is calculated as follow:

$$Var_{c0l0} = \frac{1}{(M-1)N} \sum_{x=2}^{M} \sum_{y=1}^{N} (g_i(x,y) - g_{i-1}(x+1,y))^2 \tag{2}$$

Similarly, the variance in TDI direction in different phase of adjacent frame images after mean filter can be calculated respectively. The formula is as follows:

$$Var_{cpml0} = \frac{1}{(M-m_1)N} \sum_{x=1}^{M-m_1} \sum_{y=1}^{N} (g_i(x,y) - g_{i-1}(x+m_1,y))^2 \tag{3}$$

$$Var_{cnml0} = \frac{1}{(M-m_2)N} \sum_{x=m_2+1}^{M} \sum_{y=1}^{N} (g_i(x,y) - g_{i-1}(x-m_2,y))^2 \tag{4}$$

In (3), $m_1 = 2, 3, 4$ are corresponding to the variance of phase difference 1, 2, 3 respectively. In (4), $m_2 = 0, 1, 2$ are corresponding to the variance of phase difference −1, −2, −3 respectively.

The variance in CCD direction of different phase can be calculated as follows:

$$Var_{c0lpn} = \frac{1}{(M-1)(N-n)} \sum_{x=2}^{M} \sum_{y=1}^{N-n} (g_i(x,y) - g_{i-1}(x+1,y+n))^2 \tag{5}$$

$$Var_{c0lnn} = \frac{1}{(M-1)(N-n)} \sum_{x=2}^{M} \sum_{y=n}^{N} (g_i(x,y) - g_{i-1}(x+1,y-n))^2 \tag{6}$$

In (5), $n = 0, 1, 2$ are corresponding to the variance of phase difference 1, 2, 3 respectively. In (6), $n = 0, 1, 2$ are corresponding to the variance of phase difference −1, −2, −3 respectively.

(3) Polynomial Fitting

Analysis shows the variance in different phase should satisfy some relationship. In this chapter, six-time polynomial is used to describe this relationship:

$$V = c_0 + c_1 P + c_2 P^2 + c_3 P^3 + c_4 P^4 + c_5 P^5 + c_6 P^6 = \sum_{i=0}^{6} c_i P^i \tag{7}$$

where, the coefficients of six-time polynomial are calculated using the least squares algorithm [3].

(4) Displacement Estimation

The minimum point of six-time polynomial showed in (7) is the displacement of two adjacent frame images. Since the six-time polynomial does not have numerical solution, the iterative approach is used to calculate its approximate solution.

During the motion of the satellite, the change of the displacement is a slow process. So the following iterative method is used in this chapter. Assuming the displacement calculated in the ith time is $delt_i$. The value of six-time polynomial is calculated form $delt_i - 0.005$ to $delt_i + 0.005$, and the step is 0.0001. The displacement corresponding to the minimum point in this 101 points is adopted as the displacement $delt_{i+1}$ calculated in (i + 1)th time.

B. Displacement Correction

Before executing the digital TDI, in order to make the image data of the same ground feature corresponding to the same region exactly, the image data need to be interpolated according to the displacement of row and column direction calculated above. The displacement correction along the push broom direction is shown in Fig. 7. Supposing the displacement of column direction is $delt_y$, the processing for the situation that pixel shift rate is faster than transfer rate is as follows: the first line of each image remains intact, the second line is shifted $delt_y$ downward, the third line is shifted $2 \times delt_y$ downward, and the fourth line is shifted $3 \times delt_y$ downward. After this processing, the image data will correspond to the same region exactly.

In this chapter, the seven-order Lagrange polynomial is used to shift and register the image. The calculation is as follows:

$$lx_i = \prod_{i=0, i \neq j}^{7} \frac{x \times delt_y - j}{i - j} \tag{8}$$

$$z_x(x, y) = \sum_{i=0}^{7} lx_i f(x + i - 3, y) \tag{9}$$

where, $i, j = 0, 1, 2, \ldots, 7$, $x = 4, 5, \ldots, M\text{-}4$, $f(x, y)$ is the input image.

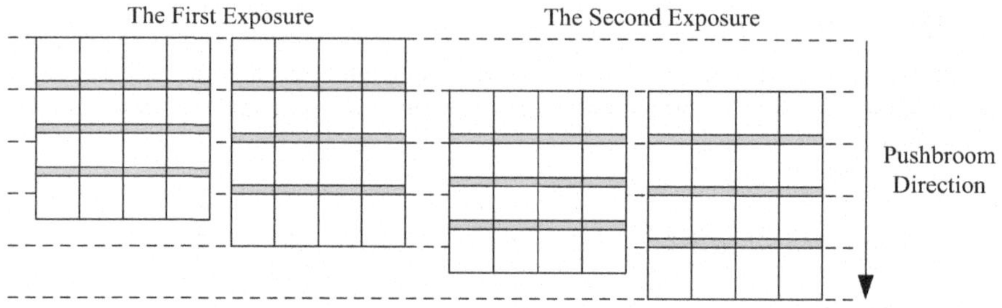

Fig. 7 The schematic diagram of displacement correction along push broom direction

From the foregoing depiction, there is a displacement between two adjacent frames of image in the horizontal direction. Supposing the displacement of this direction is $delt_x$. The displacement between the first frame image and the second frame image is $delt_x$. The displacement between the first frame image and the third frame image is $2 \times delt_x$. So the displacement correction in row direction is same as column direction. The first line of each image remains intact, the second line is shifted $delt_x$ in lateral, the third line is shifted $2 \times delt_x$ in lateral, and the fourth line is shifted $3 \times delt_x$ in lateral. The shift and register using the seven-order Lagrange polynomial is as follows:

$$ly_i = \prod_{i=0, i \neq j}^{7} \frac{y \times delt_x - j}{i - j} \tag{10}$$

$$z(x, y) = \sum_{j=0}^{7} ly_i z_x(x, y + i - 3) \tag{11}$$

where, $y = 1, 2, \ldots, N$. To make the image after processing has the same number in row direction, even extension is executed for the boundary pixel.

C. Digital TDI realization method

After the displacement correction in row and column direction completed, the digital TDI working mode can be realized by adding the data corresponding to the same scene in different frame image by using FPGA. The formula is as follows:

$$T(x, y) = \frac{1}{N} \sum_{i=1}^{N} z_{i+x-1}(N - i + 1, y) \tag{12}$$

where, N is the TDI stage, z_{i+x-1} is the image after correction.

Table 1 The SNR of different TDI stage

TDI stage	Mean value	Noise	SNR (dB)
1	12,912	236.46	34.75
4	13,177	122.02	40.50
16	12,985	68.19	45.59
32	13,015	53.38	47.76
64	12,754	42.87	49.74
128	12,836	37.55	51.06

4 Experiment Result

In this chapter, the Digital TDI operation method of array CCD is tested on an imager developed by our institute. There is a relationship between the SNR of TDI CCD and its stage. The signal of N stage TDI-CCD increase N times while the noise only increases $\sqrt{N}$ times, which indicates the SNR improve $\sqrt{N}$ [4]. Table 1 shows the Mean Value, Noise, SNR (dB) results of the images in different TDI stages. Because the image data after digital TDI is divided by N in formula (12), so the output image has the same mean value in different TDI stage. It is not difficult to find that image noise (root-mean-square error) is reduced to half of its origin when the TDI stage is increased 4 times, so the SNR is almost improved 6 dB.

5 Conclusion

In this chapter, on the basis that frame transfer area array CCD can work properly, according to the working principle of TDI CCD, a digital TDI operation mode of area array CCD based on curve fitting using FPGA is presented. This method can obviously improve the image degradation caused by the mismatch between pixel shift rate and charge transfer rate and the earth rotation. It can be broadly used in push broom imaging system using area array CCD. The experimental results show the digital TDI operation method proposed in this chapter can improve SNR of image significantly, and enhance the imaging performance of remote sensing camera.

References

1. Xue Xucheng, Li Yunfei, Guo Yongfei, "Design of analog front end of CCD imaging system", Optics and Precision Engineering, 15(8), pp. 1191–1195, 2007.
2. James R. Janesick, "Scientific Charge-Coupled Devices", Bellingham, WA: SPIE, 2001.
3. Guan Zhi, Lu Jinpu, "Fundamentals of Numerical Analysis", Higher Education Press, 2010, 7.
4. Savvas G. Chamberlain, "High Speed, Low Noise, Fine Resolution TDI CCD Imagers", Charge-Coupled Devices and Solid State Optical Sensors Proceedings, pp. 252–262, 1990.

Observation Capability and Application Prospect of GF-4 Satellite

Dianzhong Wang and Hongyan He

Abstract GF-4 satellite is the latest member of China's high-resolution earth observation system (CHEOS) and the first high resolution remote sensing satellite in geostationary orbit in the world. In this paper, its observation capabilities in spatial, temporal and spectral dimensions were introduced. To take full advantages of these capabilities five operation modes were proposed, i.e. staring mode, cruising mode, sign-in mode, tracing mode, and night mode. Each mode was illustrated by at least one civil application case together with operation recommendation. Further, two issues—data continuity and data stability, which play important roles in satellite data application were discussed. Pertinent on-orbit operation suggestions were also proposed to ensure data continuity, and evaluation models were proposed for data stability. This paper could help in rational use of the satellite and satisfying multiple observation needs.

Keywords GF-4 satellite · Observation capability · Geostationary orbit · High resolution · Remote sensing satellite

1 Introduction

China is one of the countries most affected by natural disasters, which now causes about 40 billion euros losses and more than 200 million people every year [1]. In the course of recorded history, many types of natural disasters—except volcano eruptions —have occurred in China, which include floods, droughts, meteorological, seismic, geological, maritime and ecological disasters as well as forestry and grassland fires. Natural disasters need to be effectively monitored for the development of the society. Up to the launch of GF-4 satellite, China had launched several series of land remote sensing satellite such as CBERS, ZY, HJ, SJ, RS, KZ and GF. And satellite data have

D. Wang (✉) · H. He
Beijing Institute of Space Mechanics & Electricity, Beijing, China
e-mail: wangdianzhong@gmail.com; drgnw@163.com

H. He
e-mail: hehy_cast@sina.com

H.P. Urbach and G. Zhang (eds.), *3rd International Symposium of Space Optical Instruments and Applications*, Springer Proceedings in Physics 192,
DOI 10.1007/978-3-319-49184-4_39

been widely used in flood, drought, seismic, floods monitoring and assessment [2–7]. However, almost all of the satellites mentioned above are operating in low earth orbit (LEO), which makes successive monitoring impossible for long revisit cycles. Geo-synchronous earth orbit (GEO) satellite can solve this problem [8]. However, space resolutions of previous GEO satellites were so low that applications were limited for meteorology or maritime instead of land [9]. GF-4 satellite is the highest resolution in the world, that can capture 50 m resolution multispectral images for visible and near infrared (VNIR) channel and 400 m for middle wave infrared (MWIR) at nadir. Considering that the previous GEO satellites' highest resolution is 500 m in VNIR spectrum, GF-4 has one digit improvement in resolution. At the same time, GF-4 satellite can monitor a wide spatial range including China and around, together with very high temporal resolution, and several spectral channels, this camera can be an effective tool for natural disaster monitoring and reduction.

2 Observation Capability

GF-4 satellite operates in GEO orbit which is 35,786 km away from the Earth, the camera is capable to capture VNIR multispectral of 50 m and MWIR of 400 m image at nadir, and the imaging range can satisfy a requirement of 400 km $\times$ 400 km. Specification of the camera is shown in Table 1.

Table 1 Specification of GF-4 camera

Item	Specification
Spectral range	B1: 450–900 ($\pm$15 nm)
	B2: 450–520 ($\pm$15 nm)
	B3: 520–600 ($\pm$15 nm)
	B4: 630–690 ($\pm$15 nm)
	B5: 760–900 ($\pm$15 nm)
	B6: 3.5–4.1 ($\pm$0.1 um)
Resolution at nadir	VNIR <50 m
	MWIR <400 m
Imaging range at nadir	VNIR: $\geq$500 km $\times$ 500 km
	MWIR: $\geq$400 km $\times$ 400 km
Static modulation transfer function (MTF)	B1: 0.14
	B2: 0.15
	B3: 0.15
	B4: 0.15
	B5: 0.12
	B6: 0.15
Signal noise ratio (SNR)	$\geq$46 dB(SZA* = 10°, ρ = 0.8)
	$\geq$23 dB(SZA = 80°, ρ = 0.05)
Noise equivalent temperature difference (NETD)	MWIR NETD $\leq$0.2 K (@350 K)

*SZA, solar zenith angle

3 Application Mode and Prospect

Two observation modes had been considered since the satellite was in definition phase.

3.1 Mode 1—Staring Mode

This mode means successive monitoring in short duration. As a GEO satellite, GF-4 can avoid the deficiency of LEO satellite in successive monitoring and improve regional monitoring capability, which is badly needed in disaster and emergency monitoring. In the case of successive monitoring for a given location, panchromatic channel is preferred. Despite the spectral information is single, it does not have to shift between channels and can make full use of temporal sampling capability of GEO satellite [10]. Rapid accumulation of samples makes videos for movement analysis. For 50 m spatial sampling and 5 s temporal sampling, it is expected to capture rapid changes at speed of 10 m/s or above. Accumulation of sampling also makes it practical to enlarge the dynamic range. And minus displacement within successive frames could provide information to infer more details to improve the resolution (Fig. 1). This mode could satisfy the observation requirement such as earthquake and forest fires, which usually do not change dramatically in location.

3.2 Mode 2—Cruising Mode

This mode means imaging all around as the satellite can cover. When there is no need for staring mode, GF-4 can imaging region by region as a land resource

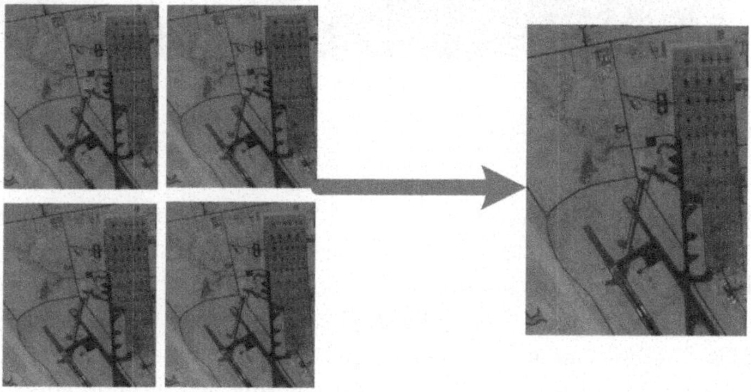

Fig. 1 Improvement of spatial resolution

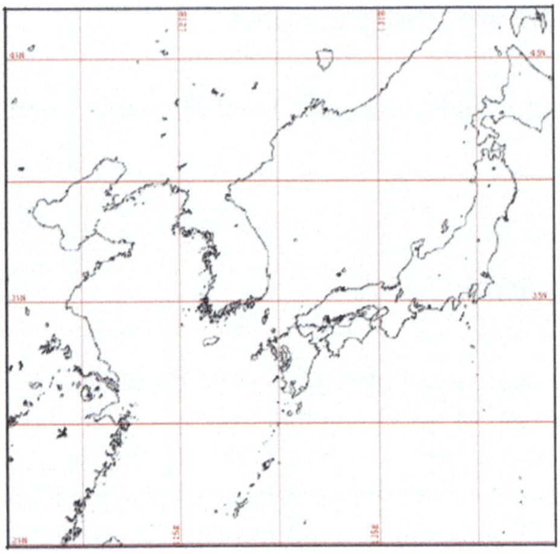

Fig. 2 Image target area of GOCI

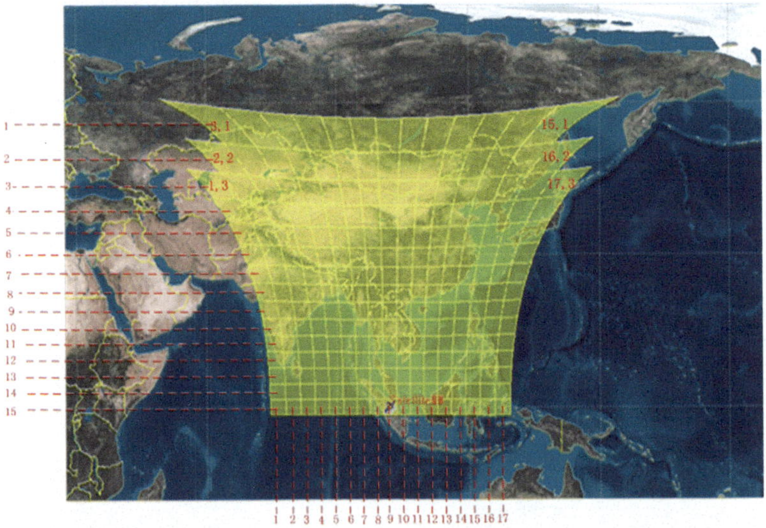

Fig. 3 Potential coverage of GF-4

satellite. According to the operation of COMS satellite, zoning makes management explicit as shown in Fig. 2 [11].

Though the major aim of imaging is China, the potential coverage of GF-4 can be separated into as many as 249 individual zones (Fig. 3). Since there is no time limit in mode 2, GF-4 can acquire full spectral information of each zone. Thus the total time relies on the span between channel altering and attitude adjusting.

During on-orbit test, both these two types of span were successively shortened to make it possible to complete observation of China all around within one work day. At the same time, three more application modes could be proposed based on the practical capabilities of the satellite.

3.3 Mode 3—Sign-in Mode

This mode means periodical monitoring for certain locations. When mode 2 was interrupted by various emergencies, there was not enough time left to observe every zone. Only regions that change frequently or be of special value should be guaranteed to visit. As the time is enough for certain locations in this mode, a full-channel observation is preferred. Among the first release of GF-4 images, there was a pair of images in Beijing, one for clear sky and the other for hazy (Fig. 4). Daily or other period cycle for regional monitoring is useful especially in long term research such as atmospheric and ecological issues.

3.4 Mode 4—Tracing Mode

This mode means successive monitoring of certain target from the original location. If the observation location of certain phenomenon changes with time, then tracing is needed instead of staring. For instance of flood monitoring, the peak of flood is not

Fig. 4 Imagery in Beijing (*Left* Clear, *Right* Hazy)

Fig. 5 Extraction of downstream of Huanghe River in GF-4 image

static and moves downstream with time. GF-4 can be assigned much time in main rivers in flood season to monitor flood situation of the watersheds (Fig. 5).

Similarly, the satellite can be assigned in certain sea routes as it had demonstrated the capability to see large seacrafts soon after launch.

3.5 Mode 5—Night Mode

MWIR channel can provide temperature information. Thanks to his channel, GF-4 can work not only at daytime but also at night. And 400 m resolution is higher than most civil IR sensors in orbit. Obviously, this is critical to improve the observation capability of emergency reaction system.

4 Influences

Among all the factors that influence the application of a certain satellite, data continuity and data stability are discussed here for their importance.

4.1 Data Continuity

Whether the terminal user can use a certain type of remote sensing data depends on its continuous span in orbit. Data continuity is especially critical for long term

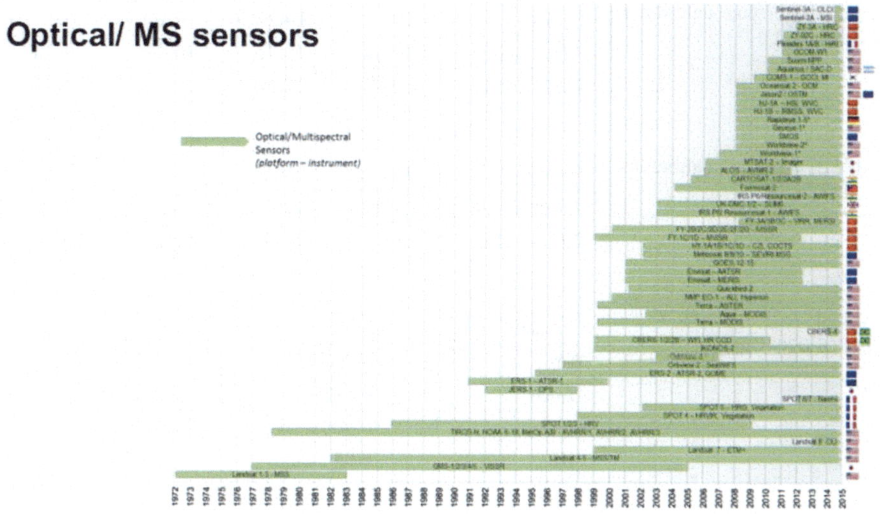

Fig. 6 World optical satellite data continuity

scientific research. Reference [14] gave a statistical comparison of world's optical satellite data continuity (Fig. 6).

Lifetime of a single satellite might be not long, while the data continuity of certain payload could be prolonged by successive launch of heritage satellite. For instance, three satellites, i.e. CBERS 01, 02, 02B together can construct data continuity of a decade scale. GF-4 was designed to have an 8-year life time, which is close to the sum of the above three satellites. Such a long life enables long term observation accumulation.

4.2 Data Stability

Data stability is an important aspect to influence data quantification and application. Reference [12] formulated gain index to evaluate radiometric performance of satellite image.

According to reference [13], absolute calibration coefficients of Pléiades 1B and SPOT 6 satellite were shown in Tables 2 and 3. There are no significant variations in each channel in their early period after launch, hence the simple gain index (SGI) and normal differenced gain index (NDGI) reached their theoretical value—1 and 0, respectively.

While two Chinese satellites, i.e. CBERS 02C satellite and ZY3 satellite were launched nearly simultaneously. The result showed apparent deviation from theoretical value in their early period after launch, as shown in Tables 4 and 5.

Table 2 Absolute calibration of Pléiades 1B

Period	Pan	Blue	Green	Red
After launch	12.04	10.46	10.47	11.32
2012–2014	12.04	10.46	10.47	11.32

Table 3 Absolute calibration of SPOT 6

Period	Pan	Blue	Green	Red
After launch	10.41	8.51	9.52	10.44
2012–2014	10.41	8.51	9.52	10.44

Table 4 SGI and NDGI of CBERS 02C satellite

Channel	SGI	NDGI
Pan	0.9623	0.0192
Green	0.8909	0.0577
Red	0.9601	0.0203
NIR	0.9539	0.0236

Table 5 SGI and NDGI of ZY3 satellite

Channel	SGI	NDGI
Pan	0.9898	0.0051
Green	0.9575	0.0217
Red	0.9213	0.0410
NIR	0.9217	0.0408

In this case, precise calibration are especially needed for further quantitative remote sensing applications.

5 Summary

GF-4 satellite is world's first high resolution remote sensing satellite in GEO orbit. It has remarkable observation capabilities in spatial, temporal and spectral dimensions. After an analysis of payload and application demands, five operation modes were proposed to meet with but not limited to urgent monitoring of earthquake, floods, typhoon and forest fire. And further calibration and enhancement processing will definitely push the application of GF-4 satellite on.

Funded by CHEOS Project: 50-Y20A07-0508-15/16

References

1. General Office of the State Council of the People's Republic of China. The 11th Five-year Plan on Comprehensive Disaster Reduction [EB/OL]. http://www.gov.cn/zwgk/2007-08/14/content-716626.htm, August 2007 (in Chinese).

2. Yang Siquan, Liu Longfei, Wang Lei, et. a1. Application Study on Assessment of Snow Disaster Monitoring with Satellite for Environment Disaster Reduction [J]. Journal of Natural Disasters, 2011, 20(6):145–150. (in Chinese).
3. Qin Xianlin, Zhang Zihui, Li Zengyuan. An Automatic Forest Fires Identification Method Using HJ-1B IRS Data[J]. Remote Sensing Technology and Application, 2010,25(5):700–706. (in Chinese).
4. Yan Nana, Wu Bingfang, Li Qiangzi, et. a1. HJ-1A/B Satellite Application on Drought Emergency Monitoring[J]. Remote Sensing Technology and Application, 2010, 25(5):675–681. (in Chinese).
5. Li Tiefeng, Xu Yueren, Pan Mao, et. a1. Study on Interpretation of Rain-induced Group Shallow Landslides Based on Multi-period SPOT-5 Remote Sensing Images [J]. Acta Scientiarum Naturalium Universitatis Pekinensis, 2007, 43(2):204–210. (in Chinese).
6. Duan Guangyao, Zhao Wenji, Gong Huili. Improved Model of Regional Flood Disaster Risk Assessment Based Oil Remote Sensing Data[J]. Journal of Natural Disasters, 2012, 21(4):57 61. (in Chinese).
7. Wang Litao, Wang Shixin, Zhou Yi, et. a1. Urgent Monitoring and Analysis on Yushu Earthquake Using Remote Sensing [J]. Journal of Remote Sensing, 2010, 14(5):1060–1066. (in Chinese).
8. T. Liu. "High Resolution Optical GEO Satellite Demonstrate Potentials in Ocean Monitoring," *Satellite Application*, pp. 70–74. December 2014. (in Chinese).
9. X. Zhang, F. Lu, "Analysis on Observation Mode Development for Foreign GEO Meteorology Satellites," *Progress on Meteorological S&T*, pp. 17–22. February 2016. (in Chinese).
10. D. Wang. "Research on Basic Observation Mode for High Resolution Satellite at Geo-stationary Orbit," *Space International*, pp. 52–54. November 2015. (in Chinese).
11. G. Kang, H. Youn, S. Choi, P. Coste. "Radiometric Calibration of COMS Geostationary Ocean Color Imager," Proc. of SPIE Vol. 6361, 636112, 2006.
12. D. Wang. "A New Method to Evaluate Radiometric Performance of Satellite Image," Symposium on New Progress and New Achievement of Remote Sensing Applications in 12th Five Years. November 2014. (in Chinese).
13. B. Cutler, L. Coeurdevey. "Pléiades 1B and SPOT6 Image Quality Status after Commissioning and 1st Year in Orbit," Session#3, Spotlight Session, JACIE 2014, March 2014.
14. K. Kuenzer, S. Deck, W. Wagner. "Remote Sensing Time Series Revealing land Surface Dynamics: Status and the Pathway ahead," In: Remote sensing time series. Berlin: Springer, pp. 1–24, 2015.

Research on Digital TDI Technology for Optimizing Sequence Remote Sensing Images Applied in an Imager with Area Array CMOS Sensor

Jiuzhe Wei, Xiaoyong Wang and Changning Huang

Abstract In recent years, the industry of small remote sensing satellites with low-cost area array CMOS imaging sensors prosper, and the trend present rising. Push-frame is the suitable working mode of the satellite with an area array imaging sensor. In order to avoid motion-blur, the situation that the integration time is insufficient of this working mode become inevitable. A digital TDI technology based on the post processing of the ground station is proposed in this paper to solve the problem above. At first, histogram equalization algorithm is conducted to the single frame image sequence with low SNR (Signal to Noise Ratio) and high overlap rate in objects on ground in order to amplify the absolute value of gray scale gradient of each point in the image along each direction; and then, SIFT (Scale Invariant Feature Transformation) algorithm is adopted to extract the feature points of each single frame image in succession; feature points matching between each single frame and forecasted standard image shall also be conducted and the error matching shall be eliminated with the RANSAC (Random Sample Consensus) algorithm to improve the registration accuracy. At last, a transformational matrix shall be calculated with the sequence image according to the coordinate relations obtained after the matching; moreover, multinomial matching registration for image fusion in order to obtain a fused image whose SNR is greatly improved. Experiments show that the matching method is accurate, the brightness and SNR of fused images were significantly improved.

Keywords Digital TDI · SIFT · Feature points matching · RANSAC · Sequence images fusion

---

J. Wei · X. Wang · C. Huang
Beijing Institute of Space Mechanics and Electricity, Beijing, China

J. Wei (✉)
Shenzhou Academy of China Academy of Space Technology, Beijing, China
e-mail: 290230371@qq.com

H.P. Urbach and G. Zhang (eds.), *3rd International Symposium of Space Optical Instruments and Applications*, Springer Proceedings in Physics 192,
DOI 10.1007/978-3-319-49184-4_40

1 Introduction

With the improvement of the semiconductor technology level in recent years, CIS (CMOS Imaging Sensor) is rapidly developed [1] and the various performance indexes and image quality of the sensor are approaching or even exceeding those of a CCD device. CIS belongs to Active Pixel Sensor. The imager with CIS has the characteristics of high reliability and integration, low cost, volume, weight and power consumption benefited from the mature standard CMOS technology.

Though being fully researched and used in the commercial and business fields, many COTS spot goods with high performance and low cost can be selected; however, TDICCDs cannot be replaced by CMOS devices in the field of low orbit and high resolution earth observation at the moment, limited by the particularity of the on-orbit-imaging condition. Research on TDICMOS sensor has been ongoing in recent years, implementation of TDICMOS sensor based on the analog domain and the digital domain proposed by Michaelis [2], Lepage [3–5], Qu [6] and Tao [7–9] etc. is depend on redesigning and upgrading of imagers' internal devices or electronics systems; although TDI functions can be implemented with these methods, some significance deficiencies of TDICCDs still exist, for instance: high customization cost, high requirements for satellite platform attitude, strict match between frame frequency and ground speed etc. These factors go against the low-cost design idea of the micro satellite where the plane array CMOS imagers are desired to be adopted.

A sequence remote sensing image digital TDI technology based on the ground station processing and applied in the plane array CMOS imager is put forward in this paper. With a post-processing algorithm, the output images with a improved SNR are synthesized by original frames with a low SNR through image registration and fusion. Imager design can be simplified and the cost of the imager can be reduced by handing over the calculation work to the ground station; moreover, this technology has a good robustness on the pointing accuracy and attitude stability error of the satellite platform, which greatly decreases the anti-requirement of the imager on the satellite platform. During the early synthesis stage, the adjusting function of a gray level transformation curve is adopted to improve the adaptability specific different illumination conditions and various features of ground objects. The improving performance of this technology on the image SNR is verified in the experiment, compared with theoretical improvement.

2 Technical Scheme

When a area array CMOS imager working on the push-frame mode which is successively framing imaging, the situation of insufficient single frame integration time will be encountered to avoid motion blur, which will cause low SNR, low brightness and poor image quality to the single frame. The digital TDI technology

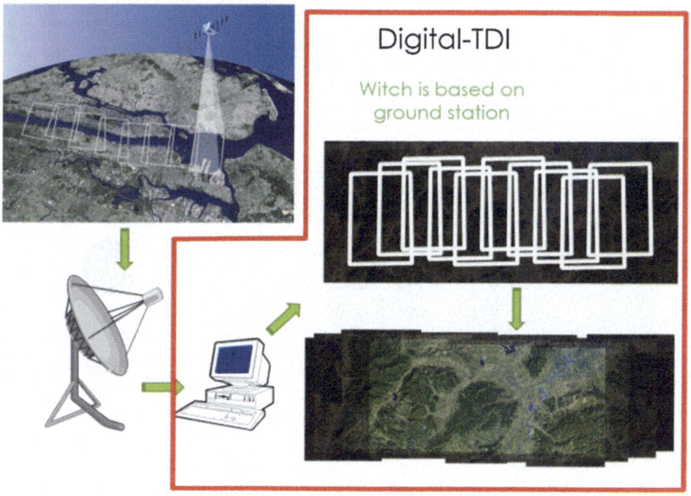

Fig. 1 Digital TDI technology process

proposed in this paper is a digital domain signal accumulation process based on groundstation processing adopted to improve the SNR and brightness of the successively framing imaging, which aims to fuse single frames with low SNR to obtain a synthetic image with improved SNR. The process of digital TDI is shown in Fig. 1.

Differing with traditional simulation domain TDI technology for electric charge accumulates the digital information of the image in the digital domain after the digital information is generated by steps of remote sensing information reading out and enlarging, A/D conversion as well as compressing and downloading etc. in the imager. As this technology is independent of the customized imager hardware, late fusion can be conducted with the presence of the original frame with a certain overlapping ratio of ground object, therefore, it is especially suitable to be applied in the low-cost light and small remote sensing load where the plane array CMOS detector is adopted. The digital TDI diagram discussed in this paper is given below.

When the sequence remote sensing image is processed, as shown in Fig. 2, displacement, rotation and even deformation of the ground object relative to the focal plane will appear; the phenomenon of change in phase position of the ground object is related with factors such as the visual angle of the imager, lens distortion and light conditions on the ground etc. As a result, matching and registration shall be done to the remote sensing image sequence during the technical construction period. The specific method is as follows: with the existing remote sensing data model as the data source, parameters such as imager on-orbit and time etc. shall be adopted to provide a forecast reference frame with the same ground characteristic target for the remote sensing sequence image. All original frames are converted with the forecast frame as reference for rotation and affinity etc., and finally, the converted original frames are fused together.

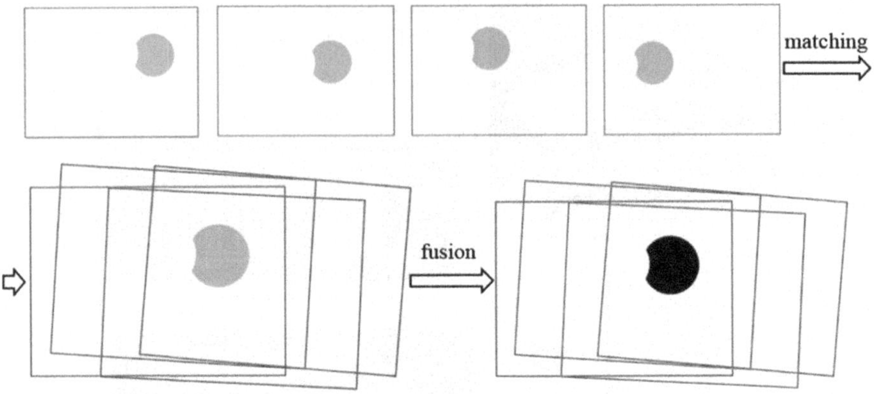

Fig. 2 Digital TDI diagram

Due to insufficient integration time, the brightness of the original frame is always not high, with low SNR, low overall gray level and centralized distribution. Histogram equalization is applied to the original frame and the scope for the gray level of the image is expanded to improve the efficiency in identifying the feature points. During the image matching period, SIFT feature extraction method is adopted in this research [10] as SIFT is stable to phase position change and affine transformation among different frames and it is also adaptive to different light conditions and visual angles; therefore, it is a method applicable for this technology.

3 SNR Analysis

SNR is an important index for measuring the quality of the image as well as one of the main evaluation indicators for TDI imagers. If the space domain noise caused by heterogeneity is not temporarily considered, the time domain noise in the system mainly comes from the background fluctuation in the detector system, internal noise from the sensor and noise from the reading circuit. As these noises are random processes independent and irrelevant with each other, the total mean squared noise of the system is the sum of the mean squared noise of various sources of noise:

$$V_{noise} = \sqrt{V_{shot}^2 + V_{inner}^2 + V_{readout}^2} \tag{1}$$

In the formula above: V_{shot}, V_{inner} and $V_{readout}$ represent the background fluctuation in the detector system, internal noise from the sensor and noise from the reading circuit. As each original frame in the digital TDI system with M overlapped frames is synthesized after being read, an accumulation effect is caused during the noise reading process. SNR of the system can be represented as below:

$$SNR_{TDI} = \frac{V_{signal-TDI}}{V_{noise\text{-}TDI}} \tag{2}$$

$$= \frac{\sqrt{M} \cdot V_{signal}}{\sqrt{V_{shot}^2 + V_{inner}^2 + V_{readout}^2}} \tag{3}$$

It is clear that the SNR of digital TDI input image to the SNR of the original frame is in direct proportion to $\sqrt{M}$.

4 Framework of the Algorithm

Original frame image sequence with a low SNR, $\{LS_k\}$, is obtained from successively framing shooting the same scenery with an imager; the imaging phase of the scenery in each frame is different with each other so that the overlapping ratio k can be assured. With the hope of applying digital TDI algorithm to the original frame sequence, the image of the captured scenery with a high SNR is fused out. The reference frame HS_0 obtained from orbit prediction is an image captured by a high-performance remote sensing satellite with the function of geographical position calibration. HS_0 with high image quality which is input as pre-knowledge after down sampling provides a benchmark for treatment such as sequence $\{LS_k\}$'s frame-frame registration, geographical position calibration and distortion elimination etc. With steps of SIFT feature extraction, feature matching, RANSAC matching filter and transformational matrix calculation etc., the original frame images are registered and accumulated according to the ground object's coordinate system of the reference frame HS_0. The flow chart of the algorithm is shown in Fig. 3.

This technology mainly consists of three parts: (1) feature point extraction with the SIFT algorithm; (2) feature matching and RANSAC algorithm screening (3) frame-frame registration and fusion.

A. *Feature points extraction with the SIFT algorithm*

The algorithm of SIFT (scale invariant feature transform) extracts local features of the image, with rotation, scale-zooming and affine transformation unchanged, changes in noises, perspectives and illumination variation stabilized; theoretically, it is suitable to be adopted as the feature extraction algorithm for digital TDI technology in this paper. By extracting the local features of the images, SIFT algorithm seeks for the extreme points, extraction points, scales and rotation invariants within the multi-scale-space. The SIFT algorithm consists of four steps [10]:

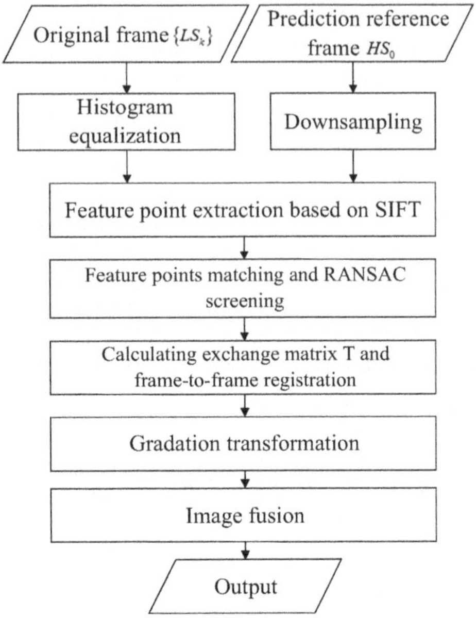

Fig. 3 Algorithm flow chart

(1) Constructing a scale space. Gauss convolution kernel is the only linear kernel to achieve the scale transformation, so a two-dimensional scale variable Gauss convolution kernel is selected to calculate the convolution with the original image.

$$L(x, y, \sigma) = G(x, y, \sigma) \cdot I(x, y) \tag{4}$$

$$G(x, y, \sigma) = \frac{1}{2\pi\sigma^2} e^{-(x^2 + y^2)/2\sigma^2} \tag{5}$$

Among above, $I(x, y)$ is an original image, (x, y) is spatial coordinates, σ is scale-space coordinates. The difference of Gaussian (DOG) scale-operator is used to build multi-scale space pyramid shown in Fig. 4.

(2) Feature points positioning. Compare between each sampling point and its adjacent points in a multi-scale space. Compare a sampling point with 8 adjacent points in same scale space and corresponding 2×9 points belong to the upper and lower adjacent scale spaces, here is totally 26 points, to find the extreme points of scale spaces and image spaces as image feature points. In order to get steady feature points, scale DOG function must be fitted, and remove the edge points generated by the edge of the DOG operator.
(3) Feature points' main direction determination. The gradient of each key point is calculated, and the gradients and directions of pixels in the neighbourhood are expressed in a histogram. Directions range of 0°–360° are divided into

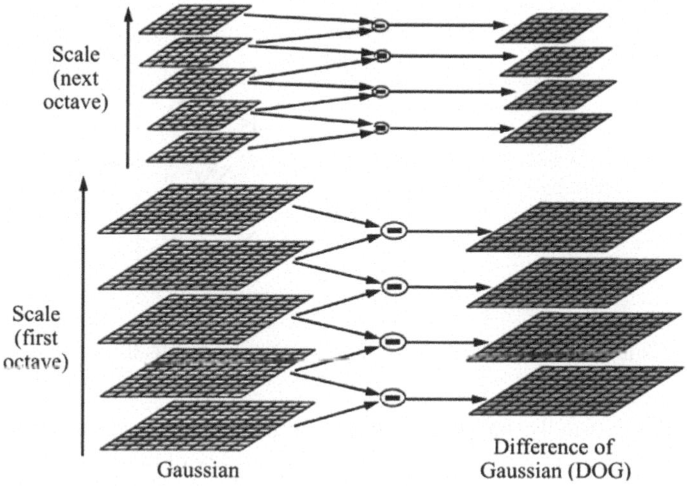

Fig. 4 Multi-scale space

36 columns. The peak direction of the histogram is used as the main direction of the feature points.

(4) Feature points description. For each feature point, there are three descriptions: location, size, and direction. The purpose of feature points description is to describe the feature points with a unique feature vector. Lowe proposed use eight directions of gradient information calculated in a 4 × 4 window which is in the scale space of the feature point to be the descriptor, which is a total of 4 × 4 × 8 = 128 dimensional vector representation.

B. *Image matching and RANSAC screening*

In this paper, the method of feature point vector included angle cosine is adopted to calculate and match the similarity of 128-dimensional feature descriptor vector which is extracted with the SIFT algorithm. Specifically, the feature descriptor vector included angle between P_k, all feature points in the image to be matched, and P_0, the feature point in the reference frame is calculated with the inverse cosine function.

$$\theta(P_0, P_k) = \arccos(\overrightarrow{des}_0 \cdot \overrightarrow{des}_k) \tag{6}$$

$\overrightarrow{des}_0$ and $\overrightarrow{des}_k$ are 128 dimensional feature descriptor vectors of normalized P_0 and P_k. Then the 128 included angles are sorted in sequence; the minimum included angle to second minimum included angle ratio is less than a certain threshold value. It is determined that P_0 and P_k are a pair of matching feature points; generally, the threshold is generally selected within 0.4–0.6 [11]; the smaller the selected threshold value is, the matching is deemed to be more strict.

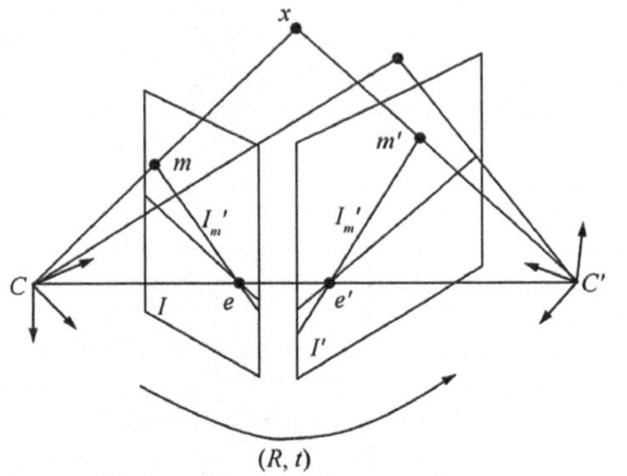

Fig. 5 Epipolar constraint

Mismatching usually happens during the feature point registration process, which will exert a great positive influence on the quality of output results; therefore, it is necessary to screen the matching points after the matching process. The random sampling consistency (RANSAC) algorithm, with which the mathematical model parameters of data are calculated according to a set of the sample data bank containing abnormal data to obtain effective sample data, is put forward by Fischler [12] in the year 1981.

$$X'^{T}FX = [x' \quad y' \quad z'] \begin{bmatrix} h_1 & h_2 & h_3 \\ h_4 & h_5 & h_6 \\ h_7 & h_8 & h_9 \end{bmatrix} \begin{bmatrix} x \\ y \\ 1 \end{bmatrix} = 0 \qquad (7)$$

In this paper, the method of epipolar constraint is used to select the feature matching points, as shown in Fig. 5. The process of the research in this paper is given below: (1) randomly selecting 8 pairs form the set of matching point pairs and substituting into the formula to calculate the initial value of *F* called basis matrix; (2) substituting the matching point pair (P_0, P_K) into the formula according to the value of *F*; if the error is less than the threshold value e, it is deemed as a pair of effective matching and this pair of matching points is included into the set of effective matching points; and the process repeats again after the value of *F* is recalculated; (3) selecting the largest consistent and effective matching points with the maximum point pairs in the set as the final effective matching point after N times of random samplings.

C. *Sequence image registration and fusion*

Images in the original frame sequence $\{LS_k\}$ are registered with the down-sampling standard frame HS_0 successively. The registration matrix T_k is calculated with the coordinate corresponding relation between the matching features of the original frame and the standard frame as the input condition. The fusion process can be conducted after each image in the original frame sequence $\{LS_k\}$ is incorporated into the standard frame HS_0 according to its own registration matrix T_k.

As the luminance difference of the ground object might be great and the number of overlapped image frames of a certain ground object captured by an orbiting imager is hard to change, therefore a gray level transformation function is incorporated before the fusion accumulation algorithm so as to meet the imaging requirements of ground object under a different luminance and further meet with the appropriate observation requirements.

5 Experimental and Analysis

The effect of the algorithm in this paper is verified in the experiment. In this experiment, the effect of the remote sensing satellite on orbit imaging is simulated by the use of window matrix, which is used to generate the sequence of the original frames. The selected experimental parameters of the simulation remote sensing satellite's orbit and performance are shown in Table 1.

The generation of the original frame to be processed is divided into the following steps.

Step.1. The size of the single frame going to be sampled and the downsampling rate of the single frame are calculated by the orbit and the performance parameters of the simulated satellite. By the known of the size of a single frame and downsampling rate, window matrix could be constructed,

Step.2. Separation distance between two frames should be calculated by the orbit altitude and the imager's frame frequency. The random error function was constructed by Gauss random number to characterize the relative displacement error of push frame to verify the robustness of the matching algorithm. The relative displacement between two frames is shown in the following:

$$V_E = \sqrt{\frac{G \cdot M}{R_E + H}} \tag{8}$$

$$d_{r-x} = f_c \cdot V_E + r_x \tag{9}$$

$$d_{r-y} = r_y \tag{10}$$

Tab.1 Parameters of the simulated imaging system

Parameter type	Orbit altitude	Standard GSD	Sampled GSD	Breadth of a frame	Frame frequency of imaging	Bit depth
Parameter value	500 km	0.5 m	1 m	1200 m × 1600 m	50 fps	8 bit

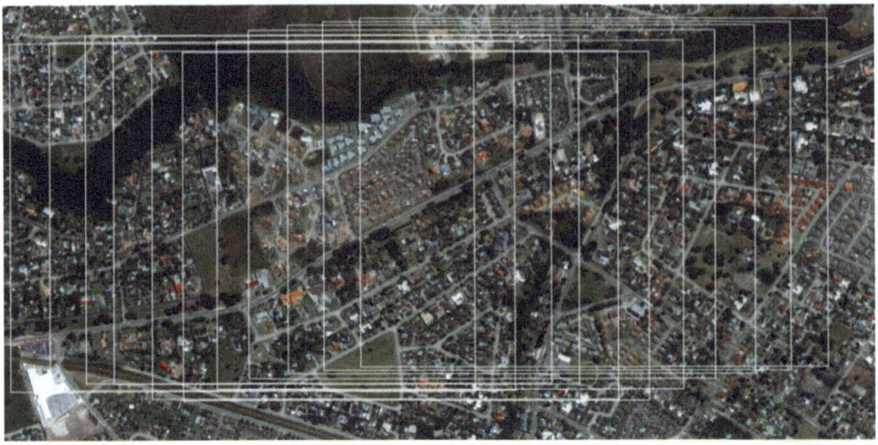

Fig. 6 Sampling windows of the spatial domain

In formulas above, V_E means the approximate relative velocity between ground and the sub-satellite points of the simulation satellites; $G \cdot M$ means the gravitational constant the earth; R_E means the radius of the earth; H means the orbit altitude of simulation satellite; d_{r-x} means the frame-frame relative displacement in ground coordinates along the push frame direction; d_{r-y} means the frame-frame relative displacement in ground coordinates along the vertical direction of push frame; f_c means the imager's frame rate; r_x means a displacement diffusion value along the push frame direction; r_y means a displacement diffusion value along the vertical direction of push frame. Figure 6 means sampling windows of the spatial domain.

Step.3. Reduce brightness of images, and add gauss white noise. Reduced the SNR to about 32 dB.

The sequence of original frames $\{LS_k\}$ going to be optimized obtained by above steps are shown below (Fig. 7)

The referential standard frame HS_0 is the original image with high quality, as shown in Fig. 8. After calculate by substituting it into the algorithm routine and screen, the feature points' matching relationship are as shown in Fig. 9; after digital TDI synthesis to the image of 9 original frames, the input effect is shown in Fig. 10.

The same area of the images with different TDI stages are picked out and compared. As shown in Fig. 11.

PSNR (peak signal to noise ratios) is the most common, the most widely used method of evaluation quality measurement. It is the logarithm between square of the value of the maximum signal and the MSE (mean square error) between the original image and the image to be processed. As formula (11) shown.

Fig. 7 Sequence of original frames $\{LS_k\}$

$$PSNR = 10 \times \log_{10}\left(\frac{255^2}{MSE}\right) \tag{11}$$

Calculating level by level the PSNR of each single original frame and level 8 digital TDI overlapped images by selecting 4 groups of different area, the calculation results are shown in Fig. 12. Judging by the calculation results, it is clear that PSNRs are decreased level by level as the level of fusion increases and that the increasing numerical relationship agrees with the theoretical calculation results.

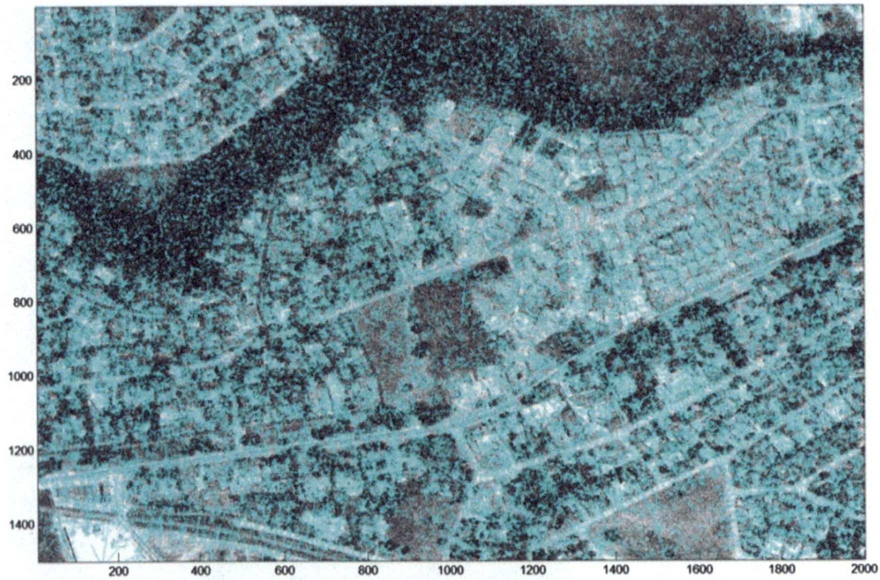

Fig. 8 Referential standard frame and the feature point extraction diagram

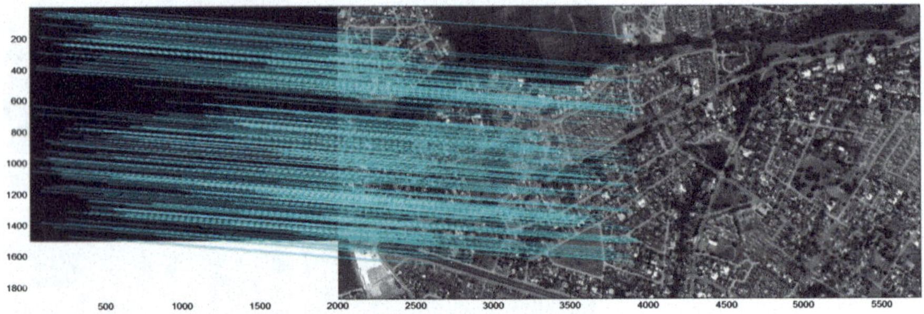

Fig. 9 Effect of feature point matching

6 Summary

This paper explained Digital-TDI technology used in low-cost CMOS imager. This technology is based on the post processing of the ground station, but not rely on special hardware design. Based on SIFT feature recognition and frame filtrated by RANSAC Frame matching method to phase change and affine transformation between different frames guaranteed stability. Therefore, the technique can be adapted to satellite platform of different attitude stability index. By experimentation, images matching accurately. Its effect of improving SNR accord with the theoretical calculation results.

Fig. 10 Synthetic effect of digital TDI

Fig. 11 Comparison of different stages

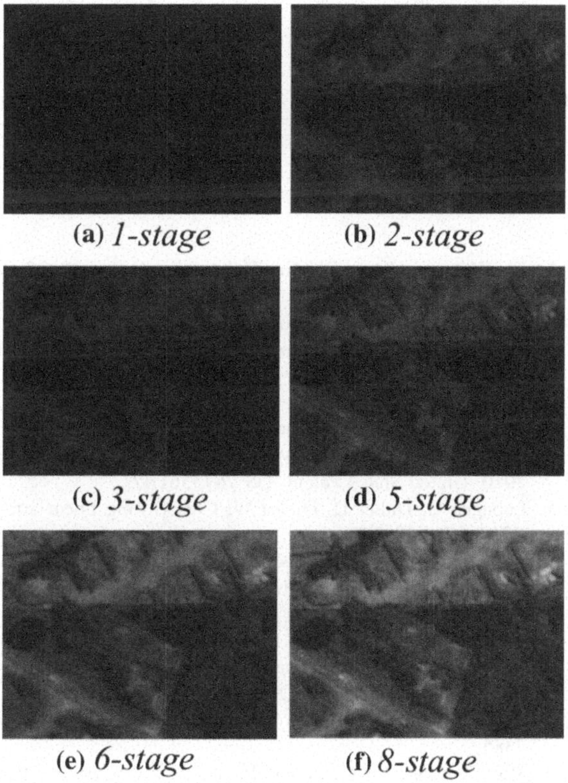

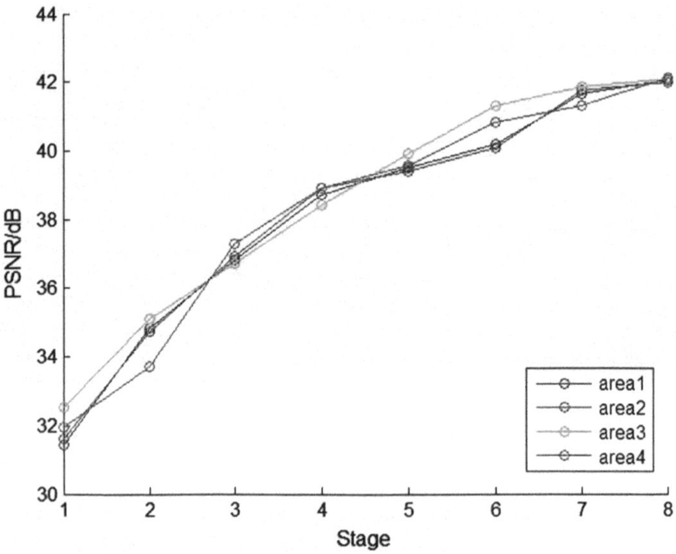

Fig. 12 Relationship between PSNR value and the digital TDI level

References

1. Liu Zhiyong, Huang Qiaolin. The Application of CMOS Camera in Small Satellites [J]. 2003. Spacecraft Recovery & Remote Sensing, (02):29–33.
2. Harald Michaelis, Ralf Jaumann, Stefano Mottola, *et al.*, 2005. "CMOS-APS sensor with TDI for high resolution Planetary Remote sensing", 2005 IEEE CCD&AIS workshop, 2005-4-25:31–34.
3. Lepage G, Bogaerts J, Meynants G. Time-delay-integration architectures in CMOS image sensors. 2009. IEEE Transactions on Electron Devices, 56(11): 2524–2533.
4. Gerald Lepage, Time delayed integration CMOS image sensor with zero desynchronization. 2010. United States Patent, US 7675561B2.
5. Lepagea G, Didier D, Dielsa W. CMOS long linear array for space application [C]. 2006. *SPIE*, 6068:606807.
6. Qu Hongsong, Zhang Ye, Jin Guang. Improvement of performance for CMOS area image sensors by TDI algorithm in digital domain [J]. 2010. *Optics and Precision Engineering*. 18 (8): 1896–1903.
7. Tao Shuping, Jin Guang. Influence Analysis on the Rolling Shutter for Time Delay and Integration in Digital Domain [J]. 2012. *Acta Optica Sinica*. 04:93–101.
8. Tao Shuping, Jin Guang, Qu Hongsong, Zhang Guixiang. Design of CMOS imaging system based on rolling TDI in digital domain [J]. *Infrared and Laser Engineering*, 2012, 09:2380–2385.
9. Tao Shuping, Jin Guang. Influence Analysis on the Rolling Shutter for Time Delay and Integration in Digital Domain. 2015. *Acta Optica Sinica*. 03:107–112.
10. Lowe D G. Distinctive image features from scale-invariant keypoints. 2004. International Journal of Computer Vision. 60(2): 91–110.

11. Fischler M A, Bolles R C. Random sample consensus: aparadigm for model fitting with applications to image analysis and automated cartography. 1981. Communications of Association for Computing Machinery. 24(6): 381–395.
12. Lowe D G. SIFT demo program (version 4) [Online], available: http://www.cs.ubc.ca/lowe/keypoints/, July 1, 2005.

A Method of Coastline Detection from High-Resolution Remote Sensing Images Based on the Improved Snake Model

Xing Kun, Zhang Bing-xian and He Hong-yan

Abstract While executing tasks such as ocean pollution monitoring, maritime rescue, geographic mapping, ship localization, cruise missile guidance and automatic navigation utilizing remote sensing images, the coastline feature should be determined. Traditional methods are not satisfactory to extract coastline in high-resolution panchromatic remote sensing image. Active contour model, also called snakes, have proven useful for interactive specification of image contours, so it is used as an effective coastlines extraction technique. Firstly, coastlines are detected by water segmentation and boundary tracking, which are considered initial contours to be optimized through active contour model. As better energy functions are developed, the power assist of snakes becomes effective. New internal energy has been done to reduce problems caused by convergence to local minima, and new external energy can greatly enlarge the capture region around features of interest. After normalization processing, energies are iterated using greedy algorithm to accelerate convergence rate. The experimental results encompassed examples in images and demonstrated the capabilities and efficiencies of the improvement.

Keywords Remote sensing images · Coastline extraction · Active contour model · Greedy algorithm

1 Introduction

Extraction of coastline from remote sensing image is of great significance for navigation, marine rescue, marine pollution monitoring, shoreline mapping, ship positioning, spacecraft position and attitude control, remote sensing image registration and cruise missile guidance. With the advent of ocean observing satellites, such as SIR-B, ESR-1 and ESR-2, coastline detection technology from SAR image is rapidly becoming a hot field in remote sensing image processing [1–3], but

X. Kun (✉) · Z. Bing-xian · H. Hong-yan
Beijing Institute of Space Mechanics and Electricity, 100094 Beijing, China
e-mail: xingkunfeixiang@aliyun.com

H.P. Urbach and G. Zhang (eds.), *3rd International Symposium of Space Optical Instruments and Applications*, Springer Proceedings in Physics 192,
DOI 10.1007/978-3-319-49184-4_41

research works about visible image are few [4–6]. There are two ways to detect coastlines: one is low level image processing technique that smoothing, segmentation and tracking are applied, the other is high level image processing technique such as multi-resolution and wavelet transform. These methods are usually used for SAR images. Unlike SAR images, visible images are formed with higher resolution, no speckle noises, large amounts of data and the gray level of water are closer to the gray level of land. So the above-mentioned methods are not strong reliability and accuracy for visible images. This paper presents a solution to the problem of extracting coastlines accurately and rapidly in high-resolution panchromatic remote sensing images. Data are from one high-resolution satellite.

Active contour models, also called snakes, were proposed by Kass et al. [7], and since then have been successfully applied in a variety of problems in computer vision and image analysis, such as edge and subjective contours detection, motion tracking and segmentation [8]. There are several key difficulties with original snakes. Many works has been proposed to improve snake models: Balloon model [9], GVF Snake [10], greedy algorithm for snake [11], ACO Snake [12] and Loop Snake [13]. However, most of the methods proposed to address problems solve only one problem while creating new difficulties. For example, GVF Snake have addressed the issue of boundary concavities, but costing a huge waste of time because of large number of GVF iterations. Another example is that of pressure forces, which can push an active contour into boundary concavities, but cannot be too strong or "weak" edges will be overwhelmed. Pressure forces must also be initialized to push out or push in, a condition that mandates careful initialization. The basic goal of this paper is to present the snake model as an effective coastline extraction technique. In order to improve motion performance, we develop better energy functions. Experiments confirm the rationality and validity.

2 Snake Model

In the original snake formulation, the best snake position was defined as the solution of a variation problem requiring the minimization of the sum of internal and external energies integrated along the length of the snake. The corresponding Euler equations, which give the necessary conditions for this minimizer, comprise a force balance equation. By introducing a temporal parameter, the force balance equation can be made dynamic. When the dynamic equation reaches its steady state, a solution to the static problem is found. We now give a brief summary of these steps. Snake is a curve that moves through the spatial domain of an image to minimize the energy functional

$$v(s) = (x(s), y(s)), s \in [0, 1] \tag{1}$$

Given a group of control points in an orderly manner $v_i = (x_i, y_i)(i = 1, 2, \cdots, n)$. The discrete form of energy functional is

$$E_{snake} = E_{\text{int}}(v) + E_{ext}(v) \tag{2}$$

where $E(v) = \sum_{i=1}^{n} E(v_i)$. The internal force E_{int} includes the first and second derivatives of v_i with respect to (x_i, y_i), called $E_{elastic}$ and $E_{bending}$.

$$E_{\text{int}}(v_i) = \alpha E_{elastic}(v_i) + \beta E_{bending}(v_i) \tag{3}$$

where α and β are weighting parameters that control the snake's tension and rigidity, respectively. The external energy E_{ext} function is derived from the image so that it takes on its smaller values at the features of interest, such as boundaries. Given a gray-level image $I(x, y)$, typical external energies designed to lead an active contour toward step edges are:

$$E_{ext} = -|\nabla I(x, y)|^2 \tag{4}$$

$$E_{ext} = -|\nabla[G_\sigma(x, y) * I(x, y)]|^2 \tag{5}$$

where $G_\sigma(x, y)$ is a two-dimensional Gaussian function with standard deviation σ, ∇ is the gradient operator, * is convolution. The internal force E_{int} discourages stretching and bending while the external force E_{ext} pulls the snake towards the desired image contour.

Its mathematical formulation makes easier to integrate image data, an initial estimated, desired contour properties and knowledge-based constraints, in a single extraction process. However, Snakes have also their limitations. First, the method can only handle topologically simple objects. The topology of the structures of interest must be known in advance since the mathematical model cannot deal with topological changes without adding extra machinery. Second, capture region is too small for the gradient of external force, which makes contours do not converge to the real edge. Third, snakes are too sensible to their initial conditions due to the nonconvexity of the energy functional and the contraction force which arises from the internal energy term. Several works have been done to address these limitations. However, the utility of such techniques is limited by performance problems. In the present paper, the method of setting initial coastlines using water segmentation addresses the topological limitations. In further steps, new internal and external energies are designed to reduce the bad effects of the contraction force and address the nonconvexity problem.

3 Initial Coastline Extraction

As can be seen from remote sensing images, the water areas are large and uniform gray scale, which show contrast with land areas. There are many approaches for image segmentation. After experimental verification, fuzzy threshold segmentation is used for pretreatment in this paper [14].

The original image is defined as X with size $M \times N$ and gray level is L. x_{mn} is the gray value of pixel (m, n). We use $\mu_X(x_{mn})$ to describe one membership value with some attribute. Image X can be represented as

$$X = \{(x_{mn}, \mu_X(x_{mn}))\} \tag{6}$$

where $0 \leq \mu_X(x_{mn}) \leq 1$, $m = 0, 1, \ldots, M-1$, $n = 0, 1, \ldots, N-1$. Membership value $\mu_x(x_{mn})$ expresses fuzzy attribute of (m, n). Supposing that gray threshold is t, $\mu_x(x_{mn})$ is defined for each pixel point as follows

$$\mu_X(x_{mn}) = \begin{cases} \frac{1}{1+|x_{mn}-\mu_0|/C}, x_{mn} \leq t \\ \frac{1}{1+|x_{mn}-\mu_1|/C}, x_{mn} > t \end{cases} \tag{7}$$

C is given by $\frac{1}{2} \leq \mu_X(x_{mn}) \leq 1$, we set $C = 255$. μ_0 is average gray value of background ($x_{mn} > t$). μ_1 is average gray value of target ($x_{mn} \leq t$). Then fuzzy membership is

$$E(X, A) = \frac{1}{MN} \sum_{m=0}^{M-1} \sum_{n=0}^{N-1} (2 \times \mu_X(x_{mn}) - 1) \tag{8}$$

where $0 < E(X, A) < 1$, we calculate $E(X, A)$ according to t. The t corresponding to the maximum of all $E(X, A)$ is the optimal threshold.

Waters are extracted preliminarily after segmentation, but there are still some noises. Basically, morphological operation is first used to smooth water's edge. In further steps, pseudo waters are removed using area threshold elimination according to area size [15]. Waters is the largest connected region in image, so other regions are lands. Then initial coastlines are extracted by using boundary tracking, which structures a closed curve connecting with image border. The control points of initial snake model are determined by choosing one point from adjacent three in the closed curve.

4 Improved Snake Model

4.1 Internal Energy

Kass used first order difference to estimate elastic force

$$E_{elastic}(v_i) = \left|\frac{dv_i}{ds}\right|^2 \approx |v_i - v_{i-1}|^2 \tag{9}$$

This approach basically causes the curve to shrink, as this is actually minimizing the distance between points. It also contributes to the problem of points bunching up on strong portions of the contour. Donna et al. [11] used another approach that $\bar{d} - |v_i - v_{i-1}|$ approximated $E_{elastic}$ through calculating the average distance $\bar{d}$ between control points. However, the average distance prevents snake points from moving. The utility of such techniques is limited. A new approach is proposed to reduce problems caused by convergence to local minima. v_i is a point on the contour. The point before and after it are v_{i-1} and v_{i+1}. Let l be the vertical through the midpoint between v_{i-1} and v_{i+1}. The elastic force is

$$E_{elastic}(v_i) = \left|v_i - v_i^l\right|^2 \tag{10}$$

where v_i^l is the point on l, as shown in Fig. 1. The approach not only removes the undesired contraction force but also retains the characteristics of equal distance distribution between control points.

Let M be the size of neighborhood of v_i at each iteration, and v_{ij} be the new control point. In order to balance the effects of each force, the value is normalized by dividing by the largest value in [0,1].

$$E_{elastic}(v_i) = \frac{\left|v_{ij} - v_i^l\right|^2}{\max\limits_{1 \le j \le M}\left\{\left|v_{ij} - v_i^l\right|^2\right\}} \tag{11}$$

Bending force using second-order central difference is given by

$$E_{bending}(v_i) = |v_{i-1} - 2v_i + v_{i+1}|^2 \tag{12}$$

Normalized form is given by

$$E_{bending}(v_i) = \frac{\left|v_{i-1} - 2v_{ij} + v_{i+1}\right|^2}{\max\limits_{1 \le j \le M}\left\{\left|v_{i-1} - 2v_{ij} + v_{i+1}\right|^2\right\}} \tag{13}$$

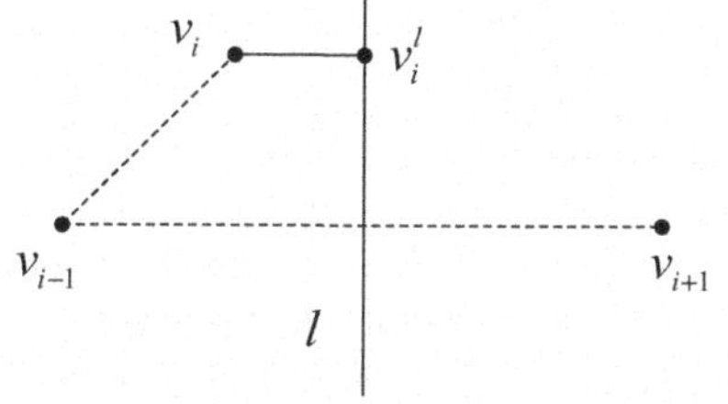

Fig. 1 Elastic force estimation approach

4.2 External Energy

It is very simple that use of the gradient magnitude as the external energy of model. However, the capture region of such techniques is limited by performance problems. If the capture region of gradient is enlarged by Gaussian smoothing, the details of edges in image will be fuzzy, which lead to accuracy decline. In Cohen [9] and Xu [10] et al. this problem is addressed by the addition of another external force term to reduce the bad effects. In both these works the computational complexity are increased and there are some trade-offs between efficiency and performance.

In order to reduce computational complexity, only a strip region centered with initial coastline is processed in the following steps. The strip region is based on snake-centered windows, which covers a small part of the total image, so running time is reduced observably. Windows are set according to requirement, such as 10×10 pixels. We combine the qualities of a good local edge detector like the Canny-Deriche extractor with a global active model. Canny edge detector is used only in the strip region of original image. We use the binary edge image to enlarge capture region of attraction forces. For each point (x, y) on snake model, (x^*, y^*) is defined as the nearest edge point in the direction of snake curves outer normal vector.

$$d(x,y) = \sqrt{(x - x^*)^2 + (y - y^*)^2} \tag{14}$$

The external energy E_{ext} is

$$E_{ext}(v_i) = k_2 \exp(-k_1 \cdot d(v_i))N(v_i) \tag{15}$$

where k_1 and k_2 are coefficient values. $N(v_i)$ is unit normal vector of v_i on snake curves.

4.3 Greedy Algorithm

In this section a greedy algorithm will be presented which allows a contour with controlled first and second order continuity to converge on an area of high image energy, in this case edges. It combines speed, flexibility, and simplicity [11]. It was compared to the original variational calculus method of Kass et al. [7] and the dynamic programming method developed by Amini ea al. [16] and found to be comparable in final results, while being faster than dynamic programming and more stable and flexible for including hard constraints than the variational calculus approach. This algorithm allows the inclusion of hard constraints as described by

Fig. 2 Greedy algorithm schematic diagram

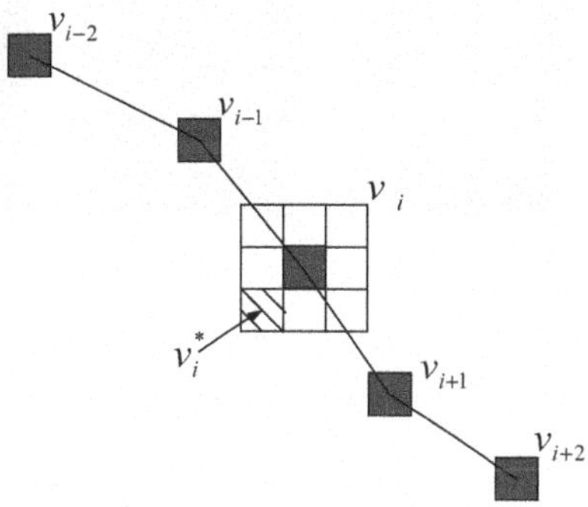

Amini et al. [16] but is much faster than their O(nm3) algorithm, being O(nm) for a contour having n points which are allowed to move to any point in a neighborhood of size m at each iteration. Figure 2 demonstrates how the algorithm works. The energy function is computed for the current location of v_i and each of its neighbors. The location having the smallest value v_i^* is chosen as the new position of v_i. v_{i-1} has already been moved to its new position during the current iteration. Its location is used with that of each of the proposed locations for v_i to compute the force term. The location of v_{i+1} has not yet been moved. Its location, along with that of v_{i-1}, is used to compute the force for each point in the neighborhood of v_i. For $i = 0$, only old values are used. For this reason v_0 is processed twice, one as the first point in the list, and once as the last point. This helps make its behavior more like that of the other points. This iterative process will stop when the number of moved points in one iterative process is less than the threshold.

5 Results and Discussion

5.1 Comparison of External Energy

In order to demonstrate the performance of external energy described in the previous section, results were given for the Sobel gradient, for the GVF (Xu et al.), for the distance potential map (Cohen et al.), and for the algorithm developed by the authors. These programs were run on one synthetic image, which was 128 × 128 pixels. Figure 3 shows the results of comparison by external energy fields. For the capture region, distance potential map and our algorithm were much larger than other algorithms. Table 1 gives the runtimes in microsecond. Using our algorithm, the speedup over GVF was observable. The gradient algorithm had the best runtime, but it had the smallest capture region. The results of distance potential map were at least as good as our algorithm, and runtime was much better, but it did not address the issue of boundary concavities.

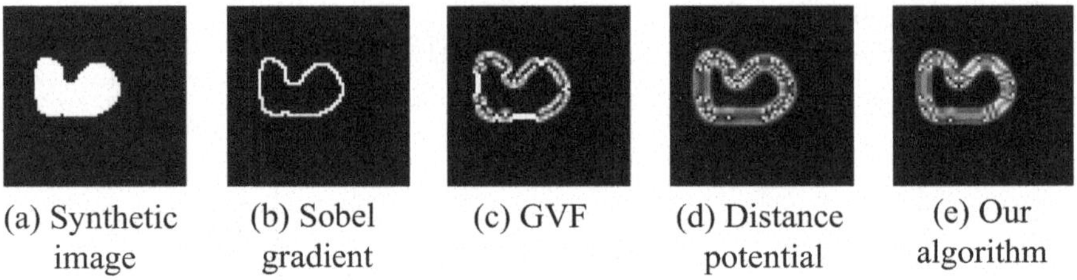

Fig. 3 Comparison of external energy fields

Table 1 Comparison of runtime by external energy methods

Algorithm	Sobel gradient	GVF	Distance potential	Our algorithm
Runtime (ms)	0.1	38	11	15

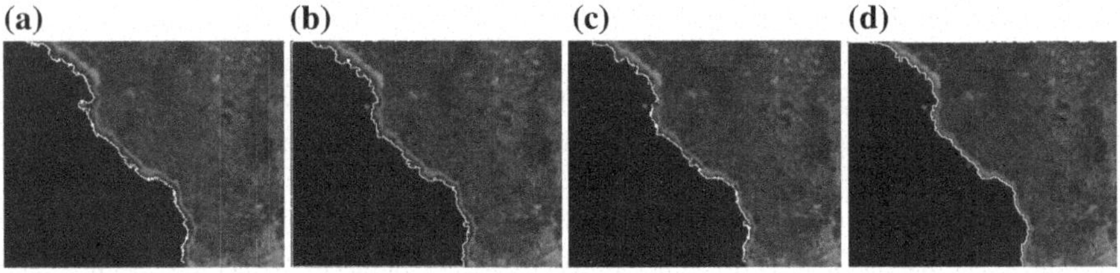

Fig. 4 Performance analysis of snake model

5.2 *Performance Analysis*

One real image tested the behavior of the algorithms when the contour spanned a region in Fig. 4 (600×500 pixels). White curves expressed the extracted coastlines. The initial coastline contour was produced by waters segmentation and boundary tracking in part (a), which is not very accurate. Part (b) shows the result of allowing the initial contour to converge to the coastlines using the variational calculus method proposed by Kass et al. Part (c) shows the result of using distance potential map as external energy. In order to enlarge capture region of the real edges, the force field was used to attract snakes. But the snakes were rigid due to the contraction force which arose from variational calculus of the internal energy term. Note that the edge points were more closely spaced on the strong portions of the contour. Part (d) shows the result of using new internal and external energies. The results achieved by all three of the methods presented were comparable.

6 Conclusion

Snakes have proven useful for interactive specification of image contours. The basic goal of this course is to present the snake model as an effective coastlines extraction technique as well as to demonstrate the capabilities and efficiencies of the improvement. Also, we aim to show some challenges in the field of snake models to motivate the research in this field. As we develop new energy functions, the power assist of snakes becomes increasingly effective. Internal energy has been done to reduce problems caused by convergence to local minima, and new external energy can greatly enlarge the capture region around features of interest. The experimental results encompass examples in images. The contours extracted by the method can accurately reflect the structure information of coastlines. The next problem we want to solve is that the runtime should be reduced.

References

1. Niedermeier A, Romaneeen E, Lehner S. Detection of coastlines in SAR Images using wavelet method. *IEEE Trans. on Geoscience and Remote Sensing*, 2000. 38(5): 2270–2281.
2. Karantzalos K G, Argialas D, Georgopoulos A. Towards automatic detection of coastlines from satellite imagery. In: Proceedings of 14th International Conference on Digital Signal Processing, Santorini, Greece, 2002(2). 897–900.
3. Lu Liming, Wang Runsheng, Li Wugao. A method of coastline extraction from synthetic aperture radar raw-data. *Journal of Software*, 2004, 16(6): 531–53.
4. Frazier P S, Page K J, Water body detection and delineation with Landsat TM data. Photogrammetric. *Engineering and remote sensing*, 2000, 66(12): 1461–1467.
5. Kevin W, Elasmar H M. Monitoring changing position of coastlines using thematic mapper imagery, an example from the Nile Delta. *Geomorphology*, 1999, 29(1–2): 93–105.
6. Jing Hao, Chen Xuequan, Gu Zhiwei. A method for coastline extraction based on edges. *Computer simulation*, 2006, 23(8): 89–93.
7. Kass M, Witkin A, Terzopoulos D. Snakes: Active contour models. *International Journal of Computer Vision*, 1987, 1(4): 321–331.
8. Li Peihua, Zhang Tianwen. Review on active contour model(Snake model). *Journal of Software*, 2000, 11(6):751–757.
9. Cohen L D, Cohen I. Finite-element methods for active contour models and balloons for 2-D and 3-D images. *IEEE Trans. on Pattern Analysis and Machine Intelligence*, 1993, 15(11): 1131–1147.
10. Xu C, Prince J L. Snakes, shapes, and gradient vector flow. *IEEE Trans. on Image Processing*, 1998, 7(3): 359–369.
11. Williams D J, Shah M. A fast algorithm for active contours and curvature estimation. *CVGIP: Image Understanding*, 1992, 55(1): 14–26.
12. Wang X N, Feng Y J, Feng Z R. Ant colony optimization for image segmentation. In: Proceedings of the 4th International Conference on Machine Learning and Cybernetics, Guangzhou, China, 2005, 5355–5360.
13. Oliveira A, Ribeiro S, Esperanca C, Giraldi G. Loop snakes: the generalized model. In: Proceedings of the 9th International Conference on Information Visualisation, Washington, DC, USA, 2005, 975–980.

14. Pal S K. Automatic graylevel thresholding through index of fuzziness entropy. *Pattern Recognition Letters,* 1983(1): 141–146.
15. Wu Fan, Wang Chao, Zhang Hong, Zhang Bo, Zhang Wei-sheng. Knowledge-based bridge recognition in high resolution optical imagery. *Journal of Electronics & Information Technology*, 2006, 28(4):587–591.
16. Amini A A, Tehrani S, Weymouth T E. Using dynamic programming for minimizing the energy of active contours in the presence of hard constraints. In: Proceedings of 2nd Int. Conf. Computer Vision, Los Alamitos, CA, USA, 1988, 95–99.

Design of High Precision Rotary Pointing Device Driven by Voice Coil Motors

Qian Cao, Ming Li and Peng-mei Xu

Abstract In the field of remote spectral sensing technology, design of high precision rotary pointing device driven by voice coil motor is presented in this paper, which can be used in many similar tracking or scanning devices. The voice coil motor has a simple structure, without any structure for reversing, and is easy to be controlled. By strictly matching the characteristics of device load and the working conditions, the features of voice coil motor is utilized efficiently. While ensuring the precision of motion, the size and mass of the devices are reduced. In the end, some analysis, calculation and experiment are presented. Comparing precision, size, mass and other parameters to some similar tracking or scanning devices in space products, this design got certain superiority.

Keywords Pointing device · Voice coil motor · Pivot · High precision

1 Introduction

With the rapid development of space optical technology and the growing requirements of space spectrum instrument, the requirements for the performance of spectrum instruments have been becoming higher and higher. The demands of light mass, compaction, low power cost and high precision for space spectrum instrument are more and more stringent. In those instruments the scan mirror is usually needed to swing regularly within a certain range of angles. Then the spectrum instruments can track targets stably when it is working in space orbit.

Voice coil motors have been used more widely on some high precision linear or rotary devices, in the field of domestic and foreign industrial control such as medical and optics, etc. But voice coil motors have never been used in any domestic space produces. The research on voice coil motors in the field of domestic space stays in laboratory.

Q. Cao (✉) · M. Li · P. Xu
Beijing Institute of Space Mechanics & Electricity, Beijing, China
e-mail: mou_lbc@163.com

H.P. Urbach and G. Zhang (eds.), *3rd International Symposium of Space Optical Instruments and Applications*, Springer Proceedings in Physics 192,
DOI 10.1007/978-3-319-49184-4_42

A Fast-steering Mirror device driven by linear voice coil motors was designed by Changchun Institute of Optics, Fine Mechanics and Physics. This system controlled the transmission direction of a laser precisely. The device has simple structure and great carrying capacity, and its steady precision is more than 2″ [1].

A precise device of scanning and positioning based on voice-coil motor was designed by Key Laboratory of Optical Calibration and Characterization, which was applied on positioning of imaging facula in the measurement of spectra. The long time working excursion of this scanning and positioning device is less than 1.5 um [2].

In this paper space tracking, pointing and scanning devices are researched, and a high precision rotary pointing device driven by voice coil motors is presented and realized in this paper.

2 Methods of Device Support

2.1 *Compare and Select*

There are following conditions for the devices in this research: the angle range of mirror swing is $<\pm 6^\circ$ during operation, the speed of swing is $<1^\circ/s$.

Rolling bearings are used usually as structural support member in some traditional rotary devices. But rolling bearing is only applied to the condition when the device rotates in complete rotations or at least the balls in bearing rotate in complete rotations. The bearings will get deviation when working at a low speed. Also the bearing's life will be seriously reduced when working long term in a small angle range of swing, because the balls in bearing are pressed unevenly.

Pivots are used as support components in this angle-limited device, which have flexible structure. Pivots do not generate any friction or friction torque. And they dose not need any lubrication. The torsion moment of the device depends on the torsion spring rate of pivots and rotation angle, and is independent of environmental factors such as temperature etc. This flexible pivotal support is especially suitable for space environment when the device works in a small angle range. The structure of flexible pivot is shown in Fig. 1. The flexible pivots type and parameters related in this paper are shown in Table 1.

2.2 *Installation of Pivot*

A certain preload should be applied to the pivots, in order to prevent them from loose in vibration when satellite launch and in long term orbit working. The amount of preload depends on the size and type of pivots, and their loads. Generally the preload of pivots is 0.01–0.03 mm, and detailed description for installing is shown in Fig. 2.

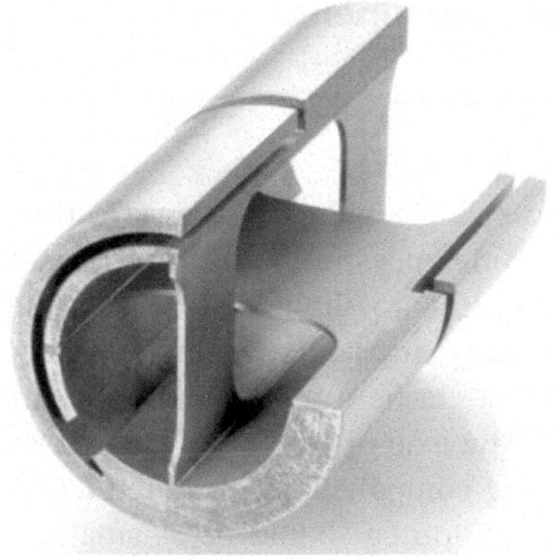

Fig. 1 Structure of flexible pivot

Table 1 Type and parameters of pivots

Type	Diameter (mm)	Torsion spring rate (Nm/rad)	Load capacity Vc (N)	Load capacity (N/mm)
5012-600	9.53	0.311	355.86	2800.11
5020-600	15.88	1.503	983.06	5075.21

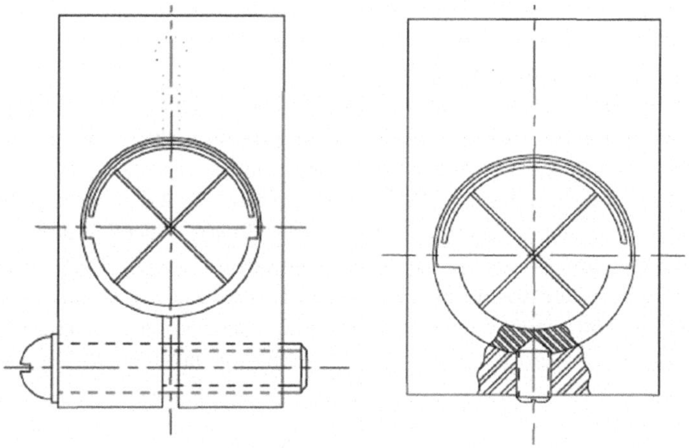

Fig. 2 Pivot installation description

The pressure to pivots which generated from the preload (0.01–0.03 mm) should be considered as loads when calculating the safety factor. The additional force to pivots which generated from the parts processing error, assembly error and the coaxial error should be considered as load also.

3 Select the Drive Components

3.1 Device Characteristics

The scan mirror needs to swing regularly within a certain range of angles in this research. There are some differences in the form of movement from the traditional devices. First, the mirror's motion is limited within a small range of angle (<±6°), while whole circle of rotating in traditional devices. Second, there is no friction torque in this device, but an elastic torque which is positively correlated with the swing angle. Third, the inertia load of the mirror under a very low speed. Therefore, the output torque of driving components is required adjustable to the elastic load, and the fluctuation of output torque should be as small as possible. Otherwise the motion accuracy of shaft in devices will be affected directly.

3.2 Compare and Select the Drive Component

Traditional brushless DC torque motor and stepper motor are circumferential structures, which have larger mass and size. It's conflict with the design principle of lightweight and compact on space products. Meanwhile the control precision of stepper motor is not enough, due to the output torque fluctuation. On the premise of meeting the requirements, the components of devices should be small, lightweight and high utilization in space products.

Rotary voice coil motor can make up the lack of DC motor and stepper motor. It has a specially angle-limited structure, also has the characteristics of torque motor. This can well meet the requirements of space products, due to its small size and light weight. The structure of rotary voice coil motor is shown in Fig. 3. Voice coil motor can drive the shaft directly. Its working principle is Ampere force (that's why it's named like a speaker). An electrified coil (conductor) placed in a magnetic field generates a force, and the size of the force is proportional to the coil current. Voice coil motor converts electrical energy into mechanical energy directly, eliminating the intermediate converting component. They have a series of advantages like small size, light weight and fast response. Voice coil motor has been widely used in lots

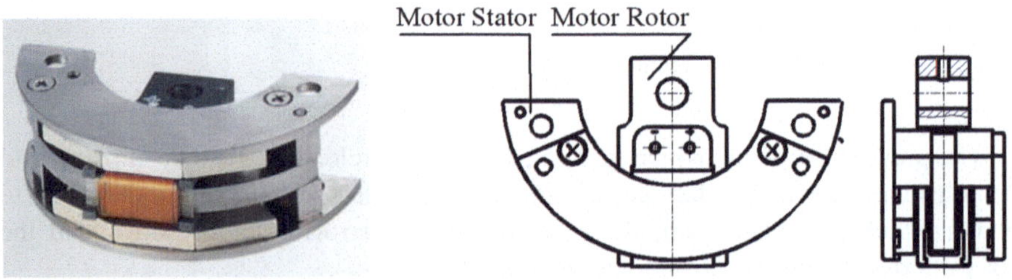

Fig. 3 Structure of rotary voice coil motor

of high precision positioning systems and high acceleration devices, such as disk positioning and optic lens positioning on the ground [3, 4].

In summary, voice coil motor is chosen as the driving component for angle-limited structure in this design, which has no reversing structure, and has advantages of simple structure, high reliability, small moment of inertia, high acceleration, fast response, no cogging effect, high precision positioning control. It is especially suitable for a system of short-stroke quick, quick-response under high-precision closed-loop servo control.

3.3 Design of Torque Margin

Torque margin comprises static torque and dynamic torque. The static torque can drive the shaft from motionless state to motion. The dynamic torque makes sure the parameters of shaft motion meeting the working conditions required.

Static torque margin of shaft:

$$\eta = \frac{M_0}{M_f} - 1 \quad (1)$$

Dynamic torque margin of shaft:

$$\eta = \frac{M_0 - 2 \times T_q}{M_f} - 1 \quad (2)$$

Note:

M_0 Maximum output torque of voice coil motor;
T_q Load torque from shaft structure moment of inertia;
M_f Maximum elastic torque of shaft structure, minimum uncertainty factor is 1.2 for spring, according to the ESA "ECSS-E-30-part3A".

4 Design of Overall

The device consists of a shaft holder, a scan mirror, a voice coil motor, a resolver and other components, as shown in Fig. 4. The scan mirror is supported by two flexible pivots, connected to the shaft holder; the motor rotor and the resolver rotor are connected to the scan mirror through two adapter structures; the motor stator and the resolver stator are fixed on the shaft holder. Then the shaft achieves the

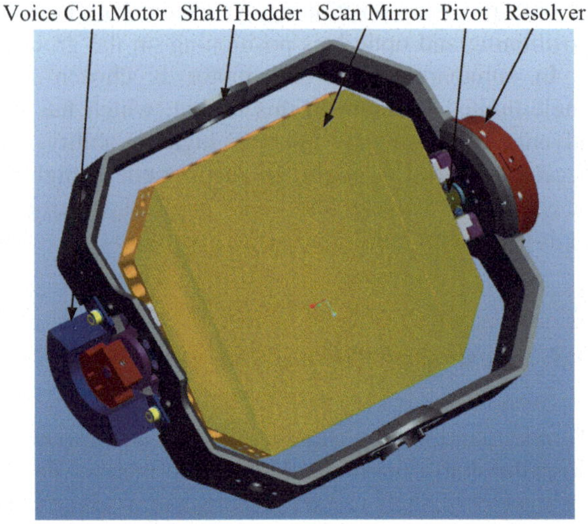

Fig. 4 A design of entirety structure

functions that the motor drives the mirror directly and the resolver feedbacks the angular position.

5 Control System Introduction

The control method is a closed-loop based on position feedback. The high-precision resolver, which electrical error is ±15″, provides the position feedback information. The control system is digitalized based on DSP servo controller. The detail control scheme is: Instructed by upper lever's parameters, DSP generates a position curve command corresponded, using the angle feedback from resolver, concurrently calculates the angle error. Then according to the control algorithm, adjust the voltage on the motor, through allocating the PWM duty cycle, the closed-loop position servo is achieved. The system function schematic is shown in Fig. 5.

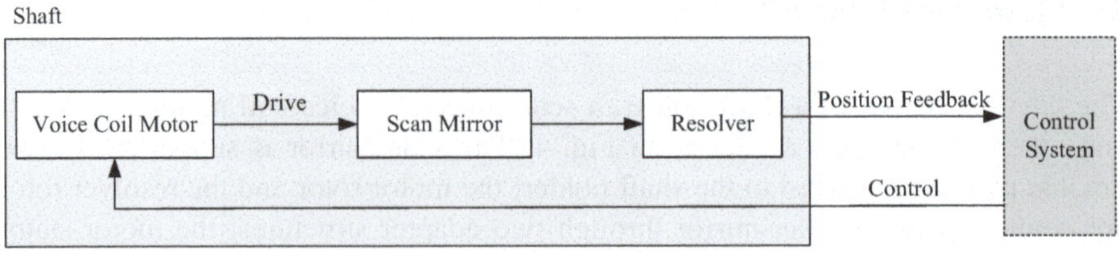

Fig. 5 System function schematic

6 Results of Test and Conclusion

Applied the above methods, we designed and realized two different tracking or pointing devices, which is shown in Fig. 6 (The inner shaft) and Fig. 7 (the inner and outer shafts). The load of devices in diagram is simulated by certain loads

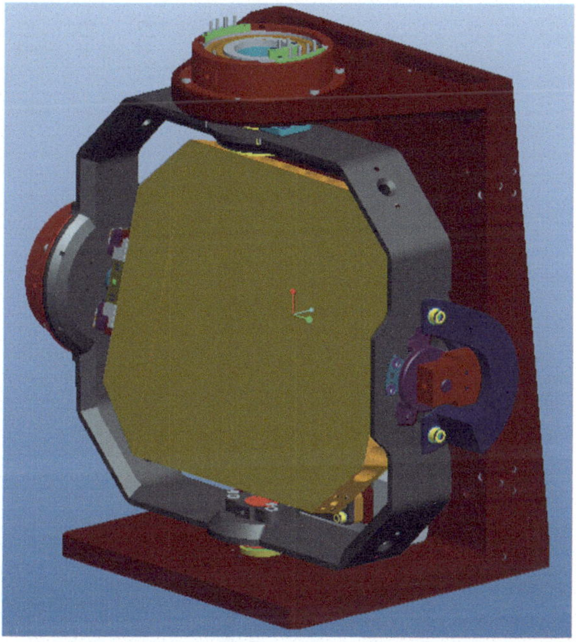

Fig. 6 A tracking device for testing

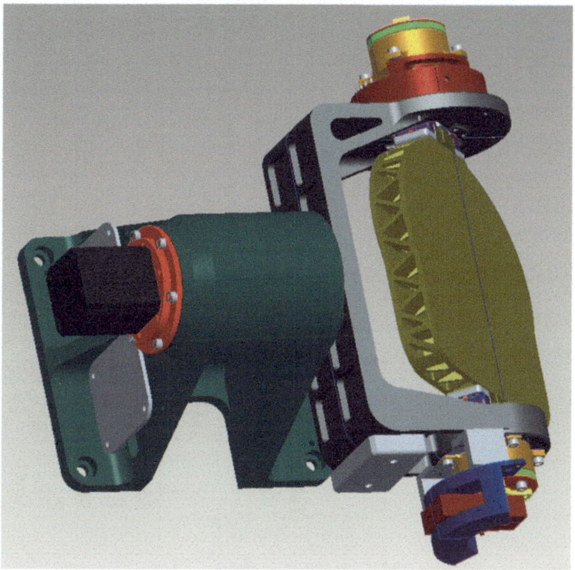

Fig. 7 A pointing device for testing

Tab.2 Key parameters and test results

	Load weight (Kg)	Load inertia (kgm^2)	Rotation range (°)	Maximum elastic torque (Nm)	Torque coefficient of motor (Nm/A)	Voice coil motor weight (Kg)	DC torque motor weight (Kg)	Electrical errors of resolver (″)	Accuracy of pointing (″)
Shaft 1	0.51	5.9×10^{-4}	±6	0.09	0.36	0.24	0.41	±15	±27.3
Shaft 2	0.71	1.2×10^{-3}	±3	0.045	0.1	0.18	0.38	±12	±12.8
Shaft 3	2.2	1.7×10^{-2}	±5	0.44	0.63	0.48	0.7	±15	±16.4

which have similar characteristics of mass or inertia respectively. Then the pointing accuracies of each device are tested. The test results are shown in Table 2.

It's shown in above data, that under the premise of ensuring the pointing accuracy, the motor weight is reduced about 40%, and the size and mass of the devices are correlate reduced, which use the method proposed in this paper.

In summary, the design of high precision rotary pointing device driven by voice coil motor got certain superiority on angle-limited structures. This method is suitable for many similar tracking and scanning devices in space products, when their working angle range is limited to $<\pm 6°$.

References

1. XU Xin-hang, WANG Bing, HAN Xu-dong, WANG Heng-Kun, LIU Ting-xia, "Design of fast-steering mirror with spherical pair supporting structure driven by voice coil actuators," Optics and Precision Engineering, J, Changchun, China: Changchun Institute of Optics, Fine Mechanics and Physics, Vol.19 No.6, pp. 1320–1324. Jun. 2011.
2. LIU En-chao, LI Xin, ZHANG Yan-na, XU Chun-feng, ZHENG Xiao-bing, "Design of Precise Scanning and Positioning Based on Voice-Coil Actuator," Journal of Atmospheric and Environmental Optics, J, Hefei, China: Key Laboratory of Optical Calibration and Characterization, Vol.9 No.5, pp. 384–390, Sep. 2014.
3. ZHANG Da-wei, FENG Xiao-mei, "The Technical Principle of Voice-Coil Actuator," JOURNAL OF NORTH UNIVERSITY OF CHINA, J, Taiyuan China: School of Mechanical Engineering, TianJin University, Vol.27 No.3, pp. 224–228. Sum No.107. 2006.
4. XING Lian-guo, ZHOU Hui-xing, HOU Shu-lin, CAO Rong-min, "Research and Application of Voice Coil Motor," Micromotors, J, Beijing China: Agricultural University, Vol.44 No.8, pp. 83–88. Aug. 2011.

Discussion on Issues in the Implementation of Spaceborne FTS

Lizhou Hou, Pengmei Xu and Bicen Li

Abstract Fourier Transform Spectrometer (FTS) has been an important analytical tool for many applications. In addition to its use in benign laboratory environment, the FTS has found great use for difficult tasks of remote sensing of the earth from satellites, and for space exploration by probes to other planets. So far many Spaceborne FTS instruments have been launched in orbit for sounding from temperature and humidity to air composition. This paper provides an overview of the issues in the implementation of Spaceborne FTS, which involves inherent limitations, practical limitations and some issues of signal chain and system integration. The inherent limitations basically include Boxcar apodization, field of view effect that is also called self-apodization. Practical limitations involve misalignment, wavefront error and beamsplitter difference. The signal chain and system integration issues contain time-delay match, flatness of electronic filter amplitude-frequency characteristic, channel effect, vibration and mechanism.

Keywords Spaceborne Fourier transform spectrometer · Remote sensing · Interferometer

1 Introduction

Fourier Transform Spectrometer (FTS) has been an important analytical tool for many applications. In addition to its use in benign laboratory environment, the FTS has found great use for difficult tasks of remote sensing of the earth from satellites, and for space exploration by probes to other planets [1]. In September 1962, the era of remote sensing measurements using spaceborne Fourier transform systems began with the flight of the Block Engineering I6T onboard the Discoverer satellite. After that many spaceborne FTSs have been put into orbit, among which the textbook

L. Hou (✉) · P. Xu · B. Li
Beijing Institute of Space Mechanics and Electricity,
No. 99 Zhongguancundong Rd., Beijing, China
e-mail: houlizhou@126.com

H.P. Urbach and G. Zhang (eds.), *3rd International Symposium of Space Optical Instruments and Applications*, Springer Proceedings in Physics 192,
DOI 10.1007/978-3-319-49184-4_43

examples include Interferometric Monitor for Greenhouse gases (IMG) onboard ADEOS, Improved Atmospheric Sounding Infrared (IASI) [2] onboard METOP-1, Michelson Interferometer for Passive Atmospheric Sounding (MIPAS) [4] onboard ENVISAT, Tropospheric Emission Spectrometer (TES) [3] onboard Aura, Atmospheric Chemistry Experiment (ACE) [4] onboard SCISAT-1, Cross-track Infrared Sounder (CrIS) [5] onboard NPP and Thermal And Near infrared Sensor for carbon Observation (TANSO) [6] onboard GOSAT. The heart of the above spaceborne FTSs is based on Michelson interferometer that was first built by Michelson in the nineteenth century for an investigation of the "motion of the medium upon the propagation of light". In the traditional Michelson configuration, beamsplitter, fixed mirror, movable mirror and its associated displacement mechanism are the main parts and they are sensitive to the tilt or vibration. Although it has obviously advantages over other dispersive spectrometers, FTS was considered high-risk sensor for space applications because of some limitations and issues in the implementation. In order to make FTS systems suitable for space missions as the above mentioned sensors did, some rigorous considerations must be taken into account not only at the outset of design effort but also in manufacturing process. This paper discussed on the limitations and issues in the implementation of Spaceborne FTS, as well as the key risk mitigations.

2 Limitation and Issues

A. *Inherent limitations*

According to the principle of operation, there exist some inherent limitation involves resolution and throughput. The simplest type of FTS based on a Michelson interferometer is sketched in Fig. 1. The incoming beam of collimated radiation is divided in two by the beamsplitter. The beam in each leg of the interferometers is reflected back toward the beamsplitter by a mirror. At the beamsplitter, each beam is again divided, one part going to the detector, the other returning to the input source. In each direction, the beams from the two legs recombine, or interfere. If the interferometer mirrors are correctly aligned, an interferogram would be obtained by moving mirror steadily. The Maximum optical Path Difference MPD is then four times of the displacement of moving mirror in this case and the spectral resolution without apodization is 1/2MPD.

the moving mirror only displace a finite distance so that the interferogram is truncated by a boxcar as shown in Fig. 2. The spectral resolution, $\Delta\lambda$, be restricted to a sinc function whose FWHM is 1.207/2MPD. It is called boxcar apodization. This is the first inherent limitation that be considered in design of FTS system.

In a real FTS system the input beam is actually divergent with a certain FOV. Consequently, when the Optical Path Difference is not zero, the fringe pattern as

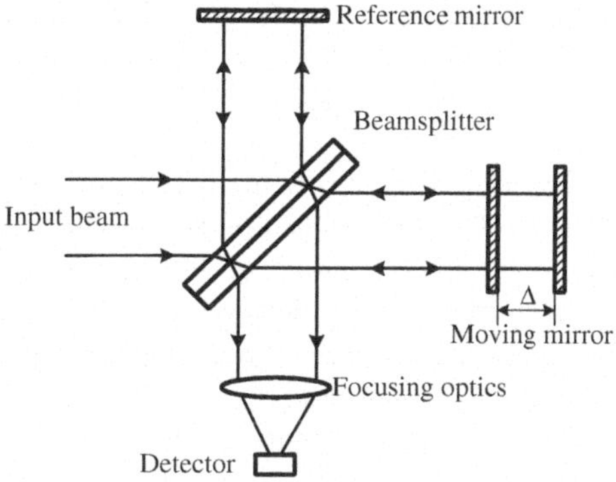

Fig. 1 Optical diagram of a basic Michelson interferometer

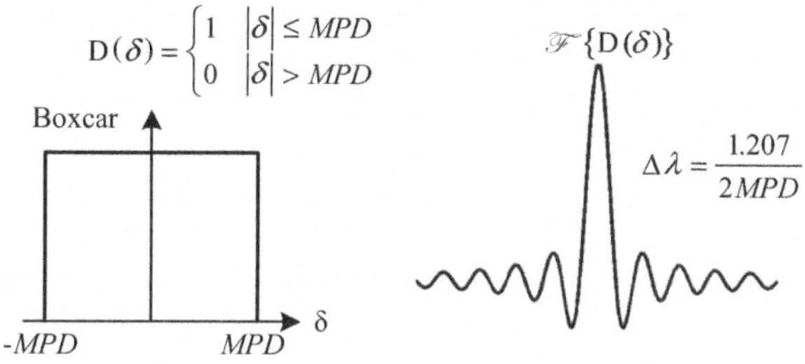

Fig. 2 Diagram of resolution of FTS

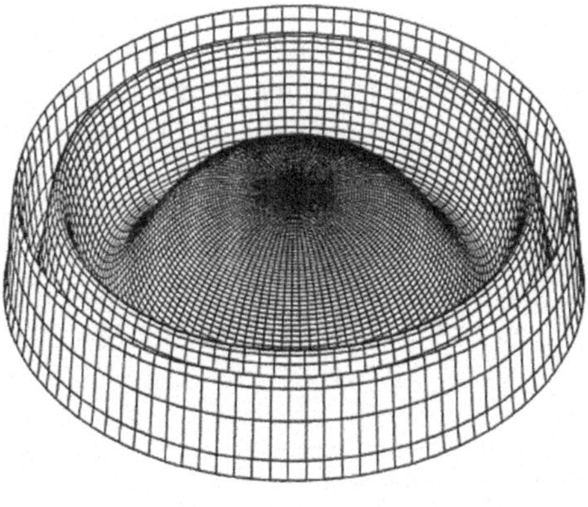

Fig. 3 Fringe pattern on the detector with a certain FOV when OPD is not zero

shown in Fig. 3 will occur on the detector. This fringe pattern results in two effects. One is to shift the apparent wavenumber by a factor close to unity. The other effect is to modulate the fringe amplitude with a sinc function, which also called self-apodization.

Obviously the self-apodization reduce the spectral resolution so that the FOV of FTS is not free to be set. An trade-off is made between FOV and spectral resolution so as to maximize the throughput of FTS for better SNR as well as not to reduce the spectral resolution significantly. There is the second inherent limitation described as $R\Omega = 2\pi$, wherein R is resolving power and Ω is solid angle of FTS.

B. *Practical limitations*

Practical limitations mainly include misalignment, wavefront error and beam-splitter. These aspects involve modulation efficiency which is one of the most important qualifications of the interferometer. As for the space application, the interferometer must undergo the launch vibration, the condition changes from on the earth to in orbit. How to maintain high modulation efficiency and keep high stability is a big challenge.

Misalignment appears in the way of tilt and shear within interferometer and detector field-of-view off-axis. For the traditional interferometer with plane mirrors tilt is the most important error to reduce modulation efficiency. And this can be resolved greatly by use of cube corner retroreflectors instead of plane mirrors or by use of Dynamic Alignment technique as shown in Fig. 4. During the movement of plane mirror the tilt of the fixed mirror is adjusted in real time by means of detecting and feeding the phase differences of metrology laser fringes so as to keep Dynamic Alignment between two mirrors. Despite of the complexity of Dynamic Alignment, it is a promising measure to maintain high modulation efficiency of interferometer especially when used in near infrared or visible band. The above mentioned CrIS made use of dynamic alignment technique and proved to be of excellent performance.

The cube corner is of great ability to be immune to tilt and was used in the sensors mentioned above except for CrIS. It is especially suitable for infrared FTS. When used in shorter wavelength band or needed to have larger clear aperture the cube corner will need to be of higher wavefront quality, which make it hard to be manufactured. The cube corner is usually an assembly adhered together by three paddles made of optical glass. For spaceborne FTS the cube corner needs to reduce weight and maintain high stability. Figure 5 shows the picture of cube corner from SAGEM. The left one is used in IASI and The right one with bipods mounting is the latest one which will be launched in this year onboard a Chinese satellite.

Shear error exist in basic Michelson interferometer with cube corner, for instance TANSO-FTS and IASI. It is derived from the apex asymmetry between two cube

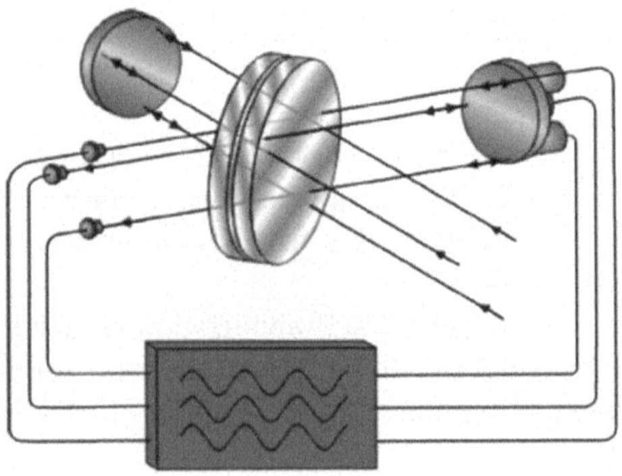

Fig. 4 Diagram of dynamic alignment interferometer

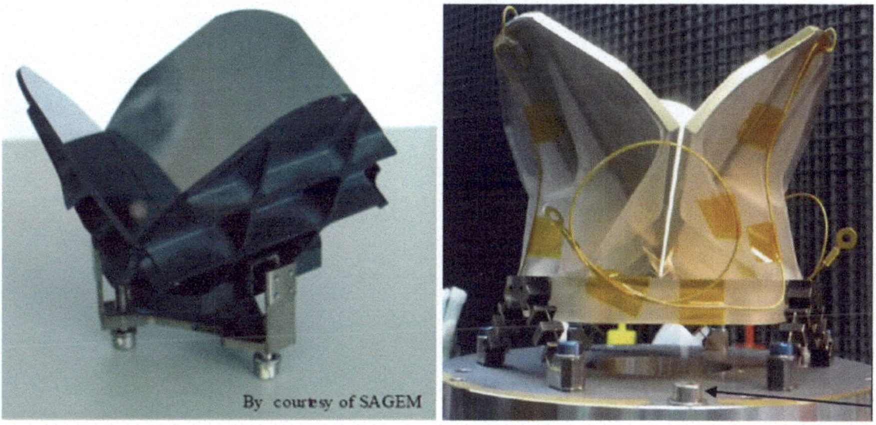

Fig. 5 Picture of cube corner from SAGEM

corners. Shear error can be eliminated by select specific interferometer configuration, such as ACE and TES configuration, in which the shear error has been compensated theoretically.

Detector field-of-view off-axis leads to the degradation and distortion of resolution. And this can be verified and realigned by monitoring of instrument line shape (ILS) of interferometer. For a well aligned detector the ILS is well symmetrical.

Wavefront error here mainly refers to the portions that cause by mounting stress of the optical components. Two aspects must be taken into account that add RTV axial pads to eliminate mounting stress and add flexible mounting nut to accommodate the thermal load variation on mounting interface between interferometer and optical bench.

Beamsplitter-compensator difference is also an important factor influence the modulation efficiency of FTS. For wide band use for example form visible to thermal range like TANSO-FTS, a self-compensated beamsplitter is a suitable chose. But it can't be used in imager FTS.

C. *Issue of time-delay match*

According to the principle of FTS, the interferogram must be sampled in terms of constant Optical Path Difference. Actually, the sampling signal driven by Optical Path Difference is generated from the co-aligned metrology laser fringes that occur simultaneously with the interferogram. The two signals must propagate synchronized along different signal chain until arrived at the ADC port at same time. Or the sampling error will occur due to the instability of velocity of the Optical Path Difference. The time-delay unmatched reduce the SNR of FTS. Figure 6 shows the detail when the time-delay unmatched considerably. Time-delay unmatched causes the SNR to be lower and thinner than the normal one.

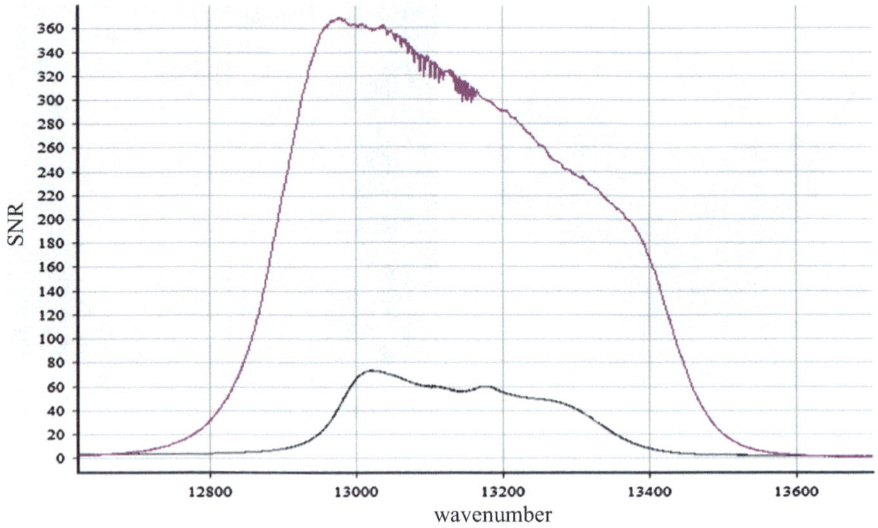

Fig. 6 SNR of a FTS versus time-delay match

Time delay match can be done by delay the sampling signal or by uniform time-sampling method [7].

D. *Issue of filter flatness*

To suppress noise in order to improve SNR, the interferogram before ADC is filtered by an anti-aliasing filter. And the frequency of interferogram is variable due to the instability of velocity of Optical Path Difference. When the amplitude-frequency characteristic of the filter is not flat enough, the amplitude of interferogram is superposed with noise. The effective method to resolve this issue is to make use of low-pass filter with higher cut-off frequency to obtain the relative flat amplitude-frequency characteristic. And at the same time the frequency of sampling signal shall to be increased to prevent aliasing.

E. *Issue of channel effect*

Channel effect, also called channel spectra, is caused by the parallel refractive optical components in the instrument. It is similar to the Fabray-Perot fringe pattern when one encounters due to a filter or a window, etc., with parallel faces somewhere in the optical path. Figure 7 shows a measured spectrum with little channel effect caused by a parallel filter substrate. Keep mind that all refractive optics in the optical path are wedged to eliminate channel spectrum.

F. *Issue of vibration and mechanism*

Micro vibration is one of the important factors that influence the interferometer. It leads to the sampling jitter which results in the degradation of SNR and even the ghost spectrum. One can choose suitable configuration for the interferometer such

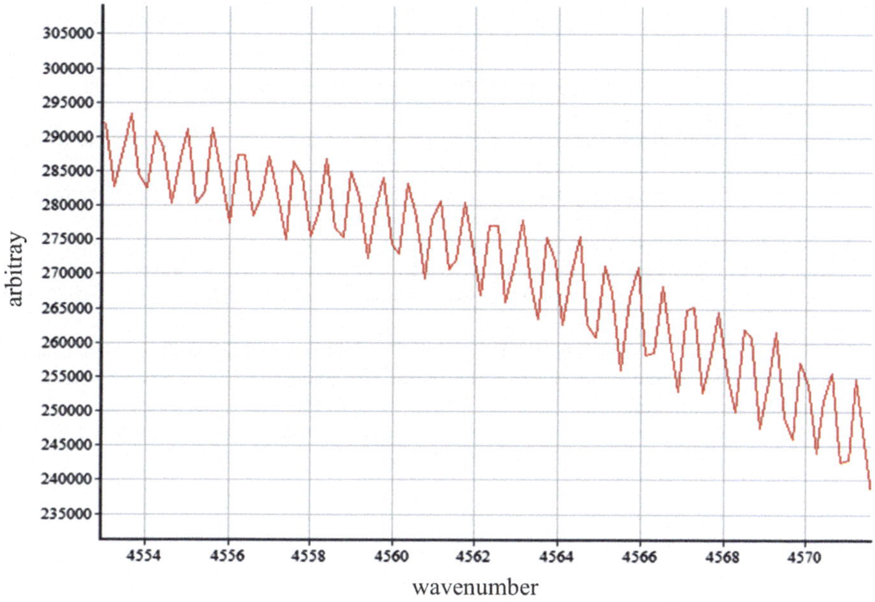

Fig. 7 Diagram of channel effect

as pendulum rotary arm with flexible pivots in ACE and TANSO to obtain the good immunity to the micro vibration. And further a vibration insulator is equipped between the FTS and its mounting base. For the mechanism, the out-of-shelf flex pivot is a common option for small range rotary in commercial FTS. But in case of spaceborne FTS, for risk-mitigation it is suggested that the flexible blade would be a good chose to undergo the launch vibration and expand the lifespan of interferometer onboard.

3 Summary

Some inherent limitations and practical limitations of spaceborne FTS have been reviewed; the key issues encountered in implementation of spaceborne FTS have been discussed as well. These limitations and issues should be rigorously not only considered at the outset of design effect of spaceborne FTS but also taken into account during the process of implementation of spaceborne FTS.

References

1. M. J. Persky, "A review of spaceborne infrared Fourier transform spectrometers for remote sensing", Rev. Sci. Instrum. 66(10), October 1995.
2. D. Blumstein & al, "IASI instrument: technical overview and measured performances" SPIE Denver 2004.
3. Reinhard Beer, "Tropospheric Emission Spectrometer Scientific Objectives & Approach, Goals & Requirements", JPL D-11294 Revision 6.0, April 1999.
4. Marc-Andre Soucy, Francois Chateauneuf, Christophe Deutsch, Nicolas Etienne, "ACE-FTS Instrument Detailed Design", Proceeding of SPIE Vol.4814 PP. 70–81, 2002.
5. K. Stumpf and J. Overbeck, "CrIS Optical System Design", Proceedings of SPIE Vol. 4486 2002.
6. A. Kuze, H. Suto, M. Nakajima, T. Hamazaki, "Thermal and near infrared sensor for carbon observation Fourier-transform Spectrometer on the Greenhouse Gases Observing Satellite for greenhouse gases monitoring", Applied Optics Vol. 48, PP. 6716–6733, 2009.
7. John C. Brasunas and G. Mark Cushman, "Uniform time-sampling Fourier Transform spectroscopy", Applied Optics, Vol. 36 No. 10 1997.

Key Performance Simulation and Analysis of Space Borne Fourier Transform Infrared Spectrometer

Bicen Li, Lizhou Hou and Pengmei Xu

Abstract With the advantages of high spectral resolution, wide spectral coverage and large throughput, space borne Fourier transform infrared spectrometer (FTS) plays more and more important role in atmospheric composition sounding. Due to the high accuracy retrieval of trace gas in atmosphere, the accuracy of observation data should be very high which must be guaranteed critically by the superior performance of gas sensor. To achieve both high signal to noise ratio (SNR) and high spectral resolution, reasonable allocation and optimization of instrument parameters are the foundation and difficulty. According to the requirements of atmospheric detection spectrometer for space application, and based on operating principle of Michelson interferometer, the SNR and spectral resolution of time modulated FTS and their influencing factors have been simulated and analyzed. An optimum design method for key performance of FTS under low input radiance of weak scene has been present. The simulation model of FTS SNR has been built, which consider satellite orbit, spectral radiometric features of surface and atmospheric composition, optical system, interferometer and its control system, measurement duration, detector sensitivity, noise of detector and electronic system and so on. Spectral resolution is defined as the full width at half maximum (FWHM) of instrument line shape function (ILS) of spectrometer. The influence factors of ILS like maximum optical path difference (MPD) and field of view (FOV) of interferometer are analyzed. The design parameters of spectrometer has been optimized combining engineering feasibility, validated and analyzed with specific sample. Theoretical analysis and simulation results indicate that this model and method could be the basis of space borne time modulated Fourier transform infrared spectrometer design.

B. Li (✉) · L. Hou · P. Xu
Beijing Institute of Space Mechanics and Electricity,
No.99 Zhongguancundong Rd., Beijing, China
e-mail: mou_lbc@163.com

L. Hou
e-mail: houlizhou@126.com

P. Xu
e-mail: 13651065845@163.com

H.P. Urbach and G. Zhang (eds.), *3rd International Symposium of Space Optical Instruments and Applications*, Springer Proceedings in Physics 192,
DOI 10.1007/978-3-319-49184-4_44

Keywords Atmospheric sounding · Space borne fourier transform spectrometer · Interferometer · Signal to noise ratio · Spectral resolution · Instrument line shape function

1 Introduction

With the growing problems of air pollution, climate change and frequent occurrence of extreme weather, people pay more attention to the research on atmospheric environment and impact of human activities, and atmospheric sounding technique is developed rapidly. Sensors used for high accuracy determination of multiple atmospheric composition should have the abilities of high spectral resolution, high sensitivity and wide spectral coverage. FTS have all these advantages and can resolve the contradiction between sensitivity and spectral resolution, which has been widely applied in infrared spectrometry and space infrared spectral radiometric measurement.

Space borne FTS can detect the characteristic spectrum of atmospheric composition, thereby the content and distribution of different composition can be obtained, like Aura/TES [1], ENVISAT-1/MIPAS [2], SCISAT-1/ACE-FTS [3], GOSAT/TANSO-FTS [4], MetOp-A/IASI [5] etc. These instruments are all timed modulated FTS whose remote sensing data have been used to study some hot issues like greenhouse effect and ozone problem. This technique has been a important means applied in many fields of atmospheric environment protection, study of weather and climate, industrial emission monitoring etc.

As key performance of FTS, SNR and Spectral resolution are crucially important to high accuracy retrieval of atmospheric composition. Based on the atmospheric spectrum detection under the scene of low radiance, analysis on influence factors of main performance of spectrometer is present in this paper. And the design parameters are optimized to meet the requirements of space application.

2 Principles of Fourier Transform Infrared Spectrometer

In Fourier transform spectroscopy, numerical computation of spectrum is by inverse fast Fourier transform (IFFT) of interferogram generated by interferometer. With the development of FTS, various types of interferometer have been invented based on principle of Michelson interferometer. The basic physical principles and interference theories for these interferometers with different configurations are in common with Michelson interferometer.

The operation principle of classical Michelson interferometer is shown in Fig. 1a. At first, the beam emitted from a light source is collimated by a collimator.

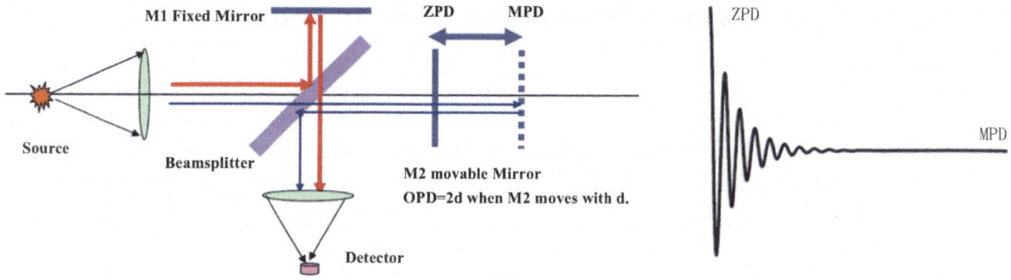

(a) Principle of Michelson interferometer **(b)** Interferogram

Fig. 1 Michelson interferometer

The parallel beam is split up into reflected beam and transmitted beam by beamsplitter. The reflected part which is then reflected by a fixed mirror (M1) and transmits the beamsplitter meets the transmitted beam which is then reflected by a moving mirror (M2) and reflected by beamsplitter. So the interference light is generated here and focused on detector. Interference intensity which is a function of optical path difference (OPD) between the two beams is called interferogram. The OPD will be *2d* by the motion of M2 mirror with the distance of *d*. The interference signal of polychromatic light is shown schematically in Fig. 1b which is generated by the motion from ZPD to MPD of M2 mirror [6].

The configuration of time modulated FTS with dual cube corners and rotary arms which is developed based on Michelson interferometer is shown in Fig. 2. This kind of FTS is very suitable for space application. Improved from classical Michelson interferometer, two hollow cube corners are used in this kind of structure so that the interference can be not sensitive to tilt and shear errors. And the moving of two rigid mounting cube corners can be differential which is realized by the dual rotary arms structure. This structure has high tolerance capacity for vibration environment.

Fig. 2 Time modulated FTS with dual cube corners and rotary arm

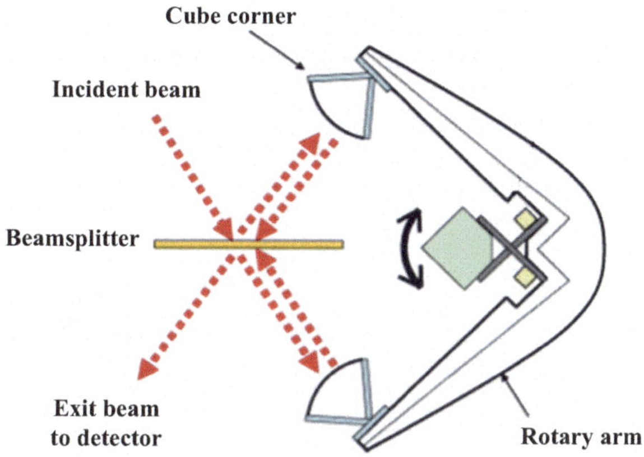

3 Analysis on Spectral Resolution and Its Influence Factors

Interference signal of FTS is a function of OPD. According to the Fourier transform relationship between double side interferogram and spectrum, true spectrum $B'(v)$ obtained by FTS is the convolution of theoretical spectrum and ILS function.

$$B'(v) = B(v) \otimes ILS(v_0, v) \tag{1}$$

$ILS(v_0, v)$ is the result measured by the FTS whose FOV is fully irradiated by a monochromatic light source.

The FWHM of ILS function is spectral resolution of FTS. As shown in Fig. 3, the main influence factors of ILS FWHM comprise MPD, FOV, off-axis and defocusing of source, aperture diffraction, laser wavelength drifting, noise of laser etc. By analysis, MPD and FOV play leading roles and other factors can be considered as small quantity. Broadening of ILS caused by MPD is independent of wavenumber, and broadening of ILS caused by FOV is directly proportional to wavenumber.

3.1 MPD

The relationship between ILS and MPD is given by

$$ILS(v_0, v) = 2L \cdot \sin c[2\pi(v - v_0)L] \tag{2}$$

According to formula (2), calculated spectrum is confined by finite moving distance of movable mirror. The first zero-crossing position of true spectrum convolved by ILS which is a sinc function is at $\delta \upsilon = 1/2L$, where $\delta \upsilon$ is un-apodization resolution or ideal spectral resolution, and L is MPD. By the influence of finite OPD, ILS is broadened and have side lobes. The relationship between FWHM of this sinc function (Δv) and L is $\Delta v = 0.60335/L$.

Fig. 3 Influence factors of FTS spectral resolution

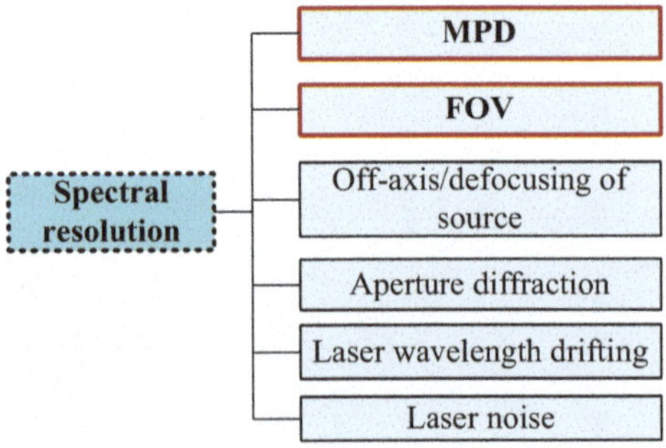

3.2 FOV

Incident light of ideal Michelson interferometer is parallel light, which should be collimated from a point light source. Whereas extended source is usually used to generate certain radiation intensity. The light comes from emitting surface which deviates from focal point of optical system will be at an angle from optical axis, so the effect of beam divergence on ILS should be considered.

$$ILS(v, v_0) = 2RT\Omega \int_{-\infty}^{+\infty} \sin c\left(\frac{v_0 x \Omega}{2}\right) \cos\left[2\pi v_0 x\left(1 - \frac{\Omega}{4\pi}\right)\right] e^{-i2\pi v x} dx \quad (3)$$

This formula describes the ILS with divergence angle of θ, where solid angle $\Omega = 2\pi(1 - \cos\theta)$.

When MPD is 2.5 cm, the ILS distortion at 6400 cm^{-1} with different FOV is simulated. The ILS simulation results, drifting and broadened wavenumber are shown in Fig. 4. ILS function is modulated as FOV increases. The peak value reduces, peak position shifts to small wavenumber, and shape of ILS is broadened.

For performance optimization of instrument, throughput AΩ should be maximized when spectral requirement is guaranteed. There is a trade-off between

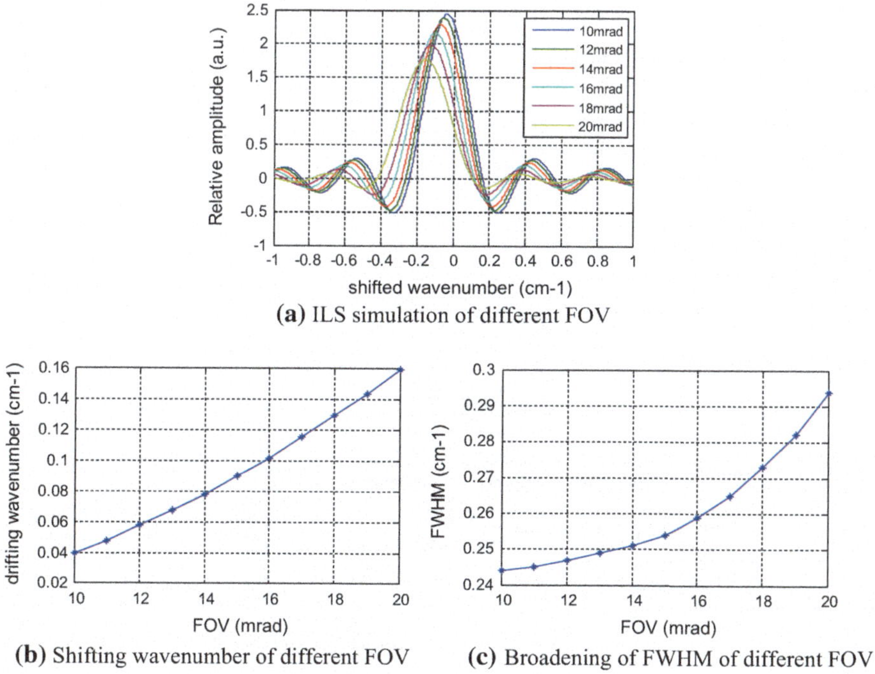

(a) ILS simulation of different FOV

(b) Shifting wavenumber of different FOV

(c) Broadening of FWHM of different FOV

Fig. 4 Analysis on influence of FOV on ILS

spectral resolution and FOV. The interference signal is dependent by the product of FOV and the Sinc function, therefore the condition of maximum signal in main lobe of Sinc function is $\Omega_{max} = 2\pi/P_{sr}$, where P_{sr} is resolution capacity,$P_{sr} = v/dv$。

Above all, MPD and maximum FOV can be determined by the requirement of spectral resolution. Then we can design the mechanical motion and optical path of interferometer based on optical path difference doubling technique.

4 Analysis on SNR and Its Influence Factors

Based on the study on principle of determination by time modulated FTS, influence factor model of SNR is established (see Fig. 5). The factors include input radiance, spectral resolution, throughput, optical efficiency, detector responsivity, duration and noise. Throughput is dependent by aperture and solid angle. Optical efficiency is expressed by transmittance of whole optical system and modulation efficiency (ME) of interferometer.

The calculation formula of SNR is:

$$SNR = \frac{B(v) \cdot \Delta v \cdot \Theta \cdot M(v) \cdot T(v) \cdot \sqrt{t} \cdot R(v)}{Noise}, \tag{4}$$

where

$B(v)$	Incident spectral radiance
Δv	Spectral interval
Θ	Throughput
$M(v)$	ME of interferometer
$T(v)$	Transmittance
t	Duration
$R(v)$	Responsivity of detector
$Noise$	Total noise

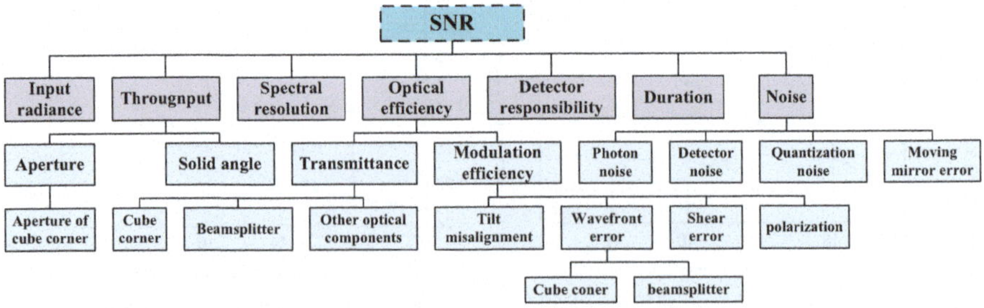

Fig. 5 Influence factors model of SNR for FTS

As the information that FTS detected is interference signal, the ratio of the light involved in interference to total signal which is ME is one of key factors affecting SNR. ME of FTS is defined as:

$$M = \frac{I_{\max} - I_{\min}}{I_{\max} + I_{\min}}, \quad (5)$$

where I_{max} and I_{min} is respectively the maximum value and minimum value of interference signal. When two beams participate in interference are identical, the ME is 1 which is the maximum value. The main factors which cause reducing of ME are tilt error, wavefront error, shear error and polarization error etc. The total ME is the product of all these modulation factors. The influence of polarization is proved to be very small. On basis of FTS with dual cube corner and rotary arm, the ME and its influence factors are discussed below.

4.1 General Angular Deviation of Cube Corner (Tilt Error)

If there is deviation of the right angle of three surface of cube corner, the emitted light and incident light are not parallel. This small angle (general angular deviation) causes that OPD deviate from theoretical value and ME decreases. The tilt modulation factor of this error is described as:

$$M_{\mathrm{T}}(\upsilon) = \left[\frac{J_1(2\pi\upsilon D\theta)}{\pi\upsilon D\theta}\right]^2, \quad (6)$$

where D is diameter of input beam, θ is general angular deviation.

When θ is 1″, the relationship between ME and beam diameter is calculated at particular wavenumbers (Fig. 6a). It can be seen that tilt modulation factor decreases along with the increasing of beam diameter, and ME decreases faster when wavenumber is larger. Because SNR is proportional to tilt modulation factor and aperture area, the optimization aperture can be obtained by calculating the product of tilt modulation factor and aperture area as a function of beam diameter. Figure 6b describes the relationship between the product and beam diameter. It can be seen that the beam diameter of about 90 mm is the optimal design value for SNR at 13,333 cm^{-1}.

4.2 General Wavefront Errors Introduced by Surface Shape

ME is also dependent on optical surface shape of interferometer. The modulation factor introduced by surface shape is:

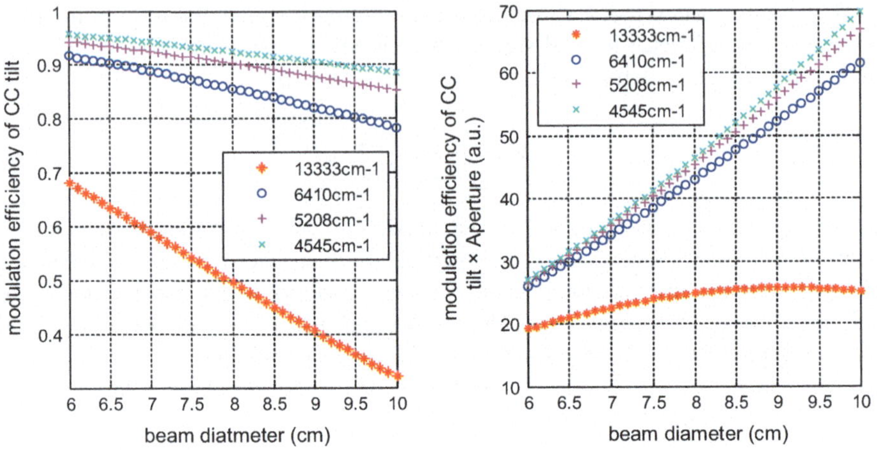

(a) Tilt modulation factor

(b) Product of tilt modulation factor and aperture area

Fig. 6 Contribution of tilt modulation factor and beam aperture to SNR

$$M = 1 - 2\pi^2 v^2 \times RMS^2, \tag{7}$$

$$RMS^2 \approx 4\sum_{1}^{m} RMS_{fi}^2 \cos^2 \alpha_i + \sum_{1}^{m} p(\alpha, n)^2 RMS_{gi}^2, \tag{8}$$

where RMS is total root-mean-square of wavefront errors, RMS_{fi} and RMS_{gi} are respectively the root-mean-square of shape errors of reflective surface and transmitting surface. The better the shape, the higher the ME.

4.3 *Shear Error of Cube Corner*

Due to mechanism clearance, vertex position of cube corner could deviate in the plane which is vertical to optical axis, which will introduce additional optical path difference. The modulation factor of cube corner shear error for particular FOV, wavenumber, OPD and shear error is given by:

$$M_s(s, \theta, v) = \frac{2J_1(\pi v \theta s)}{\pi v \theta s}, \tag{9}$$

where θ is FOV, s is the shear error.

To meet SNR requirement of FTS, all the design, manufacturing and alignment parameters which affect ME should be controlled strictly.

And the noise suppression of detector and electrical system is also very important for realizing high SNR. Analysis of noise should consider photon noise, detector noise, digitization noise and noise induced by moving mirror error etc.

Photon noise is the final limit of system SNR. Other noise sources should be restricted to the level below photon noise if possible. The sensitivity of UV, VIS or NIR detector usually can be so high that other noise can be less than photon noise. Limited by development of thermal infrared detector, detector noise exceeds all other noise sources. The photon noise spectral density of photovoltaic detectors is described as

$$I_{NSD}^{photon} = \sqrt{2 \cdot e \cdot i^{detector}}, \tag{10}$$

where e is 1.6022×10^{-19}C, $i^{detector}$ is detected current.

$$i^{dectector} = \int B(\sigma) \cdot \Theta \cdot T(\sigma) \cdot R(\sigma) d\sigma \tag{11}$$

In engineering application, intrinsic detector noise including thermal noise, dark current noise and *1/f* noise can be described as the specific detectivity, D^*. The detector noise spectral density is given by

$$I_{NSD}^{detector} = \frac{(A_d)^{1/2}}{D^*} R \tag{12}$$

where A_d is pixel area, R is responsivity.

Digitization noise which is decided by dynamic range of digital analog converter (ADC) can be reduced by using ADC of high digitalizing bit. Noises induced by moving mirror error can be restricted by increasing the moving speed stability and the flatness of frequency response of filter in signal processing unit.

5 Example

A NIR/SWIR FTS will detect total column of greenhouse gases and its variation. The requirements of this instrument are very strict. The spectral resolution should be better than 0.6 cm^{-1} for NIR channel and 0.27 cm^{-1} for SWIR channels. The SNR at weak scene of small radiance need to be better than 140–320.

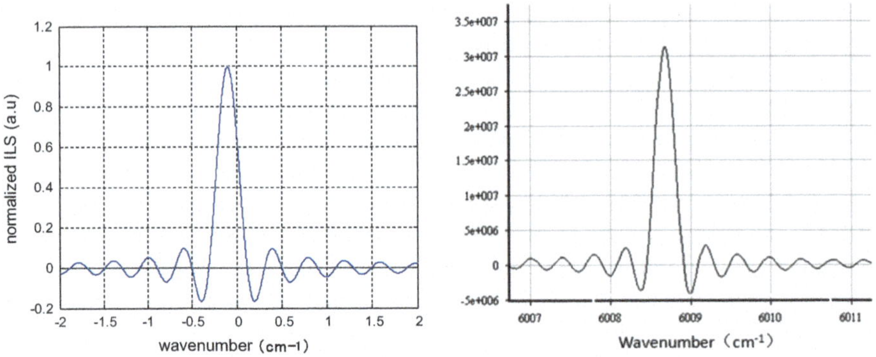

(a) B2 ILS simulation result **(b)** EM B2 ILS testing result

Fig. 7 ILS of MPD ±2.5 cm and FOV 15.8 mrad at 6008.7 cm¯1

5.1 Spectral Resolution Analysis

According to the spectral resolution requirement for gas retrieval, and based on the method discussed in Sect. 3, the MPD is ±2.5 cm, and the FOV is 15.8 mrad. The ILS simulation result of this design parameters at 6008.7 cm^{-1} is shown in Fig. 7. The FWHM of this ILS function is 0.254 cm^{-1}, which just considered MPD and FOV. The test result at 6008.7 cm^{-1} of engineering module (EM) is 0.26 cm^{-1} which is slightly larger than simulation value as other influence factors are added.

5.2 SNR Analysis

As the input conditions, surface illumination due to satellite orbit and reflective characteristics of different scene should be analyzed firstly. For the nadir observation, the reflectance of typical underlying surfaces mainly including soils, grass and so on is about 0.1–0.3 at instrument bands. According to orbit parameters of satellite, the view angle of zenith angle of 60° can cover most of low and middle latitude region. By using Modtran software, the lowest spectral radiance for this instrument is 6.5×10^{-8} to 3.4×10^{-7} W/cm^2srcm^{-1}.

The OPD of this interferometer is amplified in four times. The scan period or detection duration is 2.2 s. Substituting aperture, FOV, ME simulation results, duration, performance of detector and electronics noise into the model established in Sect. 4, the SNR simulation result of B1 channel of this instrument is shown in Fig. 8a. The simulation result agrees with the SNR test result of EM. This model is proved to be correct and reliable.

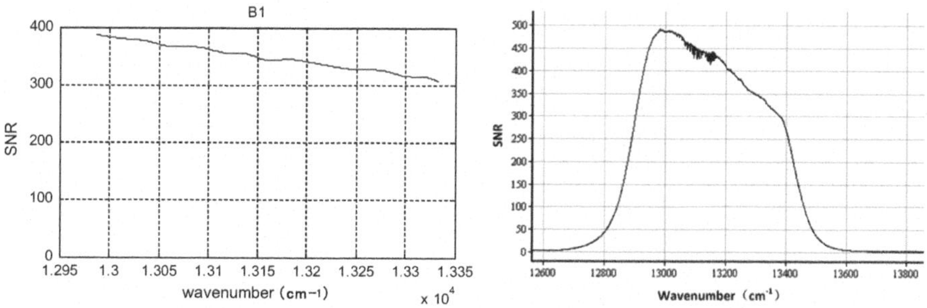

(a) B1 SNR simulation result **(b)** EM B1 SNR testing result

Fig. 8 SNR of B1 channel of green house gas monitoring instrument

6 Conclusion

Simulation and analysis for key performance of space borne FTS is the critical step for FTS design, and its accuracy is the basis of instrument engineering application. Based on the study of various parameters and its interacting principle, according to interference theory and operation principle of time modulated FTS, calculating models of some important specifications which are spectral resolution and SNR have been analyzed in detail. By comparing the simulation results with test results, the models and methods have been proved to be correct and reliable, which could be the basis of space borne Fourier transform infrared spectrometer design generally.

References

1. Reinhard Beer, "TES on the Aura Mission Scientific Objectives Measurements and Analysis Overview," *IEEE Trans. Geosci. Remote Sens.*, vol. 44, no.5, pp, 1102–1105, May 2006.
2. T. v. Clarmann, T. Chidiezie Chineke, H, Fischer, et al, "Remote sensing of the middle atmosphere with MIPAS," *Remote Sensing of Clouds and the Atmosphere VII, Proc. of SPIE,* vol. 4882, pp. 172–183, 2003.
3. Peter F. Bernath, "Atmospheric Chemistry Experiment (ACE) Mission overview," *Earth Observing Systems IX, Proc. of SPIE*, vol. 5542, pp. 146–156, 2004.
4. Akihiko Kuze, Hiroshi Suto, Masakatsu Nakajima, and Takashi Hamazaki, "Thermal and near infrared sensor for carbon observation Fourier-transform spectrometer on the Greenhouse Gases Observing Satellite for greenhouse gases monitoring", *Applied Optics*, vol. 48, no. 35, pp. 6716–6733, December 2009.
5. D. Blumstein, G. Chalon, T.Carlier, et al, "IASI instrument: Technical overview and measured performances", *Infrared Spaceborne Remote Sensing XII, Proc. of SPIE*, vol 5543, pp. 196–207, 2004.
6. Peter R. Griffiths and James A. de Haseth, "Fourier Transform Infrared Spectrometry", John Wiley & Sons., Hoboken, New Jersey, 2007.

Micro-vibration Issues in Integrated Design of High Resolution Optical Remote Sensing Satellites

Zhenwei Feng, Yufu Cui, Xinfeng Yang and Jiang Qin

Abstract With the development of remote sensing satellite technologies, and the pressing need for high resolution, high agility and low costs for remote sensing satellites, integrated design of platform and payload has been a growing tendency, which could remarkably reduce costs, volume and mass. However, compared to traditional designs, remote sensors in the integrated design are much closer to vibration sources, implying that the micro-vibration environment of remote sensors becomes even worse. Although the micro-vibration is so small that its influence to structure is negligible, it could result in considerable reduction of image quality of high resolution remote sensors. Micro-vibration has become a critical factor for high resolution image quality especially in the integrated satellites. In this article, several key solutions are proposed to mitigate the influence of micro-vibration on image quality. First, the optimization of the structure and configuration is carried out to suppress the micro-vibration transmitted to sensitive payloads. The propagation process of attenuation and amplification along vibration pathways is analysed so that the structure and configuration of satellites can be optimized, making full use of the structure to accelerate attenuation of micro-vibration. Both test and simulation show that the optimization of structure and configuration with this method is quite effective. Second, the isolator is employed to suppress the micro-vibration. The isolator is selected in system view in this article, considering the flexibility of structures, avoiding coupling with the natural frequency of the spacecraft structure. In order to improve the isolation performance, the axial, radical stiffness and damping coefficient of the isolator are optimized in system level. Simulations show that the isolator with proposed parameter is effective for micro-vibration suppression. Finally, some other issues about micro-vibration in integrated design are identified and the corresponding solutions are proposed. With the previous methods, the micro-vibration transmitted to high resolution remote

Z. Feng
Shenzhou Institute of China Academy of Space Technology, Shenzhou, China

Z. Feng · Y. Cui · X. Yang · J. Qin (✉)
Dong Fang Hong Satellite Co. Ltd, Shenzhen, China
e-mail: jiankang_53@163.com

H.P. Urbach and G. Zhang (eds.), *3rd International Symposium of Space Optical Instruments and Applications*, Springer Proceedings in Physics 192,
DOI 10.1007/978-3-319-49184-4_45

sensors has been mitigated sharply and the dynamic environment of remote sensors has been improved significantly.

Keywords Integrated design · Micro-vibration · Propagation path · Isolator

1 Introduction

Micro-vibration on spacecraft is a special kind of vibration which has tiny amplitude and high frequency, whose influence to structure is negligible. However, as to sub-meter satellites, micro-vibration has become a critical factor for high resolution image quality. With the development of satellite technologies, and the pressing need for high resolution, high agility and low costs, integrated design of satellite platform and payload has been a growing tendency.

Traditional remote sensing satellites are generally composed of propulsion module, service module and payload module. Compared to traditional designs, remote sensors in the integrated design are much closer to disturbance sources, implying that the micro-vibration environment of remote sensors becomes even worse. Analyzing the whole propagation process from momentum wheels (MW), cryocoolers and other disturbance sources to sensitive payloads, there are three possible ways to suppress the vibration transmitted to payloads: cutting down the micro-vibration of disturbance sources, accelerating the attenuation along vibration paths, lowering the sensitivity of payload. However, considering current abilities of designing, manufacturing and testing, cutting down a little bit of vibrations of disturbance sources itself could be very expensive, even if it is possible. Lowering the sensitivity of payload always contradict the demands for high resolution and high quality images. So focusing on the propagation pathways is the most feasible way currently.

In this article, several suggestions are proposed to mitigate the influence of micro-vibration on image quality for integrated design. First, the optimization design of structure and configuration is the first choice to suppress the micro-vibration transmitted to sensitive payloads. The propagation process of attenuation and amplification along vibration pathways is analyzed so that the structure and configuration of satellites can be optimized, making full use of structure to accelerate attenuation of micro-vibration. Another way to avoid degradation of image quality is to mount an isolator to cut down the micro-vibration. The isolator is selected in the system view in this article, considering the flexibility of structures, avoiding coupling with the natural frequency of the spacecraft structure and the parameters of the isolator are optimized. Finally, some other tough issues about micro-vibration in integrated design are identified and some possible solutions are proposed.

2 Optimization of Layout and Structure in Integrated Design

With the aim to cut down the vibration in the propagation path, several suggestions of layout and configuration are proposed in integrated design, making full use of structure to accelerate attenuation of micro-vibration. Although it is difficult to cut down the micro-vibration of disturbance sources itself, considering current abilities of technology and technics, it is possible to reduce the micro-vibration transmitted to payloads by optimizing the layout of disturbance sources and sensitive payloads.

2.1 *Optimization of the Layout of Disturbance Sources*

The main disturbance source of on-orbit satellites is mechanical spinning devices such as momentum wheels. The micro-vibration induced by the momentum wheels are mainly generated from its static imbalance, dynamic imbalance and bearing imperfection. Static imbalance is caused by the offset of the center mass of the wheel spin axis, while dynamic imbalance results from the misalignment of the principle axis and the rotating axis on the wheels. Bearing disturbances are caused by irregularities in balls, races and cages [1]. Micro-vibration induced by momentum wheels are determined by so many complicated factors that there are differences between momentum wheels, even they are of the same batch. The difference is the precondition of the layout optimization. Table 1 shows the transmission ratio from MW A, B, C, D to the secondary mirror when the MWs work individually.

When MW A, B, C and D work individually, the standard deviation of acceleration of bracket B is larger than that of bracket D. However, when it is transmitted to the secondary mirror, the micro-vibration from MW B is smaller than that from MW D, indicating the different transitive property of the two propagation pathways from MW B and MW D to high-resolution camera. Furthermore, it can be seen in Table 1 that the transmission ratio of the propagation from MW D to secondary mirror is the biggest while that from MW A is the smallest. So in the integrated design, if the disturbance source, which generates the biggest micro-vibration, is mounted on the propagation path of which the transmission ratio is the lowest and

Table 1 Transmission ratio from MW A, B, C, D to the secondary mirror when the MWs work individually

Subcases	From MW A to secondary mirror	From MW B to secondary mirror	From MW C to secondary mirror	From MW D to secondary mirror
Transmission ratio	0.03767	0.04065	0.04522	0.05059

the disturbance source, which generates the smallest micro-vibration, is mounted on the propagation path of which the transmission ratio is the highest, the micro-vibration transmitted to payloads would be reduced. It has been proved in the micro-vibration test of a satellite [2].

2.2 Suggestions of the Layout of Sensitive Payloads

High-resolution camera, star sensor, etc. are main sensitive payloads on-board spacecraft. The mounting position of payloads could have a great effect on their micro-vibration environment. When it comes to payloads layout design, in the premise of satisfying other constraints, the relative position relationship of sensitive payloads and disturbance sources needs to be taken into account. The relative position relationship can be divided into two aspects: the length of the propagation pathway from disturbance sources to sensitive payloads and the number of dis-continuous interface along the propagation pathways.

As shown in Table 1, the order of transmission ratio from MW A, B, C, D to high-resolution camera is MW A < MW B < MW C < MW D while the order of propagation length is MW A > MW B > MW C > MW D. It illustrates that the longer the propagation pathway gets, the more the micro-vibration attenuates, the lower the transmission ratio becomes. It can also be proved by focusing on the change of acceleration along the propagation paths, as shown in Fig. 1. Figure 1 shows accelerations along the propagation from MW A, D to the mounting position of high-resolution camera. As the length increases, the amplitude of the micro-vibration gets smaller. Therefore, if other constrains are satisfied, the layout of payloads need to be optimized to make the length of the propagation pathways as long as possible in integrated design.

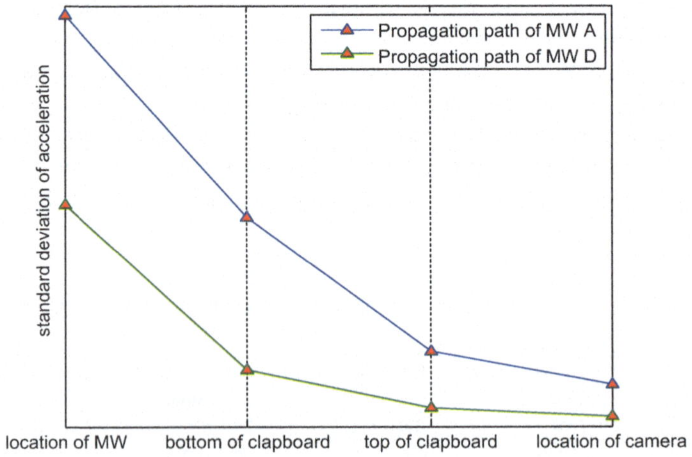

Fig. 1 Accelerations along pathways from MW A, D to the mounting position of high-resolution

Table 2 The standard deviation of acceleration above and below discontinuous interfaces

				$\times 10^{-3}$ g		
Location	Interface 1			Interface 2		
Direction	X	Y	Z	X	Y	Z
Points above the interface	2.937	3.012	1.918	2.545	2.050	1.629
Points below the interface	2.854	2.589	1.764	1.633	2.038	1.406

The number of discontinuous interface is another factor to be considered when it comes to payloads layout. The transitive property of micro-vibration transmitted between two modules is analyzed to figure out the influence of discontinuous interface, as shown in Table 2. It can be seen in Table 2 that the micro-vibration has been suppressed of three directions through discontinuous interface. So the micro-vibration environment can be improved to some extent by increasing the number of discontinuous interfaces in integrated design, taking costs, mounting space, etc. into consideration to make the best choice.

2.3 *Optimization of the Material and Size of Main Structure*

The main structure of a satellite refers to the load-carrying structure and the structures which provide mounting space for devices on-board. As to satellites employing box-plate structure as its main load-carrying structure, baseplates and clapboards of the service module and the payload module are regarded as the main structure. The material adopted by the main structure can result in great difference of local stiffness and damping coefficient. The size of the main structure, such as the thickness of the plate, can also have a significant influence on dynamic parameters

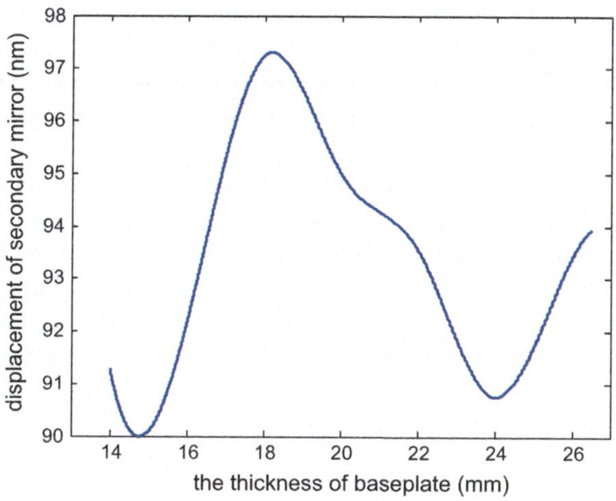

Fig. 2 The relationship of displacement of secondary mirror and the thickness of baseplate

of the structure and then it will affect the transitive property of the micro-vibration. Figure 2 shows the relationship of the stand deviation of displacement of secondary mirror and the thickness of baseplate in payload module. As shown in Fig. 2, 15 and 24 mm are the two minimum points of the curve. Therefore, the thickness around the two minimum points should be adopted in the main structure.

3 Vibration Isolation Analysis and Design

If the requirement of sensitive payloads for micro-vibration environment still could not be satisfied after adopting the above optimal methods, the isolator would be needed. As the resolution comes into sub-meter level, many spacecrafts have employed micro-vibration isolators for disturbance sources and sensitive payloads to comply with the strict mission requirement for the acquisition of high-quality images. Active-passive hybrid isolators are used in TacSat-2 lunched in 2006. WorldView-2, which was lunched in 2009, employs isolator for the CMG and the high-resolution camera. James Web space telescope, which will be lunched in 2018, adopts passive isolation techniques [3]. Some camera suppliers have supplied isolators as standard accessories.

There are four kinds of isolation techniques for isolation: passive, active, active-passive hybrid, and semi-active isolation [4]. The passive-type isolators have generally been employed in space application on account of their advantages of simplicity, low cost, and high reliability to achieve desired isolation performance. The isolator is typically designed and test in the stand-alone level. As a result, it may work well when tests alone, but when it is mounted on the satellite, its performance are usually not satisfying. Furthermore, the micro-vibration isolator is typically selected on the base of experience and a great number of experimental data, which cost large amount of money and time. What is worse is that the performance of the isolator is generally not the best using this method. In this article, the isolator is selected in the system view, considering the flexibility of the structures, avoiding coupling with the natural frequency of the spacecraft structure. In order to improve isolation performance, the axial and radical stiffness and damping coefficient of the isolator are optimized using finite element method (FEM).

The cryocooler on infrared camera produces undesirable micro-vibration disturbance during its on-orbit operation, which is one of the main sources of degradation of the image quality of high-resolution observation satellites. What is more, the propagation path from the cryocooler to high-resolution camera is short. Therefore, to avoid palpable declination of the image quality, micro-vibration disturbances induced by cryocooler operation need to be isolated.

3.1 Theoretical Analysis

Infrared camera has a much bigger mass and stiffness than that of the micro-vibration isolator and the natural frequency of the former is much higher than that of the latter. In this premise, the isolation system can be simplified into six decoupling single-degree-of-freedom isolation system. *The variables m* and x refer to the mass and displacement of the object to be isolated respectively. The variables k and c are the stiffness and damping coefficient of the isolator respectively. The variable $p(t)$ is the disturbance force while f(t) is the constraint force of the foundation—the force transmitted to the platform. According to the simplified Single-degree-of-freedom isolation system, we can establish the dynamic equation:

$$m\ddot{x} + c\dot{x} + kx = p(t) \tag{1}$$

The disturbance force is:

$$f(t) = c\dot{x} + kx \tag{2}$$

The transfer function can be obtained by Laplace transformation of (1) and (2):

$$G(s) = \frac{F(s)}{P(s)} = \frac{cs + k}{ms^2 + cs + k} \tag{3}$$

$$G(j\omega) = \frac{k + j\omega c}{(k - \omega^2 m) + j\omega c} \tag{4}$$

The natural frequency, damping ratio and frequency ratio of the isolation system are defined as follows:

$$\omega_n = \sqrt{\frac{k}{m}}, \quad \xi = \frac{c}{2\sqrt{mk}}, \quad \lambda = \frac{\omega}{\omega_n} \tag{5}$$

Then we can get the expression of transmission ratio:

$$TR = \left|\frac{F_S}{F}\right| = \sqrt{\frac{1 + (2\xi\lambda)^2}{(1 - \xi^2)^2 + (2\xi\lambda)^2}} \tag{6}$$

The isolation property of single-degree-of-freedom isolation system can be seen in Fig. 3, with frequency ratio as abscissa, transmission ratio as the ordinate, damping ratio as the parameter. As shown in Fig. 3, as the frequency ratio increases, the transmission ratio decreases. The smaller the stiffness of the isolation system is, the better the isolation performance is. It has also been proved in some experiments. However, if the stiffness is too low, the precision of installation may be impacted and so the angular displacement would increase, which is harmful to

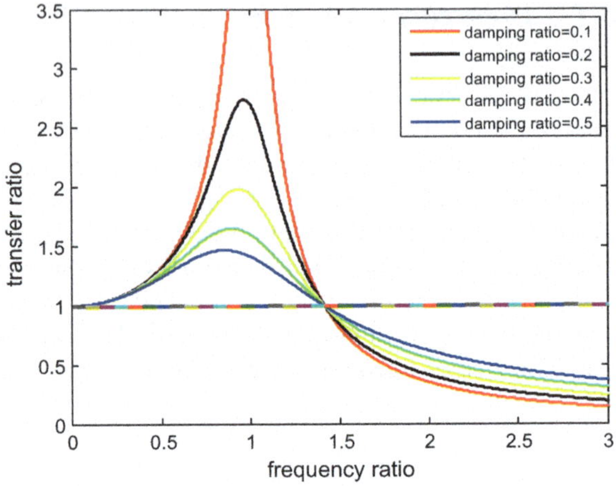

Fig. 3 Isolation property of single-degree-of-freedom isolation system

image quality. The general requirement of the resonant transmission ratio c is no greater than 3:

$$TR(\lambda = 1) = \sqrt{1 + \frac{1}{4\xi^2}} \leq 3 \Rightarrow \xi \geq 0.1768 \tag{7}$$

However, if the damping ratio is too big, the isolation performance in high frequency region could not be satisfying. Therefore, the compromise should be reached between the resonate transmission ratio and isolation performance in high frequency region.

3.2 FEM Simulation

The theoretical analysis gives the general picture of isolator parameter, but it is not accurate to simplify the satellite as rigid foundation. If the mass ratio between the foundation and the object to be isolated is no more than 20, and the natural frequency ratio is no less than 0.2, the influence of flexibility of foundation can not be neglected [5]. Considering the flexibility of foundation and coupling with other structures on-board spacecraft, FEM is a good method to take all these into account. With the FEM model of a typical high-resolution remote sensing satellite, the Bush element is employed to simulate the micro-vibration isolator.

Table 3 shows the comparison of the displacement of secondary mirror with or without an isolator. It can be seen that the isolator can reduce the micro-vibration significantly. Figures 4, 5 and 6 shows the relationship between the axial, radical stiffness and the displacement of secondary mirror of high-resolution camera. As shown in the figures, the smaller the stiffness or the damping coefficient is, the

Table 3 The comparison of the displacement of secondary mirror with or without an isolator

Subcases	*x* component	*y* component	*z* component
Without an isolator	41.4128	36.955	19.5725
With an isolator	16.5173	25.5305	15.1612

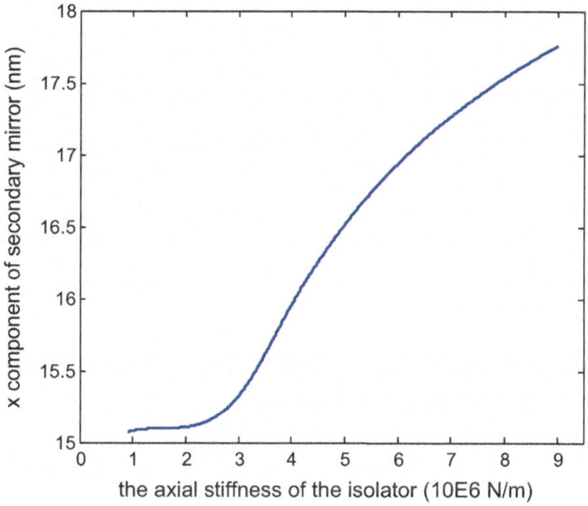

Fig. 4 Relationship between axial stiffness and the displacement of secondary mirror

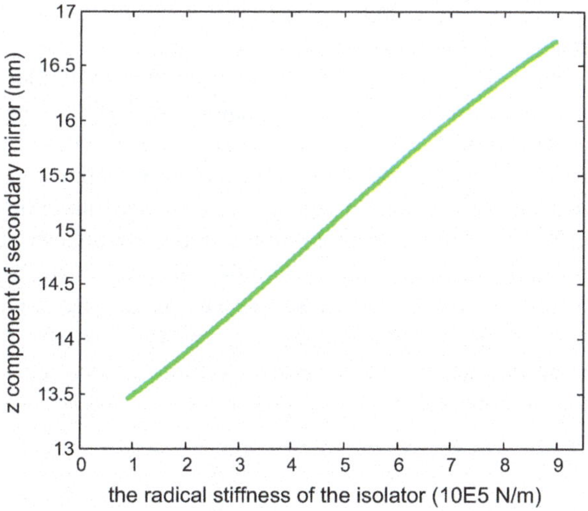

Fig. 5 Relationship between radical stiffness and the displacement of secondary mirror

better the performance of the isolator is. So if only the isolation efficiency is taken into consideration, the stiffness and damping coefficient of the isolator are recommended to be as small as possible.

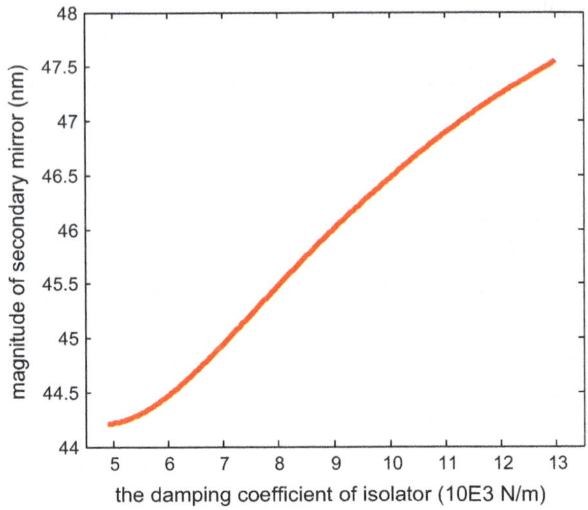

Fig. 6 Relationship between t damping coefficient and the displacement of secondary mirror

4 Other Tough Issues About Micro-vibration in Integrated Design

4.1 Micro-vibration Modeling Technique

The influence analysis of micro-vibration on sensitive payloads is complexity, involving structural subsystem, control subsystem, optical subsystem, etc. Integrated modeling method is an effective way for now. Integrated modeling method is a way to build the input-output model in system view, employing modeling tools of different majors, including structural model, optical model, control model, disturbance model, etc. Although the concept of integrated seems fresh, the techniques such as structural modeling, optical modeling, etc. have been reasonably mature, except the disturbance modeling. There are so many factors influencing the micro-vibration induced by disturbance sources such as flywheels, resulting in difficulties in building accurate disturbance models. As a result, the results achieved from FEM simulations are also not so precise.

Furthermore, the nonlinearity of the property of material and discontinuous interfaces has not been rigorously studied, so when we build the integrated model, the nonlinearity could not possibly be simulated accurately.

4.2 Micro-vibration Test Technique

In order to verify the performance of sensitive payloads under the micro-vibration disturbance, we need to do ground test. However, ground tests can not fully simulate the free boundary of on-orbit satellites, and the conditions of satellites between ground tests and on-orbit are different. For example, the solar panel is usually not mounted on the satellites in ground tests. The micro-vibration has low amplitude and so it is very sensitive to test conditions and surrounding dynamic environment. A little change of conditions or a little disturbance of surroundings would lead to great difference in micro-vibration. What is more, the effect of gravity on micro-vibration is also waiting to be studied. There are two possible solutions to these problems. One direct way is to mount acceleration sensor on the satellite. Then we can figure out the difference between the ground tests and on-orbit tests and build some correspondence between the two conditions. But the satellites carrying acceleration sensors are so less now that we cannot get enough data to analyze the difference. Another way to solve this problem is to build the FEM model according to ground tests. Then, we can modify the fixed boundary into free boundary and mount the solar panel to get a flight model, simulating the on-orbit conditions.

5 Conclusion

In this article, we proposed several suggestions to avoid degradation of image quality of high-resolution remote sensing satellites caused by micro-vibration in integrated design. Optimal design of layout and configuration is the first choice to reduce the micro-vibration transmitted to sensitive payload. Both tests and simulations have verified its efficiency. Another way to comply with the strict requirement of sensitive payload is to mount an isolator. The performance of the isolator is verified by both theoretical analysis and FEM simulations and the parameter of the isolator is optimized to get a better performance. With all these methods, the micro-vibration environment has been improved sharply and these methods can be used in subsequent integrated design satellites.

References

1. Seong-Cheol Kwon, Su-Hyeon Jeon, Hyun-Ung Oh, "Performance Evaluation of Spaceborne Cryocooler Miro-vibration Isolation System Employing Pseudoelastic SMA Mesh Washer", Cryogenics, January 2014.
2. YANG Xinfeng, BAI Zhaoguang, YANG Dong and LI Yanhui, "Study on Micro-vibration of Satellite Induced by Momentum Wheels and On-orbit Simulation Analysis", Equipment Environment Engineering, China, vol. 12, pp. 15–21, June 2015.
3. Zhang Zhenhua, Yang Lei, Pang shiwei, " Analysis of micro-vibration environment of high-precision spacecraft", Spacecraft Environment Engineering, China, vol. 26, pp. 528–534, December 2009.
4. Chunchuan Liu, Xingjian Jing, Steve Daley, Fengming Li, "Recent Advances in Micro-vibration Isolation", Mechanical Systems and Signal Processing, pp. 55–80, December 2014.
5. Wu Chengjun, "Vibration and Control in engineering" Xi'an Jiaotong University Press, January 2008.

Research on Simulation Method of Mineral Monitoring With Remote Sensing Satellites

Yue Zhang, E. Wei, Jianfeng Yin and Lixia Huang

Abstract High resolution remote sensing satellite can apply into mineral monitoring, but quantity of Chinese civil remote sensing satellites on orbit is limited. For analysing the ability and exploring the deficiency of satellites system, and improving the satellite system efficiency in mineral monitoring task, a simulation method of mineral monitoring with remote sensing satellites is proposed. Firstly the task demand analysis of mineral monitoring is developed, which is including the key observe areas and the payload parameters such as spectrum, resolution, and breadth and so on. Secondly, the satellites which can satisfy the task demand are selected, and base on the selected satellites system to build the simulation models. Thirdly, the task planning and scheduling methods is used to achieve the optimize simulation results. Finally the demand satisfaction of satellite system is assessed and the system efficiency is analysed. In this chapter, the simulation results is the parameters such as revisit time, covering time and continuous observation time of key observation areas. These parameters can be transformed to indicators which can reflect the satellite effectiveness. The contribution ratio of a single satellite to the whole satellite system is calculated and the satellite effectiveness is quantified, so the different kinds of satellites in the system can be compared with, meanwhile, the suggestion which can maximize satellites system effectiveness is provided specifically. Through the research on simulation method of mineral monitoring with remote sense satellites, the deficiency of remote sense payload and satellite system is recognized, which can provide the reference and suggestion for building optimization remote sense payload and satellite system in mineral monitoring application.

Keywords Mineral monitoring · Task planning and scheduling · Effectiveness simulation · Demand satisfaction degree

---

Y. Zhang (✉) · J. Yin · L. Huang
Institute of Spacecraft System Engineering CAST, Beijing, China
e-mail: Yzhang501@sina.com

E. Wei
School of Astronautics and Aeronautics, Harbin Institute of Technology, Shenzhen, China

H.P. Urbach and G. Zhang (eds.), *3rd International Symposium of Space Optical Instruments and Applications*, Springer Proceedings in Physics 192,
DOI 10.1007/978-3-319-49184-4_46

1 Introduction

Chinese mineral distribution has the characteristics of large area, various types and complex terrain, which makes challenges for low-cost and fast periodic monitoring. The mineral resource is wasted and produces large environment pollution problems without real-time monitoring [1, 2]. Traditional mineral monitoring with field investigation and reporting work step by step, which is not only time-consuming, expensive, but also difficult to make a timely response, therefore it is necessary to seek a new method which meets the requirements of current mine environmental monitoring. Compared with field investigation, remote sensing technology is fast, easily accessible, highly periodic and low-cost, which is play an important role in mineral monitoring task [3].

For a long time, mineral monitoring relies on foreign high-resolution remote sensing image. With the development of Chinese satellite remote sensing technology, the high-resolution satellite remote sensing data can apply into mines monitoring missions, which can reduce the cost of buying data and improve the frequency of monitoring [4–6]. However, the recent works about mineral monitoring with remote sensing satellites focus on remote sensing data management, image enhancement and calibration, information extraction and analysing [7–11]. There are few researches on performance of satellite system platform. Satellite system platform is the foundation of remote sensing data achievement, it's necessary to analyze the demand satisfaction of satellite system by effectiveness simulation.

In this chapter, the demand of mineral monitoring is analyse, including key observing areas, payload spectrum, resolution, observing swath and so on. The satellites which satisfy the demands are selected and the simulation model of these satellites are built. Based on task scheduling to optimistic satellite system, the efficiency indexes such as revisit time, coverage time and access time are calculated. Compared the simulation index value with demand index value, the satellite system performance whether can satisfy the mineral mission demand is identified. The demand satisfaction degree of satellite system is calculated at last. Not only the system deficiency is recognized, but also different system can be compared quantificationally, which can provide the reference and suggestion for building optimization remote sensing payload and satellite system in mines monitoring application.

2 Demand Analysis of Mines Monitoring

Remote sensing satellites mainly monitor the mineral exploitation status, exploitation order and geologic environment. The monitoring objects of mineral exploitation status include the position and range of mineral resource exploitation area; the mineral exploitation order monitoring is identify the situation such as

Table 1 Demand of mines monitoring

Spectrum (μm)	Important rank of spectrum	resolution	Coverage time	Revisit time	Observation swath
0.40–0.45	8	Less than 1 m	Less than 1 day	Less than 1 day	30–60 km
0.45–0.52	7				
0.52–0.59	1				
0.585–0.625	2				
0.63–0.69	3				
0.705–0.745	4				
0.770–0.895	5				
0.860–1.040	6				

exploitation without government permission, exploitation beyond the mark and change exploitation modes and mineral kinds without permission are exist; the monitoring objects of mineral geologic environment include mineral geologic disasters and pollution situation. Based on these missions, the demand for satellite and payload is shown in Table 1.

In the column of important rank of spectrum in Table 1, 1 means most important and 8 means worst important. The spectrum requirement of payload depends on detective objects. Generally, there are three optical payload configurations on satellites, which are single panchromatic spectrum payload, single multispectral payload, and panchromatic spectrum payload and Multispectral payload. Mostly satellites have panchromatic spectrum payload and Multispectral payload, and the spectrum include blue, green, red, and near inferred.

3 Effectiveness Simulation Method of Mineral Monitoring

For mineral monitoring task, an effectiveness simulation method is proposed. This method is consisted of 4 parts, which are satellite system modelling with task demand, task scheduling, efficiency index calculation, and demand satisfaction analysis. The flow of the effectiveness simulation method is shown in Fig. 1. First, based on task demand identify the observing area and satellite systems. Second, to calculate the index this can reflect the system performance, and analyze the demand satisfaction degree. When the index results can't satisfy the demand, the system is simulated using task scheduling to calculate the optimistic index. Then the demand satisfaction is analyzed once again, when the value still can't eligible, the new satellite system need to design.

A. Satellite system modelling with task demand

The orbit elements, platforms, payload parameters of global satellites on orbit and Chinese satellites which will launch in the future are collected to build one satellite and payload database. Based on the parameters of Table 1, such as resolution,

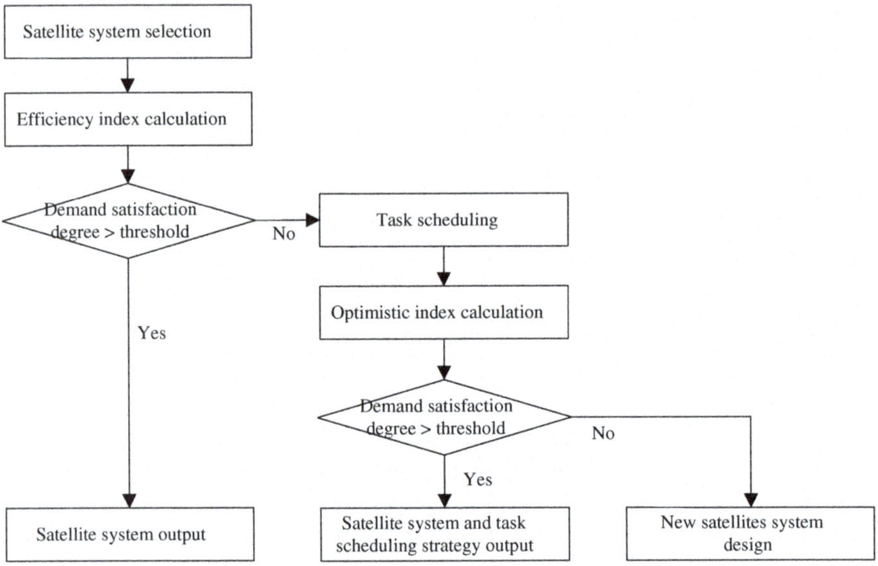

Fig. 1 Effectiveness simulation flow of mines monitoring

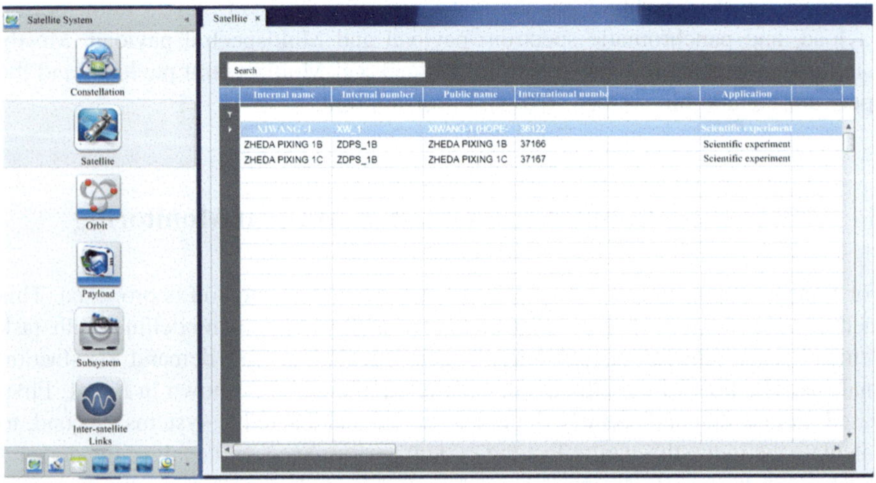

Fig. 2 Satellites and payloads database

observing swath, and spectrum, the satellites payloads which may execute the task will be selected from database. Then the simulation models of these satellites are built for effectiveness simulation (Fig. 2).

B. Task scheduling

Satellite can expand observing area by attitude maneuver and orbit maneuver, but orbit maneuver consume much more fuel than attitude maneuver, which will reduce in-orbit life of satellite sharply. Therefore, we use attitude maneuver to schedule task in the effectiveness simulation. The optimization objective of task scheduling model is minimal coverage time, and constrains are attitude maneuver ability, data storage, and electricity balance.

C. Effectiveness index calculation

For earth observing task, the efficiency indexes are revisiting time, coverage time, access number, access during time, and coverage percentage by satellite. These indexes are calculated in simulation, then using (1) to calculate the contribution ratio of satellite to the whole satellite system, which can evaluate every satellite performance in the system. The weight is changeable with different tasks.

$$C_k = \sum_{i=1}^{m} w_{ik} \frac{d_{ik}}{d_{i\max}} \tag{1}$$

where c_k is the contribution ratio of satellite k to the whole satellite system, d_{ik} is the value of index i for satellite k, d_{imax} is maximum index value in all satellites, w_{ik} is the weight of index i for satellite k, m is total number of indexes. The contribution ratio is calculated using indexes of access number, access during time, and coverage percentage.

D. Demand satisfaction analysis

Comparing the simulation index value with demand index value can identify the satellite system performance whether can satisfy the mineral mission demand. Through simulate the satellite system efficiency, the deficiency of satellite system is easily recognized, which can provide the reference and suggestion for building optimization remote sensing payload and satellite system in mineral monitoring application. The indexes of revisiting time and coverage time are used to calculate the satisfaction degree using (2) for mineral task.

$$S = \sum_{i=1}^{m} w_i \frac{(d_{ni} - d_{si})}{d_{ni}} \tag{2}$$

where S is system satisfaction degree, d_{ni} is demand value of index i, d_{si} is simulation value of index i, w_i is the weight of index i, m is total number of indexes. The contribution ratio is calculated using indexes of access number, access during time, and coverage percentage. When the value of S is level off to 1, the satellite system has better performance to satisfy the task demand.

4 Effectiveness Simulation Method Application

A. Satellites and target areas

Base on Chinese mineral distribution, three mines in northeast and one mine in west are selected in Fig. 3. Where, Mines1 located in Tonghua, Mines 2 located in Dulan, Mines 3 is located in Daqing, and Mines 4 is located in Anshan.

The agile earth observing satellite become more and more important, because it can maneuver both in pitching and rolling direction, which has boarder observing areas than traditional optical satellites. For compare with agile satellite and tradition satellites system quantificationally, two solutions of satellite systems are built for effectiveness simulation. One is consisted of three traditional optical satellites, and another is two agile earth observing satellite. The satellite parameters are shown in Table 2.

B. Effectiveness index calculation

Firstly, the indexes including access number, access during time and coverage percentage of two solutions are calculated in sub-stellar point imaging. Satellite contribution ratio of system is calculated using (1) based on the result which is shown in Table 3. The weights of access number per day, access during time per day and coverage percentage are 0.3, 0.35 and 0.35. Optical 2 has higher contribution ratio of system than the others in solution 1, and agile1 is little higher than agile 2. Therefore, the optical 1 and optical 2 are scheduled prior to improve the system efficiency.

Secondly, Using two satellite systems calculate the indexes of coverage time, revisit time. Coverage time is calculated in normal and task scheduling situation. The revisit time is calculated by considering the maximum observing range of payload,

Fig. 3 Mines areas selected in simulation

Table 2 Satellites parameters

Solution number	Solution 1	Solution2
Kind of satellite	Tradition optical satellite	Agile earth observing satellite
Quantity	3	2
Resolution	1 m	1 m
Observation swath	45 km	35 km
Panchromatic spectrum	0.45 μm–0.90 μm	0.45–0.90 μm
Multispectral	B1:0.45–0.52 μm	B1:0.45–0.52 μm
	B2:0.52–0.59 μm	B2:0.52–0.59 μm
	B3:0.63–0.69 μm	B3:0.63–0.69 μm
	B4:0.77–0.89 μm	B4:0.77–0.89 μm

Table 3 Satellite contribution ratio of system

Solution number	Solution 1			Solution2	
Satellite name	Optical 1	Optical 2	Optical 3	Agile 1	Agile 2
Access number per day	0.3	0.37	0.2	0.4	0.4
Access during times per day/s	2.8	4.3	1.8	3.9	4.1
Coverage percentage	34.41%	79.24%	28.01%	84.2%	65.31%
Contribution ratio of system	59.59%	95.69%	41.29%	96.74%	90.52%

which is equal to using task scheduling situation. The simulation results are show in Table 4, the coverage area in one day with task scheduling is shown in Fig. 4, which solution 1 is 13.6% and solution 2 is 100%. The maximum revisit time distribution is shown in Fig. 5.

C. Demand satisfaction analysis

We use revisit time and coverage time as efficiency index, because they can reflect the remote sensing satellite performance directly [12], after calculate these two efficiency index, we use (2) to calculate the demand satisfaction degree. The index weight of revisit time and coverage time both are configured to be 0.5. S1 and S2 are the demand satisfaction degree of solution 1 and solution 2. The value of S1 and S2 are −1.9792 and 0.466. When the value of S is level off to 1, the satellite system has better performance to satisfy the task demand. Therefore, solution 2 is better than solution 1 for mineral monitoring task. Base on the effectiveness simulation results, the observing performance of satellite systems is improved sharply by task scheduling, and agile earth observing satellites have better performance than tradition optical satellites in attitude maneuver ability, which effect the demand satisfaction of mineral monitoring task directly.

Table 4 Coverage and revisit time with different satellite systems

Solution number	Coverage time		Revisit time
	Sub-stellar point imaging	Task scheduling	
Solution 1	26 days	5 days 7 h	16 h
Solution 2	34 days	1 h 38 min	24 h

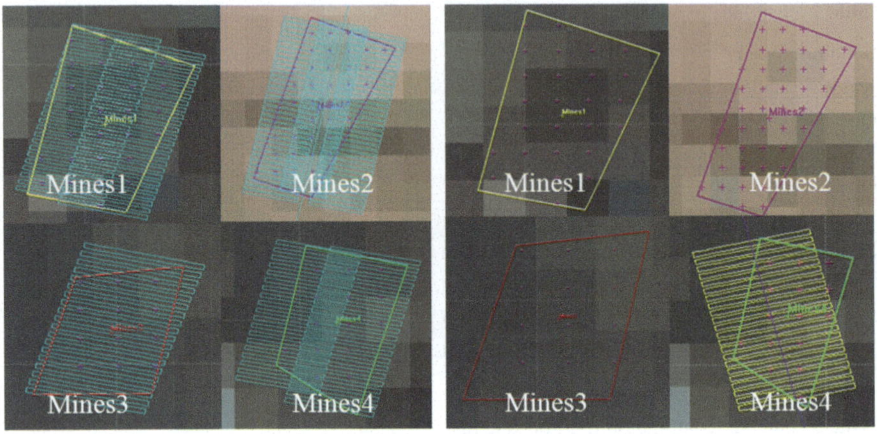

Fig. 4 Coverage areas in four mines area with task scheduling (*right* is solution 1, *left* is solution 2)

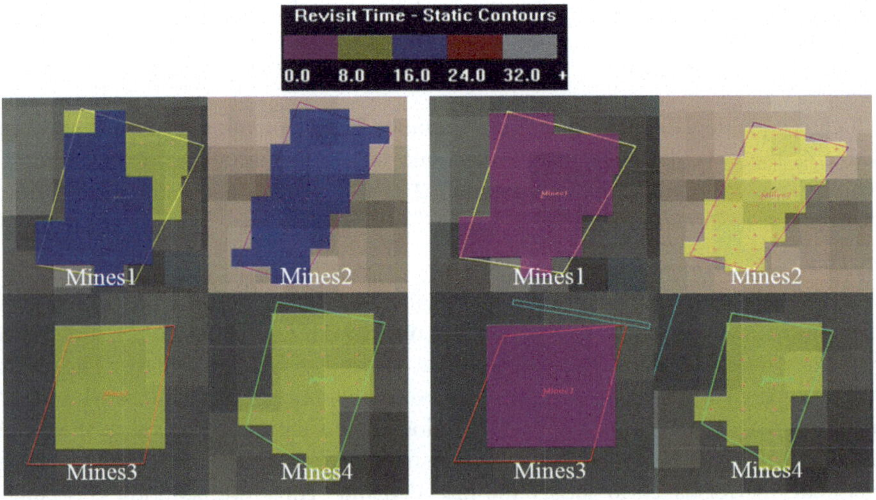

Fig. 5 Revisit time distribution in four mines area with task scheduling (*right* is solution 1, *left* is solution 2)

5 Conclusion

Through the research on simulation method of mineral monitoring with remote sensing satellite, a quantitative method of satellite effectiveness assessment is exploited based on mineral monitoring demand to remote sensing satellite system. The design solution of satellite system is assessed in the view of demand satisfaction. This method not only can be applied into mineral monitoring task, but also can assess the satellite system design solution in other remote sensing of earth observing task, which provide the foundation and suggestion for satellite system development and assessment.

References

1. Yang J Z, Qin X W, Zhang Z, et al. "Theory and Practice on Remote Sensing Monitoring of Mine," *Sinomaps Press*, 2011.
2. Qin X W, Yang J Z, Kang G F, et al. "Technology and Method Research on Remote Sensing Monitoring of Mine." *Sinomaps Press*, 2011.
3. Chen shupeng, Tong qingxi, Guo Huadong. "Information mechanism of remote sensing." *Science Press*, 1998.
4. Wang X H, Nie H F, Yang Q H, et al. "The different monitoring effects of QuickBird and SPOT-5 data in mine exploitation." *Remote Sensing for Land and Resources*, vol. 16(1), pp. 15–18. 2004.
5. Chen W T, Zhang Z, Wang Y X. "Advances in remote sensing-based detecting of mine exploitation and mine environment." *Remote Sensing for Land and Resources*, vol. 21(2), pp. 1–8. 2009.
6. Liu Z, Huang J, Shao H Y, et al. "The application of RapidEye satellite images to 1: 50000 remote sensing survey for mine exploitation." *Remote Sensing for Land and Resources*, vol. 22(1), pp. 127–129, 133. 2010.
7. LU Yunge, FAN Shuangliang, LI Chunlin. "Application of GF-2 Satellite in Remote Sensing Monitoring on Mine Exploitation in Tibet". *Spacecraft recovery & Remote Sensing*, vol. 36 (4), pp. 73–81. 2015.
8. Hao Lina. "A Remote Sensing Study on Mine Environmental Effects: A Case of the Key Ore Concentrating Area in Hubei Province". *China University of Geosciences*, 2013.
9. Chen W T, Zhang Z, Wang Y X, Zhang L. "Preliminary study on methods of geo-environment monitoring in minesites using remote sensing technique". *Geological Bulletin of China*, vol. 29(2/3), pp. 457–462. 2010.
10. Zhu Zheniian. "Study on Mine Area information extraction based on object-oriented high-resolution remote sensing image Classification and its application". *China University of Geosciences (Beijing)*, 2010.
11. LIU Zhi, HUANG Jie, SHAO Huai-yong, ect. "The Application of RapidEye Satellite Images to 1:50 000 Remote Sensing Survey for Mine Exploitation". *Spacecraft recovery & Remote Sensing*. vol. 83, pp. 127–129. 2010.
12. Wertz J R, Larson J W. "Space Mission Analysis and Design". *Springer*.1999.

Compact Spectrometers for Earth Observation

B. Snijders, L.F. van der Wal, B.T.G. de Goeij, R. Jansen, P. Toet and J.A.J. Oosterling

Abstract Driven by technology developments triggering end user's attention, the market for nano- and microsatellites is developing rapidly. At present there is a strong focus on 2D imaging of the Earth's surface, with limited possibilities to obtain high resolution spectral information. More demanding applications, such as monitoring trace gases, aerosols or water quality still require advanced imaging instruments, which tend to be large, heavy and expensive. In recent years TNO has investigated and developed several innovative concepts to realize advanced spectrometers for space applications in a more compact and cost-effective manner. This offers multiple advantages: a compact instrument can be flown on a much smaller platform (nano- or microsatellite); a low-cost instrument opens up the possibility to fly multiple instruments in a satellite constellation, improving both global coverage and temporal sampling; a constellation of low-cost instruments can provide added value to the larger scientific and operational satellite missions. Application of new technologies allowed us to reduce the instrument size significantly, while keeping the performance at a sufficient level. Low-cost instruments may allow to break through the 'cost spiral': lower cost will allow to take more development risk and thus progress more quickly. This may lead to a much faster development cycle than customary for current Earth Observation instruments. This new development approach is demonstrated using the most advanced design of a hyperspectral imaging spectrometer (named 'Spectrolite') as an example. Several different novel design and manufacturing techniques were used to realize this compact and low-cost design. Laboratory tests as well as the first preliminary results of airborne measurements with the Spectrolite breadboard will be presented. The design of Spectrolite offers the flexibility to tune its performance (spectral range, spectral resolution) to a specific application. Thus, based on the same basic system design, Spectrolite offers a range of applications to different clients. To illustrate this, we present a mission concept to monitor NO_2 concentrations over urban areas at high spatial resolution, based on a constellation of small satellites.

B. Snijders (✉) · L.F. van der Wal · B.T.G. de Goeij · R. Jansen · P. Toet · J.A.J. Oosterling
TNO, PO Box 155, 2600 AD Delft, The Netherlands
e-mail: bart.snijders@tno.nl

H.P. Urbach and G. Zhang (eds.), *3rd International Symposium of Space Optical Instruments and Applications*, Springer Proceedings in Physics 192,
DOI 10.1007/978-3-319-49184-4_47

Keywords Earth observation · Imaging spectrometer · Constellation · Air quality · Spatial resolution · Temporal resolution · Spectral resolution · Freeform mirrors · 3D-printing · Small-sats · Airborne test · Modelling · Data assimilation

1 Introduction

The market for nano- and microsatellites is developing rapidly, largely driven by the continued interest of the commercial sector in nano- and microsatellite applications.

Over the past decades TNO has a leading role in developing high-end spectrometers for remote sensing, with a focus on air quality monitoring. This experience has now been aimed at the design and realisation of a very compact instrument, fit for nano- and microsatellites, with a performance that comes close to the performance of existing larger instruments.

2 Obtaining Spectral Information

The strong increase in the number of smaller satellites for Earth Observation applications is mainly focused on imaging the Earth's surface with medium or high resolution. Possibilities to obtain spectral information are still limited. Therefore, more demanding applications, such as monitoring trace gases, aerosols, or water quality still require advanced spectral imaging instruments, which are large, heavy and expensive. Given our strong background in the design of high-grade atmospheric monitoring instruments, we were challenged to see if we could create a break-through in this area.

In 2013 we started to investigate different innovative designs to realize advanced spectrometers for space applications in a very compact and cost-effective manner. Our R&D efforts were motivated by the following ambition:

- A compact instrument can be flown on a much smaller platform (nano- or microsatellite);
- A low-cost instrument opens up the possibility to fly multiple instruments in a satellite constellation, improving both global coverage and temporal sampling (e.g. multiple overpasses per day to study diurnal processes);
- In this way a constellation of low-cost instruments may provide added value to the larger scientific and operational satellite missions (e.g. the Copernicus Sentinel missions);
- A low-cost instrument may also allow us to break through the 'cost spiral': lower recurring cost will allow us to take more risk in the development and thus progress more rapidly. This may lead to a much faster development cycle than customary for current Earth Observation instruments, which may lead to opening up the worldwide application of this type of air quality monitoring.

3 Spectrolite

The above considerations will be illustrated using the most mature design of a hyperspectral imaging spectrometer (recently named 'Spectrolite') as an example. During the conceptual development of Spectrolite we applied the following principles:

- Make stand-alone system that is optimized for a specific task (i.e. a single observable phenomenon); this keeps the development complexity relatively small and ensures that tough requirements for different observable phenomena are not mingled, keeping the cost for development and production low.
- Make a modular system that can easily be adopted for different observable phenomena (e.g. air quality, land use or water quality) without a full redevelopment and qualification; this should allow the system to be built in series, allowing the development cost to be shared over multiple modules.

Based on these constraints Spectrolite ended up as an all-reflective, off-axis design, consisting of only 8 optical components, including 4 free-form mirrors and a flat grating in the spectrometer. The all-reflective approach has a number of advantages:

- A favourable thermal stability due to the all-aluminium design (important for a low-mass satellite, with limited possibilities for thermal isolation and stabilisation);
- The design can be copied unchanged for other wavelength ranges between 270 and 2400 nm (with tuning of the grating and the detector only);
- Fewer optical elements without the need for correction of chromatic aberration and the application of diamond turned freeform optics;
- More design freedom regarding stray light baffling due to the off-axis nature of such a design;
- Little or no ghost images from reflections of detector and optical surfaces.

3.1 Requirements for Spatial and Temporal Resolution

The spatial resolution, characterised by the Ground Sampling Distance (GSD), has been improved further for every generation of imaging spectrometers developed by TNO. Especially for a cost-efficient concept the balancing of the GSD is a key element in the specification of the instrument concept; this requires a trade-off between spatial resolution, data volume, mission cost and user needs. An typical case of information needed by end-users is the emission of the NO_2 sources in urban areas. In Fig. 1 an overview picture of a densely populated area with typical NO_2 emissions of industry, domestic and traffic is given.

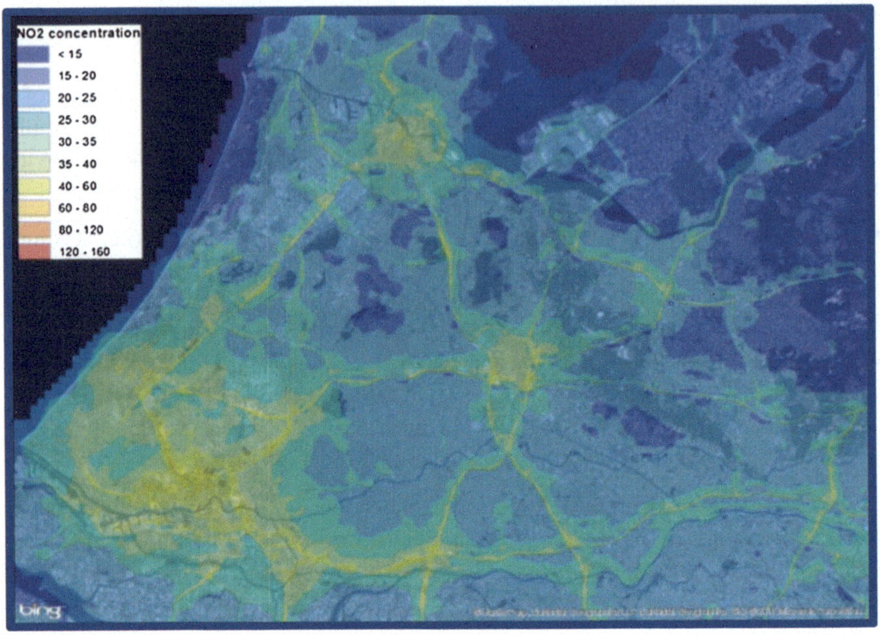

Fig. 1 NO_2 emissions of the western part of the Netherlands, generated from an emission database (Sander Jonkers, TNO 2015)

Figure 2 provides a resampled picture, simulating observation with an satellite instrument with 1 km GSD; many of the details of emissions are clearly visible.

In Fig. 3 a resampling with 10 km GSD is applied; many of the details are lost here. This analysis has led to the very preliminary choice of 1 km GDS; this number has to be further optimised by more detailed analyses.

Synthesis of different types of measurements has shown the dynamics of NO_2 variations, see Fig. 4.

The graph shows that NO_2 concentrations can vary strongly over a few hours, and the satellite measurement (in this case by the OMI instrument) only provides a single sample of this variation per day. Extension to 3 measurements over the period of daylight hours will definitely improve capturing the daily variation, and improve distinguishing the different source signatures. The Spectrolite approach with its low-cost platform enables sampling the air quality multiple times per day. Further analysis has to be performed to derive the optimal configuration of Spectrolite instruments, in combination with other sources of air quality data, for accurate measurement of daily variations of NO_2.

The temporal and spatial frequencies of NO_2 variations decrease as a function of the distance to the source, so nominally the variations on ground level are larger than higher in the atmosphere. In this way high resolution measurements can be used to further reduce the uncertainties in the vertical distribution of NO_2.

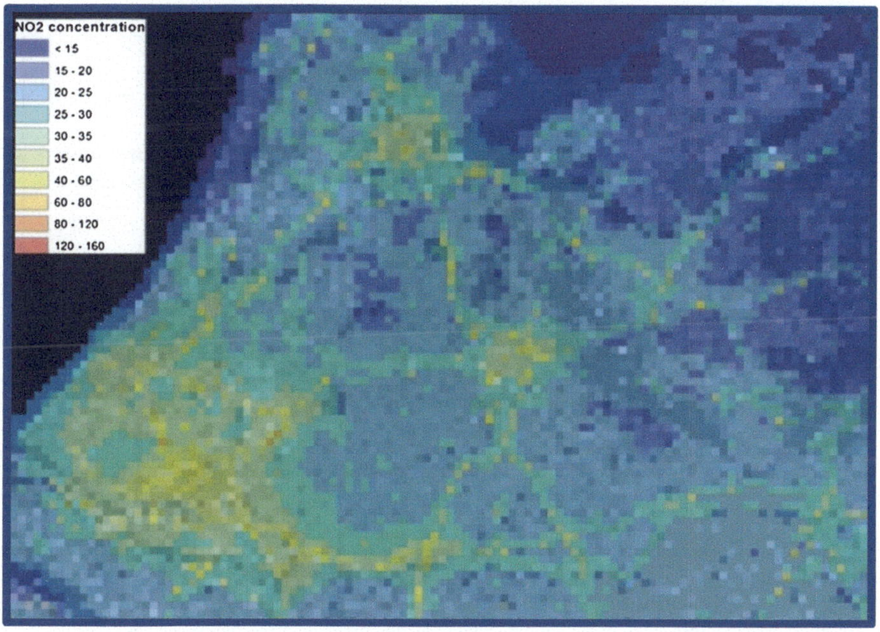

Fig. 2 NO_2 emissions, resampled with 1 km GSD

Fig. 3 NO2 emissions, resampled with 10 km GSD

A constellation of two of more Spectrolite instruments can be used to observe the atmosphere simultaneously under different angles, improving the accuracy of the vertical profile measurement further.

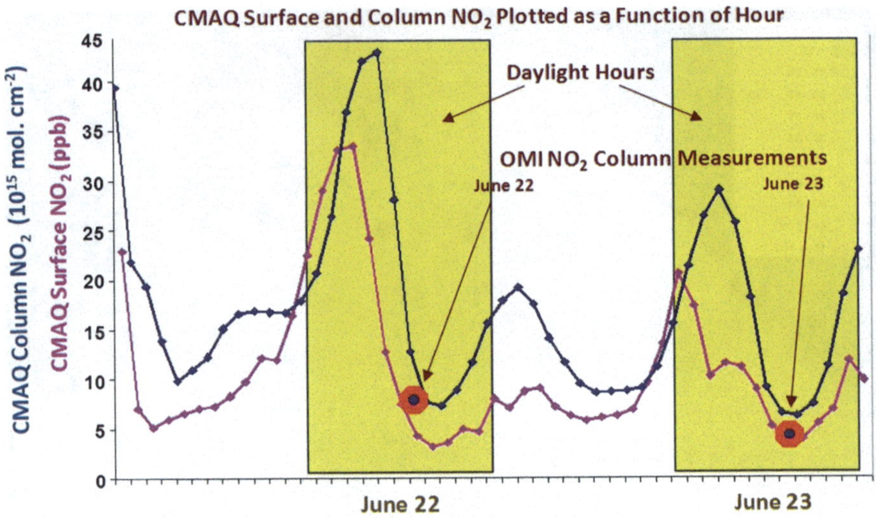

Fig. 4 Ground-based validation of EOS-Aura OMI NO_2 vertical column data in the midlatitude mountain ranges of Tien Shan (Kyrgyzstan) and Alps (France), D. Ionov, YM. Timofeyev, VP. Sinyakov, VK. Semenov, F. Goutail, JP. Pommereau, EJ. Bucsela, EA. Celarier, M. Kroon, J. Geophys. Res., 2008

3.2 *Spectrolite Breadboard*

In order to demonstrate that this design philosophy actually works, a Spectrolite breadboard was developed and built. The breadboard was optimized for the detection of NO_2 concentrations in the atmosphere.

For this breadboard the following requirements were set:

- Spectral range: 320–495 nm
- Spectral resolution: <0.5 nm
- Spatial resolution: 0.1° (1 km GSD @ 600 km)
- Field of view: 60° ($\sim$700 km swath @ 600 km)
- SNR for NO_2 measurement: 500 over urban areas

Figure 5 shows a 3D drawing of the designed Spectrolite breadboard, including the optical path (Ref. [1]).

The design is very compact and fits in a volume of $10 \times 20 \times 20$ cm^3, as shown in Fig. 6, due to this small volume, Spectrolite easily fits in a standardized 6U or 12U volume CubeSat structure, while leaving sufficient room for an additional instrument and the supporting hardware (electronics, communication, attitude and thermal control, etc.).

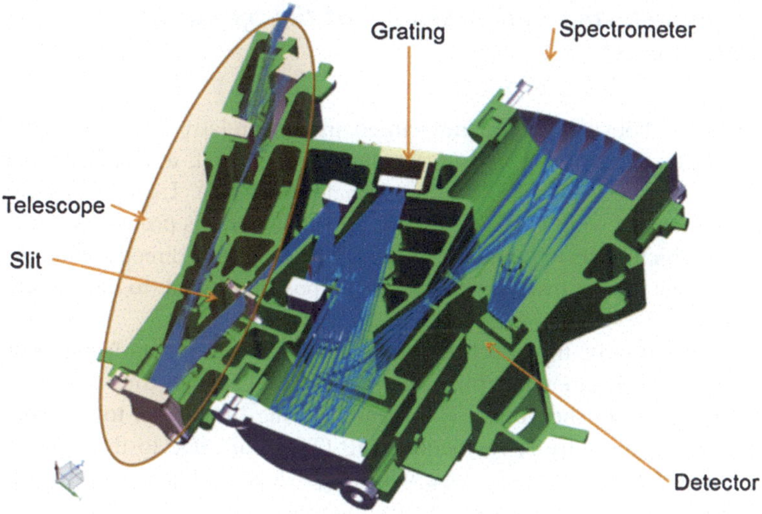

Fig. 5 3D drawing of Spectrolite, showing the housing (*green*), all components (*white*) and the optical path (*blue*)

Fig. 6 Schematic, showing how Spectrolite fits into the standardized mechanical structure of a large CubeSat

3.3 Manufacturing and Assembly of the Spectrolite Breadboard

The Spectrolite housing was manufactured using a relatively novel method for space hardware: 3D printing and investment casting. First, a 3D wax model was printed, from which a ceramic cast was made (Ref. [2]). Using this cast, three aluminium housings were made. The spatial accuracy of this process is in the order of 0.1 mm, which is insufficient for mounting the mirrors directly without alignment. Therefore the mirror interfaces were post-machined to an accuracy of 20 μm.

The four mirrors of the spectrometer (see Fig. 5) were manufactured at TNO using Single Point Diamond Turning with a slow tool servo. After this the mirrors were lightly hand polished only. The manufacturing turned out to be more challenging than e.g. the mirrors for Sentinel 5P (Tropomi), due to the more difficult mounting and the 'aggressive' shape with large slopes. This required a very low turning speed, which in turn required very high thermal stability of the cutting tool (<10 nm over a period of minutes or more). Also the effective aperture was only ~1 mm smaller than the edge of the mirrors.

4 Verification of the Spectrolite Breadboard

A number of tests and analyses have been performed to verify the performance of the spectrometer part of the Spectrolite breadboard. The main performance parameters that were verified were:

- The (RMS) spatial spot size
- The (RMS) spectral spot size
- The polarization sensitivity
- Co-registration

A detailed description of these verification activities falls without the scope of this paper. Here we only present the results of the spectral spot size verification.

4.1 Spectral Spot Size Verification

The spectral spot size was measured using a HgAr spectral line source coupled into a small integrating sphere. The sphere was placed in front of the 'telescope' objective, so that the slit is overfilled in spectral direction, but it was not possible to illuminate the entire slit length. Therefore a number of measurements were performed at several spatial slit positions. Figure 7 shows a single detector frame from

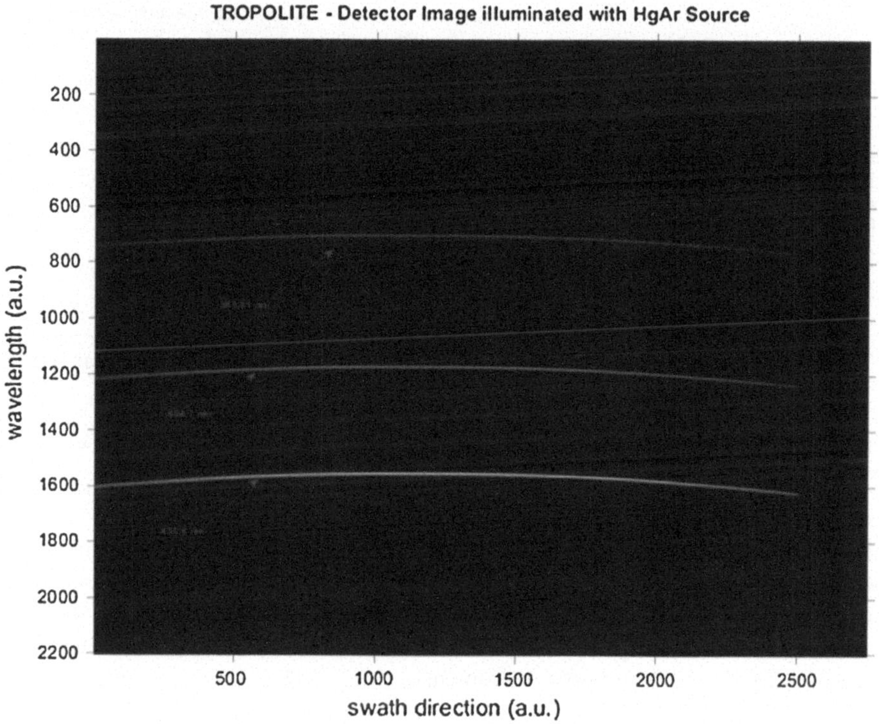

Fig. 7 Detector frame; the three lines correspond to 365, 405 and 436 nm (from *top* to *bottom*)

the spectral measurements. The three lines correspond to 365, 405 and 436 nm from top to bottom. A sub frame around each spectral line was selected and the position and height of the peak in each column (i.e. spatial position) was determined. With the results, the 'smile' was characterized and the peak heights were normalized.

The spectral resolution varies between ~0.2 and ~0.33 nm and is best for the highest wavelengths. The small gradient in spatial direction might be caused by a non-optimal alignment of the detector. The spectral resolution measurement was confirmed by looking closely at the 365 nm line. In reality this is a double line, with a spacing of 0.47 nm. This doublet could be clearly resolved by the spectrometer, as shown in Fig. 8.

4.2 *Spectrolite Airborne Test Campaign with DLR*

In the first months of 2016 the Spectrolite spectrometer breadboard was combined with an off-the-shelve imaging system to provide an instrument for the DLR

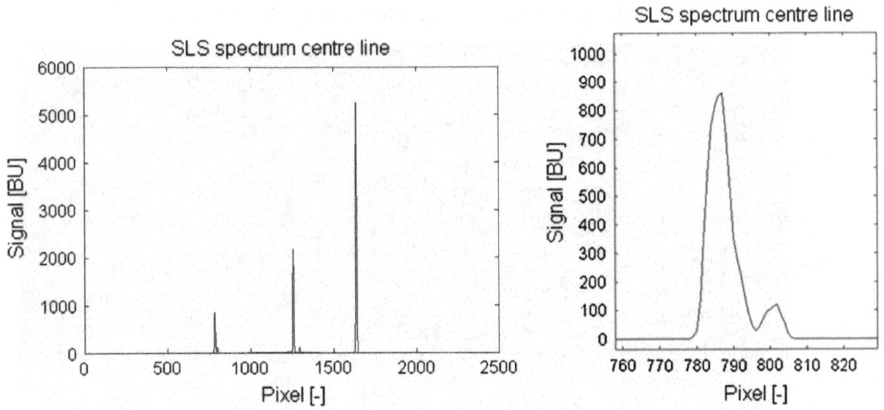

Fig. 8 Cross cut through the HgAr source spectrum (*left*) and zoom in on the double peak at 365 nm (*right*)

airborne measurement campaign over Berlin. Basic thermal control and data acquisition systems were added, and the instrument system was integrated in the airplane. The system performed well during the measurements; a first analysis of the observations show good correlation with NO_2 concentrations simultaneously measured by the other airborne instruments.

4.3 *Spectrolite Consortium Activities*

Initiated by the positive results of our efforts to design high-grade, compact spectrometers for Earth Observation from SmallSats, several Dutch space companies decided to join TNO and establish a consortium that intends to provide ‘end-to-end’ Earth Observation solutions with nano- and microsatellites, aiming at the production of small series. In the long run the portfolio of the consortium will include:

- Mission development
- Instrument/platform design
- Data retrieval/processing
- Launch arrangement
- Ground segment
- Modelling/data assimilation
- Customized data service

Currently the Spectrolite consortium consists of the following organizations: Airbus Defence & Space Netherlands, Innovative Solutions in Space (ISIS), Science & Technology (S&T) and TNO.

5 Discussion and Conclusions

We have proposed and realized different innovative designs to manufacture compact, high-grade spectrometers for space applications in a cost-effective manner. To demonstrate the feasibility of these designs, we realised the Spectrolite breadboard. Spectrolite is based on an all-reflective, off-axis design, consisting of only 8 optical components, including 4 freeform mirrors and a flat grating in the spectrometer.

The Spectrolite housing was manufactured using a relatively novel method for space hardware: 3D printing and investment casting. The combination of 3D printing of the wax model and investment casting provides a large design freedom. The 3D printing of the wax model allows the casting to have a very complex shape, while the casting itself allows the design of accurate, lightweight and stiff structures. When producing multiple housings in small series, 3D printing + investment casting becomes more cost-effective than traditional milling. Cost savings of more than 20% can be achieved for series of 10 or more.

The manufacturing of the freeform mirrors turned out to be more challenging than expected, due to the more difficult mounting and the 'aggressive' shape with large slopes. This required a very low turning speed, which in turn required very high thermal stability of the cutting tool (<10 nm over a period of minutes or more). The mirrors and the grating were assembled on manufacturing tolerance; the assembly took approximately 1.5 days in total. The alignment of the detector was done in approximately one day. The spectral resolution varies between $\sim$0.2 and $\sim$0.33 nm and is best for the highest wavelengths.

By realizing the Spectrolite breadboard, we have successfully demonstrated the feasibility to design and manufacture very compact, high-grade spectrometers in a cost-effective manner. Given their size and weight these spectrometers can be flown on very small platforms, e.g. nano- and microsatellites. Moreover, they can easily be 'tuned' to observe different phenomena (e.g. air quality, land use or water quality) without requiring full redevelopment and qualification. This offers the opportunity to build these spectrometers in series and share the development among multiple modules.

An airborne test has demonstrated the correct performance of the Spectrolite instrument system.

Initiated by the positive results of our efforts to design high-grade, compact spectrometers for Earth Observation from SmallSats, a number of Dutch space companies decided to join TNO and establish a consortium that intends to provide 'end-to-end' Earth Observation solutions with nano- and microsatellites, aiming at the production of small series.

The consortium will first focus on a SmallSat mission to monitor NO_2 over densely populated areas. In order to pinpoint local sources and characterize their source signatures better, the consortium intends to supply NO_2 data at high spatial resolution to a variety of cities/regions/countries multiple times per day using a constellation of nano- or microsatellites (Ref. [3]).

References

1. Volatier J-P., Baümer S., Kruizinga B. and Vink R., "Case study: TROPOLITE", Proc. SPIE 9131, 91310L, May 1, 2014.
2. Oosterling J.A.J., "Advanced Manufacturing Methods for Systems-of-Microsystems Nanospacecraft", TNO technical note for ESA, ref. TNO-ESA-AM-00, 2015.
3. Goeij B.T.G. de, Groot Z. de, Rotteveel J., Valk N. van der and Otter G., "Using small satellites and instruments for an operational air quality service", Proc. 10th Symposium on Small Satellites for Earth Observation, Berlin, 2015.

Radiometric Calibration of the GOME-2 Instrument

Gerard Otter, Niels Dijkhuizen, Amir Vosteen, Sanneke Brinkers, Bilgehan Gür and Pepijn Kenter

Abstract The Global Ozone Monitoring Experiment-2 [1] (GOME-2) represents one of the European instruments carried on board the MetOp satellite within the ESA's "Living Planet Program". Consisting of three flight models (FM's) it is intended to provide long-term monitoring of atmospheric ozone and other trace gases over a time frame of 15–20 years, thus contributing valuable input towards climate and atmospheric research and providing near real-time data for use in air quality forecasting. The ambition to achieve highly accurate scientific results requires a thorough calibration and characterization of the instrument prior to launch with highly specialised sources [2]. These calibration campaigns were performed by TNO in Delft in the Netherlands, in the "Thermal Vacuum Calibration Facility" of the institute. Due to refurbishment and/or storage of the instruments over a period of a few years, several re-calibration campaigns were necessary. These re-calibrations provided the unique opportunity to study the effects of long term storage and build up statistics on the instrument as well as the calibration methods used. During the re-calibration of the second flight model a difference was found in the radiometric calibration output, which was not understood initially. In order to understand the anomalies on the radiometry, a deep investigation was performed using numerous variations of the setup and different

G. Otter (✉) · N. Dijkhuizen · A. Vosteen · S. Brinkers
TNO, Delft, The Netherlands
e-mail: gerard.otter@tno.nl

N. Dijkhuizen
e-mail: niels.dijkhuizen@tno.nl

A. Vosteen
e-mail: amir.vosteen@tno.nl

S. Brinkers
e-mail: Sanneke.brinkers@tno.nl

B. Gür · P. Kenter
S&T, Delft, The Netherlands
e-mail: bilgehan.gur@tno.nl

P. Kenter
e-mail: kenter@stcorp.nl

H.P. Urbach and G. Zhang (eds.), *3rd International Symposium of Space Optical Instruments and Applications*, Springer Proceedings in Physics 192,
DOI 10.1007/978-3-319-49184-4_48

sources. The major contributor was identified to be a systematic error in the alignment, for which a correction was applied. Apart from this, it was found that the geometry of the sources influenced the results. Based on the calibration results combined with a theoretical geometrical hypothesis inferred that the on-ground calibration should mimic as close as possible the in-orbit geometry.

Keywords GOME-2 · Calibration · METOP · Radiometry · Spectrometer · UV · VIS · NIR · Stray-light

1 Introduction

1.1 The Instrument

The GOME-2 instrument is a scanning grating spectrometer flying in low earth orbit LEO. The spectral bands are continuous in a range from 240 to 790 nm with a spectral resolution ranging from 0.2 (240–400 nm) to 0.4 nm (400–790 nm). Additionally two polarization sensitive prism spectrometers are implemented measuring s and p polarization of the incoming light in the range of 310–1100 nm.

The instrument is equipped with a calibration unit with three different sources; a hollow cathode lamp used to produce a spectrum with atomic lines that is used to monitor the wavelength calibration in-orbit, a quartz tungsten halogen lamp which gives a black body like spectrum to monitor changes of the detector etalon and a sun diffuser which is illuminated by the sun while in the eclipse, this is used to do the radiometric calibration.

1.2 Calibration Approach

The GOME-2 instrument uses an extensive on-ground calibration in combination with on-board sources in a dedicated calibration unit. The on-ground calibration delivers key-data that is used in the on-ground processor to convert the raw data of the instrument into physical units. The on-board unit is used to monitor changes of the instrument while in orbit and to alter calibration key-data if necessary.

The instrument is calibrated on-ground for radiometry, polarization, spectral response, field of view, detector response and stray-light. The instrument is placed into a vacuum chamber (<10–5 mBar) and cooled down to its nominal in-orbit temperature. In this way the in-orbit conditions are simulated except for exact temperature gradients and gravity.

The position of the instrument in the chamber is fixed so that only nominal angles of incidence in the earth and sun port can be measured. Setups in ambient are used to determine the instrument response to different sun incidence angles on the on-board diffuser and to determine the scan mirror angle dependence.

1.3 Storage Approach

The MetOp mission consist of three satellites which were built in sequence. The instruments were all calibrated prior to integration on the satellite. After the first MetOp launch in 2006 the GOME-2 instruments were dismounted from their satellites and stored at the instrument prime contractor's premises.

For each next launch the corresponding instrument would again have to be calibrated before mounting it to the satellite. This was intended to be a delta campaign with respect to the original calibration to confirm stability.

1.4 Observation During Storage

After a storage period of 6 years the calibration measurements of the second flight model were repeated. For most of the over 100 key-data the differences with the previous calibration were not significant, proving the stability of the instrument and its calibration [3]. However some differences were more fundamental and forced a deeper investigation of the measurement method and the instrument design. In particular for the radiometric calibration large differences were found.

2 Radiometric Calibration

2.1 Setup

The GOME-2 instrument has an earth viewing port and a sun port. The radiometric responses of both ports was to be absolutely calibrated. The response of the earth port is referenced to as radiance response and the response from the sun port is referenced to as the irradiance response. A NIST calibrated FEL lamp is used for this absolute calibration.

For the irradiance measurements (Fig. 1) the FEL lamp directly illuminates the instrument sun diffuser in the sun port. For the radiance measurements (Fig. 2) the FEL lamp illuminates an external diffuser to convert the FEL irradiance into a radiance (and simulate the earth as a diffuser). This diffuser is also calibrated in the absolute radiometric calibration facility at TNO. In both cases the distance between either external or instrument diffuser and the FEL lamp were varied.

As the FEL lamp is essentially a point source, it does not only illuminate the diffuser but also the surrounding. Therefore the clean room walls, ceiling and floor are all of black material. In addition to this, dedicated baffling is performed.

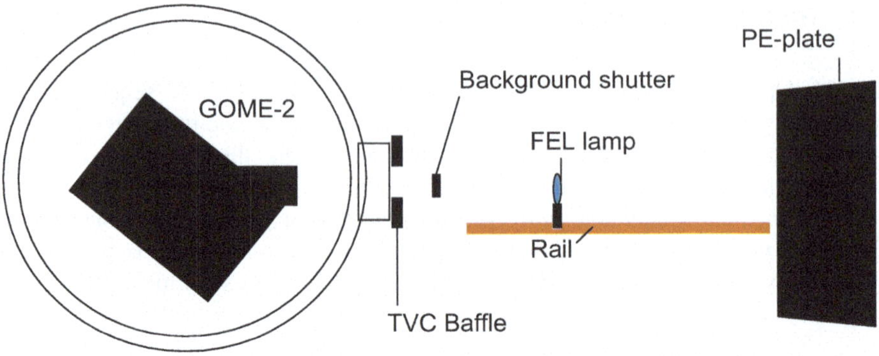

Fig. 1 Scheme of the setup for absolute irradiance calibration (*side view*). The FEL lamp illuminate the on-board diffuser directly. To avoid stray-light a baffle is placed on the vacuum chamber flange and behind the FEL lamp a black plate is placed under an angle such that the little reflection coming from the plate is reflected away from the instrument

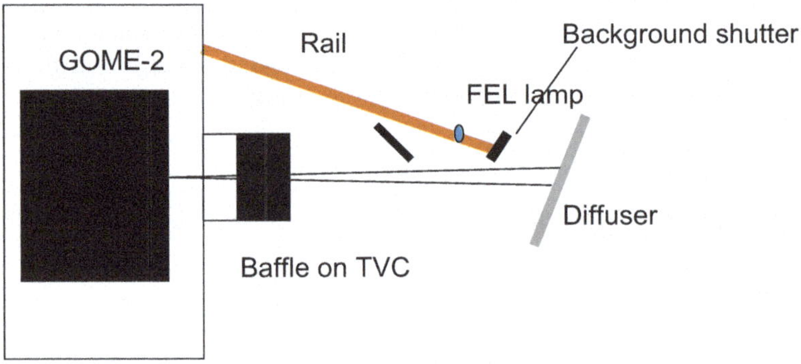

Fig. 2 Scheme of the setup for absolute radiance calibration (*top view*). The FEL lamp is illuminating an external calibrated diffuser. A baffle is place between the lamp and the vacuum chamber port to avoid stray-light

2.2 *Radiance Measurements*

The first radiance measurements with the FEL lamp showed differences of several percent with the previous calibration. To confirm these results, measurements were repeated and again resulted in different values. It was found that the differences in the radiance measurements could be explained by a small alignment error, which was first corrected and later prevented by more stringent alignment procedures.

A small error was left between the measurements at different distances between FEL lamp and diffuser. This error could be explained if the size of the instrument detector footprint on the diffuser is taken into account. In the analyses a homogenous illumination of the diffuser was assumed. As the FEL lamp resembles a point

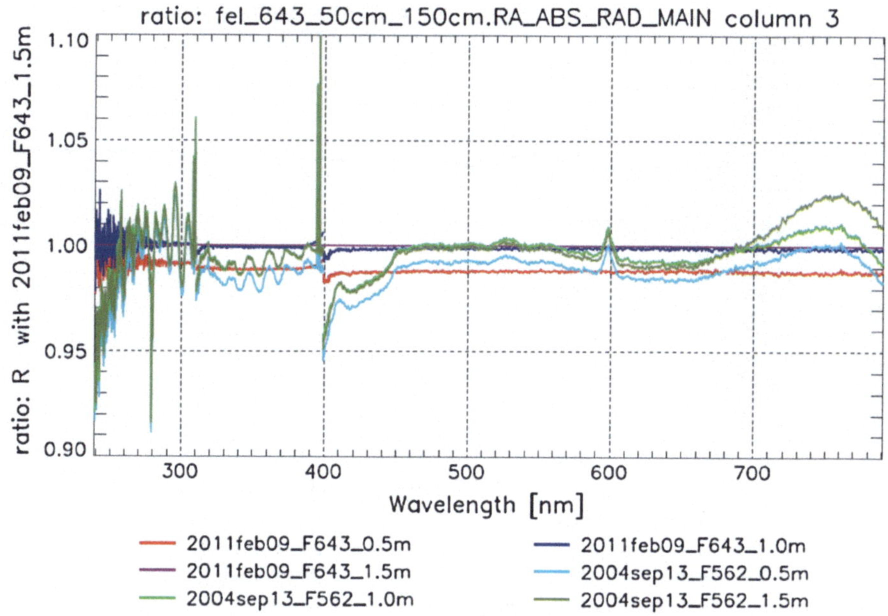

Fig. 3 Comparing different radiance measurements. Results are very comparable. The strong wavelength dependent behaviour of the 2004 measurements are due to a difference in detector etalon which is not fully corrected

source this is only true at a far enough distance from the diffuser w.r.t. the size of the instrument detector footprint. Considering the field of view of 3 degrees (in one direction) and a distance of ~2.5 m from the instrument, the footprint becomes about ~13 cm. At the outer edges of the footprint the distance to the lamp is therefore about 0.85% larger than at the centre, leading to a 1.7% lower irradiance at that point.

The assumed homogenous illumination leads therefore to a significant systematic error of ~1% at 0.5 m distance. The data for the longest distance was used for the calibration key-data because at that distance the illumination is most homogenous (Fig. 3).

2.3 *Irradiance Measurements*

The irradiance response measurements showed differences which were much larger and inconsistent. Unfortunately the irradiance measurements were carried out at a single distance, except for one case. Even though the results also indicated an alignment error similar to the radiance measurements, the order of magnitude of this error was improbably large and was not consistent with the other measurements. Therefore a more deep investigation was performed (Fig. 4).

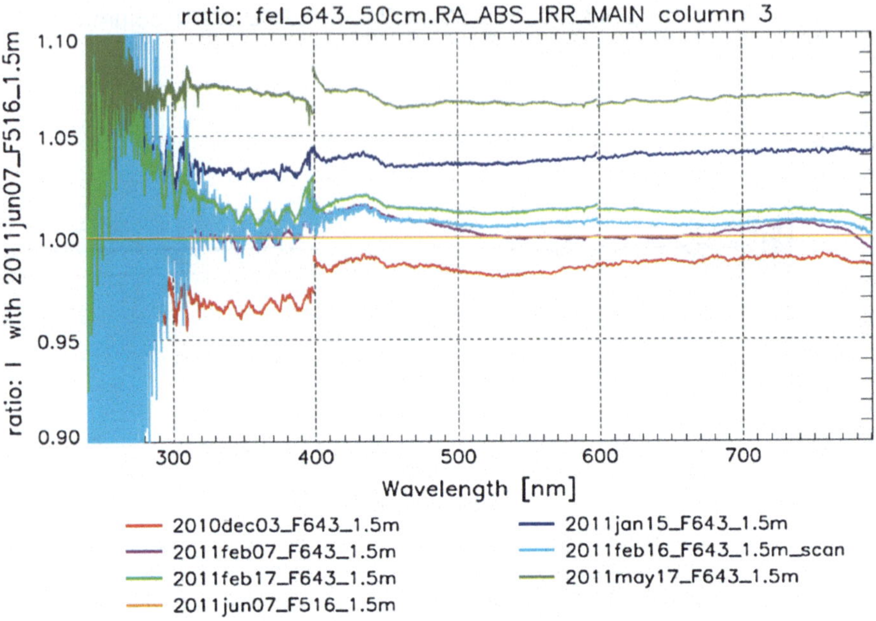

Fig. 4 Comparing different irradiance measurements. Results show large differences

3 Investigation of Irradiance Response Anomaly

3.1 Setup Analyses

As stated previously the radiometric anomaly was found to be in the irradiance measurements [4]. Therefore this setup is described here in greater detail.

The FEL irradiance setup is relatively simple: the source is placed at a known distance from the instrument diffuser. The instrument itself is inside a thermal vacuum chamber (TVC). The light enters the TVC via a vacuum window. Just in front of this window there is a baffle to block all unwanted light into the vacuum chamber to minimize stray-light. In practice this means that only the diffuser and the sun port baffle will be illuminated.

Behind the lamp a dark plastic plate (PE-plate) is placed at an angle w.r.t. the axis diffuser source. This is done to avoid specular reflections towards the diffuser. In some of the test cases the PE-plate was removed or replaced by a mirror.

The FEL lamp itself is placed upon a rail which is used to adjust the distance between lamp and diffuser. This rail itself has a high accuracy of a few micrometres. The distance between FEL and diffuser is between 1.5 and 2.5 m. This distance is determined by triangulation of the sun port baffle and the lamp alignment jig using a theodolite. The accuracy of this technique is well within a few millimetres.

Fig. 5 Picture of the FEL setup for irradiance. As can be seen the whole clean room is illuminated by the lamp even though the walls are *black*

Most of the light of the FEL lamp will illuminate the environment and not the instrument diffuser. This means that there will be a considerable amount of environmental stray-light. To characterize this a background measurement was performed using a background shutter that only blocks the direct light. Preferably the shutter should be close to the lamp, however this would influence the temperature of the lamp and therefore invalidate the NIST calibration. The background shutter is therefore placed further from the lamp blocking more of the environmental stray-light. The background level measured in this way is about 3 orders of magnitude lower than the exposure measurements so it is assumed that the influence of the position of the shutter is small. A number of potential sources of the deviations were investigated; namely environmental stray-light, instrument stability and alignment error (Fig. 5).

3.2 Environmental Stray-Light

The influence of environmental stray-light was investigated by changing the baffling behind the FEL lamp. Three case were investigated:

1. The default configuration with the PE-plate.
2. The PE-plate replaced with a mirror at an angle such that it views towards a dark area inside the clean room at a long distance.

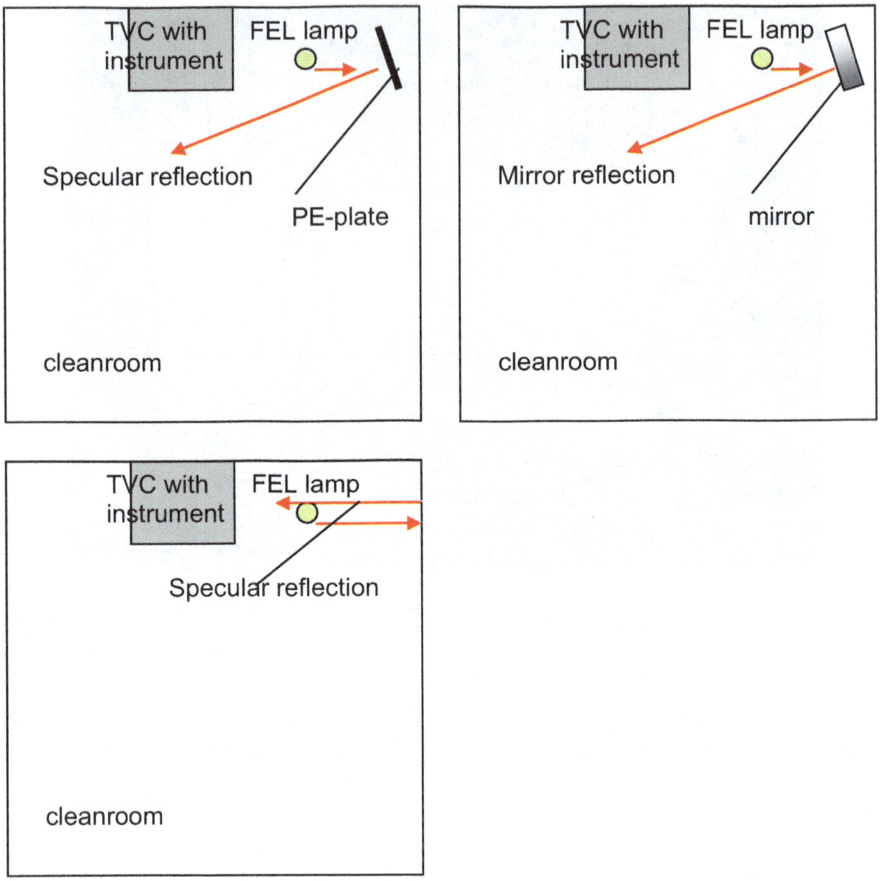

Fig. 6 Three different situations of environmental stray-light that were investigated. *Top left* the default situation with PE-plate. *Top right* the situation were the PE-plate was replaced by a mirror. *Bottom left* the situation were no extra measures for stray-light were taken to avoid stray-light from the wall behind the lamp

3. No PE-plate or mirror, in this case the clean room wall will generate a specular reflection towards the instrument diffuser.

The measurement results reproduced very well (within 1%). It was therefore concluded that the influence of the environmental stray-light was very well under control (Figs. 6 and 7).

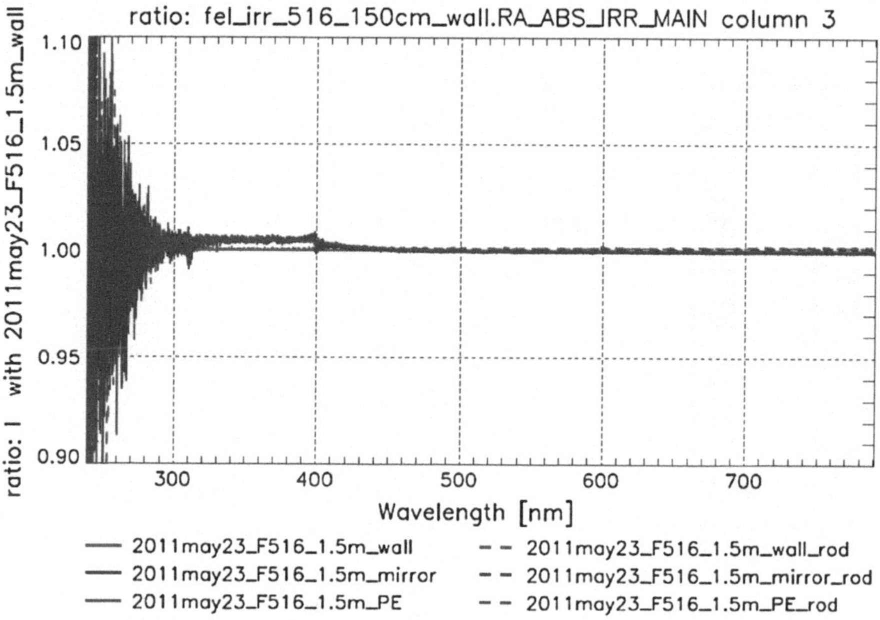

Fig. 7 Comparison of the different environmental stray-light conditions. There were no significant differences found

3.3 *Instrument Stability*

The instrument stability was tested by performing environmental cycles while leaving the setup intact. The instrument was switched off and on and sent through thermal cycling and vacuum cycling. The irradiance response after any of these environmental changes showed no significant deviation (within 1%). It was therefore concluded that the instrument was stable and was not the cause for the anomaly.

3.4 *Alignment Error*

The distance measurement using triangulation was considered very accurate based on the accuracy of the theodolite. However after doing many repetitions of the measurement it was found that only after re-alignment of the setup major difference in irradiance response were observed. On one occasion, when the instrument was back in ambient condition with the setup still aligned, the alignment was checked using a ruler. A difference was found of no more than 5 mm. A misalignment of 5 mm could not explain the differences in the irradiance response, for that a distance error of more than few centimetres was needed.

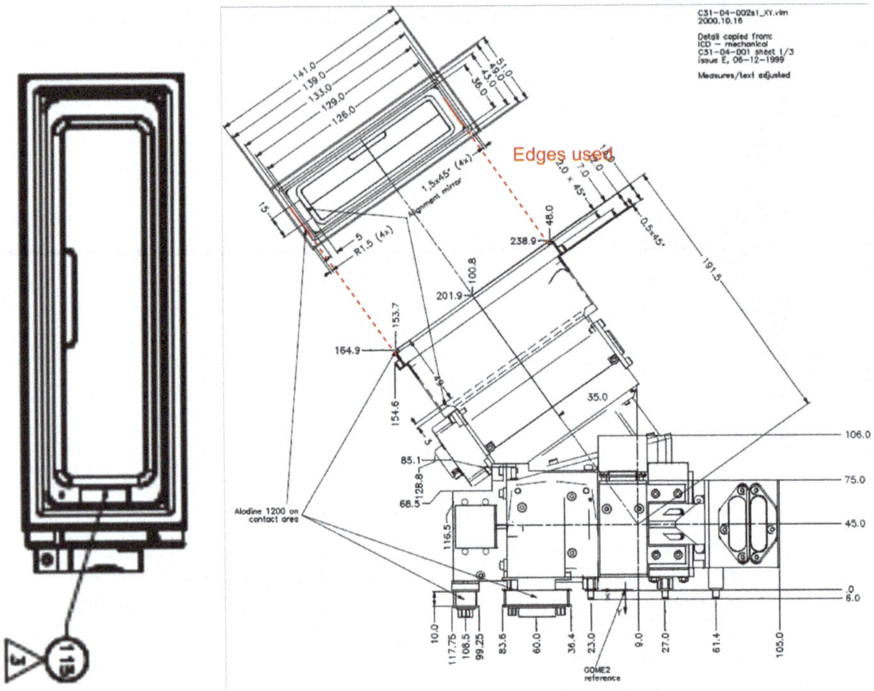

Fig. 8 Drawing of the GOME-2 calibration unit. On the *left* the view *inside* the baffle as seen by the FEL lamp and the theodolite used for alignment. On the *right* the complete unit

As the differences in irradiance response of the instrument were wavelength independent, an alignment error was still considered the most likely cause. Focus was therefore given to the target on the instrument used for the triangulation. The edges of the instrument sun port baffle have a well-known separation distance to each other and to the diffuser. Theoretically they pose a very accurate target. However the baffle is black and placed into a dark vacuum chamber.

The mechanical lay-out setup is shown in Fig. 1 and drawings of the calibration unit with the sun port in Fig. 8. As can be seen in when viewing the baffle (top left view) from outside the chamber there are many edges visible close to the target edges. A distance error of 129/126 − 1 = 2.4% is introduced when the nearest edges are mistakenly used for triangulation. The theodolite is typically placed at a distance of ~3 m (typical), which can lead to an distance error of ~7 cm. The FEL lamp is calibrated at a distance of 50 cm. The point-source like behaviour is used to determine the absolute irradiance at other distances (using the r^2 rule). With the lamp at 150 cm this leads to an error in irradiance response of 9.5% (5.7% with the lamp at 250 cm). This makes it plausible that the deviations in irradiance response found are indeed from alignment errors.

3.5 The Absolute Irradiance Response

After reaching the conclusion that the deviations were caused by alignment errors, the question remained to find the true irradiance response of the instrument. Fortunately the FEL source was not the only source used for radiometric calibration. Additionally a sun simulator was used for both radiance and irradiance measurements.

The sun simulator is a highly collimated source which in theory has no distance dependence. Due to inhomogeneity of the beam some minor distance dependent effect can be observed but this will only lead to negligible errors.

The sun simulator is not absolutely calibrated but can be used for relative measurements. The ratio between the instrument response of the earth port and the sun port to the sun simulator was determined. This ratio is referred to as the instrument BSDF. As the radiance response of the instrument to both sources (FEL and sun simulator) was consistent, the correctly aligned irradiance results using the FEL lamp could be chosen.

During the investigation of the irradiance anomaly with the FEL lamp, the distance between diffuser and lamp was varied. In these measurements the FEL lamp seemed to not behave like a point source. Given the geometrical configuration

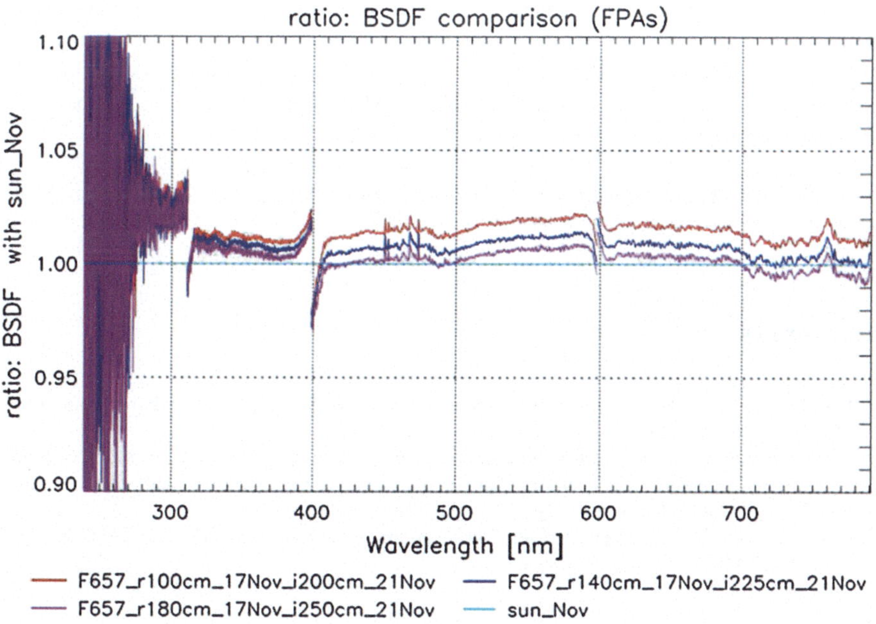

Fig. 9 Comparison of the instrument BSDF determined with the FEL and the sun simulator. For the FEL different distances between lamp and diffuser were used for both the radiance and irradiance measurements. Measurements at the longest distances for both radiance and irradiance show the best resemblance with the sun simulator results

of the setup this was not expected. It was found that the measurement at the longest distance resembled best the sun simulator measurements. The explanation was found in considering that the divergence of the light beam illuminating the diffuser changes with distance. The divergence of the sun is about 0.5° and therefore the source should have a similar divergence. Considering the size of the diffuser and the FEL lamp a distance of about 1.8 m would be required.

Additionally, the baffle structure in front of the diffuser will give some stray-light. If the divergence of the illumination is similar to the sun, the stray-light will resemble the in-orbit conditions. If, however, the FEL lamp is placed too close to the sun port, the effective divergence of the beam is larger and more stray light is produced by the baffle. This was proven by a measurement with the sun simulator were first the baffle and diffuser was illuminated and then only the diffuser. A change of 0.5–1% in signal was observed. For a more divergent source as the FEL this is most likely higher (Fig. 9).

4 Conclusion

An anomaly was observed for the irradiance response of the GOME-2 instrument after recalibration. After an extensive investigation it was found that the anomaly observed was caused by alignment error and stray-light from the calibration unit baffle.

For the investigation different setup configuration were used to determine the true value of the irradiance response of GOME-2. This shows that for accurate absolute radiometric calibration a redundant method is required.

Acknowledgements We would like the acknowledge the GOME-2 teams of EUMETSAT, ESA and Selex Galileo for their support during the investigation.

References

1. J. Callies et al, GOME-2 ozone instrument onboard the European METOP satellites, Proc. SPIE 5549, 60 (2004).
2. K. Smorenburg et al, Slit function measurement optical stimulus, Proc. SPIE 4881, 511 (2003).
3. P. Kenter, GOME-2 Calibration Comparison FM2-1 vs. FM2-2 Calibration Results, MO-TN-TPD-GO-0088 issue 1, GOME-2 calibration datapack 2012.
4. G. Otter, GOME-2 Calibration Radiometric anomaly investigation, MO-TN-TPD-GO-0094 issue 2, GOME-2 calibration datapack 2012

Sentinel-3A: First Flight Results of Its Optical Payload

Jean-Loup Bézy, Jens Nieke, Johannes Frerick, Constantin Mavrocordatos and The S3 team

Abstract Sentinel-3 is a Global Land and Ocean monitoring mission developed in the frame of the joint EC/ESA Copernicus initiative. Carrying a suite of cutting-edge instruments, Sentinel-3 will measure systematically Earth's oceans, land, ice and atmosphere to monitor and understand large-scale global dynamics. It will provide essential information in near-real time for ocean and weather forecasting. Four identical Sentinel-3 satellites are planned that will collectively provide a 20-year period of continuous observations. The Sentinel-3A (S3A) satellite was launched on the 16th of February 2016. Sentinel-3B satellite is under development and will be launched in 2017. Sentinel-3C and D satellites are in preparation as replacement for the A and B units. In this paper we will describe the S3A optical payload and summarize the first commissioning phase results.

Keywords Copernicus · Sentinel-3 · Ocean Colour · Sea Surface Temperature

1 Introduction

Sentinel-3 is a Global Land and Ocean monitoring mission developed in the frame of the joint EC/ESA Copernicus initiative, previously known as Global Monitoring for Environment and Security (GMES) [1, 2]. Carrying a suite of cutting-edge instruments, Sentinel-3 will measure systematically Earth's oceans, land, ice and atmosphere to monitor and understand large-scale global dynamics. It will provide essential information in near-real time for ocean and weather forecasting.

Four identical Sentinel-3 satellites are planned that will collectively provide a 20-year period of continuous observations. The Sentinel-3A (S3A) satellite was launched on the 16th of February 2016. Sentinel-3B satellite is under development and will be launched in 2017. Sentinel-3C and D satellites are in preparation as replacement units for the A and B.

J.-L. Bézy (✉) · J. Nieke · J. Frerick · C. MavrocordatosThe S3 team
European Space Agency, European Space Research and Technology Centre,
Keplerlaan 1, 2200 AG Noordwijk ZH, The Netherlands
e-mail: Jean-Loup.Bezy@esa.int

H.P. Urbach and G. Zhang (eds.), *3rd International Symposium of Space Optical Instruments and Applications*, Springer Proceedings in Physics 192,
DOI 10.1007/978-3-319-49184-4_49

In this paper we will describe the S3A optical payload and summarize the first commissioning phase results.

2 Satellite and Orbit

Sentinel-3 features a 3 axis stabilized satellite flying in a frozen sun-synchronous orbit, with an orbit repeat cycle longer than 20 days as required for the topography mission and a local time at descending node located at 10:00 h as desired for the optical instruments while keeping continuity with Envisat. With a 1250 kg mass and 1.1 kW power consumption, Sentinel-3 fits the Vega land Rocket launchers. The satellite is designed for a minimum lifetime of 7.5 years.

The Sentinel-3 mission is based on a constellation of two satellites, both orbiting the Earth at an altitude of 814.5 km. This configuration offers for some of the instruments global coverage within one to two days. An overview about the S3 orbit parameters is given in Table 1.

Table 1 Sentinel-3 satellite orbit parameters

Orbit type	Repeating frozen sun synchronous low Earth orbit
Repeat cycle	27 days (14 + 7/27 per day)
Local solar time	10:00 a.m. at descending node
Average altitude	814.5 km
Inclination	98.65°
S3B satellite	Identical orbit to S3A but flown 140° out of phase with S3A

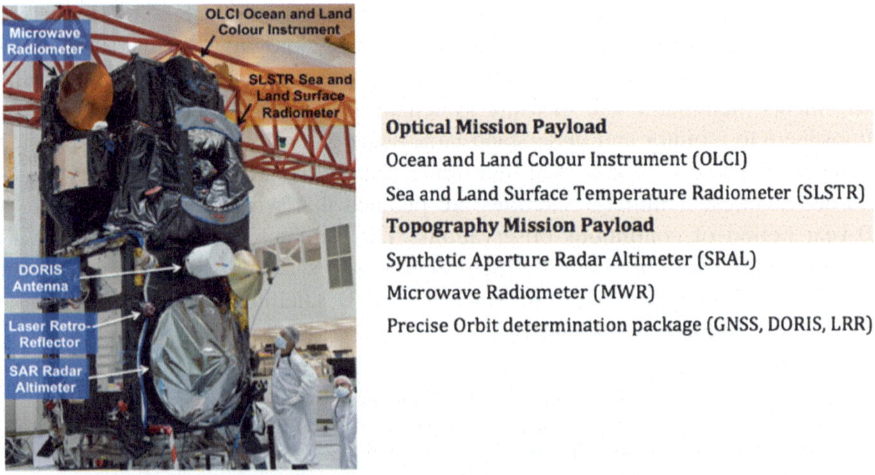

Fig. 1 The Sentinel-3A satellite fully equipped during on-ground testing at Thales Alenia Space, France, in December 2014

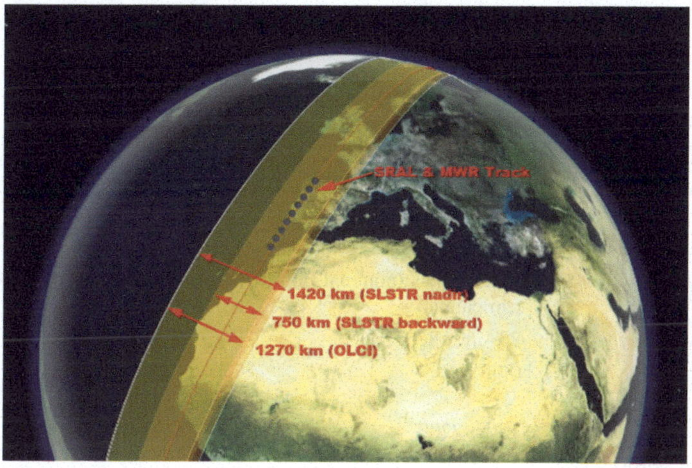

Fig. 2 Ground swath of OLCI and SLSTR (nadir and oblique) and the centrally nadir located footprint of the SRAL and MWR

Sentinel-3 satellite carries five instruments: two optical instruments and an altimetry topography payload package. Figure 1 shows the S3A satellite fully equipped during on-ground testing.

The swaths of the optical instruments are collocated on Earth. Whereas the OLCI and SLSTR nadir field-of-views have a slight offset to the west of the nadir position, the SLSTR oblique view is centred on the nadir ground track of the satellite. Also the footprints of SRAL and MWR are directed to nadir, on the satellite ground track (see Fig. 2).

3 Ocean and Land Colour Imager

The Ocean and Land Colour Imager (OLCI) [3] is based on a heritage design from ENVISAT's Medium Resolution Imaging Spectrometer (MERIS) [4, 5]. OLCI has 21 spectral channels, compared to the 15 of MERIS, an observation geometry optimised to minimise the negative impact of sun-glint, and a spatial resolution of 300 m over all surfaces. As a unique feature, the spectral channels can be changed in width and position by ground command to support spectral calibration campaigns. The key performances of the OLCI are listed in Table 2.

OLCI is a large swath imaging dispersive spectrometer with high spectral and radiometric accuracy. It is a nadir-looking sensor operated in a push-broom mode. The instrument consists of five identical cameras sharing the large field of view of 68.6°. The cameras are mounted on a temperature controlled optical bench in a fan-shaped configuration. The cameras view the Earth through five depolarizing windows. The spectrometer within each camera consists of collimator optics, a diffraction grating and imager optics. A two-dimensional frame-transfer CCD array

Table 2 OLCI key performance requirements

Swath width	1280 km
Spatial resolution	300 m
MTF	>0.3
Spectral range	390–1040 nm
Spectral sampling interval	1.25 nm
Spectral resolution	1.5 nm (FWHM)
Spectral channels	21 channels (nominal Earth View), 45 channels (spectral campaigns)
Absolute accuracy	2% in reflectance
Relative spectral accuracy	0.05%
SNR	150…2200
Polarisation sensitivity	<0.3%

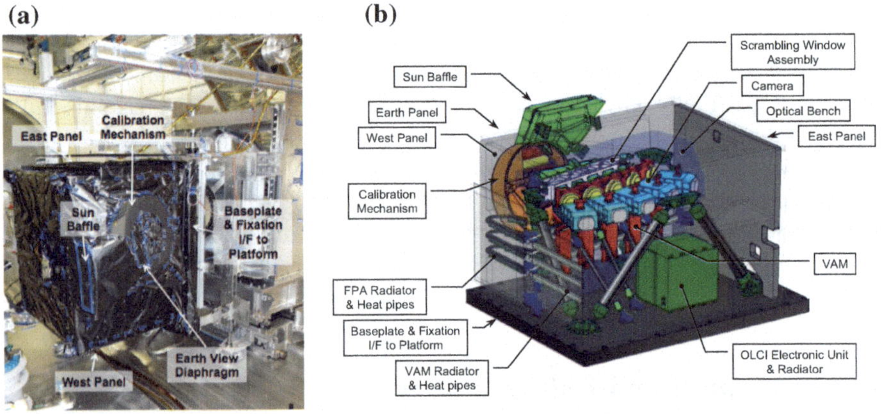

Fig. 3 **a** OLCI-A in the instrument test configuration inside the vacuum chamber at Thales Alenia Space, France, **b** OLCI design configuration overview

(MERIS heritage back-illuminated CCD55-20) receives the spectrally dispersed image of the spectrometer entrance slit. The OLCI-A PFM under test configuration and the instrument optical configuration are indicated in Fig. 3.

In contrast to the MERIS instrument, OLCI employs an asymmetric swath with respect to the satellite ground-track in order to avoid direct solar reflection at sea surface (sun-glint). The amount of tilt is defined by the need to minimize the maximum Observation Zenith Angle (OZA) at the outer border of the swath and at the same time guaranteeing global coverage. Figure 4 shows the across-track tilt of the overall field of view of 12.6°, resulting in a maximum OZA slightly above 55°.

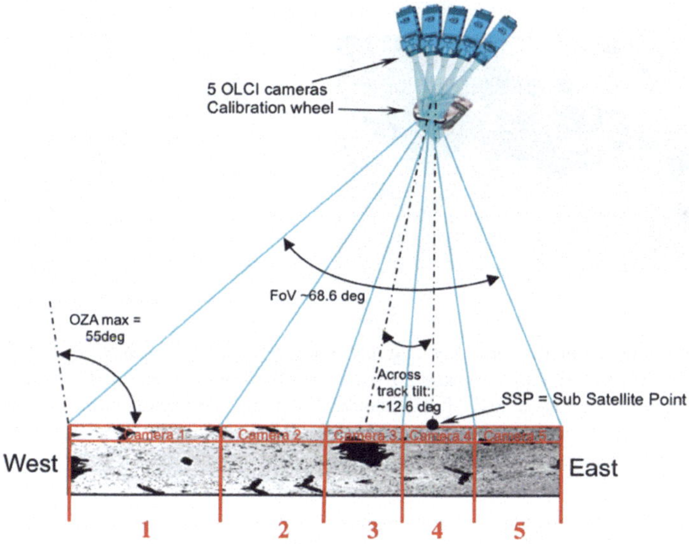

Fig. 4 OLCI's 5 camera modules arranged in asymmetrical viewing geometry to avoid the sun reflection upon sea surface

3.1 First Results from OLCI

After the successful S3A Launch and Early Orbit (LEOP) and the Spacecraft-In-Orbit-Verification Phases (SIOV), the calibration-validation phase of the OLCI instrument started. The end of this Commissioning Phase is marked through the OLCI In-Orbit Commissioning Review (IOCR) which was held after 5 month in orbit. During the IOCR an in-depth assessment of the OLCI CAL/VAL activities (functional, performance, product verification and validation) was provided by all involved experts and centers confirming the overall excellent performance. In most aspects (radiometric, spectral and geometric), OLCI-A is performing well and data have been successfully assessed (see Fig. 5). Deeper analyses concerning some of the radiometric performance are still ongoing and are planned to be finalized by the end of the year 2016.

4 Sea and Land Surface Temperature Radiometer

The Sea and Land Surface Temperature Radiometer (SLSTR), that uses heritage concepts taken from ENVISAT's Advanced Along Track Scanning Radiometer (AATSR) [6], will determine global sea-surface temperatures to an accuracy of better than 0.3 K. SLSTR is a dual view (near-nadir and backward views) conical imaging radiometer with a swath of 1440 km at nadir and of 740 km in the backward view. The dual view enables robust atmospheric correction over the

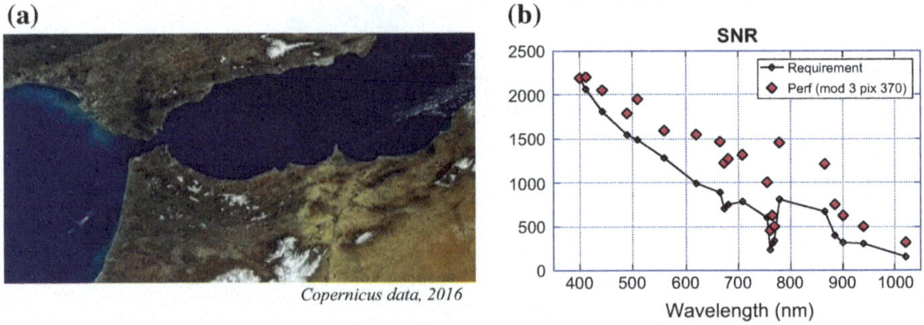

Fig. 5 **a** Featuring southern Spain, Portugal and North Africa, this is one of the first images from OLCI. The image was taken on 1st March 2016 and clearly shows the Strait of Gibraltar between the Atlantic and Mediterranean. Swirls of sediment and algae in the seawater can be seen along the southwest coast of Spain and along the coast of Morocco. **b** Fully compliant OLCI-A SNR performance (central pixel performance shown) scaled from sun calibration signal to typical water radiance

swath. SLSTR measures in nine spectral channels with two additional channels optimised to monitor fire. The SLSTR has a spatial resolution in the visible and shortwave infrared channels of 500 m and 1 km in the thermal infrared channels. The key performances of the SLSTR are listed in Table 3.

The nadir and backward views are generated by separate scanners, allowing a wider swath than possible with the single conical scan of the original AATSR design (500 km) (see Fig. 6). The Focal Plane Assemblies are common to the nadir and backward views and are addressed alternatively from both views via a flip mirror. The SWIR/MWIR and TIR detectors are HgCdTe elements (photovoltaic and photoconductive elements) cooled to about 89 K by active Stirling coolers. The VNIR detectors are Silicon diode operated at about 265 K. The SLSTR optical design was reported in [7].

The instrument includes two on-board calibration reference black bodies and one visible calibration source (sun illuminated diffuser) for accurate and stable in-flight calibration.

Table 3 SLSTR key performance requirements

Swath width	Nadir view: 1400 km Oblique view: 740 km
Spectral channels	VNIR: 0.555, 0.659, 0.859 μm SWIR: 1.38, 1.61, 2.25 μm TIR: 3.74, 10.85, 12.0 μm Fire: 3.74, 10.85 μm
Spatial resolution	VNIR/SWIR: channels: 0.5 km TIR and fire: 1 km
Radiometric resolution	NEΔT: 30 mK (LWIR) NEΔT: 50 mK (MWIR) SNR: 20 (VNIR-SWIR)
Radiometric accuracy	VNIR-SWIR: 2% TIR: 0.1 K (goal) Fire: 3 K

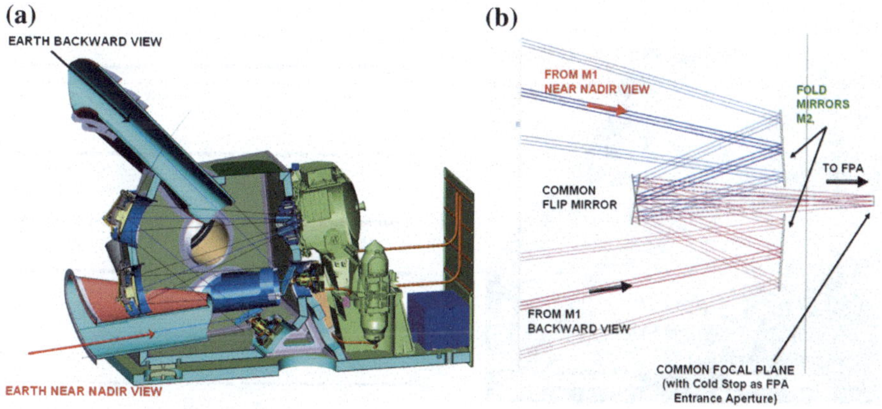

Fig. 6 **a** SLSTR, consisting of two baffles, two telescopes (near nadir and backward view), each with a scan mirror, an off-axis parabolic mirror and a folding plane mirror. **b** A plane flip mirror is used to switch the beam between the two telescopes, focalizing at the entrance diaphragm of the Focal Plane Assembly (FPA)

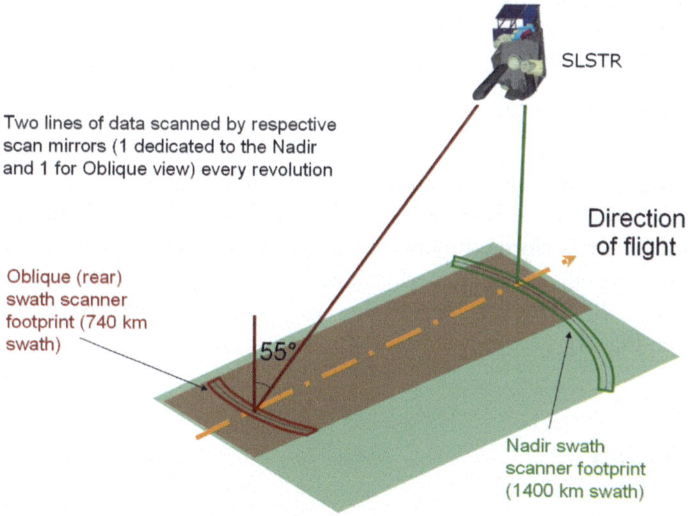

Fig. 7 Outline sketch of the SLSTR near nadir and oblique viewing geometry highlighting the asymmetric nadir swath with respect to the nadir point

All pixels are acquired with an Observation Zenith Angle (OZA) of less than 56° (see Fig. 7), so that only the OZA dependency of the sea surface emissivity needs to be considered, while variations due to salinity, temperature and wind speed can be considered negligible. Further, the geometry provides a path length ratio between the two views (atmospheric optical thickness ratio) larger than 1.54 in order to maintain the same retrieval algorithms quality achieved by the algorithms applied to the data of the predecessors instruments.

(a) (b)

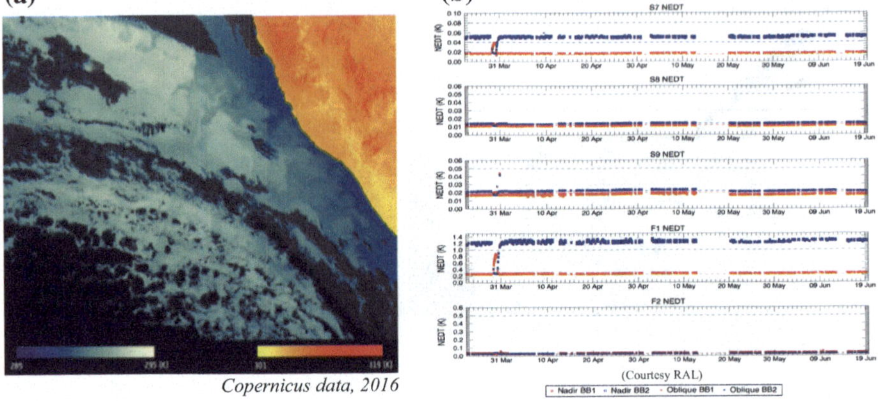

Fig. 8 **a** SLSTR-A thermal-infrared channels depicting thermal signatures over a part of western Namibia and the South Atlantic Ocean. The Namibian land surface is shown in *red–orange* colours, corresponding to a brightness temperature range 301–319 K. The *blue* colours over the ocean correspond to a temperature range of 285–295 K, **b** Noise-equivalent Delta temperature (NEDT) of SLSTR-A channel S7 (3.7 μm), S8 (10 μm) and S9 (12 μm) matching very well with specification and indicating initial long term stability

4.1 First Results from SLSTR

SLSTR-A went in operations shortly after the successful S3A LEOP and SIOV. During the calval phase of SLSTR all functional, performance, product verification and validation were performed by involved experts and centers, confirming—as for OLCI-A—the overall excellent performance. In all aspects (radiometric, spectral and geometric), SLSTR-A is performing well and data have been successfully assessed (see Fig. 8).

5 Conclusion

With the successful launch of Sentinel-3A, a new era for the Copernicus Services has started offering data over oceans and lands with unprecedented coverage. Together with Sentinel-3B, its twin satellite scheduled for launch in 2017, and later on with the replacement Sentinel-3C and D units, a 20-year period of continuous observations is guaranteed. Among the five instruments embarked, the OLCI and SLSTR optical payload ensure the continuity of the ENVISAT mission with very much improved performance. During the calibration-validation phase functional, performance, product verification and validation were performed confirming the overall excellent performance of the optical payload.

References

1. Donlon, C. et al, "The Global Monitoring for Environment and Security (GMES) Sentinel 3 mission", *Remote Sensing of Environment*, Vol. 120, p. 37–57, 2012
2. EC Copernicus site (http://copernicus.eu)
3. J. Nieke, et al., "Ocean and Land Colour Imager on Sentinel-3," in Optical Payloads for Space Missions, *John Wiley & Sons*, chapter 10, 2015
4. M. Rast, and J.-L. Bézy, "The ESA Medium Resolution Imaging Spectrometer MERIS: A review of the instrument and its mission", *International Journal of Remote Sensing*, 20, 1681–1702, 1999
5. J.-L.Bézy, et al., "Medium Resolution Imaging Spectrometer for ocean colour on board Envisat," in Optical Payloads for Space Missions, *John Wiley & Sons*, chapter 4, 2015
6. T. Llewellyn-Jones, M.C. Edwards, C.T. Mutlow, A.R. Birks, I.J. Barton and H. Tait, "AATSR: Global Change and Surface Temperature Measurements from ENVISAT", *ESA Bulletin* 106, February 2001
7. Coppo, P., B. Ricciarelli, F. Brandani, J. Delderfield, M. Ferlet, C. Mutlow, G. Munro, T. Nightingale, D. Smith, S. Bianchi, P. Nicol, S. Kirschstein, T. Hennig, W. Engel, J. Frerick & J. Nieke, "SLSTR: a high accuracy dual scan temperature radiometer for sea and land surface monitoring from space," *J. Mod. Opt.* 57(18), 1815–1830, 2010

Erratum to: 3rd International Symposium of Space Optical Instruments and Applications

H. Paul Urbach and Guangjun Zhang

Erratum to:
H.P. Urbach and G. Zhang (eds.), *3rd International Symposium of Space Optical Instruments and Applications*, Springer Proceedings in Physics 192, DOI 10.1007/978-3-319-49184-4

The original version of the book was inadvertently published with spell errors in the authors' names. "Xuhie Huang" in Chapter "Optical Design of an Aperture-Divided MWIR Imaging Polarimeter" has to be changed to "Xujie Huang" and "Qiao-yun Fa" in Chapter "In-Orbit Calibration Method for Sun Sensor Based on Sun Ephemeris and Star Sensor" has to be changed to "Qiao-yun Fan". The erratum book has been updated with these changes.

The updated online versions of these chapters can be found at
http://dx.doi.org/10.1007/978-3-319-49184-4_7;
http://dx.doi.org/10.1007/978-3-319-49184-4_25

H.P. Urbach and G. Zhang (eds.), *3rd International Symposium of Space Optical Instruments and Applications*, Springer Proceedings in Physics 192,
DOI 10.1007/978-3-319-49184-4_50

Index

H.P. Urbach and G. Zhang (eds.), *3rd International Symposium of Space Optical Instruments and Applications*, Springer Proceedings in Physics 192,
DOI 10.1007/978-3-319-49184-4

The manufacturer's authorised representative in the EU is Springer Nature Customer Service Centre GmbH, Europaplatz 3, 69115 Heidelberg, Germany. If you have any concerns regarding our products, please contact ProductSafety@springernature.com

Printed and bound by CPI Group (UK) Ltd, Croydon, CR0 4YY
30/09/2026
02225572-0018